Bioanalysis

Advanced Materials, Methods, and Devices

Volume 7

Series editor

Tuan Vo-Dinh
Fitzpatrick Institute for Photonics
Duke University
Durham, NC, USA

More information about this series at http://www.springer.com/series/8091

Manabu Tokeshi
Editor

Applications of Microfluidic Systems in Biology and Medicine

Springer

Editor
Manabu Tokeshi
Division of Applied Chemistry, Faculty
of Engineering
Hokkaido University
Sapporo, Japan

ISSN 2364-1118 ISSN 2364-1126 (electronic)
Bioanalysis
ISBN 978-981-13-6231-6 ISBN 978-981-13-6229-3 (eBook)
https://doi.org/10.1007/978-981-13-6229-3

Library of Congress Control Number: 2019935498

This Springer imprint is published by the registered company Springer Nature Singapore Pte Ltd.
The registered company address is: 152 Beach Road, #21-01/04 Gateway East, Singapore 189721, Singapore

Preface

About 30 years have passed since the concept of μTAS was proposed by Andreas Manz in 1990. In the mid-1990s, as the human genome project was being promoted, many researchers were working on the development of microfluidic "electrophoresis" devices. From around 2000, development of devices integrating chemical analysis, bioassay, etc. has increased. Then, this research field called μTAS, lab-on-a-chip, microfluidic device, etc. has greatly expanded and developed. Currently, μTAS field continues to expand; new concepts such as digital microfluidics, nanofluidics, μPAD (microfluidic paper-based analytical device), and organ-on-a-chip which were originally unexpected have been proposed; and not only to analytical chemistry but also to various fields such as medical diagnosis and biological applications has being carried out. Especially, medical and biological applications are developing rapidly, and there are striking things. However, the published scientific papers on these applications are enormous, and it is hard for experts in this field to read all of them.

This book focuses on state-of-the-art microfluidic research in medical and biological applications. The top-level researchers in this research field explain carefully and clearly what can be done by using microfluidic devices. Beginners in the field — undergraduates, engineers, biologists, medical researchers—will easily learn to understand microfluidic-based medical and biological applications. Because a wide range of topics are summarized here, it also helps experts to learn more about fields outside their own specialties. The book covers many interesting subjects, including protein separation, protein crystallization, cell separation, single-cell analysis, cell diagnosis, point-of-care testing, immunoassay, and regenerative medicine. Readers will be convinced that microfluidic devices have great potential for medical and biological applications.

I would like to thank Shinichi Koizumi and Asami Komada at Springer for their help in pulling the book together. The publication of this book would have been impossible without their help. Finally, I would like to express my sincere thanks to the authors for all of the time and efforts they spent writing their chapters.

Sapporo, Japan
August 2018

Manabu Tokeshi

Contents

Chapter 1
Acoustofluidic Blood Component Sample Preparation and Processing in Medical Applications

Maria Antfolk and Thomas Laurell

Abstract Recent developments of bulk acoustofluidic technology (BAW – bulk acoustic wave) in biomedical applications is described in this chapter. The basic principles for setting up an acoustic standing wave in a microchannel in 1 or 2 dimensions in the transversal direction to flow is outlined. BAW acoustofluidics is a preferred solution as compared to SAW based acoustofluidics due to the relatively higher acoustic energies that can be accomplished in BAW systems. This in turn lends BAW technology to perform cell manipulation based handling in a sufficiently high flow through format that can fulfill many biomedical and bioanalytical applications. Several unit operations for BAW based cell handling have today reached a level of maturity where these are being integral components in cytometry and cell processing instrumentation. Most of these applications are still realized at an analytical level and have not yet reached process scale or therapeutic scale throughput. However, intense developments are in progress to also reach into this domain of larger scale processing since the performance and label free operation offered from BAW systems would significantly impact current bioprocess industry and clinical practice. The importance of having full control of the buffer systems used is discussed since poorly matched buffers/fluids, with respect to the acoustic properties (acoustic impedance), may significantly impact the processing outcome as a consequence of acoustically driven fluid relocation. Also, the challenge of manipulating smaller bioparticles, e.g. bacteria, is discussed and strategies to tackle the fact that the inherent acoustic streaming in acoustic standing wave based microfluidics may be counteracting the desired alignment of cell/particles defined by the acoustic standing wave. A focus is put on applications in blood component processing, where unit operations such as cell separation (WBC, RBC, WBC subpopulations, CTC and platelets), buffer exchange and concentrating cell samples have become important modalities in cell based microfluidics. The current state of diagnostic BAW applications such as blood plasma separation, circulating tumor cell (CTC) isolation, rapid hematocrit determination and bacteria enrichment/purification in sepsis are discussed.

M. Antfolk · T. Laurell (✉)
Department of Biomedical Engineering, Lund University, Lund, Sweden
e-mail: thomas.laurell@bme.lth.se

M. Tokeshi (ed.), *Applications of Microfluidic Systems in Biology and Medicine*, Bioanalysis 7, https://doi.org/10.1007/978-981-13-6229-3_1

Keywords Acoustofluidics · Acoustophoresis · Cell separation plasmapheresis · CTC · Sepsis · Hematocrit

1.1 Introduction

Acoustofluidics uses ultrasound to separate and handle cells either in continuous flow, using bulk acoustic waves (BAW) [1–3] or standing surface acoustic waves (SSAW) [4], or in a batch mode using acoustic trapping. The methods rely to a majority on the formation of standing acoustic waves to handle and process cells based on their size, density, and/or compressibility.

Acoustophoresis is most often operated in a label-free mode which is advantageous when isolating cells that are not susceptible to surface marker based separations, but has also been combined with affinity beads to target specific cell types. In a continuous flow mode, devices depending on BAW can be operated at relatively high flow rates compared to other microfluidic methods, and where SSAW devices have not shown the same throughput. Devices based on SSAW, however, are fabricated in softer materials such as PDMS, while BAW devices generally rely on more rigid materials such as silicon or glass in order to efficiently support a standing wave formation defined solely by the geometry of the microfluidic compartment. BAW systems are commonly operated by exciting the entire chip structure at an oscillation frequency that coincides with a fundamental resonance mode of a microfluidic compartment. By designing the resonance cavity as a microchannel, particles or cells can be focused in well-defined and reproducible positions/streamlines [5]. In a SAW device the microchannel is placed between a pair of IDTs on a piezoelectric substrate. When actuating the substrate the SAWs propagate in opposite directions on the substrate surface and leak into the liquid of the microchannel. The interference between the two propagating fields causes pressure fluctuations in the liquid resulting in the formation of acoustic pressure nodes and antinodes [6]. Both types of acoustophoresis offer opportunities for a number of different unit operations such as cell separation, sample concentration, alignment and buffer exchange (Fig. 1.1).

1.1.1 Acoustofluidics

When a particle is suspended in an ultrasound standing wave field, the particle is subjected to both a primary and secondary radiation force, as well as a Stokes' drag force induced by the acoustic streaming. The primary acoustic radiation force originates from scattering of the standing wave on a particle and affects the particles position within the microchannel. The secondary acoustic radiation force is due to interactions of the scattered waves from two particles and affects the particle-particle

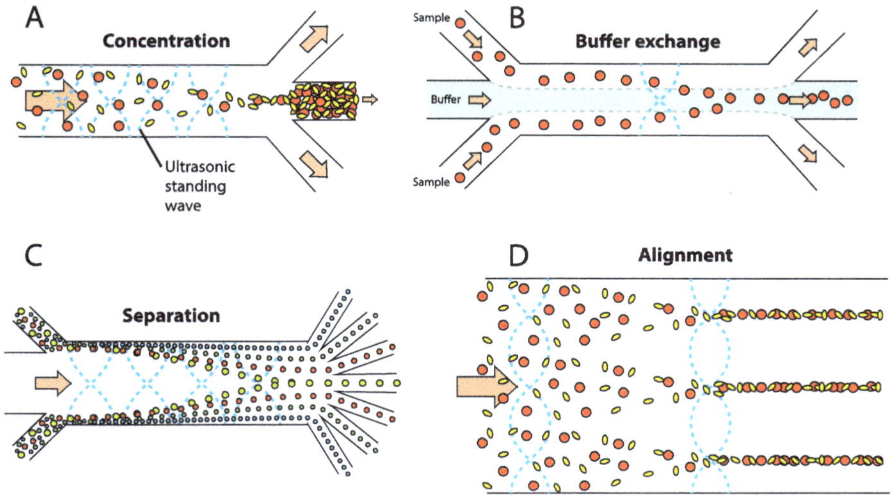

Fig. 1.1 Schematic of different cell handling unit operation achievable with acoustophoresis

relative position. This force is commonly orders of magnitude smaller and is only significant on very short particle-particle distances, *i.e.* at very high particle concentrations. Most acoustofluidic devices are operated below this critical particle concentration where these secondary forces are negligible. The fundamental theory on acoustic standing wave forces on particles have been further described by King [7], Yoshioka and Kawashima [8], Gorkov [9] and Nyborg [10], among others.

1.1.1.1 Primary Acoustic Radiation Force

The primary acoustic radiation force (F_{rad}) is responsible for moving the cells within the microchannel, to the node or anti-node of the standing wave. The expression for the primary acoustic radiation force for a plane standing acoustic wave, on a spherical particle with a radius much smaller than the wavelength can be written, Eq. 1.1:

$$F^{rad} = -\nabla U^{rad} = -\nabla \frac{4\pi}{3} a^3 \left[f_1 \frac{1}{2} \kappa_0 p_{in}^2 - f_2 \frac{3}{4} \rho_0 v_{in}^2 \right] \tag{1.1}$$

$$f_1(\tilde{\kappa}) = 1 - \tilde{\kappa} \quad \text{and } \tilde{\kappa} = \frac{\kappa_p}{\kappa_0}$$

$$f_2(\tilde{\rho}) = \frac{2(\tilde{\rho} - 1)}{2\tilde{\rho} + 1} \quad \text{and } \tilde{\rho} = \frac{\rho_p}{\rho_0}$$

U^{rad} denotes the acoustic radiation potential, a the particle radius, f_1 and f_2 are the monopole and dipole scattering coefficients respectively, κ_0, ρ_0, κ_p and ρ_p are the

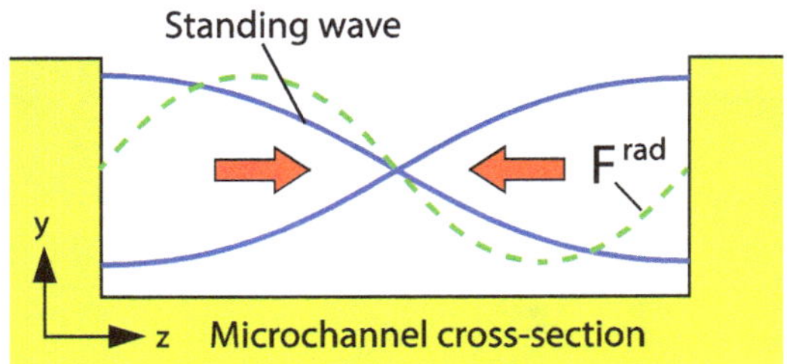

Fig. 1.2 Schematic of acoustic pressure field (solid blue line) and acoustic radiation force (dashed green line) in the channel cross-section. Red arrows indicate direction of F_{rad} for particles with a positive acoustic contrast factor, Φ

compressibility and density of the fluid and particle, respectively, $\langle p_{in} \rangle$ and $\langle v_{in} \rangle$ are the pressure and velocity field time averages. When considering an ideal one-dimensional standing wave (Fig. 1.2), the expression for the acoustic radiation force can be simplified to the following [11]; Eq. 1.2.

$$F_z^{rad} = -\partial_z U^{rad} = 4\pi\phi(\tilde{\kappa}, \tilde{\rho}) k a^3 E_{ac} \sin(2kz) \tag{1.2}$$

$$E_{ac} = \frac{p_a^2}{4\rho_0 c_0^2}$$

$$\phi(\tilde{\kappa}, \tilde{\rho}) = \frac{1}{3}\left[\frac{5\tilde{\rho} - 2}{2\tilde{\rho} + 1} - \tilde{\kappa}\right]$$

where $k = 2\pi/\lambda$, and $\phi(\tilde{\kappa}, \tilde{\rho})$ is the acoustic contrast factor, E_{ac} the acoustic energy density, z the position of the particle in the direction of the channel cross-section (Fig. 1.2), p_a the pressure amplitude, ρ_0 and c_0 the density and the speed of sound in the medium respectively. It should be noted that E_{ac} is proportional the square of the voltage applied to the piezo electric element.

The acoustic radiation force-induced movement of cells and particles in a standing wave is balanced by the Stokes' drag force such that $F_{\text{rad}} = F_{\text{drag}}$. When these forces are balanced the velocity, u_{rad}, of the particle can be derived according to Eq. 1.3, where η is the fluid viscosity.

$$u_{\text{rad}} = \frac{2\Phi}{3\eta} a^2 k_y E_{\text{ac}} \sin(2k_y y) \tag{1.3}$$

It can be seen that the particle velocity is dependent on the square of the particle radius and hence acoustophoretic separation of particles of the same physical properties are strongly dependent of the particle size. Using Eq. 1.3 the unknown acoustic contrast factor for a particle or cell with a known radius can be estimated by comparing the voltage needed to translate the particle to the center outlet under constant flow rate versus the corresponding voltage for a reference particle where both the contrast factor and radius are known.

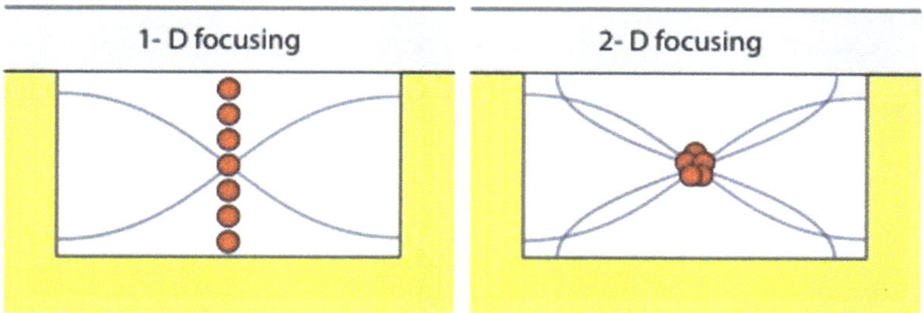

Fig. 1.3 Schematic of one-dimensional and two-dimensional acoustophoretic focusing

The sign of the acoustic contrast factor, Φ, depends on the density and compressibility of the particles in relation to the suspending medium, and determines if the particle will move towards the pressure node or anti-node. This factor is typically positive for cells suspended in physiological fluids such as plasma, phosphate buffered saline (PBS), or cell culture medium, and negative for particles such as air bubbles or oil particles in the same medium. It should however be noted that the acoustic contrast factor of a cell is dependent of the acoustophysical properties of the buffer conditions and by designing the suspending medium accordingly a cell can also display a negative acoustic contrast [12]. The acoustic contrast factor of a cell in a given medium can be calculated if the density and compressibility of the cell is known. Although density data many times can be found in the literature, the compressibility is scarcely found. To alleviate this shortcoming, Cushing et al. recently presented a method to measure the compressibility for specific cell types [13], which opens the route to a simplified way to tailor make acoustophoresis buffers for optimized separation performance.

1.1.1.2 Two-Dimensional Focusing

Early acoustophoresis based particle separation experiments relied on focusing the particles in one dimension (Fig. 1.3, left) [5, 14]. The parabolic flow prole in the microchannels will have an impact on the separation performance when operated in this mode. This is due to the fact that particles suspended in different parts of the channel will have different retention time in the device and, thus, experience the acoustic standing wave field increasingly longer when the particles are positioned nearer the walls or the top and bottom of the channel. A smaller particle initially positioned closer to the wall and especially top and bottom of the channel may, thus, be able to move to the same final position as a larger particle initially positioned closer to the channel center. As the retention time in the force field for the smaller particle will be longer it will make up for the fact that this particle will experience a smaller acoustic radiation force and thus move slower.

In order to improve the resolution of the separation, the particles can be pre-aligned in two dimensions, both horizontally and vertically, to ensure that they all are positioned in the same flow vector and thereby have the same retention time in

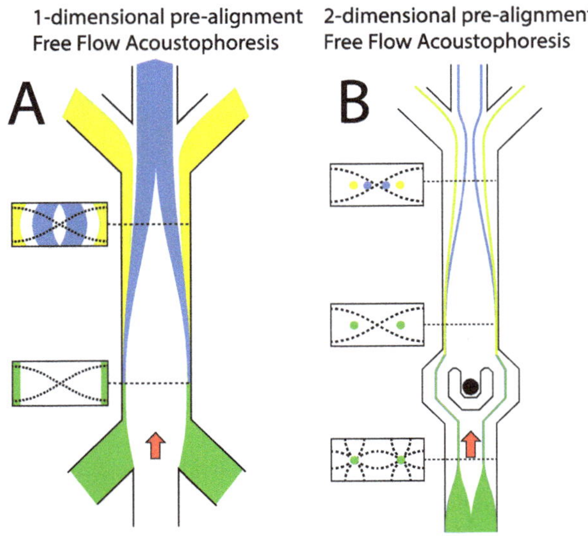

Fig. 1.4 Schematic of the impact of separation performance in free flow acoustophoresis using either (**a**) one-dimensional pre-alignment or (**b**) two-dimensional pre-alignment

the separation microchannel (Fig. 1.3). Particles pre-aligned in 1-D along the sidewall, Fig. 1.4a, will display a broadened distribution at the outlet depending on the initial vertical distribution of the particles in the acoustophoresis channel during one-dimensional separation, whereas particles that are pre-aligned in two-dimensions before the acoustophoresis separation step, Fig. 1.4b, will display a minimized dispersion at the separation outlet. Two-dimensional pre-alignment has been used to improve the sorting efficiency [15–17].

Two-dimensional focusing has also been used to improve the ability to concentrate cells and particles in continuous flow using acoustophoresis [18]. The two-dimensional focus positions the cells in the fastest moving fluid regime when passing the outlet region. In the outlet region where the channel widens, the channel width no longer matches the wavelength of the ultrasound. In this region resonances occur in several directions, creating an spurious acoustic standing wave field that may divert the particles from their original trajectories to end up in the waste outlet. Also, acoustically induced vortices appear at the edges of the outlet flow splitter which may divert one-dimensionally focused particles from the central stream line, *i.e.* if the particles move in the slower moving fluid regime past this region they may spend enough time there to be diverted. If they are, however, two-dimensionally focused to be positioned in the fast-moving central fluid regime they will pass this region rapidly and will not divert from their original trajectories.

1.1.1.3 Acoustic Streaming

An acoustic standing wave in a microchannel will not only exercise a primary radiation force on particles suspended in the acoustic standing wave field, it will

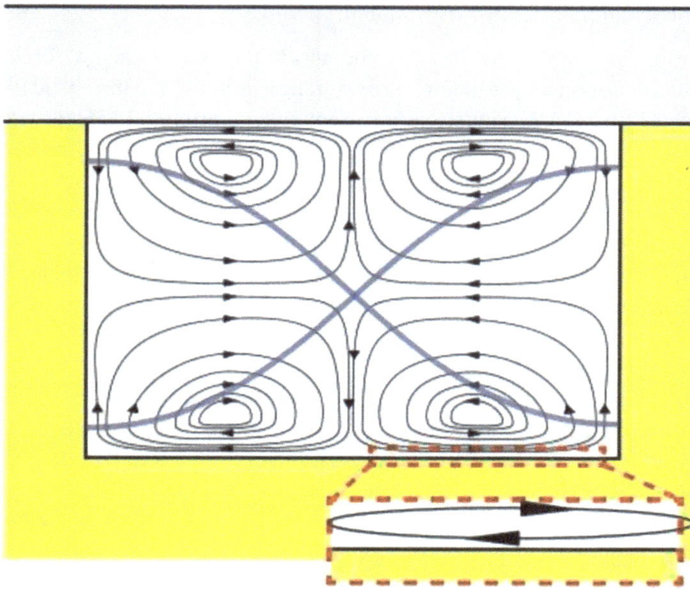

Fig. 1.5 Schematic of Rayleigh streaming in the cross-section of a microchannel actuated at a λ/2 resonance. Lower insert illustrates the boundary driven Schlichting streaming that induce the Rayleigh streaming rolls

also induce the acoustic streaming, driven by the shear stress near rigid walls in the acoustic boundary layers of the standing wave [19]. This phenomenon was originally described by, and named after, Lord Rayleigh to Rayleigh streaming [20]. Later, the theory has been extended by Schlichting [21], Nyborg [22], Hamilton [23], and Muller [24], among others.

The Rayleigh streaming seen in acoustofluidic devices is driven by the viscous dissipation of acoustic energy into the boundary layer of the fluid along the channel walls. The dissipation in the boundary layer next to the wall is relatively large in comparison with the bulk dissipation because of the steep velocity gradient that is formed perpendicular to the channel wall due to the non-slip boundary condition, dictating that the velocity by the wall decreases to zero. The boundary layer δ_v in most acoustofluidic devices is less than 1 μm thick, harboring the steep velocity gradient. The viscous dissipation results in a momentum flux close to the channel wall, which in turn results in the formation of boundary layer vortices called Schlichting streaming (Fig. 1.5). This boundary layer streaming will in turn generate the large streaming vortices seen in the bulk of the fluid, the Rayleigh streaming. The acoustic streaming is particularly pronounced when $\lambda >> h >> \delta_v$, where λ is the ultrasound wavelength and h is the microchannel height (perpendicular to the direction of propagation of the standing wave) [19].

1. *One dimensionally actuated Rayleigh streaming*

In common acoustofluidic devices the motion in the acoustic force field of particles (with properties similar to polystyrene) larger than the critical particles diameter of ≈2 μm (BAW acoustophoresis actuated at around 2 MHz) is dominated by the acoustic radiation force. The motion of smaller particles is, however, dominated by the Stokes' drag-induced acoustic streaming [25, 26]. At some positions in the microchannel the resulting motion of the Stokes' drag-induced acoustic streaming counteracts the motion induced by the primary acoustic radiation force. For larger particles, this presents no problem as the primary acoustic radiation force is larger which keeps the focused particles in the nodal plane of the acoustic standing wave. For smaller particles, however, this disables focusing of the particles when the acoustic streaming drags the particles away from the nodal plane at the channel top and bottom. The transition from acoustic radiation force-induced to acoustic streaming-induced drag force motion is easily understood when considering that the primary acoustic radiation force scales with the cube of the particle radius while the acoustic streaming-induced drag force scales with the radius. Thus, as the particle radius decreases the primary radiation force decreases faster than the streaming-induced drag force.

2. *Two dimensionally actuated Rayleigh streaming*

When actuating the microchannel in two dimensions by two orthogonal resonances the generated acoustic streaming velocity field is fundamentally different compared to a single standing wave actuation [27]. The two-dimensionally actuated streaming velocity field does not counteract the primary radiation force in the same way as the classic Rayleigh streaming does. When actuated in only one dimension, sub-micrometer particles are caught in the characteristic quadrupolar structure of the streaming flow counteracting the focusing on these smaller particles (Fig. 1.5). When simultaneously actuating in two dimensions the streaming pattern instead takes the form of a single vortex (Fig. 1.6 (left)), no longer counteracting the focusing of the smaller particles but instead rotating them and under the influence of the primary radiation force they take on a spiraling trajectory towards the channel center (Fig. 1.6 (right)). The occurrence of the circular streaming is dependent of the oscillation phase shift between the side walls and top/bottom walls. Using this finding Antfolk et al. reported that the critical particle diameter for acoustic particle focusing could be reduced below 0.5 μm in diameter (for polystyrene-like particles) when actuating the microchannel in two dimensions [27].

1.1.1.4 Size-Insensitive Separation

Most successful acoustophoresis separation experiments are dependent mostly on the particle size differences, but the effect of the density and compressibility can also be seen for example when Grenvall et al. [28] managed to separate monocytes from granulocytes even though they are very similar in size.

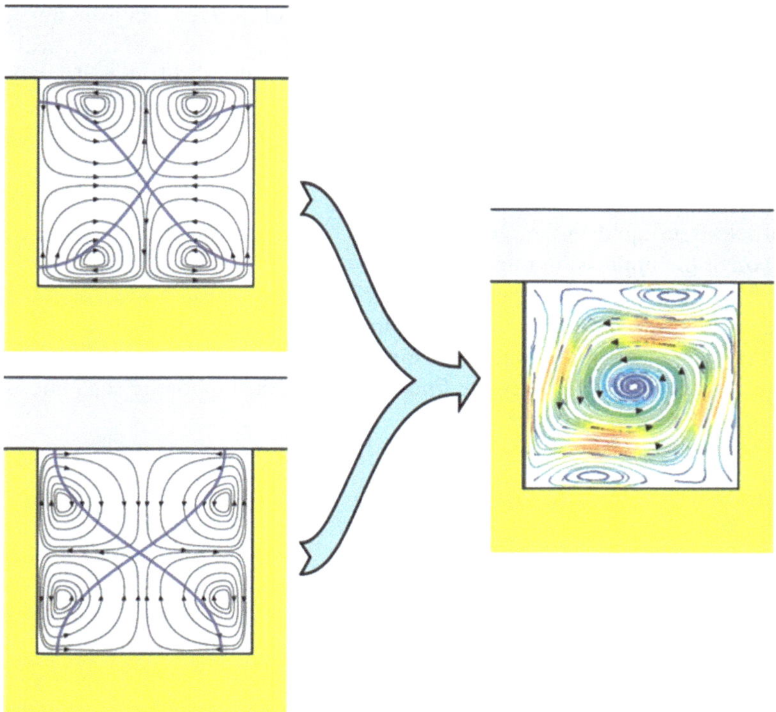

Fig. 1.6 Superimposing horizontal and vertical λ/2 resonance modes and Rayleigh streaming (left) in a square cross-section channel yields a particle trajectory that is driven by acoustic streaming and the radiation force towards the channel center (right). (Adapted from Ref. [27])

In light of this, Augustsson et al. [12] recently reported on an innovative method, iso-acoustic focusing, to perform size-insensitive acoustophoretic separation of cells in inhomogeneous density liquid flow. The cells can be separated based on their acoustic impedance that in turn depends on their density. Using this device, they showed that the acoustic impedance of the breast cancer cell line MCF7 differs from that of the different WBC populations. This indicates the possibility of a microfluidic separation method less dependent on size that will not be biased towards only recovering the larger cancer cells.

1.1.1.5 Medium Switching

Another acoustofluidic phenomenon that must be accounted for is the medium switching effect. Similar to particles, whole fluids volumes can also be acoustically manipulated. As most acoustofluidic cell separation experiments utilizes a particle-free liquid in the microchannel centre into which the separated particles can be focused from the side streams it is important to match to acoustic impedance of the two separate carrier fluids accordingly. If two fluids of different acoustic impedance stream alongside each other in an λ/2 acoustic standing wave field there will be an acoustic force acting at the fluid interface such that the fluid of higher acoustic impedance will relocate to the acoustic pressure node. This may have profound impact on the outcome of acoustophoresis experiments. *E.g.* if the acoustic impedance of the outer particle streams is higher than the central particle-free fluid the entire particle fluid stream will relocate to the pressure node in the centre and drive the particle-free liquid to the channel sides [29]. Thus, no separation has occurred even though the impression might be that all target particles are now focused in the centre. The medium switching effect might be seen for example when attempting to separate cells from plasma if the acoustic impedance of the cell free centre liquid is not properly matched. Medium switching occurs at very modest differences in acoustic impedance (<1%) why the fluids used in acoustophoretic separations with striated flows have to be properly tuned.

1.1.1.6 Biocompatibility

Acoustofluidics have been shown to be a gentle method to cells, not affecting the viability or other cell functions [30, 31]. When separating platelets from peripheral blood progenitor cells Dykes et al. reported that the platelets still retained their capacity to become activated after the acoustic separation, and furthermore, that there were significantly fewer activated platelets in the acoustophoretically separated sample compared to control samples obtained from standard buffy coat platelet concentrates. It was also shown that the blood progenitor cells retained their colony-forming ability after the acoustic separation [32]. Similarly, Urbansky et al. showed a preserved clonogenic capacity after separating hematopoietic progenitor cells in a BAW device [33]. Further evidence that acoustic cell handling is gentle was reported by Hultström et al. in studies of acoustic cell trapping where the viability and proliferation was unaffected after acoustic trapping of COS-7 cells for 30–75 min [34].

1.2 Blood Components

Applications of acoustophoresis has been dominated by the life science field and much effort has been put into the separation of different cell types and/or blood components from each other.

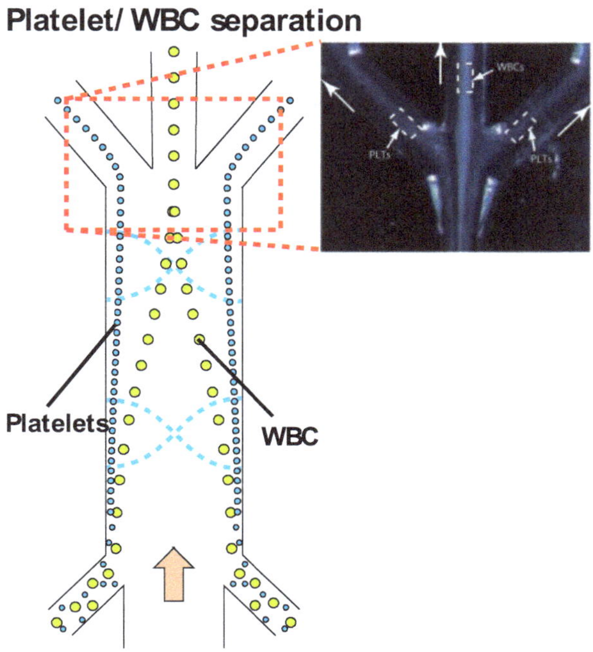

Fig. 1.7 Schematic of platelet separation from WBC in an apheresis product

1.2.1 Platelets

As platelets display a very modest acoustophoretic mobility, several groups have reported approaches to separate platelets from other blood cells by migrating these other cells to the pressure node, leaving the platelets in the original streamline. Dykes et al. removed platelets from peripheral blood progenitor cell products by focusing the WBCs into a clean central buffer stream while the platelets followed the plasma stream along the walls of a $\lambda/2$ resonator with a central and a single side outlet configuration (Fig. 1.7). A depletion efficiency of 89% of the platelets and a leukocyte recovery of 98% at a flow rate of 20 μL/min using a BAW device was reported [32].

Using a SAW device, Nam et al. separated platelets from whole blood with a recovery or 74.1% and a removal of 99.9% of the RBCs and WBCs. The device could, however, only be operated at a flow rate of 0.25 μL/min [35]. Chen et al. later proposed a parallel plate BAW device configured to perform an h-filter type separation which could be operated in an impressive high-throughput mode of 10 mL/min but with a modest separation performance, removing about 80% of the RBCs and WBCs while recovering around 80% of the platelets [36].

1.2.2 WBC Sub-populations

A more challenging blood component separation is fractionation of different white blood cell populations. Here, Grenvall et al. managed to separate lymphocytes, granulocytes, and monocytes into three different populations. Using a BAW device, they showed that 86.5% of the lymphocytes could be recovered with a purity of 95.2% and 68.4% of the granulocytes were recovered with a purity of 98.5%, and 83.1% of the monocytes could be recovered with a purity of 25.5% at a sample flow rate of 8 μL/min and a cell throughput of 8000 cells/min [28].

When different cell types display overlapping acoustophoretic mobilities and are not possible to separate based on their acoustophysical properties, affinity beads directed towards a specific cell surface marker can be used. As the microbeads bind to their target cell the bead cell complex display a different acoustic fingerprint and will migrate at a different velocity as compared to the unbound cells, thereby enabling acoustophoretic separation, analogous to magnetic bead based separation. Using a BAW device Lenshof et al. utilized anti-CD4 affinity beads that gave the target cell a higher acoustic mobility compared to the unlabeled lymphocytes of a similar cell size. In this way 65 ± 22% of the CD4+ T cells could be recovered with a purity of 87 ± 12% at a sample flow rate of 30 μL/min and a cell concentration of 10^7 cells/mL [37]. Similarly, Urbansky et al. also utilized affinity beads to separate CD8 + T cells from peripheral blood progenitor cells in a BAW device. The mean purity of the separated sample was 91.8 ± 8% and a median separation efficiency of 63% (range 15.1% to 90.5%) could be obtained at a sample flow rate of 60 μL/min and a cell concentration of 10^7 cells/mL [33].

Utilizing a BAW device (Fig. 1.8) Urbansky et al. separated the mononuclear cell (MNC) fraction of the WBCs, including monocytes and lymphocytes, from RBCs in 100x diluted blood samples. By optimizing the acoustic impedance of the medium, thereby altering the relative acoustic mobility of the MNCs compared to the RBCs, a high separation efficiency could be reported, where 99.4 ± 0.4% of the collected cells where MNCs with a contamination of 0.8 ± 0.6% of RBCs. This corresponded to at 390.6 ± 169.8-fold enrichment of the MNC fraction and a recovery of 68% [38].

1.2.2.1 Integrated Acoustic Device for Monocyte Subpopulation Isolation

Nivedita et al. proposed an integration of a spiral microfluidic device for Dean flow and a BAW acoustic device utilizing microbubbles to size-selectively trap cells. The device was used to separate a subpopulation of monocytes that were larger than 18 μm in diameter from 10x diluted whole blood. At a high flow rate of 1.1 mL/min they were able to obtain a pure sample of only monocytes [39].

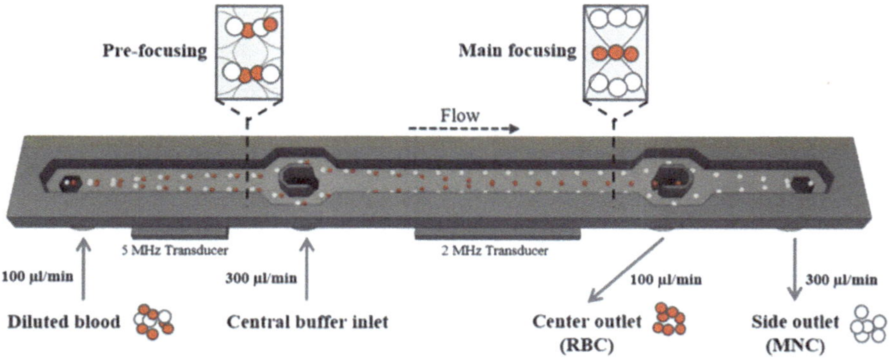

Fig. 1.8 Acoustic device for separating mononuclear cells from red blood cells. (Reprinted from [38], Copyright (2017) The Author(s) licensed under CC BY 4.0)

1.2.3 Plasma and Plasma Proteins

Lenshof et al. demonstrated a way to obtain cell-free plasma from whole blood by acoustically focus and sequentially removing the concentrated blood cells focused in the center of a BAW device while retaining as much plasma as possible. At a sample inlet flow rate of 80 µL/min plasma with only $3.65 * 10^9$ cells/L could be produced, which is below the recommended maximum threshold of $6 * 10^9$ cells/L for plasma transfusions set by the Council of Europe [40]. Later, the same platform was integrated with an immunoaffinity-capture platform for the detection of PSA from the plasma separated from whole blood. Spiked PSA could be detected in the separated plasma at clinically significant levels of 1.7–100 ng/mL [41]. The presented blood plasma isolation method fulfilled clinical standards for diagnostic purposes, and if possible to integrate with an analytical step, the development may become a valuable contribution in the field of clinical diagnostics.

In a similar way Tenje et al. used a BAW device to wash RBCs from plasma in an attempt to produce blood plasma protein-free RBC concentrates important for blood transfusion to patients suffering from IgA-sensitivity. While recovering 97% of the RBCs they showed an almost complete removal of IgA with a concentration of less than 0.25 µg/mL in the cell fraction, which is below the limit of detection and also in accordance with the European guidelines. The RBC sample was diluted 20x before processing and the sample inflow rate was 100 µL/min [42]. It should be noted that it was critical for the separation to balance the acoustic impedance of the central buffer versus the sample and ensure that medium switching did not occur. Although showing a good performance in RBC isolation and background elimination, the method still needs to show its value in terms of upscaling and throughput to meet clinically relevant applications.

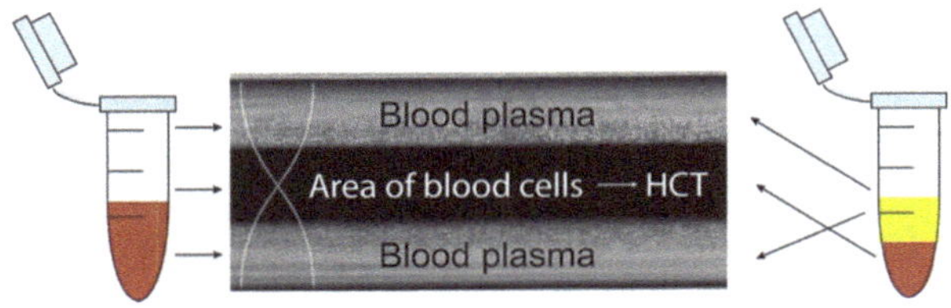

Fig. 1.9 Simple acoustic hematocrit measurement with optical readout. (Reprinted from [44], Copyright (2018), with permission from Elsevier)

1.2.4 RBC Separation

Most acoustophoresis devices operate at flow rates around 1–100 μL/min. At the lower end of this range the flow rate might not be clinically relevant for processing a blood sample. Especially, if the target cells are less abundant in the blood sample. In an attempt to show a high throughput device Adams et al. fabricated a parallel plate BAW device that could be operated at an impressive flow rate of 16.7 mL/min. At this flow rate, they showed that they could separate 2 μm and 10 μm diameter particles with a recovery of around 80% of the 10 μm particles and a contamination of 10% of the 2 μm particles [43]. When using this set-up to separate RBCs from whole blood into PBS a recovery of 95% was reported at a flow rate of 16.8 mL/min. It is however unclear if this separation actually occurred or if the recovery of cells at the target outlet resulted as an effect of fluid relocation due to the differences in acoustic impedance between PBS and whole blood as previously reported by Tenje et al. [42]. No data was reported showing that the blood plasma background was eliminated and since only the cell recovery and not the wash efficiency was documented the actual separation was not concluded in the paper.

1.2.5 Hematocrit Measurements

Hematocrit measurements are commonly made by cell counting or centrifugation followed by estimation of the packed cell volume. Using a simple BAW device, Petersson et al. showed that acoustic hematocrit measurements were possible in just 2 s by acoustically focusing the blood cells in the channel center, separating it from the plasma. The hematocrit could be detected by optically monitoring the cell-focused area in the microchannel center with an error of less than 3% [44]. (Fig. 1.9). The method was tested on a limited number of healthy volunteers and the full clinical value of the method still needs to be clarified by benchmarking a broader population of healthy and sick individuals compared to the golden standard.

1.3 Circulating Tumor Cells

Circulating tumor cells (CTCs) are among the most targeted rare cells found in blood. Shed from a primary tumor they travel with the blood circulation to remote tissues where they potentially can form metastases. Found in quantities as low as 1–10 CTC/mL they are challenging to detect but have a great diagnostic value and interest to both clinicians and researchers. CTCs are sought after both for their prognostic and diagnostic values, *e.g.* as a marker for the disease progression and response to different treatments [45].

1.3.1 Cancer Cell Separation Based on a Positive Acoustic Contrast Factor

In an early approach, Augustsson et al. [17] used a BAW device to separate spiked tumor cells from WBCs based on the fact the that the tumor cells display a larger average size. With the same configuration as in Fig. 1.8, the device could recover 85.4% of the cancer cells with a contamination of 0.7% of WBCs at a sample flow rate of 70 μL/min. Later Antfolk et al. presented a BAW device with a simplified fluidic setup based on the same separation principle. Simplification was achieved by removing the center cell-free fluid inlet when performing the separation. In a first channel section the cells were pre-aligned in two dimensions in two parallel streams $\lambda/4$ from the channel side walls, before entering the separation zone that operated in a $\lambda/2$ mode (and analogous to the separation in Fig. 1.8), where the tumor cells were separated from the WBCs and collected in the central outlet. This device had three obvious advantages.; (1) It facilitated the flow control of the device, which will generate higher recoveries and purities of target cells; (2) It would also more easily lend itself to integration with other unit operations, keeping the fluidic connections to a minimum.; and (3) The need to match the acoustic impedances of fluids, to avoid acoustic impedance mismatch when using multiple inlet streams, was eliminated. Using this device, spiked cancer cells could be separated from WBCs with a recovery of $86.5 \pm 6.7\%$ of the cancer cells with $1.1 \pm 0.2\%$ contamination of WBCs at a sample flow rate of 100 μL/min. In addition to the mentioned advantages, this device also made it possible to concentrate the recovered cells about 2.5-fold instead of diluting them as devices relying on hydrodynamic pre-focusing often do, a fact that is especially important for rare cell processing [16]. The original platform presented by Augustsson et al. [17] was later improved to facilitate the processing of clinical-scale samples of 10 mL of 2x dilute RBC-lysed blood spiked with 50 MCF-7 breast cancer cells in $\approx$ 2 h. In a high-recovery mode the cancer cell recovery was

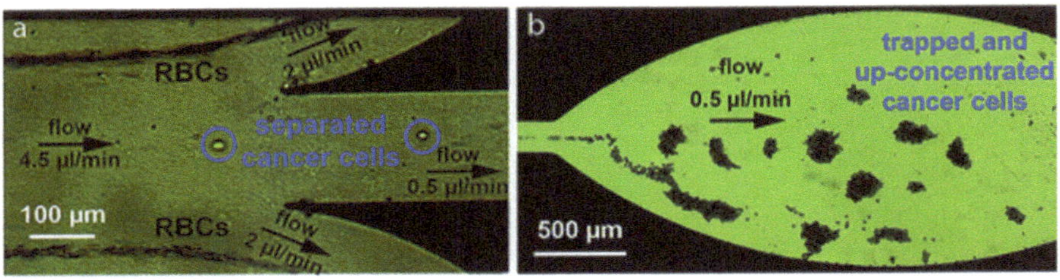

Fig. 1.10 Separation and subsequent trapping of lung cancer cells A549 from RBCs. (Reprinted from [47], Copyright (2015) The Author(s) licensed under CC BY 3.0)

86 ± 2.3% with a WBC contamination of 0.57 ± 0.14%, and in a high-purity mode the cancer cell recovery was 62 ± 8.7% with a WBC contamination of 0.1 ± 0.07%. This corresponded to cancer cell enrichment factors of 162 ± 55 and 1445 ± 1811, respectively for the high-recovery and high-purity modes and demonstrated a clinical potential for CTC enumeration based on acoustophoresis [46].

Antfolk et al. also presented an analogous integrated BAW device to simultaneously separate and concentrate spiked cancer cells from WBCs. It was reported that MCF7 cells could be separated from WBCs with a recovery of 91.8 ± 1.0% and a contamination of 0.6 ± 0.1% WBCs with a simultaneous 23.8 ± 1.3-fold concentration of the cancer cells. Correspondingly, DU145 cells could be recovered with an efficiency of 84.1 ± 2.1% with 0.2 ± 0.04% contamination of WBCs while concentrating the cancer cells 9.6 ± 0.4-fold at a sample flow rate of 80 μL/min. This might reduce the need for a separate concentration step, commonly performed by centrifugation, before sample analysis and cancer cell enumeration which greatly reduces the risk of losing rare target cells during this step.

In an attempt to take on-chip sample preparation further, Iranmanesh et al. [47] presented a BAW device that simultaneously separated spiked cancer cells (lung cancer cell line A549) from RBCs, trapped and concentrated the recovered cancer cells and allowed for on-chip fluorescence image analysis of the trapped cells. 85.5% of the cancer cells could be recovered with a purity of 100%, and the cells could be concentrated 130 times in the trap (Fig. 1.10). The sample flow rate was, however, only 4.5 μL/min making rare cell processing very time consuming in its current state. It should also be noted that the experiments were made on an RBC suspension without WBCs and the main challenge in CTC separation from blood is to discriminate against the WBC background that has more similar acoustophysical properties to CTCs than RBCs do.

Using a tilted-angle SSAW device, Li et al. [48] separated spiked cancer cells from WBCs at a sample flow rate of 20 μL/min, *i.e.* >4 h for processing a 5 mL clinical sample. While removing 99% of the WBCs 60–80% of the cancer cells could be recovered. However, the system was tuned to operate at a high CTC recovery, 90%, (based on the cell line data) when running three metastatic breast cancer, leading to a challenging background of 10% remaining WBCs. This should be compared to the golden standard, CellSearch, that display a WBC depletion of about log 3 to 3.5.

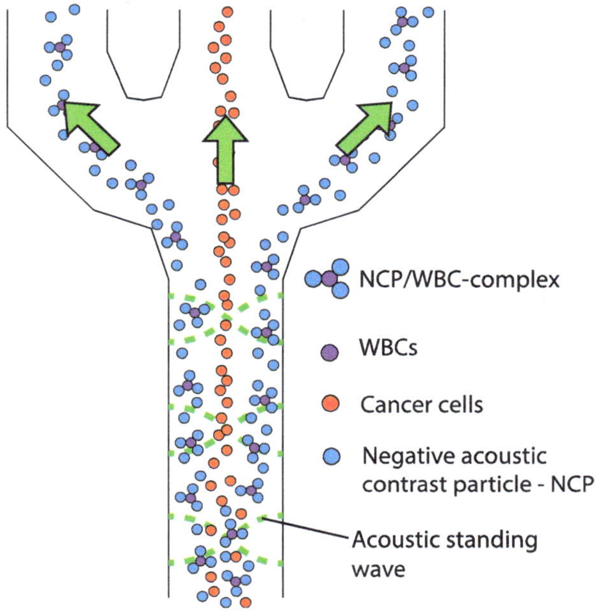

Fig. 1.11 Schematic of acoustophoretic negative selection using antibody activated microbeads having a negative acoustic contrast factor (NCP). NCP microbeads bind to WBCs, forming a complex that displays a net negative acoustic contrast factor, moving towards the pressure antinode in the microchannel, while cancer cells (red), having a positive acoustic contrast factor, focus in the pressure node in the center of the channel

1.3.2 Cancer Cell Separation Based on a Negative Acoustic Contrast Factor

Faridi et al. used anti-EpCAM coated microbubbles to show that they could separate microbubble-bound cells (HCT 116, a colon cancer cell line) from unbound cells. The microbubbles display a negative acoustic contrast factor, *i.e.* they move towards the pressure antinode in the acoustic standing wave field. At a sample flow rate of 3 μL/min they showed a 75% separation efficiency [49].

Using PDMS-based negative contrast factor particles (NCP), Cushing et al. showed that they could separate cancer cells from WBCs bound to the anti-CD45 activated negative contrast particles. A separation efficiency of 99% and a cancer cell recovery of 85–90% and a contamination of 1–2% WBCs, spiking 1000 MCF-7 or DU145 cells per 100,000 WBCs, was demonstrated (Fig. 1.11) [50].

1.3.3 Cancer Cell Concentration

Rare cells, after isolation, are often dilute since they originate from large volumes of sample. When handling dilute cell suspensions containing few target cells, or small sample volumes produced by microfluidic processing, ordinary concentration techniques such as centrifugation might not be practically usable due to inherent losses

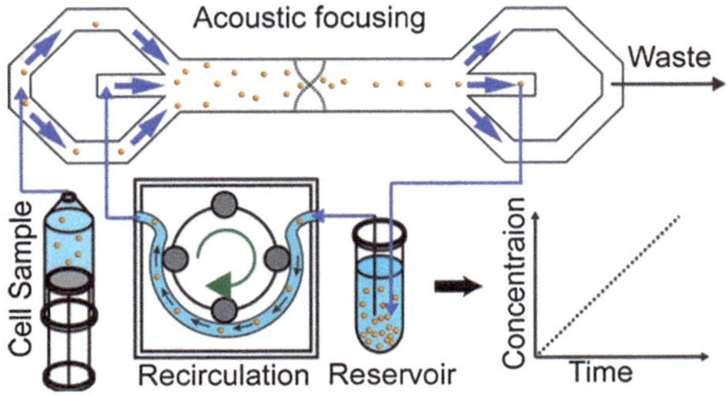

Fig. 1.12 Acoustophoresis BAW concentration device operated in a recirculating mode. (Reprinted with permission from [51]. Copyright 2015 American Chemical Society)

linked to centrifugation, i.e. no formation of a cell pellet or adsorption to centrifugation tube walls. To this end, BAW acoustophoresis has been used to concentrate cells and particles in order to facilitate subsequent analysis.

Nordin and Laurell showed that sequential concentration steps and two-dimensional focusing was crucial to achieve a high concentration factor in a continuous flow BAW device. The sequential concentration steps were used to multiply a modest concentration factor achieved in each step to a high overall concentration factor. The two-dimensional focusing allowed the particles to all be positioned in the flow regime with the highest flow rate which reduced the time spent in the acoustically chaotic trifurcation regions to a minimum, preventing them from being deflected from their established flow trajectories and into the waste outlets. Using these findings dilute cancer cell suspensions could be concentrated 195.7 ± 36.2-fold with a recovery of 97.2 ± 3.3% at a sample input flow rate of 200 μL/min [18]. Jakobsson et al. took the concentration possibilities even further and using a recirculating BAW device they showed that RBCs could be concentrated a staggering 1166 ± 110 times with a 98.7% recovery, and DU145 cells and MCF7 cells could be concentrated 817 ± 125 times with a recovery of 90.2%, and 519 ± 115.7 times with a recovery of 81.7%, respectively (Fig. 1.12) [51].

1.3.4 Integrated Acoustic Devices for CTC Processing

Acoustophoresis systems have also been integrated with other microfluidic devices to add even more to the usability of the devices. Kim et al. presented an integration of a BAW concentration device and a dielectrophoresis (DEP) trapping device, enabling simultaneous concentration, trapping and subsequent analysis of target cells. As a consequence of the cell concentration step, the device also translated a high flow rate into a lower flow rate adapted for the in-line integrated DEP trapping chip that was dependent of a low flow rate to efficiently trap cells. Without the concentration step, the low flow rate limit of the DEP trapping chip would be very time consuming and in principle be non-applicable to clinical samples. When integrating the two devices a higher sample flow rate could be used. The integrated system could process samples with a tenfold higher flow rate as compared to the DEP trapping chip alone, while still recovering over 90% of the cells at a sample input flow rate of 40 μL/min [52].

An acoustic separation step was later added to the integrated rare cell concentration and DEP-trapping system. The totally integrated CTC system efficiently eliminated the risk of cell losses in each of the unit operations by enabling direct on-chip tumor cell separation, concentration, trapping and labelling prior to identification and enumeration. Prostate cancer cells (DU145) spiked into a sample of the peripheral blood mononuclear cell (PBMC) fraction, with a concentration corresponding to their concentration in whole blood, were efficiently separated and trapped at a recovery of 76.2 $\pm$ 5.9% of the cancer cells and a minute contamination of 0.12 $\pm$ 0.04% PBMCs while simultaneously enabling a 20x volumetric concentration at a sample flow rate of 80 μL/min [53].

1.4 Bacteria

1.4.1 Passive Bacteria Processing

The smaller size of bacteria compared to eukaryotic cells make them more challenging to process using microfluidics as the active forces often scales with the size of the cell or particle. Many attempts of bacteria processing in microfluidic devices thus aim to isolate them by actively removing the larger blood cells, leaving the bacteria unaffected in the plasma fraction.

Ai, Sanders and Marrone [54] used SAW to separate *E. coli* spiked into pre-isolated peripheral blood mononuclear cells (PBMCs). In the target outlet 95.65% bacteria and 3.92% WBCs could be found, while 7.24% of the cells collected in the waste outlet were found to be bacteria and 91.48% were WBCs. The sample was processed with a flow rate of 0.5 μL/min. Ohlsson et al. [55] proposed a system combining a BAW device with an acoustic trapping device for rapid bacteria separation from whole blood in sepsis samples. In the acoustophoresis chip blood cells where focused and subsequently removed (plasmapheresis) leaving the bacteria in the cell free plasma which were acoustically trapped and concentrated in the down stream trapping unit (Fig. 1.13). The trapped and enriched bacteria were subsequently released directly into a PCR microchip for analysis. With a sample flow rate of 80 μL/min they processed a blood sample of 1 mL (diluted to 70% whole blood) in less than 2 h while recovering ~11% of the bacteria and a ~0.1% RBC contamination, resulting in a detection limit of 1000 bacteria/mL blood.

Using a glass BAW device, Ngamsom et al. separated the bacteria *S. typhimurium*, spiked at a rate of 10^7 CFU/mL, from 2x diluted horse blood. By focusing the RBCs in the channel center at a sample flow rate of 10 μL/min, they

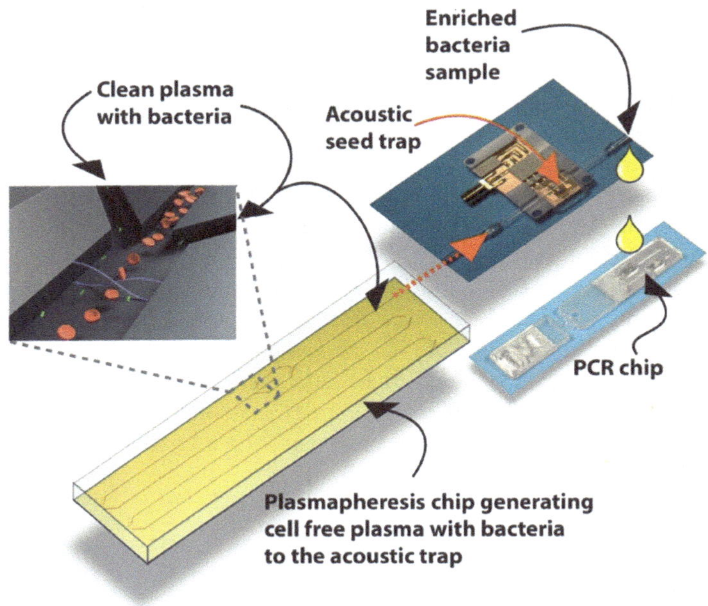

Fig. 1.13 Schematic of the components assembling the integrated sepsis diagnostics system presented by Ohlsson et al. [55], where an initial acoustophoretic removal of blood cells produced a clean plasma (plasmapheresis) containing bacteria that continued to an acoustic trapping unit where bacteria were enriched and washed prior to release into a disposable PCR microchip followed by DNA amplification and bacteria identification

could deplete around 99.8% of the RBCs while recovering around 10% of the bacteria. (Fig. 1.14) [56] Presumably, the low recovery of the bacteria stems not from an acoustic mobility velocity overlap but rather from a mismatch of the acoustic impedances of the sample and the buffer solution where the phenomena medium switching (mentioned above, see also [42]) occurs rather than a separation. In the case of medium switching the sample fluid will, if it has a higher acoustic impedance than the central buffer, entirely switch place with the central buffer rather than to just allow the cells to acoustophoretically migrate across the stream lines into the clean buffer. This phenomenon would explain for the fact that 90% of the bacteria ends up together with the RBCs rather than staying at the sides even though they are significantly smaller than the horse RBCs and thus have a much lower acoustic mobility, ensuring that they would transition to the central stream much later than the RBCs under conditions where no medium switching occurred.

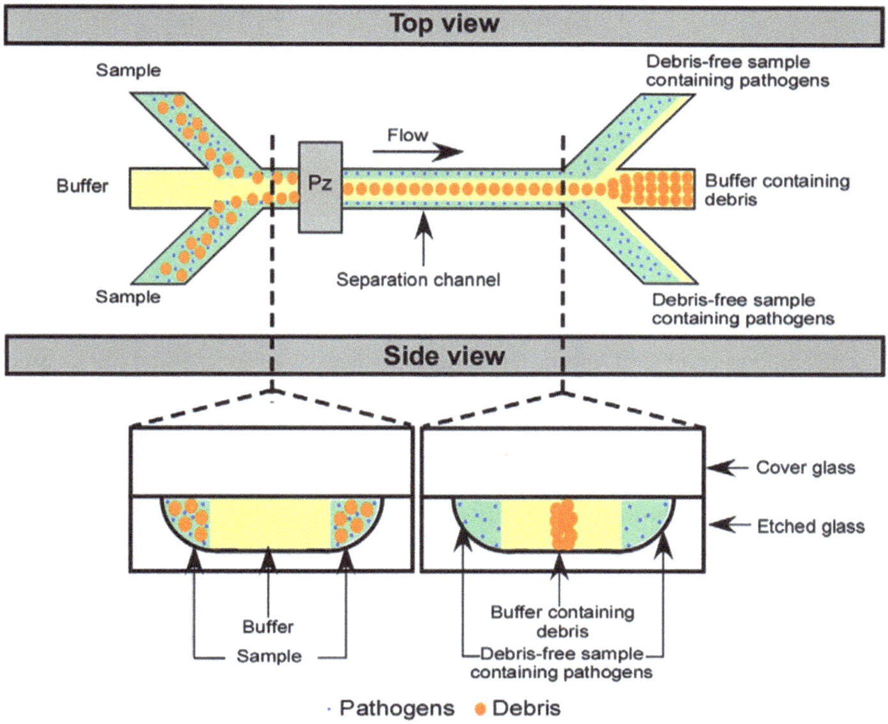

Fig. 1.14 Acoustofluidic device for passive separation of bacteria from blood by depletion of the RBCs. (Reprinted from [56], Copyright (2016), with permission from Elsevier)

1.4.2 Active Bacteria Processing

Although often more challenging, some attempts have been made to process bacterial samples by actively moving them with external forces. Acoustophoresis devices that focus particles in one dimension and are operated at a frequency of around 2 MHz handles larger cells efficiently. They can, however, not be efficiently used to handle particles and rare cells smaller than about 2 μm in diameter, practically meaning that most bacteria cannot be actively processed by acoustophoresis. While the acoustophoresis-induced motion of larger particles is dominated by the acoustic radiation force, the motion of smaller particles is instead dominated by the acoustic streaming-induced drag force. Larger particles can, thus, be focused while smaller particles instead are trapped in the streaming vortices inherent with the acoustic field. When striving to solve this problem, Antfolk et al. found that while simultaneously exciting two orthogonal resonances an acoustic streaming velocity field was formed that no longer counteracted the primary radiation force, which thereby allowed sub-micrometer particles to be focused (Fig. 1.6, section *Two dimensionally actuated Rayleigh streaming).* The device was also used to focus

E. coli achieving a focusability of 0.95 ± 0.35. Simulations and experiment showed that the streaming velocity field was dominated by a large centered flow roll instead of the four characteristic Rayleigh vortices seen when actuating the channel in one dimension (Fig. 1.6) [27]. This achievement opens up the acoustophoresis field against many microbiology applications not previously possible.

1.5 Conclusions

Acoustophoresis has been extensively used for processing biological samples with a special emphasis on blood sample handling and processing. The method can be used for separation, concentration/enrichment, buffer exchange, or alignment and has been found useful for blood component separation, cancer cell separation, or hematocrit measurements amongst other processes. It has also been demonstrated that the method can be used to process samples of a clinically relevant size in a relevant time. Acoustophoresis is, however, limited to the size of particles that can be actively processed, due to the acoustic streaming drag force. This limit has been lowered through recent developments and is likely to be lowered more in the future. Caution also has to be made to ensure that the different liquids used are clearly matched in their acoustic impedances to avoid medium switching to occur, where two striated fluids relocates in the acoustic field instead of just the intended cells or particles. Even so, when considering these concerns during the experimental design, acoustophoresis has proven to be a useful method with the ability to integrate with other microfluidic or macroscale methods, why it can be anticipated to see acoustophoresis solutions become an integral part of future cell processing systems as well as bioanalytical and clinical diagnostic systems.

References

1. Lenshof A, Magnusson C, Laurell T (2012) Acoustofluidics 8: applications of acoustophoresis in continuous flow microsystems. Lab Chip 12:1210–1223. https://doi.org/10.1039/c2lc21256k
2. Jakobsson OJE, Grenvall C, Nordin M et al (2014) Acoustic actuated fluorescence activated sorting of microparticles. Lab Chip 14:1943–1950. https://doi.org/10.1039/c3lc51408k
3. Jakobsson O, Antfolk M, Laurell T (2014) Continuous flow two-dimensional acoustic orientation of non-spherical cells. Anal Chem 86:6111–6114
4. Gedge M, Hill M (2012) Acoustofluidics 17: theory and applications of surface acoustic wave devices for particle manipulation. Lab Chip 12:2998. https://doi.org/10.1039/c2lc40565b
5. Laurell T, Petersson F, Nilsson A (2007) Chip integrated strategies for acoustic separation and manipulation of cells and particles. Chem Soc Rev 36:492–506. https://doi.org/10.1039/b601326k
6. Li S, Ding X, Mao Z et al (2015) Standing surface acoustic wave (SSAW)-based cell washing. Lab Chip 15:331–338. https://doi.org/10.1039/c4lc00903g
7. King LV (1934) On the acoustic radiation pressure on spheres. Proc R Soc Lond Ser A-Math Phys Sci 147:212–240

8. Yoshioka K, Kawashima Y (1955) Acoustic radiation pressure on a compressible sphere. Acustica 5:167–173
9. Gor'kov LP (1962) On the forces acting on a small particle in an acoustical field in an ideal fluid. Sov Phys Dokl 6:773–775
10. Nyborg WL (1967) Radiation pressure on a small rigid sphere. J Acoust Soc Am 42:947–952
11. Bruus H (2012) Acoustofluidics 2: perturbation theory and ultrasound resonance modes. Lab Chip 12:20–28
12. Augustsson P, Karlsen JT, Su H-W et al (2016) Iso-acoustic focusing of cells for size-insensitive acousto-mechanical phenotyping. Nat Commun 7:11556. https://doi.org/10.1038/ncomms11556
13. Cushing K, Garofalo F, Magnusson C et al (2017) Ultrasound characterization of microbead and cell suspensions by speed of sound measurements of neutrally buoyant samples. Anal Chem 89(17):8917–8923. https://doi.org/10.1021/acs.analchem.7b01388
14. Petersson F, Åberg LB, Swärd-Nilsson A-MK, Laurell T (2007) Free flow acoustophoresis: microfluidic-based mode of particle and cell separation. Anal Chem 79:5117–5123. https://doi.org/10.1021/ac070444e
15. Antfolk M, Magnusson C, Augustsson P et al (2015) Acoustofluidic, label-free separation and simultaneous concentration of rare tumor cells from white blood cells. Anal Chem 87:9322–9328. https://doi.org/10.1021/acs.analchem.5b02023
16. Antfolk M, Antfolk C, Lilja H et al (2015) A single inlet two-stage acoustophoresis chip enabling tumor cell enrichment from white blood cells. Lab Chip 15:2102–2109. https://doi.org/10.1039/C5LC00078E
17. Augustsson P, Magnusson C, Nordin M et al (2012) Microfluidic, label-free enrichment of prostate cancer cells in blood based on acoustophoresis. Anal Chem 84:7954–7962. https://doi.org/10.1021/ac301723s
18. Nordin M, Laurell T (2012) Two-hundredfold volume concentration of dilute cell and particle suspensions using chip integrated multistage acoustophoresis. Lab Chip 12:4610–4616. https://doi.org/10.1039/c2lc40629b
19. Wiklund M, Green R, Ohlin M (2012) Acoustofluidics 14: applications of acoustic streaming in microfluidic devices. Lab Chip 12:2438–2451. https://doi.org/10.1039/c2lc40203c
20. Rayleigh L (1884) On the circulation of air observed in Kundt's tubes, and on some allied acoustical problems. Philos Trans R Soc Lond A 175:1–21. https://doi.org/10.1080/15265161.2011.596400
21. Schlichting H (1932) Berechnung obener periodischer Grenzschichtströmungen. Phys Z 33:327–335
22. Nyborg WL (1958) Acoustic streaming near a boundary. J Acoust Soc Am 30:329. https://doi.org/10.1121/1.1909587
23. Hamilton MF, Ilinskii YA, Zabolotskaya EA (2003) Acoustic streaming generated by standing waves in two-dimensional channels of arbitrary width. J Acoust Soc Am 113:153–160
24. Muller PB, Rossi M, Marín ÁG et al (2013) Ultrasound-induced acoustophoretic motion of microparticles in three dimensions. Phys Rev E 88:23006. https://doi.org/10.1103/PhysRevE.88.023006
25. Barnkob R, Augustsson P, Laurell T, Bruus H (2012) Acoustic radiation- and streaming-induced microparticle velocities determined by microparticle image velocimetry in an ultrasound symmetry plane. Phys Rev E 86:56307. https://doi.org/10.1103/PhysRevE.86.056307
26. Muller PB, Barnkob R, Jensen MJH, Bruus H (2012) A numerical study of microparticle acoustophoresis driven by acoustic radiation forces and streaming-induced drag forces. Lab Chip 12:4617–4627. https://doi.org/10.1039/c2lc40612h
27. Antfolk M, Muller PB, Augustsson P et al (2014) Focusing of sub-micrometer particles and bacteria enabled by two-dimensional acoustophoresis. Lab Chip 14:2791–2799
28. Grenvall C, Magnusson C, Lilja H, Laurell T (2015) Concurrent isolation of lymphocytes and granulocytes using prefocused free flow acoustophoresis. Anal Chem 87(11):5596–5604. https://doi.org/10.1021/acs.analchem.5b00370

29. Deshmukh S, Brzozka Z, Laurell T, Augustsson P (2014) Acoustic radiation forces at liquid interfaces impact the performance of acoustophoresis. Lab Chip 17:3394–3400. https://doi.org/10.1039/c4lc00572d
30. Burguillos MA, Magnusson C, Nordin M et al (2013) Microchannel acoustophoresis does not impact survival or function of microglia, leukocytes or tumor cells. PLoS One 8:e64233
31. Wiklund M (2012) Acoustofluidics 12: biocompatibility and cell viability in microfluidic acoustic resonators. Lab Chip 12:2018–2028. https://doi.org/10.1039/c2lc40201g
32. Dykes J, Lenshof A, Åstrand-Grundström I-B et al (2011) Efficient removal of platelets from peripheral blood progenitor cell products using a novel micro-chip based acoustophoretic platform. PLoS One 6:e23074. https://doi.org/10.1371/journal.pone.0023074
33. Urbansky A, Lenshof A, Dykes J et al (2016) Affinity-bead-mediated enrichment of CD8+ lymphocytes from peripheral blood progenitor cell products using acoustophoresis. Micromachines 7:101. https://doi.org/10.3390/mi7060101
34. Hultström J, Manneberg O, Dopf K et al (2007) Proliferation and viability of adherent cells manipulated by standing-wave ultrasound in a microfluidic chip. Ultrasound Med Biol 33:145–151. https://doi.org/10.1016/j.ultrasmedbio.2006.07.024
35. Nam J, Lim H, Kim D, Shin S (2011) Separation of platelets from whole blood using standing surface acoustic waves in a microchannel. Lab Chip 11:3361–3364. https://doi.org/10.1039/c1lc20346k
36. Chen Y, Wu M, Ren L et al (2016) High-throughput acoustic separation of platelets from whole blood. Lab Chip 16:3466–3472. https://doi.org/10.1039/C6LC00682E
37. Lenshof A, Jamal A, Dykes J et al (2014) Efficient purification of CD4+ lymphocytes from peripheral blood progenitor cell products using affinity bead acoustophoresis. Cytom Part A 85:933–941. https://doi.org/10.1002/cyto.a.22507
38. Urbansky A, Ohlsson P, Lenshof A et al (2017) Rapid and effective enrichment of mononuclear cells from blood using acoustophoresis. Sci Rep 7:17161. https://doi.org/10.1038/s41598-017-17200-9
39. Nivedita N, Garg N, Lee AP, Papautsky I (2017) A high throughput microfluidic platform for size-selective enrichment of cell populations in tissue and blood samples. Analyst 142:2558–2569. https://doi.org/10.1039/C7AN00290D
40. Lenshof A, Ahmad-Tajudin A, Järås K et al (2009) Acoustic whole blood plasmapheresis chip for prostate specific antigen microarray diagnostics. Anal Chem 81:6030–6037. https://doi.org/10.1021/pr8007545.(27
41. Ahmad-Tajudin A, Petersson K, Lenshof A et al (2013) Integrated acoustic immunoaffinity-capture (IAI) platform for detection of PSA from whole blood samples. Lab Chip 13:1790–1796. https://doi.org/10.1039/c3lc41269e
42. Tenje M, Lundgren MN, Swärd-Nilsson A-M et al (2015) Acoustophoretic removal of proteins from blood components. Biomed Microdevices 17:95. https://doi.org/10.1007/s10544-015-0003-5
43. Adams JD, Ebbesen CL, Barnkob R et al (2012) High-throughput, temperature-controlled microchannel acoustophoresis device made with rapid prototyping. J Micromech Microeng 22:75017. https://doi.org/10.1088/0960-1317/22/7/075017
44. Petersson K, Jakobsson O, Ohlsson P et al (2018) Acoustofluidic hematocrit determination. Anal Chim Acta 1000:199–204. https://doi.org/10.1016/J.ACA.2017.11.037
45. Cohen SJ, Punt CJA, Iannotti N et al (2008) Relationship of circulating tumor cells to tumor response, progression-free survival, and overall survival in patients with metastatic colorectal cancer. J Clin Oncol 26:3213–3221. https://doi.org/10.1200/JCO.2007.15.8923
46. Magnusson C, Augustsson P, Lenshof A et al (2017) Clinical-scale cell-surface-marker independent acoustic microfluidic enrichment of tumor cells from blood. Anal Chem 89:11954–11961. https://doi.org/10.1021/acs.analchem.7b01458
47. Iranmanesh I, Ramachandraiah H, Russom A, Wiklund M (2015) On-chip ultrasonic sample preparation for cell based assays. RSC Adv 5:74304–74311. https://doi.org/10.1039/C5RA16865A

48. Li P, Mao Z, Peng Z et al (2015) Acoustic separation of circulating tumor cells. Proc Natl Acad Sci U S A 112:4970–4975. https://doi.org/10.1073/pnas.1504484112
49. Faridi MA, Ramachandraiah H, Iranmanesh I et al (2017) Microbubble activated acoustic cell sorting. Biomed Microdevices 19:23. https://doi.org/10.1007/s10544-017-0157-4
50. Cushing K, Undvall E, Ceder Y et al (2018) Reducing WBC background in cancer cell separation products by negative acoustic contrast particle immuno-acoustophoresis. Anal Chim Acta 1000:256–264. https://doi.org/10.1016/J.ACA.2017.11.064
51. Jakobsson O, Oh SS, Antfolk M et al (2015) Thousand-fold volumetric concentration of live cells with a recirculating acoustofluidic device. Anal Chem 87:8497–8502. https://doi.org/10.1021/acs.analchem.5b01944
52. Kim SH, Antfolk M, Kobayashi M et al (2015) Highly efficient single cell arraying by integrating acoustophoretic cell pre-concentration and dielectrophoretic cell trapping. Lab Chip 15:4356–4363. https://doi.org/10.1039/C5LC01065A
53. Antfolk M, Kim SH, Koizumi S et al (2017) Label-free single-cell separation and imaging of cancer cells using an integrated microfluidic system. Sci Rep 7:46507. https://doi.org/10.1038/srep46507
54. Ai Y, Sanders CK, Marrone BL (2013) Separation of Escherichia coli bacteria from peripheral blood mononuclear cells using standing surface acoustic waves. Anal Chem 85:9126–9134. https://doi.org/10.1021/ac4017715
55. Ohlsson PD, Evander M, Petersson K et al (2016) Integrated acoustic separation, enrichment and microchip PCR detection of bacteria from blood for rapid sepsis diagnostics. Anal Chem 4:9403–9411. https://doi.org/10.1021/acs.analchem.6b00323
56. Ngamsom B, Lopez-Martinez MJ, Raymond J-C et al (2016) On-chip acoustophoretic isolation of microflora including S. typhimurium from raw chicken, beef and blood samples. J Microbiol Methods 123:79–86. https://doi.org/10.1016/J.MIMET.2016.01.016

Chapter 2
Microfluidic Technologies and Platforms for Protein Crystallography

Masatoshi Maeki and Manabu Tokeshi

Abstract Protein crystallization and its three-dimensional structure analysis is indispensable for understanding the protein function in the body and life phenomenon. Three dimensional structure of protein also plays important role for drug discovery and it have been already used to design new drug. To determine the three dimensional protein structure, protein crystallization conditions: concentration of protein, kinds and concentration of precipitant, buffer, pH, temperature, and additives must be optimized. In addition, high-diffraction quality protein crystals are needed to determine the protein three-dimensional structure at high resolution. However, optimization of the protein crystallization condition and preparation of high quality protein crystals require the labor intensives and trial-and-error. Microfluidics can provide the solution for the problems of traditional protein crystallography. A lot of microfluidic based technologies and platforms have been developed to utilize their unique characteristics. In this chapter, microfluidic technologies and platforms for protein crystallography is summarized. In particular, the application of microfluidics for high-throughput protein crystallization condition screening, controlling of protein crystal growth, and on-chip X-ray diffraction experiment using microfluidic devices are overviewed.

Keywords Protein crystallography · Droplet microfluidics · Normally-closed valve · Crystal growth · X-ray crystal structure analysis

2.1 Introduction

Biological sample is valuable and expensive due to the difficulty of sample preparation acquired from human patients, animals, cells, and bacteria. The acquired sample contains many impurities, which have to remove by appropriate pretreatment methods for each clinical diagnosis, assay, or experiment. In the pretreatment step,

M. Maeki (✉) · M. Tokeshi
Division of Applied Chemistry, Faculty of Engineering, Hokkaido University, Sapporo, Japan
e-mail: m.maeki@eng.hokudai.ac.jp

M. Tokeshi (ed.), *Applications of Microfluidic Systems in Biology and Medicine*, Bioanalysis 7, https://doi.org/10.1007/978-981-13-6229-3_2

the target biological molecules isolate from the impurities using separation techniques such as, centrifuge, electrophoresis, and chromatography. Generally, the purification step requires multiple trials with different separation modes [1, 2]. In the case of protein purification, affinity chromatography is the first choice for tag-fusion proteins. After the affinity chromatography, ion exchange chromatography and gel filtration chromatography are considered to obtain the high purity protein sample. When we need a large amount or high concentration of target protein sample, a scale scale-up of purification volume is required, because protein cannot be amplified unlike the nucleic acids. In particular, preparation of membrane protein is more complicated compared with soluble protein. From these reasons, preparation of the high purity biological sample is labor-intensive and time-consuming process.

Microfluidic technologies and platforms can offer many advantages in biological applications [3–6]. Reduction of sample consumption, high-throughput screening, shortening detection time, distinguishing fluid characteristic are remarkable features of microfluidic devices, and it facilitate the cell separation and single cell analysis [7–9], biological assays [10–12], pharmaceuticals [13–15], drug screening [16–18]. These are essential and powerful features for the biological applications, especially for protein analysis, because the high purity protein sample is valuable and expensive.

In the field of protein analysis, structure biology is a key research area to elucidate the relationship between the protein structure and the protein function [19–21]. Protein can function by precisely folding into its native "three-dimensional" structure. Conversely, misfolded proteins not only lose a function but also cause of the disease like an amyloidosis [22]. Determination of the protein three-dimensional structure provides us the essential information including protein-protein interaction, active site of protein, and protein-ligand binding site. The protein structural information is indispensable for the structure based drug design and development of molecularly-targeted therapy [23–25]. To determine the protein three-dimensional structure, X-ray crystallography [26–28], nuclear magnetic resonance spectroscopy (NMR) [29], and cryo-electron microscopy (cryo-EM) [30] are gold standard analytical methods. Although these are powerful analytical methods for protein structure and dynamics measurement, 80 ~ 90% of protein structures have been determined by X-ray crystallography.

In the protein X-ray crystallography, the preparation of high quality protein crystals is remaining problem to determine the precise three-dimensional protein structure. After the purification step, screening of the protein crystallization conditions is carried out by the conventional protein crystallization method, such as the hanging drop vapor diffusion method, sitting drop vapor diffusion method, microbatch method, and liquid-liquid interface diffusion method, as shown in Fig. 2.1 [31]. At the primary screening, the kinds and concentration of precipitant solution, concentration of protein solution, kinds, pH, and concentration of buffer solution, crystallization temperature, and additives are optimized to obtain protein crystals. We can roughly predict the suitable crystallization condition from the homology of proteins, which have been recorded in the Protein Data Bank (PDB). However, protein crystals do not always appear the similar crystallization condition.

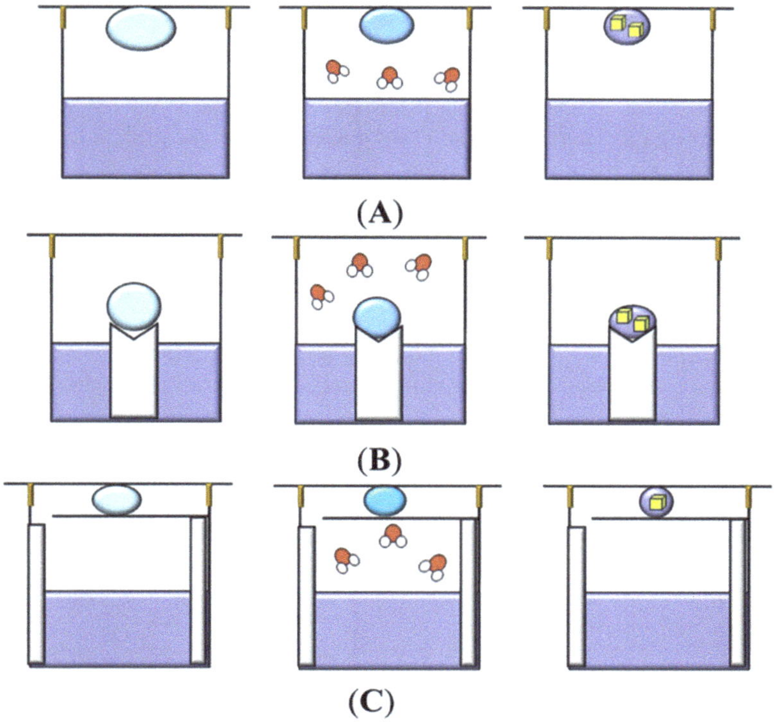

Fig. 2.1 Schematic illustrations of the conventional protein crystallization methods. Vapor diffusion method. A drop containing unsaturated precipitant and protein solution is placed in a crystallization well containing a precipitant solution. The well is sealed with grease to prevent evaporation and to allow vapor equilibration of the droplet and the reservoir. Equilibration of water vapor from the protein-containing droplet to the reservoir solution causes the protein solution to reach a supersaturation level, where nucleation and initial growth occur. (**A**) Hanging drop vapor diffusion method. (**B**) Sitting drop vapor diffusion method. (**C**) Sandwich drop vapor diffusion method. (Reprinted with permission from Ref. [31] with permission from MDPI)

There-fore, a comprehensive trial is unavoidable at the primary screening experiment. Typically, 1 ~ 2 μL of protein solution is needed for one trial, and thus, a several hundred microliter valuable protein solution is required for the comprehensive crystallization condition screening.

Then the secondary screening is conducted based on the results of primary screening to obtain high quality protein crystals. Although optimization of the crystallization condition is significant, the control of protein crystal growth is indispensable factor for obtaining the high quality of protein crystals. Growth control of the crystal regardless of the proteins or inorganic materials, at the ground environment is attracted in many researchers for improving the molecular packing. Formation of the natural convection is the major reason to disrupt the ordered molecular packing of crystal at the ground environment. For improving the quality of crystals, the microgravity, electric field, and magnetic field are used to control the crystal growth environment, even though specialized experimental apparatus are required [32–37].

Microfluidic technologies and platforms can resolve the abovementioned problems using its unique characteristics [38–41]. Reduction of the protein consumption is easily achievable by downsizing of the device dimension. Only nano to picoliter volume of protein solution is needed for one trial, and the microfluidic device enables the high-throughput screening of protein crystallization condition [42–54]. In addition, the "micro-sized crystallization space" allow the control of protein crystal growth different form the conventional crystallization environment [55–61]. In this chapter, microfluidic techniques and platforms for protein crystallography to accelerate the field of structural biology are summarized and the applications of microfluidic devices combined with the recent synchrotron facilities are also overviewed.

2.2 Principle of the Protein Crystallization and X-Ray Diffraction Measurement

In brief, the protein crystallization experiment constructs: screening of protein crystallization condition, preparation of high quality protein crystal, screening of cryoprotection condition, and X-ray diffraction measurement. In the case of conventional methods, a micropipette is used to prepare the protein crystallization screening setups. Generally, 1 μL of protein solution and 1 μL of precipitant solution is mixed to prepare the crystallization solution, and the droplet of crystallization solution is placed into a well of microplate. The precipitant works to decrease the solubility of proteins. The microplate setting the crystallization droplet is stored at the appropriate temperature until forming the protein crystals. For example, the hanging drop crystallization method is the most widely used for primary screening. Characteristics of the method is that supersaturation of the crystallization solution is gradually increased by the water-vapor diffusion from the crystallization droplet to precipitant solution in the reservoir during incubation. Diffusion rate of the water-vapor and the time necessary for reach the equilibrium can control by the mixing ratio of protein and precipitant solutions. When the appropriate kinds and concentration of precipitant solution is selected, protein crystals form in the droplet by increasing supersaturation of crystallization solution. However, protein crystals do not form at the suitable crystallization condition. Therefore, the optimization of protein crystallization condition is the first barrier of the protein crystallization.

A nucleation of protein crystal occurs at the high supersaturation condition by adding the precipitant solution. However, the moderate supersaturation condition is preferable for the crystal growth, because the high supersaturation condition lead to increase lattice defects. This is a major dilemma of protein crystallization process to control the nucleation and crystal growth [62, 63].

X-ray diffraction experiment is also essential part of the protein crystallography. Proteins are crystallized with hydrated water molecules, the percentage of water molecules occupy 50% w/w in average. Thus, manipulation of the protein crystals for the X-ray diffraction experiment difficult due to the fragility of protein crystals. In addition, cryoprotection process to prevent the deterioration of protein crystals during the X-ray diffraction experiment is also important for protein crystallography. Current synchrotron facilities enabled X-ray diffraction data collection from the micrometer-sized crystals by high energy and photon flux X-ray beam [64, 65]. However, the high energy X-ray beam caused the radiation damage during the measurement process. To avoid the X-ray radiation damage, protein crystal undergoes the cryoprotection and is measured at 100 K in cold nitrogen gas stream [66, 67]. Optimization of cryoprotection condition and damage-less cryoprotection procedure are indispensable for high quality X-ray diffraction data. Finally, the protein three dimensional structure is determined from the X-ray diffraction data using a PC.

A variety of the microfluidic devices have been developed for protein crystallography, which realize the effective crystallization condition screening, growth control of the protein crystals, on-chip X-ray diffraction measurement, membrane protein crystallography, time-resolved protein crystallography. Basically, microfluidic devices for protein crystallography was miniaturization of the conventional technique. However, the applications of microfluidic devices for protein crystallography have been expanding by the advances of the synchrotron facilities.

2.3 Microfluidic Devices for High-Throughput Screening of Protein Crystallization Condition

Screening of protein crystallization condition is the one of the most significant problems in protein crystallography. Screening for hundreds or thousands of protein crystallization condition is necessary to prepare diffraction quality protein crystals. In the case of the conventional protein crystallization screening, a robotic system equipped with an automatic liquid dispenser is employed for high-throughput screening. For example, mosquito® supplied from TTP LabTech Ltd. is a major automated dispenser [56]. The pipetting range of sample is tens of nano litter to micro litter, and the typical microplate is useable for the screening experiment. However, these kinds of automated dispenser are astonishingly expensive, and require a constant running cost.

In comparison with the conventional methods, microfluidic devices provide many advantages for protein crystallization screening. Figure 2.2 shows the droplet-based high-throughput screening system of protein crystallization condition [43]. Ismagilov's group first reported the droplet-based protein crystallization system. Protein, precipitant, buffer, and additives (PEG) were separately introduced into the microchannel, and droplets containing the all reagent at the defined concentration were continuously generated by shearing the oil flow stream. The composition of

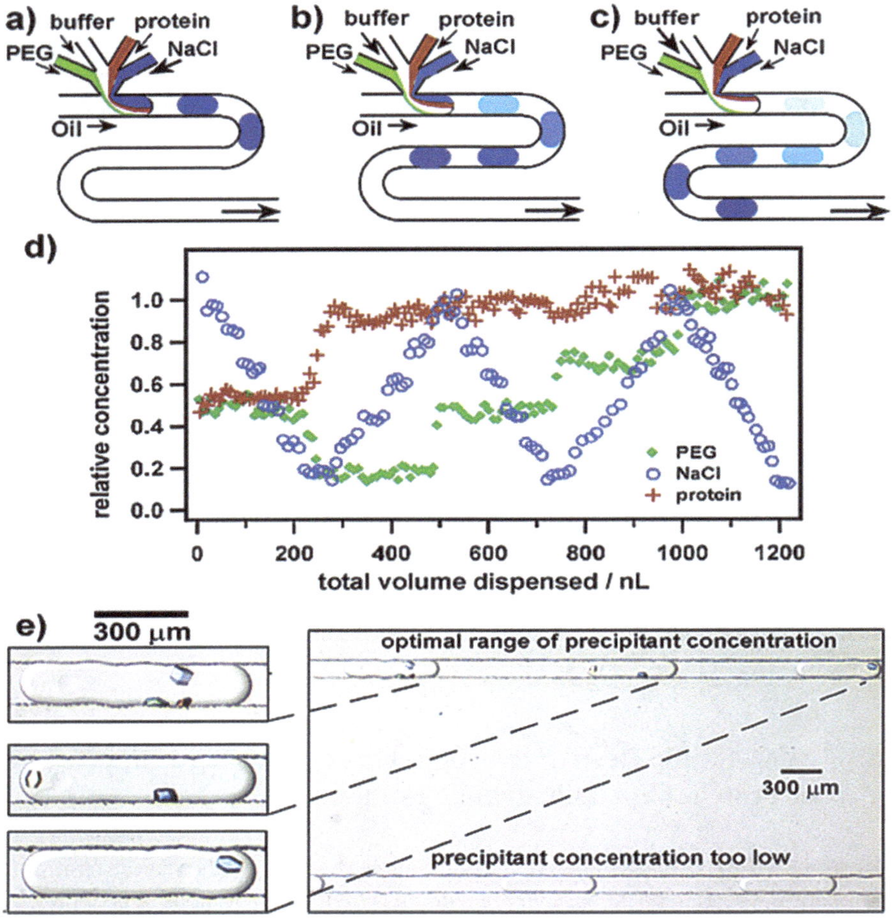

Fig. 2.2 Droplet-based high-throughput protein crystallization condition screening system. (**a–c**) A schematic illustration of generation process of microdroplets including protein, NaCl, PEG, and buffer. When the flow rate of NaCl is decreased and the flow rate of buffer solution is increased, the supersaturation condition of the droplet is decreased. (**d**) Characterization of the idea shown in (**a–c**). Flow rate of oil was 12 nL/s. Flow rates of aqueous solutions of PEG, salt, lysozyme, and acetate buffer were varied between 0.8 and 5 nL/s. The total flow rate of the aqueous solutions was set at 15 nL/s. (**e**) Photographs of lysozyme crystals appeared in the droplets of variable composition. (Reprinted with permission from Ref. [43] with permission from The American Chemical Society)

each reagent was defined by initial concentration of solutions and flow rates. When the flow rate of NaCl was gradually decreased as shown in Fig. 2.2a–c, tens of droplets containing different concentration of NaCl were automatically generated in several seconds. The droplet volume was picoliter to nanoliter or-der and controlled by the flow conditions. Figure 2.2d shows relative concentration of each reagent within the droplets. Protein crystals formed at the optimal crystallization condition ranges (Fig. 2.2e).

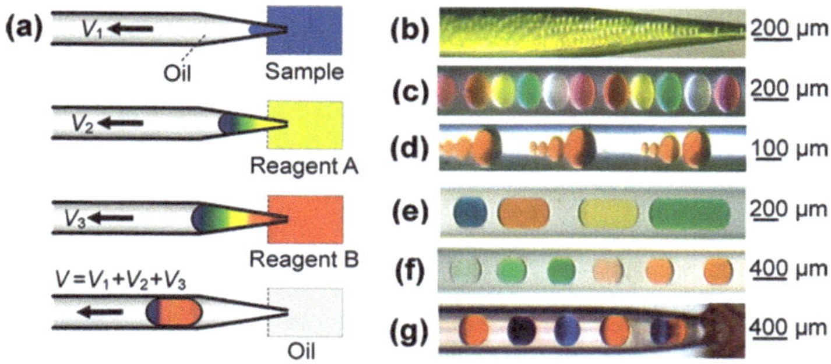

Fig. 2.3 (**a**) Schematic illustration of droplet formation process. (**b**) Continuous droplet generation of 20 pL volume at a throughput of 4.5 s per droplet by sequentially aspirating a 20 pL fluorescence solution and 80 pL of oil at flow rates of 2 and 8 nL/min. (**c**) Continuous generation of 1.6 nL droplets containing five different dyes. (**d**) Droplets with different volumes of 20, 40, 160, and 1000 pL. (**e**) Droplets generation with different volumes (2.5, 4.5, 5.5, and 8.0 nL) and different kinds of dyes. (**f**) Droplets of 2.5 nL volume with concentration gradient formed by diluting the green and red dye solutions. (**g**) 2.5 nL droplets formed by sequentially introducing red dye, water, and blue dye with different mixing ratios. The mixing ratios were 1.0:1.5:0, 0.5:1.5:0.5, and 0:1.5:1.0. (Reprinted with permission from Ref. [68] with permission from The American Chemical Society)

An automated microfluidic droplets formation platform named DropLab has been developed by Fang's group. Schematic illustrations of droplets formation using DropLab system is shown in Fig. 2.3 [68]. The minimum droplet volume using the DropLab was ~ 20 pL and the operation speed was 4.5 s per droplet. The concentration gradient between droplets was able to form by changing the dispensing ratio of each reagent. Figure 2.4 shows the protein crystallization condition screening and the optimization experiments using the DropLab system. In the case of crystallization condition screening, 50 droplets with 12 nL containing different precipitant and concentration solution in each were arrayed in a 10 cm long capillary within 22.5 min. As shown in Fig. 2.4a–c, lysozyme crystals were appeared at the suitable precipitant condition. Then, the optimization of the mixing ratio between the 100 mg/mL lysozyme solution and precipitant #1 (1.0 M NaCl and 25% PEG 6000 in 0.1 M sodium acetate buffer (pH 4.6)) were carried out to define the best crystallization condition (Fig. 2.4d–e). Forty droplets array with different mixing ratios of lysozyme to precipitants #1 were generated similarly with precipitant screening.

The droplets which are 5 kinds of mixing ratios of lysozyme to precipitant with 4:1, 2:1, 1:1, 1:2, and 1:4, were prepared in a 10 droplet sequence. The best lysozyme crystals were obtained at the mixing ratio of 1:2 (4 nL lysozyme and 8 nL precipitant #1).

A microwell-based protein crystallization platform is first reported by Quake's group in 2002 [47]. In the case of microwell-based crystallization platform, introduction of sample solutions is carried out by pneumatic valve operation [69]. They fabricated the protein crystallization device enabled 144 parallel condition screening. The microfluidic device was made from PDMS and sealed with glass substrate.

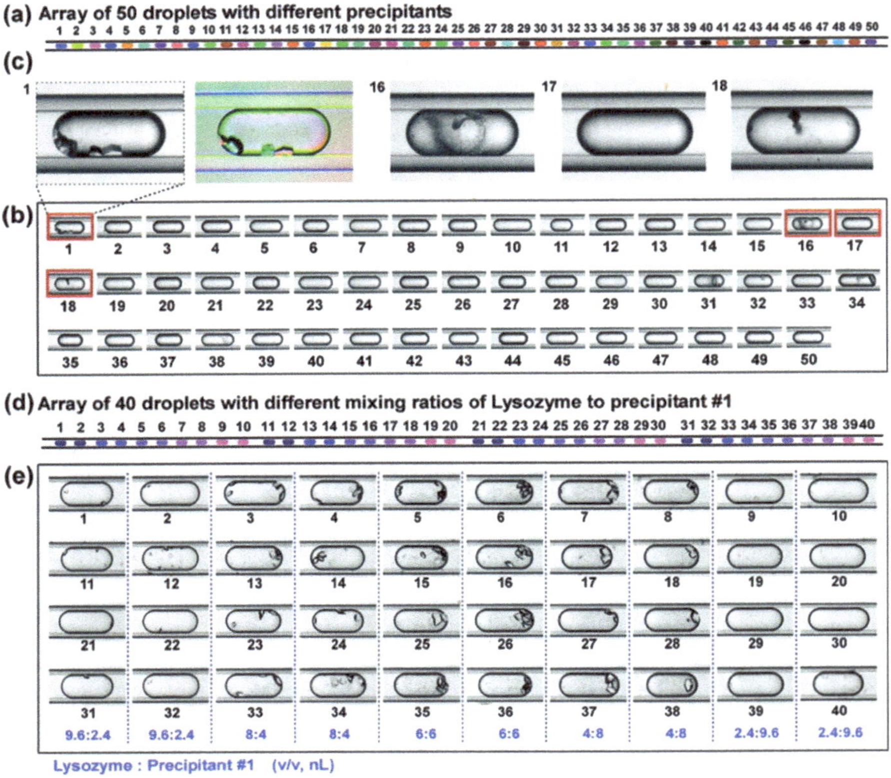

Fig. 2.4 Lysozyme crystallization screening using DropLab. (**a**) Schematic illustration of droplet array for lysozyme crystallization condition screening using 50 different precipitants. (**b**) Photographs of 50 droplets containing lysozyme and 50 different precipitants in the droplet array of one capillary after a 20 h incubation. (**c**) Enlarged views of droplets containing precipitant #1, #16, #17, and #18 in (**b**). Lysozyme crystals were obtained in droplet #1. The second image of droplet #1 is a polarized microphotograph. (**d**) Schematic illustration droplet array for optimization of mixing ratio of protein to precipitant #1. (**e**) Photographs of 40 droplets (4 duplicates) in the droplet array with different mixing ratios of 4:1, 2:1, 1:1, 1:2, and 1:4 between lysozyme and precipitant #1. Droplet volume was 12 nL. (Reprinted with permission from Ref. [68] with permission from The American Chemical Society)

Different volume ratio of protein solutions and precipitant solutions were filled separately to microwells via the normally closed valve. After introducing each solution, the valves at the middle section (mixing valve) were opened to mix the protein and precipitant solutions by interface diffusion. This crystallization method is called as "free interface diffusion" in the field of protein crystallography. The protein and precipitant solutions were gradually mixed by molecular diffusion, thus the supersaturation of the solution can control by the opening time of mixing valve. This type of microfluidic device enabled comprehensive protein crystallization condition screening by only pipetting the protein and precipitant solutions at each

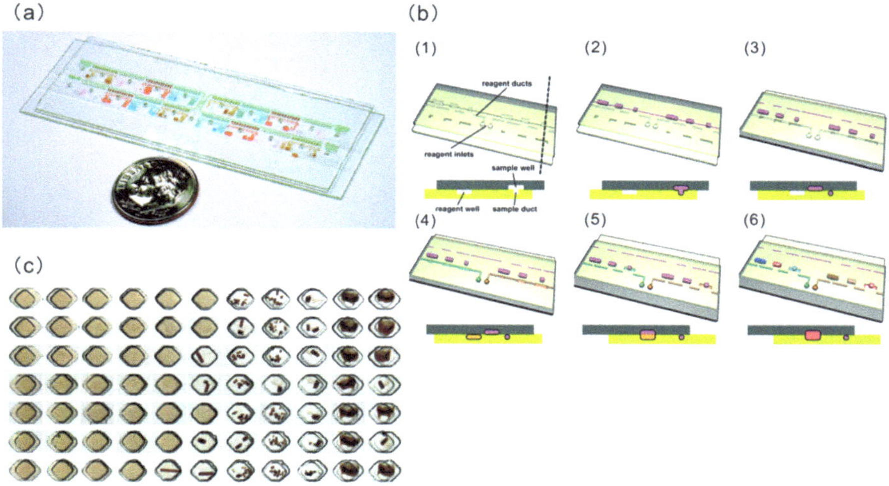

Fig. 2.5 (**a**) Photograph of SlipChip. (**b**) Schematic illustrations of reagent-loading procedure. (**c**) Screening of the photosynthetic reaction center using SlipChip. 11 trials using the same precipitant at different concentration ratios were demonstrated. (Reprinted with permission from Ref. [45] with permission from The American Chemical Society)

inlet. The optimal protein and precipitant concentration are simultaneously explored by the different volume ration of microwell sets and mixing time of the solutions. Six different proteins such as, lysozyme, bacterial primase catalytic core domain, type-II topoisomerase, thaumatin, xylanase, and glucose isomerase were crystalized using the microfluidic device. In addition, a thaumatin crystal collected from the micro wells (5 nL volume) after appearing the crystal were analyzed by X-ray diffraction experiment and a high-resolution diffraction pattern is observed. This result suggests that microwell platform was able to grow a diffraction quality protein crystal as well as screening of the protein crystallization condition.

A unique protein crystallization condition screening device was reported from Ismagilov's group [45, 70]. The microfluidic device was named as "SlipChip". Figure 2.5 shows the schematic illustration of the SlipChip and its operating photographs. Protein solution was loaded into the microchannel (bottom side) and microwells (upper side) followed by slipped to divide the microchannel and microwells (Fig. 2.5a–c). Then, precipitant solution was loaded into the other microchannel and microwells, and the substrates are additionally slipped to mix the solutions. There were also fabricated different sized microwells to realize the high-throughput protein crystallization screening. Basically, the crystallization principle of SlipChip was free-interface diffusion like microwell-based device mentioned above. They applied SlipChip to determine the crystal structure of glutaryl-CoA dehydrogenase and dihydrofolate reductase/thymidylate synthase. These proteins are diffracted at the resolution of 1.6 Å and 1.9 Å, for glutaryl-CoA dehydrogenase and dihydro-folate reductase/thymidylate synthase, respectively. SlipChip

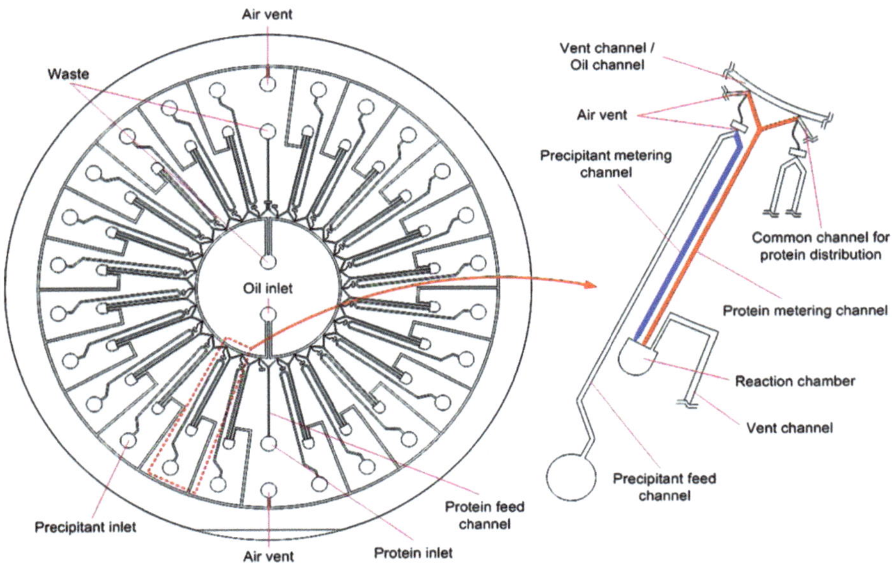

Fig. 2.6 Schematic illustrations of centrifugal microfluidic device with 24 crystallization array. The device has 24 parallel microstructure units. Each unit comprises a protein metering structure, a precipitant metering structure, and a crystallization chamber. All protein metering structures are connected via a common channel to the sample inlet, and each precipitant metering structure is connected via a feed channel to an individual inlet. Outlets of both the protein metering channel and the precipitant channel are connected to the crystallization chamber. (Reprinted with permission from Ref. [72] with permission from The American Chemical Society)

provide us the simple operation for protein crystallization screening, although the mass productivity of device is lower than that of PDMS-based microfluidic devices.

Centrifugal microfluidic devices are also used for protein crystallography [71, 72]. Wang et al. demonstrated the 24 parallel protein crystallization experiment with one operation using a single microfluidic device (Fig. 2.6). The microfluidic device was made from PDMS microchannel and glass substrate. They tested the protein crystallization using lysozyme, xylanase, lipase B, glucose isomerase, and thermolysin. A protein solution and precipitant solutions were introduced each chamber by centrifugal and capillary force. Each chamber set of protein and precipitant solution were connected with microchannels. Therefore, the microfluidic device was able to employ the vapor diffusion crystallization method. The supersaturation of protein solution was gradually increased by water vapor diffusion similar with the conventional hanging drop vapor diffusion method. For the lysozyme crystallization, the microfluidic device got lysozyme crystals at 81 different conditions. This result indicates the hits rate of the vapor diffusion chip is same as the conventional hanging drop vapor diffusion method (76 hits). However, the vapor diffusion chip allowed the rapid experimental setup (6–8 min for vapor diffusion chip, and 20–30 min for the conventional hanging drop vapor diffusion method) and reduction of sample consumption (6 nL/trial for vapor diffusion chip, and 2 μL/trial for the conventional vapor diffusion method).

2.4 Protein Crystal Growth in a Microspace Environment

Microfluidic device is useful for not only the protein crystallization condition screening, but also the preparation of high quality protein crystal. When the optimized protein crystallization is founded by the primary screening, improving the quality of protein crystal is addressed by additional optimization of crystallization condition screening. Typically, high diffraction quality protein crystals are not necessarily obtained even with the crystallization condition is optimized. The high diffraction quality crystal is defined as a good molecular packing single crystal (not multiple or aggregated crystals), and it lead to obtain high quality X-ray diffraction data. Crystallization environment is significant factor to prepare the ordered molecular packing protein crystal. In the case of the ground-based protein crystallization experiment, the natural con vection is the main factor of the formation of disordered crystal. From these reasons, protein crystallization experiments at a special environment: a microgravity (at the International Space Station), a magnetic field, or an electric field have been reported [33, 34]. These environments make it possible to suppress the natural convection and the offer a diffusion-controlled protein crystal growth environment. Microspace environments: the microdroplet and the microwell are alternative ground-based crystallization platforms of the special crystallization environment. In this section, microfluidic platforms for preparation of high diffraction crystal is introduced.

Advantages of the microspace environment on the protein crystallography are producing one single crystal in a microdroplet or microwell (single crystallization) and controlling protein crystal shape suitable for X-ray crystal structure analysis [55–57, 73, 74]. Single crystallization is a technique for preventing an aggregation of protein crystals to obtain high quality X-ray diffraction data. When a number of nucleus or crystals formed in the crystallization droplet, each protein crystal cannot grow enough to obtain good X-ray diffraction data due to the consumption of protein molecules at the nucleation and early stage of crystal growth. In addition, one single crystal must be exposed by X-ray beam to collect the "single crystal X-ray diffraction data". From the viewpoint of X-ray data collection, stacked or aggregated protein crystals: a needle-like crystal and a cluster-like crystal lowers the diffraction quality and complicates the crystal structure analysis after the diffraction data collection. In the case of these types of crystals, crystal growth control plays important roles to prevent staking of each crystal.

Maeki et al. reported droplet-based protein single crystallization method. They focused on the effect of droplet size on the protein crystallization and elucidated the critical droplet size for generating one single crystal in the microdroplet both theoretically and experimentally [56]. Figure 2.7 shows the concept of the single protein crystallization method based on the microdroplet. After first nucleation of a protein crystal in the droplet, the protein crystal grow in competition with further nucleation until the supersaturation of the solution reaches the metastable region. Thus, a one single crystal was generated in a droplet smaller than the critical droplet size. Conversely, a large size droplet made it possible continuously nucleation rather

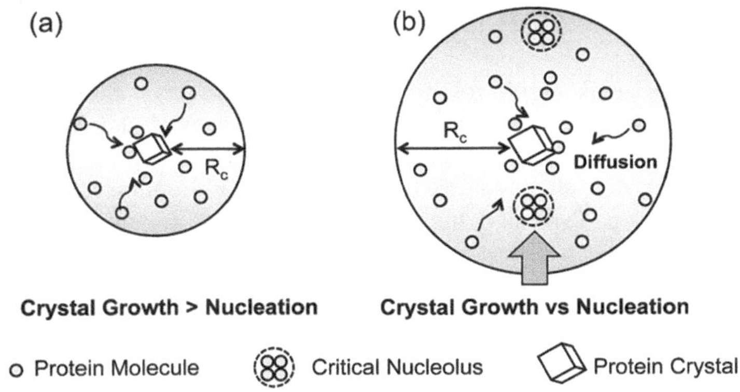

Fig. 2.7 Droplet-based high-throughput protein crystallization condition screening system. Concept of the single crystallization method based on droplet microfluidics. (**a**) Droplet size < critical size (*Rc*). (**b**) Droplet size > *Rc*. The critical size was estimated from the diffusion coefficient, the initial concentration of protein, and the consumption rate of protein in solution. (Reprinted with permission from Ref. [56] with permission from Wiley-VCH)

than the crystal growth after first nucleation. The microdroplet single nucleation method is based on the mass balance of protein molecules and Fick's laws of protein diffusion. The critical droplet (*Rc*) size was able to calculate from the diffusion coefficient of protein molecules (*D*), the initial concentration of protein in the droplet (C_O), and the consumption rate of protein by crystal growth (*q*), as shown in the following equation:

$$Rc = (6DC_0/q)1/2$$

To proof of concept, four model proteins: lysozyme (14 kDa), thaumatin (22 kDa), glu cose isomerase (179 kDa, four subunits), and ferritin (440 kDa, 24 subunits) were used for crystallization experiments. The *Rc* values were theoretically calculated as 600, 600, 450, and 200 μm for lysozyme, thaumatin, glucose isomerase, and ferritin, respectively. In the case of lysozyme and thaumatin, single crystallization enabled using 200 and 360 μm sized droplets, regardless of the supersaturation condition. On the other hand, multiple crystals formed in the 500 μm sized droplet after first nucleation at the high supersaturation condition. After the first nucleation, protein concentration gradient formed in the microdroplet, regardless of the droplet size, and it have been confirmed by in situ Raman spectroscopy. In other words, high concentration of protein remains in the area far away from the crystal. The protein concentration gradient gradually disappeared by molecular diffusion due to the depletion of protein molecules around the formed crystal. In the case of small-sized droplet, protein molecules could rapidly diffuse and incorporated to the formed crystal compared with the large sized droplet. However, in the case of large sized droplet, although protein molecules diffuse to the formed crystal, the time to necessary to reach the protein molecules to the crystal became long compared with

Fig. 2.8 Photographs of protein crystals unsuitable for single crystal X-ray crystallography grown without microseeds. (left) Photoactive yellow protein (PYP) and (right) cytochrome bo_3 oxidase (cyt bo_3). (Reprinted with permission from Ref. [58] with permission from The Royal Society of Chemistry)

the small sized droplet. For this reason, further nucleation enables in the region, which protein molecules remains at high concentration. For the ferritin crystallization, molecular diffusion of ferritin assumed slower than that of the small molecular weight proteins: lysozyme and thaumatin. Therefore, multiple crystals form in the droplet like the lysozyme and thaumatin crystallization using the large sized droplet. These results suggest that a strategy for design of microfluidic devices to control the protein crystallization behavior.

Stacking or aggregating of protein crystals are serious problems in the protein crystallography (Fig. 2.8). Typically, the aggregations lower the diffraction quality and complicates the structure determination process. To avoid the aggregation of protein crystals, crystallization control based on the phase diagram is useful technique both the conventional methods and microfluidic methods. The aggregation occurs by the nucleation onto the surface of protein crystals formed in the crystallization droplet (or well). Thus, protein crystals incubated under the suitable crystallization (supersaturation) condition can avoid the aggregations.

Kenis et al. reported a microseeding method using a microfluidic device [58]. Figure 2.9 shows a microfluidic approach for the protein crystallization experiment using microseed. The microfluidic device composed four layers including PDMS layers (50 μm for one layer) and cyclic olefin copolymer layers (50 μm for one layer). The total thickness of the microfluidic device was 200 μm, which expected good X-ray transmission efficiency. The microfluidic device enabled 24 crystallization trials using 6 different concentration of diluted microseed solutions at once. The solutions: the protein-precipitant solution and the microseed solutions were introduced into the 24 microwells via the normally closed valves and mixed to transport the microseeds to the crystallization solution under the metastable region (Fig. 2.9b). Under the metastable region, protein crystal can grow without further nucleation. Figure 2.9c–d shows the photographs of the crystals of (c) photoactive yellow protein (PYP, soluble protein) and (d) cytochrome bo_3 oxidase (cyt bo_3, membrane

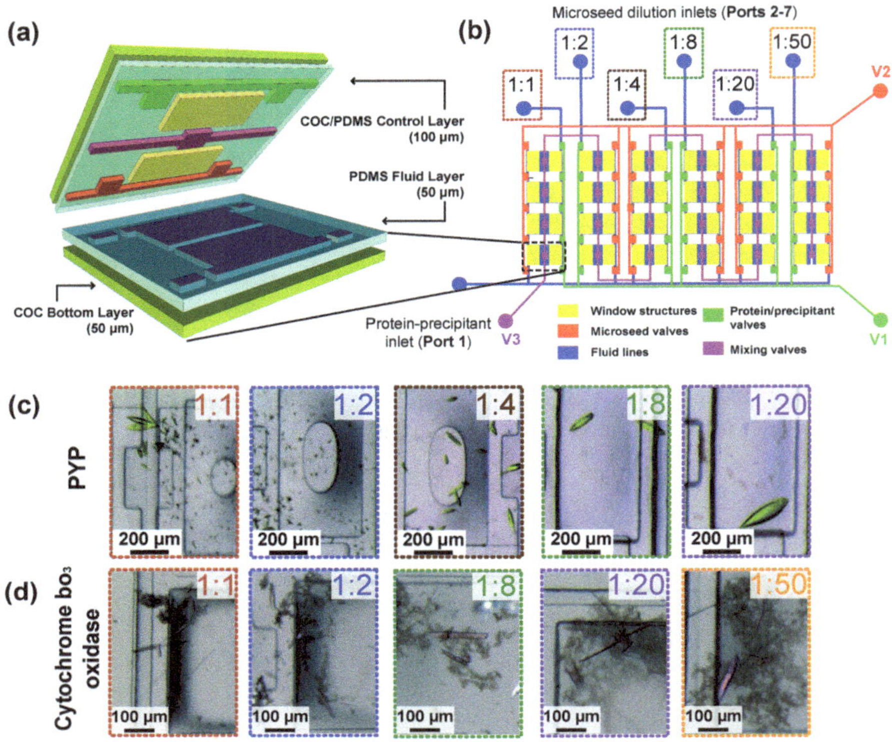

Fig. 2.9 (**a**) 3D perspective view of the microfluidic device with normally closed pneumatic valves. The microfluidic device comprised of four layers: an impermeable cyclic olefin copolymer top layer bonded to an PDMS control layer containing valves, a PDMS fluid layer to contain protein, precipitant and seed solutions, and a COC bottom layer. (**b**) Schematic illustration of a 24-well array chip used for microseeding. Fluid layer is represented in blue, and the valve lines, V1, V2 and V3, are colored based on their function. In the case of microseeding experiments, different microseed dilutions and a pre-mixed protein–precipitant solution were introduced into the device from inlets 2–7 and inlet 1, respectively, prior to on-chip mixing. (**c** and **d**) Results of the dilution ratio screening of the seed solutions. At greater seed dilutions (lower seed concentration), PYP crystals (**c**) grew into fewer, larger crystals and cyt bo_3 crystals (**d**) grew into fewer, thicker crystals. (Reprinted with permission from Ref. [58] with permission from The Royal Society of Chemistry)

protein). In comparison with the conventional crystallization method shown in Fig. 2.8, the aggregations of protein crystals were able to prevent by the microseeding method using microfluidic device. When the high concentration of microseed solution was used, the small sized crystals formed in the crystallization well. Conversely, 200 and 100 μm sized crystals for PYP and cyt bo_3 were obtained at the low concentration of microseed solution. At the microseeding process, the competition of dissolution of microseeds and growth of microseeds occurs in the crystallization solution and therefore, the appropriate dilution rate of microseed solution allows the crystal growth of one or a few protein crystals in the solution.

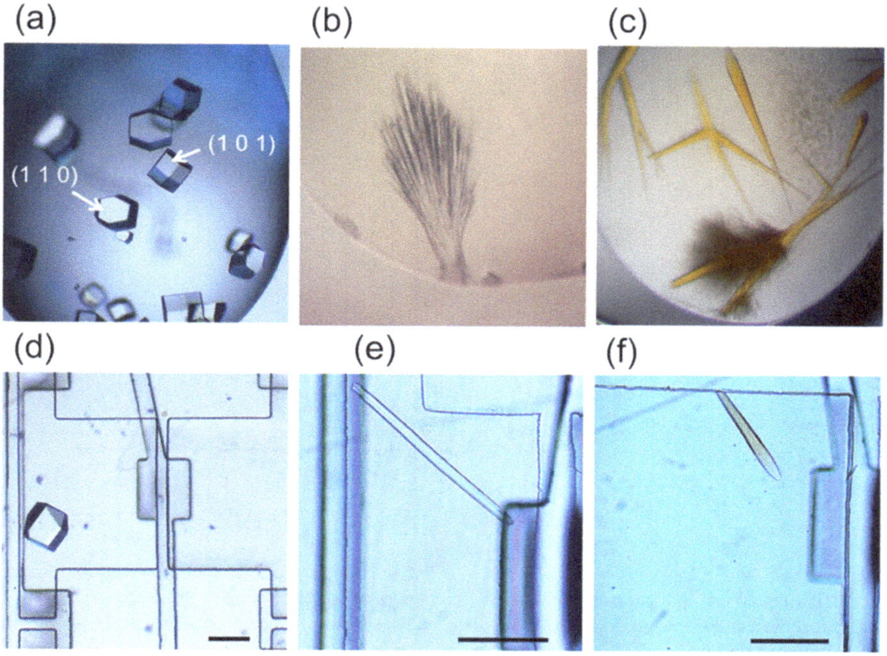

Fig. 2.10 Photographs of protein crystals formed by the (**a**) microbatch method, (**b**, **c**) hanging drop vapor diffusion method, and the (**d–f**) microfluidic chip method. (**a**, **d**) Lysozyme crystal, (**b**, **e**) PsGK crystal, and (**c**, **f**) CPR–HO complex crystal. The scale bars represent 200 μm. (Reprinted with permission from Ref. [55] with permission from The Royal Society of Chemistry)

They also optimized the microfluidic system with separated microchannels for protein, precipitant, and microseed solutions to precisely control the mixing of three solutions. By the improved microfluidic device, the microseeds solution was transported to the crystallization solution (mixture of protein and precipitant solution) at preferable time. The resolution of bo_3 was improved by the microfluidic approach from 12 Å to 10 Å. Although, the diffraction resolution was not enough to determine the three-dimensional structure, the microfluidic approach expects to become a non-invasive and high throughput optimization method to obtain high diffraction quality protein crystal after the first protein crystallization condition screening.

Maeki et al. found the difference of protein crystal growth behavior depended on the depth of crystallization chamber and demonstrated that the reforming of aggregated protein crystal [55]. They also employed the microseeding method using the microfluidic devices with normally closed valves. Lysozyme, glucokinase from *Pseudoalteromonas* sp. AS-131 (PsGK), and NADPH- cytochrome P450 oxidoreductase-heme oxygenase complex (CPR-HO complex) as model proteins. Figure 2.10 shows the photographs of protein crystals formed by the (a) microbatch method, (b, c) hanging drop vapor diffusion method, and the (d–f) microfluidic

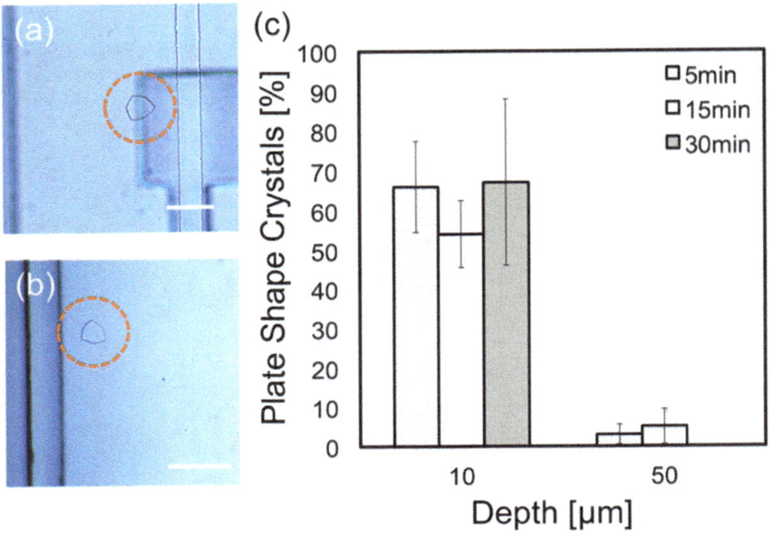

Fig. 2.11 (**a**, **b**) Photographs of the plate-shaped PsGK crystals formed in the microfluidic device with 10 µm deep crystallization chambers. (**c**) Percentage of the plate-shaped PsGK crystals formed under different mixing times. The scale bars represent 100 µm. (Reprinted with permission from Ref. [55] with permission from The Royal Society of Chemistry)

devices. This result indicates that the microseeding method coupled with the microfluidic device was a universal protein crystallization control technique. In addition, they demonstrated the shape of PsGK crystal was dramatically changed by the depth of the crystallization chamber. Although, the road-shaped PsGK crystal formed in 50 µm depth microfluidic device, the plate-shaped PsGK crystal formed in 10 µm depth microfluidic device, as shown in Fig. 2.11. Mixing time (5–30 min) between the crystallization solution and the microseed solution did not affect the crystal shape. The three-dimensional structure of PsGK have been determined by the road-shaped crystal formed by the conventional hanging drop vapor diffusion method. Then the plate shaped PsGK crystal was also measured by the X-ray beam at synchrotron facility. The plate shaped PsGK crystal diffracted at a resolution limit of 2.8 Å and the crystallographic parameters: the space group and the lattice constant were same as the rod-shaped PsGK crystal. For the X-ray diffraction measurement, the plate-shaped crystal is easy to irradiate the X-ray beam on the targeted crystal, because the irradiation area of crystal is larger than that of the rod-shaped crystal. The thick cubic-shaped crystal provides the high diffraction intensity and ideally is the best for the X-ray single crystallography. Theoretically, the plate-shaped crystal can grow to the cubic-shaped crystal by the additional macroseediong method. Therefore, the reforming method followed by the macroseeding may offer the solution for the preparation of high diffraction quality crystal.

2.5 On-Chip Protein Crystal Structure Analysis and Other Applications

X-ray diffraction measurement is the last experimental step on the protein crystallography. When the X-ray diffraction data is obtained, the three-dimensional protein structure is calculated using a personal computer or a super computer. The size of protein crystals obtained by the microfluidic device is typically smaller than 200 μm due to the limitation of crystallization space (microdroplet or microwell). Thus, a synchrotron facility is widely employ for the X-ray diffraction measurement. Generally, X-ray diffraction measurement is carried out at the cryogenic condition to reduce the radiation damage. The room-temperature measurement also enables, if the diffraction data collected from the different protein crystals is able to merge each other. Advances in X-ray free electron laser (XFEL), the serial femtosecond X-ray crystallography at room temperature will be indispensable technique for elucidating the protein structures and dynamics [75–78]. In any cases, optimization and selection of the quality of protein crystals, characteristics of microfluidic device (material and thickness), and the characteristic of X-ray beam (photon flux and beam spot size) requires to collect high quality diffraction data.

Microfluidic devices for both of the on-chip X-ray measurements at cryogenic and room temperature conditions have been reported in several papers. For the on-chip X-ray measurement, the suitable substrate materials should be selected to reduce the background signal. In addition, the thickness of the microfluidic device also affects the X-ray diffraction intensity. Ismagilov et al. demonstrated that *in situ* X-ray analysis using microdroplets collected into the glass capillary with 200 μm inner, and 180 μm outer diameters [79]. Microdroplets containing thaumatin crystal were directly exposed X-ray in 10 s beam at BioCARS station 14BM-C and 14ID-B at the Advanced Photon Source at Argonne National Laboratory (USA). Thaumatin crystals diffracted to 1.8 Å resolution and the crystallographic parameters are the same as the reported thaumatin crystal prepared by the conventional crystallization method. The background signal from the glass capillary is enough to low, because they used 10 μm thickness-wall glass capillary. Teflon or PFA (perfluoroalkoxy alkane) capillary (I. D. 200 μm, O. D. 360 μm, 80 μm wall thickness) was also acceptable for the device material to determine the crystallographic parameters: lattice constant and space group (BL 07 in the SAGA Light Source, Japan) (Fig. 2.12) [80]. However, the background signal at the small angle was slightly higher than that of thin-wall glass capillary. These results suggest that the thinner microfluidic device is desirable for obtaining high signal to noise ration X-ray diffraction data by on-chip measurement.

X-ray compatible COC-PDMS hybrid microfluidic device was developed by Kenis's group [52]. They evaluated the effect of substrate materials and thickness on the X-ray transmission and confirmed that the COC-PDMS device showed the transmission factor was more than 90%. Then, they carried out X-ray measurement

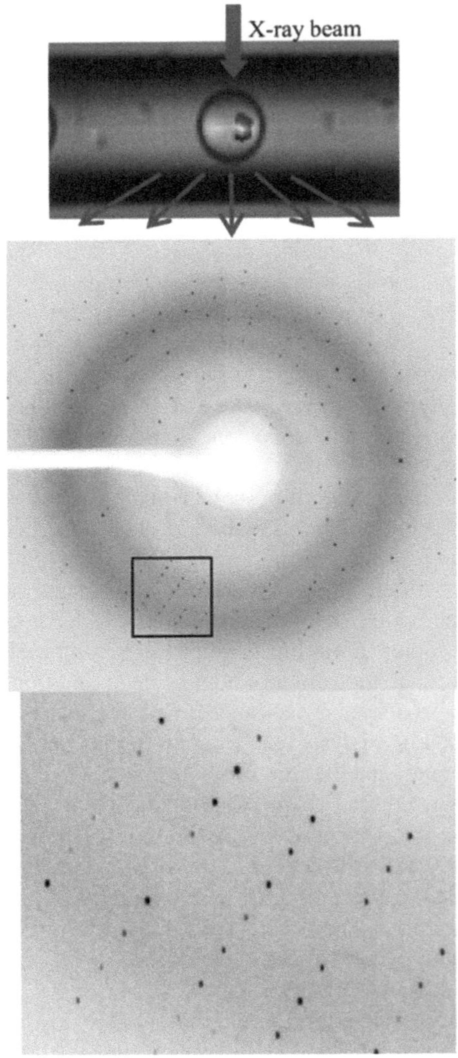

Fig. 2.12 (top) Lysozyme crystal formed in the microdroplet within capillary. (middle and bottom) X-ray diffraction patters obtained by directly irradiation of X-ray to the lysozyme crystal within the capillary. (Reprinted with permission from Ref. [80] with permission from The Japan Society for Analytical Chemistry)

using lysozyme crystal at room temperature in the oscillation range from −5° to +5° (10° per one lysozyme crystal, 1° oscillation step with 1 s exposure). Diffraction data from multiple lysozyme crystals were merged to obtain a complete data set. The diffraction data was compared with the lysozyme crystal prepared by the conventional crystallization method over the resolution range from 50 Å to 1.55 Å. The lysozyme crystal prepared by the conventional method underwent cryoprotection to obtain a complete diffraction data set. The quality of the diffraction data using the microfluidic device was compatible with the conventional crystallization and analytical method. Maeki et al. first demonstrated the integration of protein

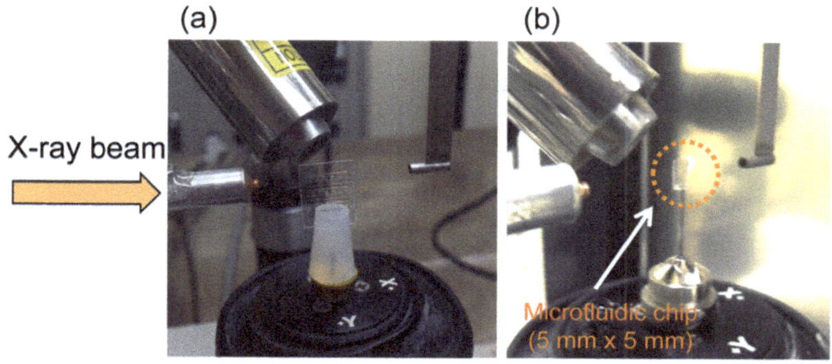

Fig. 2.13 Comparison of the cooling situation of the microfluidic chip by cold nitrogen gas. (**a**) Photograph of the in situ X-ray diffraction experiment using a standard microfluidic chip at 277 K. (**b**) Photograph of a microfluidic chip that was cut to a size of 5 × 5 mm^2. This microfluidic chip was used for the X-ray diffraction experiment at 100 K. (Reprinted with permission from Ref. [81] with permission from The American Chemical Society)

crystallization to on chip X-ray diffraction procedure including cryoprotection [81]. Figure 2.13 shows the photographs of the X-ray diffraction experiment setup: at (a) 277 K and (b) 100 K (cryogenic condition). They also used the COC-PDMS microfluidic (240 μm thickness) device with normally closed pneumatic valves to mix the cryoprotectant after appearing lysozyme crystals in microwells. The standard microfluidic device (3 × 4 cm^2) was not completely placed in the cold nitrogen stream and the crystals formed in the device could not apply the flash cooling. In contrast, the microfluidic device optimized for the cryoprotection enabled the flash cooling process and obtained complete diffraction data set from one protein single crystal (oscillation range: 0° to 90°, 1° oscillation step with 30 s exposure at BL 07 in the SAGA Light Source, Japan). They can avoid the X-ray radiation damage by the cryoprotection and the lysozyme crystal diffracted to 1.5 Å resolution and the three-dimensional structure was determined from one single crystal. The microfluidic device with normally closed valves made it possible the step-wise cryoprotection to reduce the osmotic shock at the cryoprotection. The stepwise cryoprotection induced the improvement of diffraction data quality.

Perry et al. developed a graphene-based microfluidic device to reduce the background signal. Figure 2.14 shows schematic illustrations of fabrication and operation procedures [82]. Unlike the polymer-based materials, the thickness of single layer graphene is sub-nanometer. Therefore, the signal to noise ratio is expected extremally increase, because of the decreasing of the background signal. They confirmed that the 100 μm thick COC layer increased the attenuation and background scatter [83]. Typically, micro crystals shower the low diffraction intensity compared with large-sized crystals, thus the graphene-based microfluidic device is useful for the micro-crystallography. In addition, the graphene-based microfluidic device can be applied not only for the X-ray crystallography and also the other analytical methods including small angle X-ray diffraction (SAXS) and cryo-transmission electron microscopy (cryo-TEM) to analyze the biological targets.

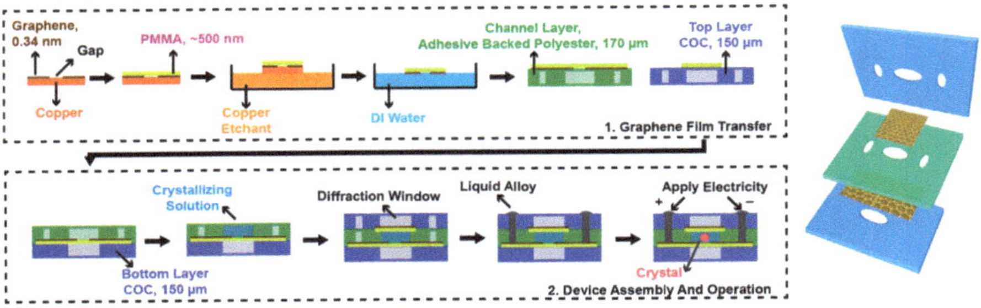

Fig. 2.14 Schematic illustrations of the device fabrication and operation procedures. A graphene membrane on copper is coated with a layer of PMMA, and then released from the copper substrate by etching. The graphene-PMMA film is floated on the surface of water and transferred to an adhesive polyester support layer. This layer is adhered to a COC layer containing the cut-out pattern for the microfluidic channels. Assembled device is held together by the adhesive layers defining the window structures. (Reprinted with permission from Ref. [82] with permission from MDPI)

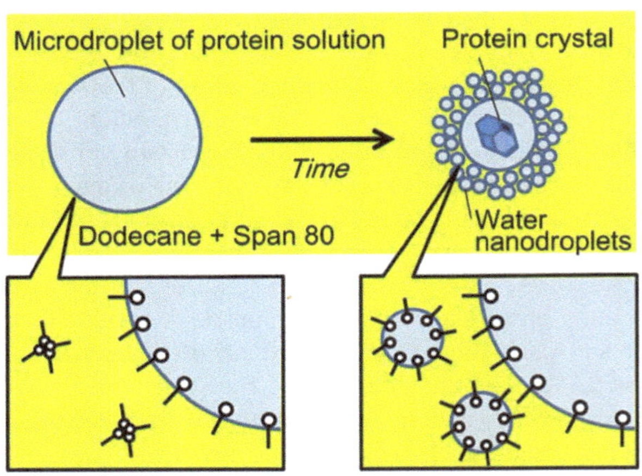

Fig. 2.15 Concept of microdroplet-based protein crystallization using spontaneous emulsification. (Reprinted with permission from Ref. [84] with permission from The Royal Society of Chemistry)

Fukuyama et al. reported a droplet-based vapor diffusion protein crystallization method [84]. They focused on the spontaneous emulsification to increase the supersaturation in microdroplets. Figure 2.15 shows the concept of microdroplet-based protein crystallization using spontaneous emulsification. They used lysozyme as a model protein and microdroplets containing a crystallization solution: a mixture of lysozyme and precipitant were dispersed into a continuous phase (Dodecane + Span 80). The supersaturation in the microdroplets are gradually increased by the spontaneous emulsification. When a 60 mM Span 80 solution was used as the continuous phase, the microdroplet was shrunk 180 μm to 100 μm within 35 min. The increasing rate of supersaturation was able to control by the concentration of Span 80. Vapor diffusion method is most widely used protein crystallization method in the conventional methods. The supersaturation is increased to reach the equilibrium to reservoir solution with increasing the incubation time. Therefore, the crystallization condition is explored in wide supersaturation range. The microdroplet-based vapor diffusion method provides many advantages in protein crystallography by reducing the protein sample consumption and high-throughput crystallization condition screening compared with the typical microdroplet-based protein crystallization method.

2.6 Summary

This chapter summarized that the microfluidic techniques and platforms for protein crystallography. Microfluidics has a great potential in the protein crystallographic applications: high-throughput protein crystallization condition screening, low amount of sample consumption, high quality single crystal preparation, and on-chip X-ray diffraction measurement. A number of nano or picoliter-scale trials are prepared by simple operation compared with the conventional crystallization methods. In addition, the quality of protein crystals is able to improve using the microfluidic device and environment without special apparatus. On-chip X-ray crystallography using micro and nano protein crystals will be main research topic, because of the recent advances of synchrotron facility. In particular, microfluidic devices coupled with XFELs certainly accelerate the protein three dimensional structure analysis. Novel microfluidic-based techniques and devices for protein crystallography will be developed along with the progress of synchrotron facility and diffraction data and statistical processing.

References

1. Nilsson J, Ståhl S, Lundeberg J, Uhlén M, Per-Å (1997) Affinity fusion strategies for detection, purification, and immobilization of recombinant proteins. Protein Expr Purif 11:1–16
2. Arnau J, Lauritzen C, Petersen GE, Pedersen J (2006) Current strategies for the use of affinity tags and tag removal for the purification of recombinant proteins. Protein Expr Purif 48:1–13
3. Maeki M, Kimura N, Sato Y, Harashima H, Tokeshi M (2018) Advances in microfluidics for lipid nanoparticles and extracellular vesicles and applications in drug delivery systems. Adv Drug Deliv Rev 128:84–100
4. Mazaafrianto D, Maeki M, Ishida A, Tani H, Tokeshi M (2018) Recent microdevice-based Aptamer sensors. Micromachines 9(5):202
5. Gong MM, Sinton D (2017) Turning the page: advancing paper-based microfluidics for broad diagnostic application. Chem Rev 117:8447–8480
6. Osaki T, Takeuchi S (2017) Artificial cell membrane systems for biosensing applications. Anal Chem 89:216–231
7. Armbrecht L, Dittrich PS (2017) Recent advances in the analysis of single cells. Anal Chem 89:2–21
8. Wu J, Chen Q, Lin JM (2017) Microfluidic technologies in cell isolation and analysis for biomedical applications. Analyst 142:421–441
9. Hao SJ, Wan Y, Xia YQ, Zou X, Zheng SY (2018) Size-based separation methods of circulating tumor cells. Adv Drug Deliv Rev 125:3–20

10. Kaminski TS, Garstecki P (2017) Controlled droplet microfluidic systems for multistep chemical and biological assays. Chem Soc Rev 46:6210–6226
11. Kurita R, Niwa O (2016) Microfluidic platforms for DNA methylation analysis. Lab Chip 16:3631–3644
12. Strohmeier O, Keller M, Schwemmer F, Zehnle S, Mark D, von Stetten F, Zengerle R, Paust N (2015) Centrifugal microfluidic platforms: advanced unit operations and applications. Chem Soc Rev 44:6187–6229
13. Gutmann B, Cantillo D, Kappe CO (2015) Continuous-flow technology-a tool for the safe manufacturing of active pharmaceutical ingredients. Angew Chem Int Ed Eng 54:6688–6728
14. Suryawanshi PL, Gumfekar SP, Bhanvase BA, Sonawane SH, Pimplapure MS (2018) A review on microreactors: reactor fabrication, design, and cutting-edge applications. Chem Eng Sci 189:431–448
15. Shi HH, Xiao Y, Ferguson S, Huang X, Wang N, Hao HX (2017) Progress of crystallization in microfluidic devices. Lab Chip 17:2167–2185
16. Shembekar N, Chaipan C, Utharala R, Merten CA (2016) Droplet-based microfluidics in drug discovery, transcriptomics and high-throughput molecular genetics. Lab Chip 16:1314–1331
17. Dittrich PS, Manz A (2006) Lab-on-a-chip: microfluidics in drug discovery. Nat Rev Drug Discov 5:210–218
18. Sesen M, Alan T, Neild A (2017) Droplet control technologies for microfluidic high throughput screening (muHTS). Lab Chip 17:2372–2394
19. Venkatakrishnan AJ, Deupi X, Lebon G, Tate CG, Schertler GF, Babu MM (2013) Molecular signatures of G-protein-coupled receptors. Nature 494:185–194
20. Surade S, Blundell TL (2012) Structural biology and drug discovery of difficult targets: the limits of ligandability. Chem Biol 19:42–50
21. Nooren IMA, Thornton JM (2003) Diversity of protein-protein interactions. EMBO J 22:3486–3492
22. Knowles TP, Vendruscolo M, Dobson CM (2014) The amyloid state and its association with protein misfolding diseases. Nat Rev Mol Cell Biol 15:384–396
23. Lonsdale R, Ward RA (2018) Structure-based design of targeted covalent inhibitors. Chem Soc Rev 47:3816–3830
24. Wang SH, Yu J (2018) Structure-based design for binding peptides in anti-cancer therapy. Biomaterials 156:1–15
25. Ferreira LG, Dos Santos RN, Oliva G, Andricopulo AD (2015) Molecular docking and structure-based drug design strategies. Molecules 20:13384–13421
26. Blundell TL, Jhoti H, Abell C (2002) High-throughput crystallography for lead discovery in drug design. Nat Rev Drug Discov 1:45–54
27. Liu W, Wacker D, Gati C, Han GW, James D, Wang D, Nelson G, Weierstall U, Katritch V, Barty A, Zatsepin NA, Li D, Messerschmidt M, S.b. Boutet GJ, Williams JE, Koglin MM, Seibert C, Wang STA, Shah S, Basu R, Fromme C, Kupitz KN, Rendek I, Grotjohann P, Fromme RA, Kirian KR, Beyerlein TA, White HN, Chapman M, Caffrey JCH, Spence RC, Stevens VC (2013) Serial femtosecond crystallography of G protein-coupled receptors. Science 342:1521–1524
28. Rosenbaum DM, Rasmussen SG, Kobilka BK (2009) The structure and function of G-protein-coupled receptors. Nature 459:356–363
29. Hong M, Zhang Y, Hu F (2012) Membrane protein structure and dynamics from NMR spectroscopy. Annu Rev Phys Chem 63:1–24
30. Cheng Y, Grigorieff N, Penczek PA, Walz T (2015) A primer to single-particle cryo-electron microscopy. Cell 161:438–449
31. Russo Krauss I, Merlino A, Vergara A, Sica F (2013) An overview of biological macromolecule crystallization. Int J Mol Sci 14:11643–11691
32. McPherson A, DeLucas LJ (2015) Microgravity protein crystallization. NPJ Microgravity 1:15010

33. Kundrot CE, Judge RA, Pusey ML, Snell EH (2001) Microgravity and macromolecular crystallography. Cryst Growth Des 1:87–99
34. Pareja-Rivera C, Cuéllar-Cruz M, Esturau-Escofet N, Demitri N, Polentarutti M, Stojanoff V, Moreno A (2016) Recent advances in the understanding of the influence of electric and magnetic fields on protein crystal growth. Cryst Growth Des 17:135–145
35. Li F, Lakerveld R (2018) Electric-field-assisted protein crystallization in continuous flow. Cryst Growth Des 18:2964–2971
36. Koizumi H, Uda S, Fujiwara K, Tachibana M, Kojima K, Nozawa J (2015) Crystallization of high-quality protein crystals using an external electric field. J Appl Crystallogr 48:1507–1513
37. Sazaki G, Yoshida E, Komatsu H, Nakada T, Miyashita S, Watanabe K (1997) Effects of a magnetic field on the nucleation and growth of protein crystals. J Cryst Growth 173:231–234
38. Shang L, Cheng Y, Zhao Y (2017) Emerging droplet microfluidics. Chem Rev 117:7964–8040
39. Hibara A, Fukuyama M, Chung M, Priest C, Proskurnin MA (2016) Interfacial phenomena and fluid control in micro/nanofluidics. Anal Sci 32:11–21
40. Beebe DJ, Mensing GA, Walker GM (2002) Physics and applications of microfluidics in biology. Annu Rev Biomed Eng 4:261–286
41. Di Carlo D (2009) Inertial microfluidics. Lab Chip 9:3038–3046
42. Li L, Ismagilov RF (2010) Protein crystallization using microfluidic technologies based on valves, droplets, and slipchip. Annu Rev Biophys 39:139–158
43. Zheng B, Roach LS, Ismagilov RF (2003) Screening of protein crystallization conditions on a microfluidic chip using nanoliter-size droplets. J Am Chem Soc 125:11170–11171
44. Maeki M, Yamaguchi H, Tokeshi M, Miyazaki M (2016) Microfluidic approaches for protein crystal structure analysis. Anal Sci 32:3–9
45. Li L, Du W, Ismagilov R (2010) User-loaded slipchip for equipment-free multiplexed nanoliter-scale experiments. J Am Chem Soc 132:106–111
46. Li L, Mustafi D, Fu Q, Tereshko V, Chen DL, Tice JD, Ismagilov RF (2006) Nanoliter microfluidic hybrid method for simultaneous screening and optimization validated with crystallization of membrane proteins. Proc Natl Acad Sci U S A 103:19243–19248
47. Hansen CL, Skordalakes E, Berger JM, Quake SR (2002) A robust and scalable microfluidic metering method that allows protein crystal growth by free interface diffusion. Proc Natl Acad Sci U S A 99:16531–16536
48. Hansen CL, Classen S, Berger JM, Quake SR (2006) A microfluidic device for kinetic optimization of protein crystallization and in situ structure determination. J Am Chem Soc 128:3142–3143
49. Heymann M, Opthalage A, Wierman JL, Akella S, Szebenyi DM, Gruner SM, Fraden S (2014) Room-temperature serial crystallography using a kinetically optimized microfluidic device for protein crystallization and on-chip X-ray diffraction. IUCrJ 1:349–360
50. Shim J-u, Cristobal G, Link DR, Thorsen T, Jia Y, Piattelli K, Fraden S (2007) Control and measurement of the phase behavior of aqueous solutions using microfluidics. J Am Chem Soc 129:8825–8835
51. Talreja S, Kim DY, Mirarefi AY, Zukoski CF, Kenis PJA (2005) Screening and optimization of protein crystallization conditions through gradual evaporation using a novel crystallization platform. J Appl Crystallogr 38:988–995
52. Guha S, Perry SL, Pawate AS, Kenis PJ (2012) Fabrication of X-ray compatible microfluidic platforms for protein crystallization. Sensors Actuators B Chem 174:1–9
53. Khvostichenko DS, Schieferstein JM, Pawate AS, Laible PD, Kenis PJ (2014) X-ray transparent microfluidic chip for mesophase-based crystallization of membrane proteins and on-chip structure determination. Cryst Growth Des 14:4886–4890
54. Liang YR, Zhu LN, Gao J, Zhao HX, Zhu Y, Ye S, Fang Q (2017) 3D-printed high-density droplet array chip for miniaturized protein crystallization screening under vapor diffusion mode. ACS Appl Mater Interfaces 9:11837–11845

55. Maeki M, Yamazaki S, Pawate AS, Ishida A, Tani H, Yamashita K, Sugishima M, Watanabe K, Tokeshi M, Kenis PJA, Miyazaki M (2016) A microfluidic-based protein crystallization method in 10 micrometer-sized crystallization space. CrystEngComm 18:7722–7727
56. Maeki M, Teshima Y, Yoshizuka S, Yamaguchi H, Yamashita K, Miyazaki M (2014) Controlling protein crystal nucleation by droplet-based microfluidics. Chem Eur J 20:1049–1056
57. Yamaguchi H, Maeki M, Yamashita K, Nakamura H, Miyazaki M, Maeda H (2013) Controlling one protein crystal growth by droplet-based microfluidic system. J Biochem 153:339–346
58. Schieferstein JM, Pawate AS, Varel MJ, Guha S, Astrauskaite I, Gennis RB, Kenis PJA (2018) X-ray transparent microfluidic platforms for membrane protein crystallization with microseeds. Lab Chip 18:944–954
59. Zhang S, Gerard CJJ, Ikni A, Ferry G, Vuillard LM, Boutin JA, Ferte N, Grossier R, Candoni N, Veesler S (2017) Microfluidic platform for optimization of crystallization conditions. J Cryst Growth 472:18–28
60. Ildefonso M, Candoni N, Veesler S (2011) Using microfluidics for fast, accurate measurement of lysozyme nucleation kinetics. Cryst Growth Des 11:1527–1530
61. Grossier R, Magnaldo A, Veesler S (2010) Ultra-fast crystallization due to confinement. J Cryst Growth 312:487–489
62. Vekilov PG (2010) Nucleation. Cryst Growth Des 10:5007–5019
63. Vekilov PG (2007) What determines the rate of growth of crystals from solution? Cryst Growth Des 7:2796–2810
64. Moukhametzianov R, Burghammer M, Edwards PC, Petitdemange S, Popov D, Fransen M, McMullan G, Schertler GF, Riekel C (2008) Protein crystallography with a micrometre-sized synchrotron-radiation beam. Acta Crystallogr D Biol Crystallogr 64:158–166
65. Yamamoto M, Hirata K, Yamashita K, Hasegawa K, Ueno G, Ago H, Kumasaka T (2017) Protein microcrystallography using synchrotron radiation. IUCrJ 4:529–539
66. Berejnov V, Husseini NS, Alsaied OA, Thorne RE (2006) Effects of cryoprotectant concentration and cooling rate on vitrification of aqueous solutions. J Appl Crystallogr 39:244–251
67. Pflugrath JW (2004) Macromolecular cryocrystallography--methods for cooling and mounting protein crystals at cryogenic temperatures. Methods 34:415–423
68. Du W-B, Sun M, Gu S-Q, Zhu Y, Fang Q (2010) Automated microfluidic screening assay platform based on droplab. Anal Chem 82:9941–9947
69. Thorsen T, Maerkl SJ, Quake SR (2002) Microfluidic large-scale integration. Science 298:580–584
70. Li L, Du W, Ismagilov RF (2010) Multiparameter screening on slipchip used for nanoliter protein crystallization combining free interface diffusion and microbatch methods. J Am Chem Soc 132:112–119
71. Wang L, Sun K, Hu X, Li G, Jin Q, Zhao J (2015) A centrifugal microfluidic device for screening protein crystallization conditions by vapor diffusion. Sensors Actuators B Chem 219:105–111
72. Li G, Chen Q, Li J, Hu X, Zhao J (2010) A compact disk-like centrifugal microfluidic system for high-throughput nanoliter-scale protein crystallization screening. Anal Chem 82:4362–4369
73. Maeki M, Yamaguchi H, Yamashita K, Nakamura H, Miyazaki M, Maeda H (2012) A method for generating single crystals that rely on internal fluid dynamics of microdroplets. Chem Commun (Camb) 48:5037–5039
74. Maeki M, Yamaguchi H, Yamashita K, Nakamura H, Miyazaki M, Maeda H (2011) Analysis of kinetic behavior of protein crystallization in nanodroplets. Chem Lett 40:825–827
75. Chapman HN, Fromme P, Barty A, White TA, Kirian RA, Aquila A, Hunter MS, Schulz J, DePonte DP, Weierstall U, Doak RB, Maia FR, Martin AV, Schlichting I, Lomb L, Coppola N, Shoeman RL, Epp SW, Hartmann R, Rolles D, Rudenko A, Foucar L, Kimmel N, Weidenspointner G, Holl P, Liang M, Barthelmess M, Caleman C, Boutet S, Bogan MJ, Krzywinski J, Bostedt C, Bajt S, Gumprecht L, Rudek B, Erk B, Schmidt C, Homke A, Reich C, Pietschner D, Struder L, Hauser G, Gorke H, Ullrich J, Herrmann S, Schaller G, Schopper F, Soltau H, Kuhnel KU, Messerschmidt M, Bozek JD, Hau-Riege SP, Frank M,

Hampton CY, Sierra RG, Starodub D, Williams GJ, Hajdu J, Timneanu N, Seibert MM, Andreasson J, Rocker A, Jonsson O, Svenda M, Stern S, Nass K, Andritschke R, Schroter CD, Krasniqi F, Bott M, Schmidt KE, Wang X, Grotjohann I, Holton JM, Barends TR, Neutze R, Marchesini S, Fromme R, Schorb S, Rupp D, Adolph M, Gorkhover T, Andersson I, Hirsemann H, Potdevin G, Graafsma H, Nilsson B, Spence JC (2011) Femtosecond X-ray protein nanocrystallography. Nature 470:73–77

76. Spence JCH (2017) XFELs for structure and dynamics in biology. IUCrJ 4:322–339
77. Suga M, Akita F, Hirata K, Ueno G, Murakami H, Nakajima Y, Shimizu T, Yamashita K, Yamamoto M, Ago H, Shen JR (2015) Native structure of photosystem II at 1.95 A resolution viewed by femtosecond X-ray pulses. Nature 517:99–103
78. Tosha T, Nomura T, Nishida T, Saeki N, Okubayashi K, Yamagiwa R, Sugahara M, Nakane T, Yamashita K, Hirata K, Ueno G, Kimura T, Hisano T, Muramoto K, Sawai H, Takeda H, Mizohata E, Yamashita A, Kanematsu Y, Takano Y, Nango E, Tanaka R, Nureki O, Shoji O, Ikemoto Y, Murakami H, Owada S, Tono K, Yabashi M, Yamamoto M, Ago H, Iwata S, Sugimoto H, Shiro Y, Kubo M (2017) Capturing an initial intermediate during the P450nor enzymatic reaction using time-resolved XFEL crystallography and caged-substrate. Nat Commun 8:1585
79. Zheng B, Tice JD, Roach LS, Ismagilov RF (2004) A droplet-based, composite PDMS/glass capillary microfluidic system for evaluating protein crystallization conditions by microbatch and vapor-diffusion methods with on-chip X-ray diffraction. Angew Chem Int Ed Eng 43:2508–2511
80. Maeki M, Yoshizuka S, Yamaguchi H, Kawamoto M, Yamashita K, Nakamura H, Miyazaki M, Maeda H (2012) X-ray diffraction of protein crystal grown in a nano-liter scale droplet in a microchannel and evaluation of its applicability. Anal Sci 28:65–68
81. Maeki M, Pawate AS, Yamashita K, Kawamoto M, Tokeshi M, Kenis PJ, Miyazaki M (2015) A method of cryoprotection for protein crystallography by using a microfluidic chip and its application for in situ X-ray diffraction measurements. Anal Chem 87:4194–4200
82. Sui S, Wang Y, Dimitrakopoulos C, Perry S (2018) A graphene-based microfluidic platform for electrocrystallization and in situ X-ray diffraction. Crystals 8
83. Sui S, Wang Y, Kolewe KW, Srajer V, Henning R, Schiffman JD, Dimitrakopoulos C, Perry SL (2016) Graphene-based microfluidics for serial crystallography. Lab Chip 16:3082–3096
84. Fukuyama M, Akiyama A, Harada M, Okada T, Hibara A (2015) Microfluidic protein crystallisation controlled using spontaneous emulsification. Anal Methods 7:7128–7131

Chapter 3
Application of SERS-Based Microfluidics for *In Vitro* Diagnostics

Jinhyeok Jeon, Namhyun Choi, Joung-Il Moon, Hao Chen, and Jaebum Choo

Abstract In this chapter, we describe SERS-based immunoassay techniques using various types of microfluidic platforms. SERS is a highly sensitive detection modality, and microfluidic platforms provide many advantages such as automatic sampling and reduced sample volume. Therefore, the integration of SERS with microfluidic platforms offers wide applications in chemical or biological analysis. These novel SERS-based microfluidic platforms provide a powerful clinical tool for highly sensitive *in vitro* diagnostics.

Keywords Surface-enhanced Raman scattering (SERS) · Microfluidics · *In vitro* diagnostics

3.1 Introduction

The development of fast and sensitive immunoassays for clinical samples has been a critical issue in the area of early diagnosis of food-borne illness or infectious disease [12, 22, 51]. Many efforts have been employed to develop sensitive detection methods for specific biomarkers in human serum, including enzyme-linked immunosorbent assay (ELISA) [42, 53], immunochemiluminescence assay (ICMA) [60], and radioimmunoassay (RIA) [46], ELISA is the most widely used screening tool for validating candidate protein markers, and various ELISA kits have been commercialized for quantitative identification of specific biomarkers. In many cases, however, this approach does not meet the requirements for accurate assessment of a specific biomarker at low concentrations. In ELISA assays, quantification of a marker is achieved through observing enzyme-mediated color changes but a discernible color change of a target cannot be induced at low concentrations of a marker

J. Jeon · N. Choi
Department of Bionano Technology, Hanyang University, Ansan, South Korea

J.-I. Moon · H. Chen · J. Choo (✉)
Department of Chemistry, Chung-Ang University, Seoul, South Korea
e-mail: jbchoo@cau.ac.kr

M. Tokeshi (ed.), *Applications of Microfluidic Systems in Biology and Medicine*, Bioanalysis 7, https://doi.org/10.1007/978-981-13-6229-3_3

and consequently a microplate reader fails to distinguish the color change. Likewise, RIA allows rapid and inexpensive screening of a clinical sample but the use of radioisotopes and scintillation fluids restricts its application for practical detection of a specific biomarker in human serum. ICMA is one of the most popular immunoassay techniques in clinical laboratories but it does not satisfy the low limit of detection (LOD) for several hormone-related biomarkers such as sexual hormones (estradiol and testosterone) or thyroid stimulating hormone (TSH). Recently, mass spectrometry was adopted for more accurate and confident diagnostics in the clinical laboratory [47, 48]. However, despite a low detection capability of less than 10 pg/mL this technique has several analytical disadvantages including long sample preparation steps and the need for expensive instrumentation. For these reasons, mass spectrometry is not considered suitable for routine clinical applications. Thus, there is an urgent need for new technologies that can achieve efficient and sensitive detection of target biomarkers in serum for use in routine clinical diagnostics.

Recently, surface-enhanced Raman scattering (SERS)-based assay platforms have been increasingly considered as promising formats for the detection of various biomarkers due to their high sensitivity and multiplex detection capability [17, 21]. When SERS nano tags are used as detection probes, the Raman scattering signals are greatly enhanced at active junctions known as "hot spots" as a result of electromagnetic and chemical enhancement effects. Such enhancements have shown great promise in overcoming the sensitivity problems inherent in the fluorescence or luminescence detection of ELISA or ICMA assays. Moreover, SERS-based detection methods do not require the culturing or amplification steps inherent to bacteria colony counting and PCR assays. To date, many different types of biomarkers including proteins [8, 10, 11, 28, 29], viruses [2, 20, 40], and bacteria [13, 26, 44, 57, 63] have been studied using SERS-based assays.

The last decade has also seen great progress in the development of microfluidic technology for use in chemical and biomedical sciences [43, 54, 61]. These developments have been driven by its advantages over conventional macroscale analytical methods, including reduced consumption of reagents and mixing time, high analytical throughput, facile automation, and improved production conversion [34, 35, 37]. It has long been recognized that the system used for detecting the progress of a reaction is a key factor in determining the applicability of a microfluidic system. Because of the extremely small volume in a microfluidic channel, a highly sensitive detection method is essential for monitoring the progress of a chemical or biological reaction. Consequently, the development of SERS-based microfluidic platforms has recently attracted significant attention in the biomedical sciences. The integration of SERS, a highly sensitive detection modality, with microfluidic platforms, which have many advantages over microscale methods [3, 23, 24, 32], offers significant promise for chemical and biological experimentation. There are two primary types of flow regimes in microfluidic platforms: continuous and segmented (or droplet) flows. In this chapter we introduce various types of SERS-based microfluidic platforms for the rapid and sensitive detection of biomarkers in serum.

3.2 Multiplex Immunoassay Using Array-Embedded Microfluidic Channel

A gold-patterned microarray chip has been extensively used as a SERS-based immunoassay platform but it has some technical problems [6, 16, 19, 28, 29]. First, it is not easy to optimize the homogeneous distribution of target biomarkers on gold wells because the sequential processes including antibody/antigen immobilization and washing steps are difficult to control manually using a micropipette. Second, the repetition of washing steps required to remove non-specific binding proteins makes this assay platform inconvenient. Finally, a tedious manual dilution process is required for the immunoassay of various concentrations of target biomarkers.

To resolve this problem, M. Lee and co-workers [30] developed a programmable and fully automatic gold array-embedded gradient microfluidic chip that integrates a gradient microfluidic device with gold-patterned microarray wells. Serial dilution of the antigen marker can be achieved in a stepwise manner using microfluidic concentration gradient generators with N cascade-mixing stages [27–29, 39]. In this channel, the desired concentrations can be achieved by controlling the volumetric mixing ratios of two merging solutions in each stage. In addition, 30-μm gold-patterned microarray wells (5 × 5) were embedded onto a glass substrate for antibody immobilization on the surface. Use of this novel gold microarray-embedded gradient microfluidic channel avoids the tedious manual dilution process, and highly accurate immunoanalysis can be achieved with automatic flow controls. The utility of this platform was demonstrated by quantitative immunoassay of the alpha-fetoprotein protein marker. The total assay time, including incubation, washing steps and detection, was less than 60 min. Figure 3.1 shows the layout of a gold array-embedded gradient chip for SERS-based immunoassay.

However, during the clinical sample analyses, the method described above is affected by nonspecific binding of non-target species. To resolve this problem, K. K. Reza and co-workers [41] developed a platform that utilizes a combination of alternating current electrohydrodynamic (ac-EHD)-induced surface shear forces to remove nonspecific molecules from the electrode surface. The use of ac-EHD surface shear forces improves analyte transport across an antibody-functionalized capture domain and simultaneously displaces nonspecific binding molecules from the electrode surface [45, 52]. This capability enables rapid SERS immunoassays that outperformed traditional pressure driven flow based assays. To demonstrate its applicability, a microfluidic device containing five individual microchannels, each with an array of asymmetric electrode pairs, was constructed. In a typical immunoassay, a complex biological sample containing multiple protein biomarkers and the solution containing detection antibody conjugated to SERS nano tags are driven through the devices sequentially under an ac-EHD field. In this study, simultaneous capture and multiplexed detection was performed for human epidermal growth factor receptor 2 (HER2), mucin 1, cell surface associated (MUC1), epidermal growth factor receptor (EGFR), and mucin 16, cell surface associated (MUC16)

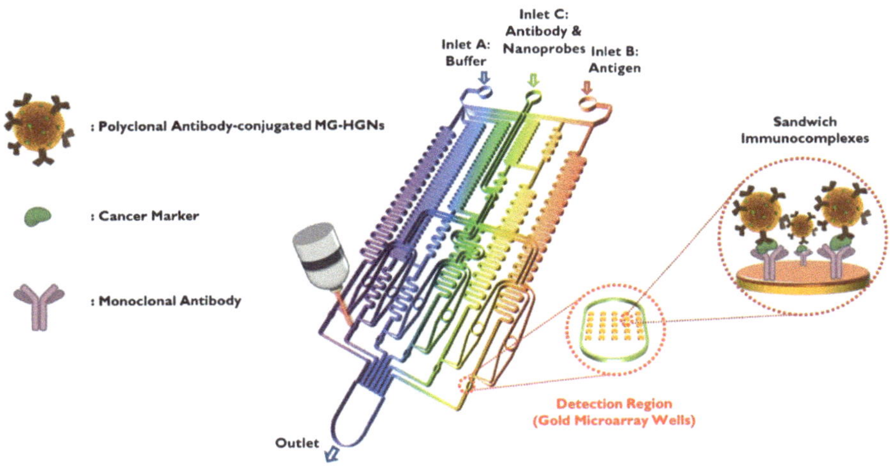

Fig. 3.1 Layout of a gold array-embedded gradient chip for the SERS-based immunoassay. The illustrations in the enlarged circles represent the formation of sandwich immunocomplexes on the surface of 5 × 5 round gold wells embedded in the gradient channel. (Reprinted with permission from Lee et al. [30]. Copyright (2012) The Royal Society of Chemistry)

proteins that are overexpressed in breast, lung, and ovarian cancer, respectively. Considering the extensive multiplexing ability of SERS immunoassays, the presented approach with the ability to simultaneously detect four individual cancer biomarkers within each channel can potentially detect a broad panel of cancer biomarkers. Figure 3.2 shows a schematic illustration of the microfluidic channel for the multiplex detection of protein biomarkers using ac-EHD induced SERS detection.

For high throughput detection of multiple biomarkers, L. Wu and co-workers [59] developed a SERS-assisted 3D barcode microfluidic system. Figure 3.3 shows a schematic illustration of this immunoassay protocol. Multiple samples were dispensed into parallel channels using a programmable syringe pump. Different antigens in the sample were captured by the corresponding antibodies immobilized on the surface of the channels. Next, a mixture of SERS nanoprobes was dispensed into the channel and bound onto corresponding antigens to form immunocomplexes. For quantitative evaluation of multiple biomarkers, a 3D barcode containing the spatial and spectroscopic information was acquired from the variation of Raman peak intensities (Fig. 3.3d, right). This array-embedded microfluidic technology could be successfully used for the simultaneous detection of multiple biomarkers.

Wang et al. [56] developed a microfluidic device with multiple channels for the duplex detection of pathogen antigens as illustrated in the photograph and scheme in Fig. 3.4. Microfluidic channels were connected with three inlets and three outlets to enable the simultaneous detection of multiple targets simultaneously. Each channel contained gold patterns that could be used for the immobilization of specific antibodies. Herein, NYscFv-350 antibody was immobilized on the surface in the

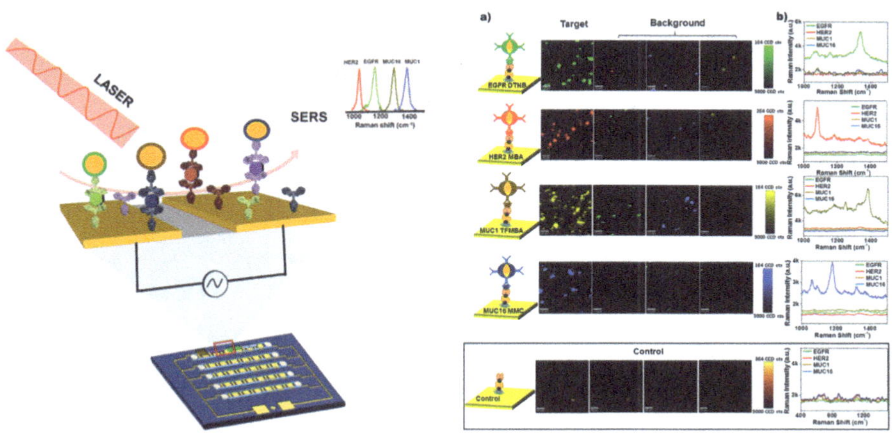

Fig. 3.2 Schematic illustration of multiplexed protein biomarker detection using ac-EHD induced SERS-immunoassay. (**a**) Schematic illustration of typical immunoassay and false-color SERS images, and (**b**) corresponding SERS spectra, for specific target capture and background response from serum samples containing equal concentrations (10 pg mL^{-1} for each antigen) of EGFR, HER2, MUC1, and MUC16 antigens under ac-EHD field (f = 1 kHz and Vpp = 100 mV). Control experiments were performed using devices functionalized without any capture antibody. Scale bar is 10 μm. (Reprinted with permission from Reza et al. [41]. Copyright (2016) John Wiley & Sons, Inc.)

first channel, and NYscFv-030 antibody was bound on the surface in the second channel. In the third channel, NYscFv-350 and 030 antibodies at a 1:1 ratio were immobilized to capture both antigens. Immunoreactions were performed, and detection antibody-conjugated silica-coated SERS nano tags were employed for readout. The demonstrated duplex detection of 350 and 030 paves the way for a powerful analytical platform with potential applications for high throughput multiplex detection of biomarkers.

3.3 Magnetic Bead-Based Immunoassay in Microfluidic Channel

Magnetic beads have been extensively used for immunoassays because they can be manipulated and separated using a magnetic field, strongly facilitating assay protocols [50]. When magnetic beads are used as a substrate, capture antibodies are immobilized on the surface of the beads for use as mobile substrates. Their large surface-to-volume ratio provides efficient capture of target antigens [36]. Antibody-conjugated SERS nano tags are bound to target antigens captured by the magnetic beads to form sandwich immunocomplexes [5, 58, 62]. Here, SERS nano tags are used as detection probes for the highly sensitive quantification of target antigens. The sandwich immunocomplexes can be immobilized on the wall of a microtube

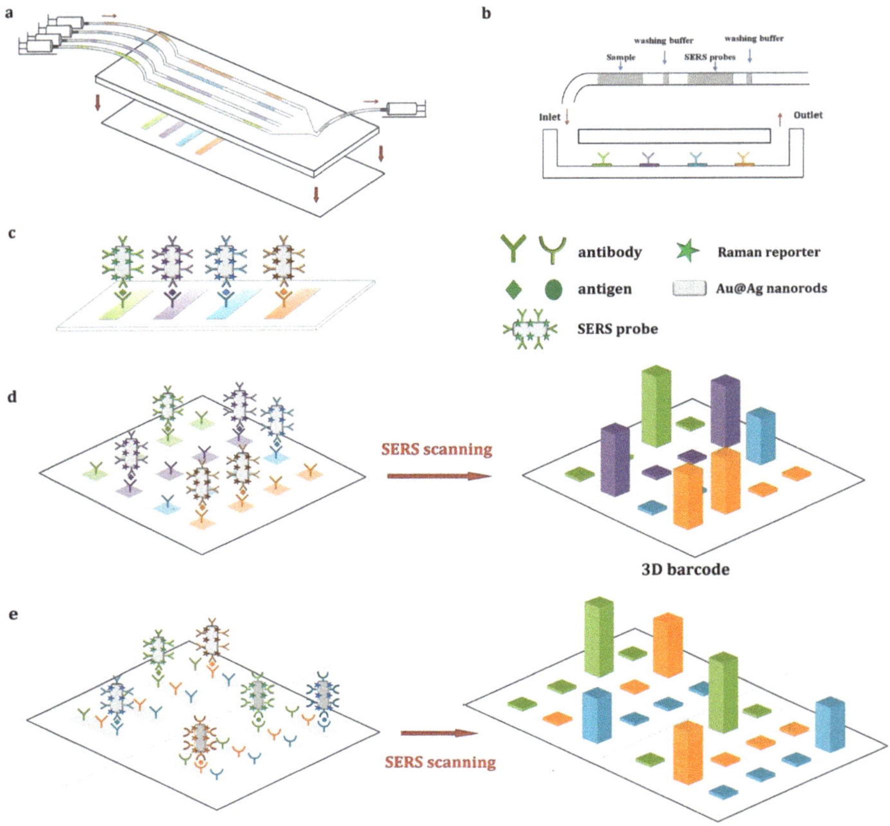

Fig. 3.3 (**a**) Schematic illustration of the microfluidic system. The syringes are connected to microfluidic channels through polyethylene (PE) tubes and controlled by programmable microfluidic pumps. The upper PDMS mold was bonded with the antibody barcode substrate to obtain the microfluidic platform. (**b**) Side view of the microfluidic system. The PE tube is preloaded with samples and reagents, with each kind of fluid separated by air. The samples sequentially pass the detection zones which are patterned with different antibodies in a process (known as "bubble-based sample delivery.") (**c**) The principle of sandwich immunoassays. The "antibody barcode" was used to separate and capture different antigens. The SERS probes can be further captured by the corresponding antigens, forming a sandwich structure. The different colours represent different kinds of antibodies, antigens, and Raman reporters. (**d**) Multiplex immunoassay using the 3D barcode chip (each channel was patterned with one antibody). The SERS spectrum in each unit of the hybridization array is acquired and the measurement results are presented as a 3D barcode. The different colors represent the Raman fingerprint of different reporters and the bar height reflects the SERS intensity. (**e**) Multiplex immunoassay using the 3D barcode chip (each channel is patterned with multiple antibodies). Different antigens in different samples can be identified according to the 2D spatial information and the characteristic Raman peaks, both of which are included in the 3D barcode. (Reprinted with permission from Wu et al. [59]. Copyright (2015) John Wiley & Sons, Inc.)

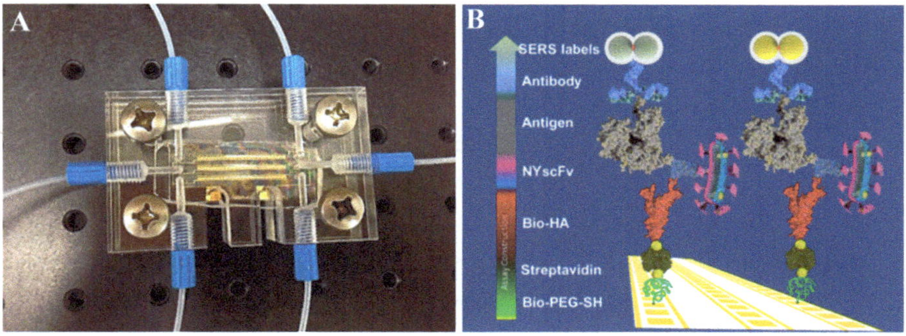

Fig. 3.4 (**a**) Photograph of the 3-channel microfluidic device containing gold patterns and (**b**) schematic illustration of the SERS biosensor platform with NYscFv for duplex antigen detection. (Reprinted with permission from Wang et al. [56]. Copyright (2014) American Chemical Society)

using a magnetic bar, and quantitative analysis can be performed by measuring the characteristic Raman signals of a Raman reporter. This SERS-based immunoassay overcomes the slow immunoreaction resulting from diffusion-limited kinetics on two-dimensional substrates because of the large effective surface area of the magnetic beads. Nonetheless, this assay technique is inconvenient due to problems associated with multiple manual washing steps, tedious manual handling of samples, and difficulties in controlling assay conditions. To resolve these problems, SERS-based microfluidic platforms were developed for more efficient analysis of biomarkers. A highly accurate and reproducible immunoassay is possible by maintaining a continuous flow and homogeneous mixing conditions in a microfluidic channel.

Chon et al. [9] developed a gradient microfluidic channel to automatically achieve serial dilutions of a target marker. This device has subsequently been used to automatically serially dilute a target sample with buffer [4, 49]. For the formation of sandwich immunocomplexes in a microfluidic channel, functionalized HGNs and magnetic beads with antibodies were introduced into the channel. Next, minisolenoids [19] were integrated into the microfluidic device for immobilization of different concentrations of immunocomplexes. Here, minisolenoids were placed close to each microfluidic channel to generate a magnetic field gradient to trap magnetic beads in a flowing stream. The solenoids can be switched on or off, and the field intensity can be tuned on demand. Finally, the SERS signal for sandwich immunocomplexes immobilized on each microfluidic channel was measured using a confocal Raman microscope. Consequently, a SERS-based immunoassay can be automatically performed in a microfluidic channel. With this novel technique, the tedious manual dilution process is eliminated and rapid and sensitive immunoanalysis can be achieved. Figure 3.5 demonstrates the layout of the SERS-based optofluidic sensor and its working scheme.

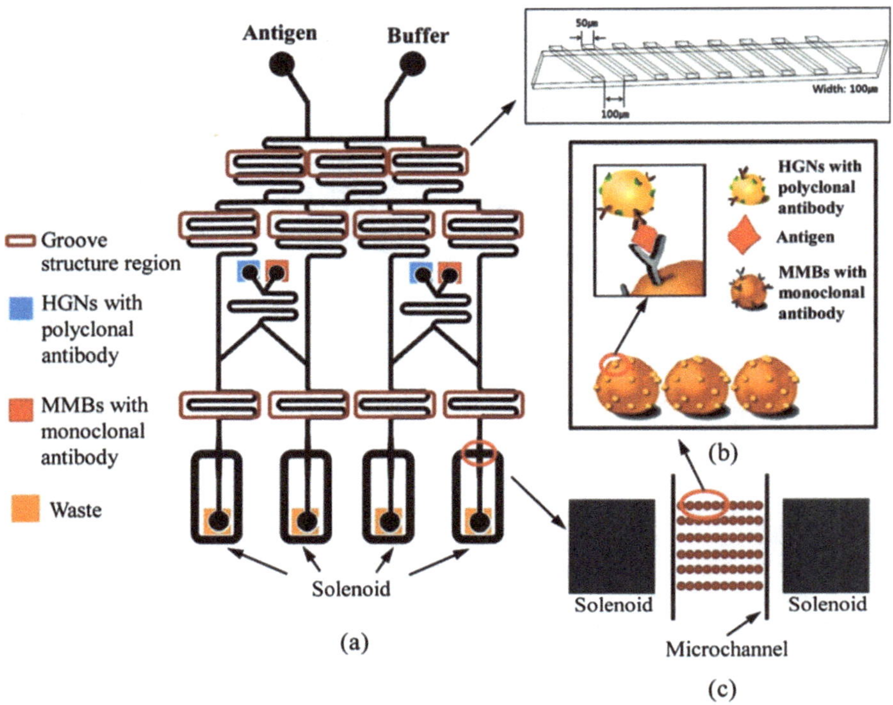

Fig. 3.5 (**a**) Layout of a SERS-based gradient optofluidic sensor integrated with solenoids. (**b**) Formation of a sandwich immunocomplex between HGNs and magnetic beads. (**c**) Trapping of sandwich immunocomplexes in a microfluidic channel. (Reprinted with permission from Wang et al. [58]. Copyright (2010) American Chemical Society)

Li et al. [31] improved the detection sensitivity using magnetic focusing and plasmonic coupling techniques. For this purpose, NiFe core-Au shell nanoparticles (NPs) were synthesized by successive reduction of $HAuCl_4$ using a seeded growth approach. Using tunable magnetic core-shell NPs, effective magnetic focusing could be achieved via an effective coupling between magnetic NPs and gold NPs in a microfluidic platform. Figure 3.6 displays a schematic illustration of a SERS-based microfluidic platform for magnetic focusing immunoassays.

During the process of magnetic focusing [33, 55], magnetic particles accumulate on the bottom of the channel near the edge of the magnetic bar. This aggregation induces a high density of electromagnetic "hot spots" from the nanogaps between NiFe core-Au shell NPs and gold NPs. Figure 3.7 also shows the signal enhancement effect according to the size of Au NPs. The peak intensity increases with the size of Au NPs, and is maximized at a diameter of 60 nm. This result also shows good agreement with theoretical simulation results for the electromagnetic field enhancement in Fig. 3.7c.

Gao et al. [15] developed a novel SERS-based magnetic sensor for highly sensitive detection of the anthrax biomarker poly-γ-D glutamic acid (PGA) [25] in

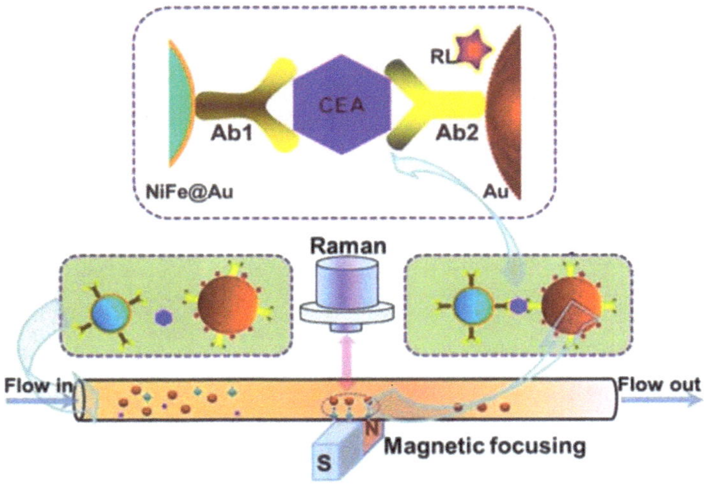

Fig. 3.6 Illustration of SERS detection of cancer biomarker CEA using functional nanoprobes consisting of Au-coated NiFe magnetic nanoparticle (NiFe@Au), Ab1 (capture antibody), Ab2 (detection antibody) and RL (Raman label). (Reprinted with permission from Han et al. [19]. Copyright (2015) American Chemical Society)

human serum. For safe, sensitive, and rapid detection of a hazardous material in human serum, they designed and fabricated a solenoid-embedded dual channel microfluidic device to perform the PGA immunoassay in an automatic manner. Figure 3.8 illustrates the schematic design of the microfluidic device used in this study. The device consists of two parallel channel compartments: one for PGA sensing and the other for control measurements. The sensing channel assays serum for various concentrations of the PGA trace whereas the control channel only detects PGA-free serum as an external standard. Both channels are also composed of three compartments in the vertical direction. In the first compartment of the sensing channel, serum including target trace PGA antigens and anti-PGA-conjugated magnetic beads were introduced into the channel via two inlets. In the second compartment, PGA-conjugated AuNPs were introduced into the channel via the inlet located in the middle part of the device. Here, PGA antigens and PGA-conjugated AuNPs underwent competitive reaction with the antibodies on magnetic beads under flow conditions. As shown in this figure, a staggered herringbone groove-shaped mixer was incorporated into the channel to improve mixing efficiency. The third compartment acted to trap and detect magnetic immunocomplexes. To achieve effective trapping of the magnetic beads in the stream, two yoke-type mini-solenoids were arranged at the end of each channel. After trapping the immunocomplexes, PGA-free serum was introduced from the inlet located in the middle part of the device to wash out unbound PGA antigens and AuNPs. Finally, the SERS signal was measured by focusing the laser beam within each microfluidic channel. For the control measurements, the assay and SERS detection were performed using the same procedure. In

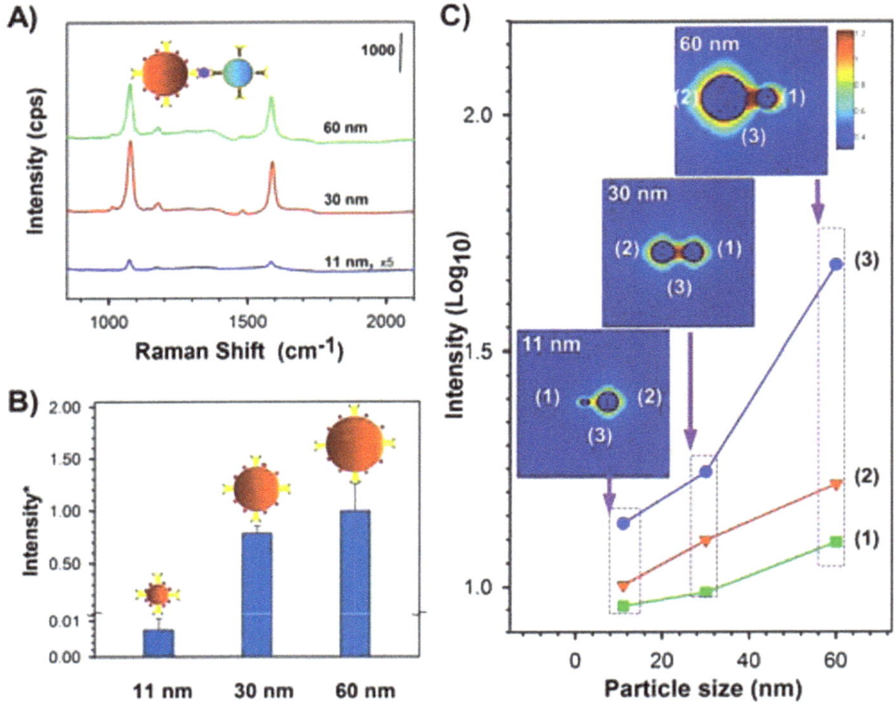

Fig. 3.7 (**a**) SERS spectra of the sandwich complex of Ab1 conjugated NiFe@Au NPs-CEA antigen-Ab2 conjugated Au NPs; (**b**) Plot of normalized peak intensity over total surface area at 1076 cm^{-1} vs Au NP size. (**c**) Theoretical simulation of a dimer model for the sandwich complex in terms of E-field enhancement around a small particle (area 1, squares), large particle (area 2, triangles), and in the center (area 3, circles); insert: contour plots external to the dimers for 11–27 nm (bottom), 30–27 nm (middle), and 60–27 nm (top) pairs. (Reprinted with permission from Han et al. [19]. Copyright (2015) American Chemical Society)

this study, the external standard values for PGA-free serum were measured each time using the control microfluidic channel, which greatly improved reliability by minimizing the influence of most experimental variables.

3.4 Segmented Flow-Based Immunoassay in Microfluidic Channel

Gao et al. [14] developed a SERS-based microdroplet sensor for the rapid and sensitive detection of hazardous materials. This sensor is composed of two droplet compartments, one for the on-chip synthesis of fresh Ag NPs and the other for droplet merging and SERS detection. Silver ions were nucleated and grown to larger size silver nanoparticles in droplets, and then each droplet was synchronously

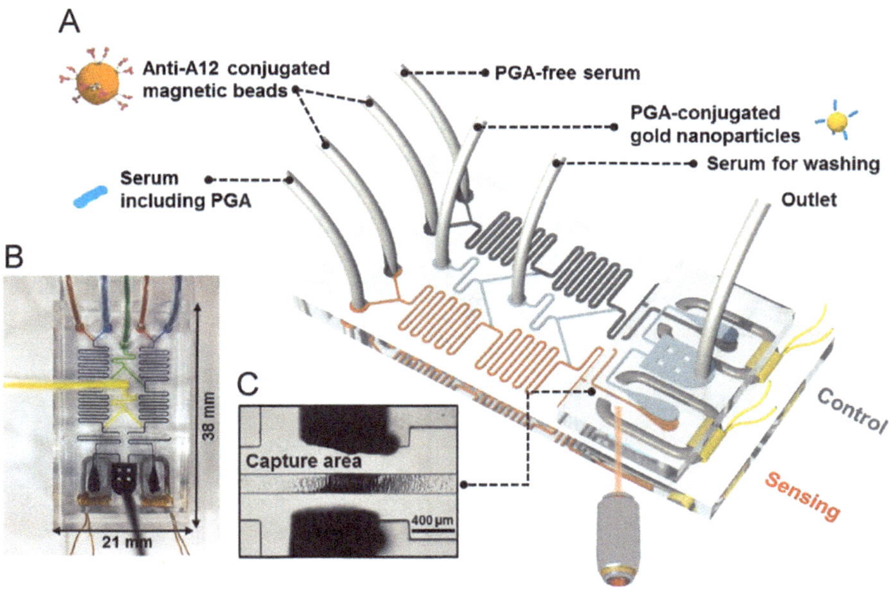

Fig. 3.8 (**a**) Schematic illustration of the solenoid-embedded dual channel microfluidic sensor for SERS-based competitive immunoassay. The sensor is composed of two parallel channels: one for PGA sensing (gray) and the other for control (red). (**b**) Optical images of the solenoid chip filled with four different colours of inks. (**c**) Photograph of the capture area for magnetic immunocomplexes. (Reprinted with permission from Wang et al. [55]. Copyright (2015) Elsevier Inc.)

merged with a droplet containing target analytes for SERS detection. As shown in Fig. 3.9, a high mixing efficiency among silver nitrate solution, distilled water, and hydroxylamine hydrochloride/sodium hydroxide solutions was achieved in the first compartment. Droplet formation resulted from induced shear forces at the interface between the two different phases. After formation of droplets, aqueous samples in each droplet were mixed by transport through the winding channels. Synthesizing Ag NPs within droplets separated by an immiscible fluid environment allowed for easy isolation of the reactants from the surrounding environment, thus avoiding contamination. In the second compartment, each droplet was synchronously merged with a droplet containing target analytes for SERS detection. The combination of a droplet merging system with in situ Ag NP synthesis for on-line SERS detection is expected to be a powerful analytical tool for the fast and reproducible bioanalysis.

Recently, Choi et al. [7] reported a fully integrated SERS-based microdroplet platform for the automatic immunoassay of specific antigen. In this study, a novel integrated microfluidic system that includes droplet generation, transport, mixing, merging, and splitting modules was designed and fabricated. The device allows efficient immunoreactions to be achieved through sequential droplet generation, transport, and merging [1, 16, 18, 38], while wash-free immunoassays are realized through the droplet splitting. This defines a conceptually new multifunctional

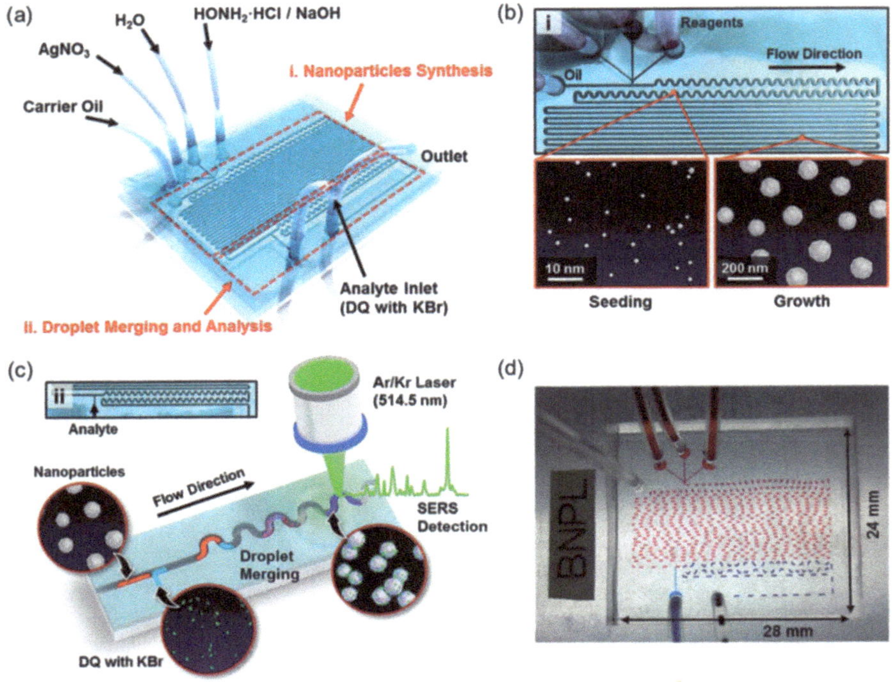

Fig. 3.9 (**a**) Schematic illustration of an integrated microdroplet channel for SERS detection of DQ. The channel is composed of two compartments. The first is the synthetic section for Ag NP synthesis (i) and the second is for droplet merging and SERS analysis (ii). (**b**) Expanded view of the first compartment (i): nucleation of silver ions and large silver nanoparticles. (**c**) Expanded view of the second compartment (ii): intersection for droplet merging and SERS detection. (**d**) Optical images of the entire microdroplet channel filled with two different colors (red and blue) of ink. (Reprinted with permission from Jang et al. [25]. Copyright (2014) The Royal Society of Chemistry)

microfluidic platform capable of performing complex multistep immunoassays in nanoliter volumes for the safe and sensitive detection of hazardous materials. To validate this approach, the fraction 1 (F1) antigen for *Yersinia pestis*, which is of significant interest to the defense and security communities because of its potential use as a biological weapon, was chosen as a target protein marker. Figure 3.10 illustrates the schematic design of the integrated microfluidic channel used in the study.

The device consists of six compartments. In the first compartment, aqueous solutions of antigens and antibody-conjugated SERS nano tags are supplied from two central inlets, and carrier oil is supplied orthogonally. Droplet formation results from shear forces at the interface between the aqueous and oil phases (top image in Fig. 3.10b). In the second compartment, antigens and detection antibody-conjugated SERS nano tags in each droplet are efficiently mixed by transport through multiple winding channels. Since the antibody–antigen reactions in each droplet are

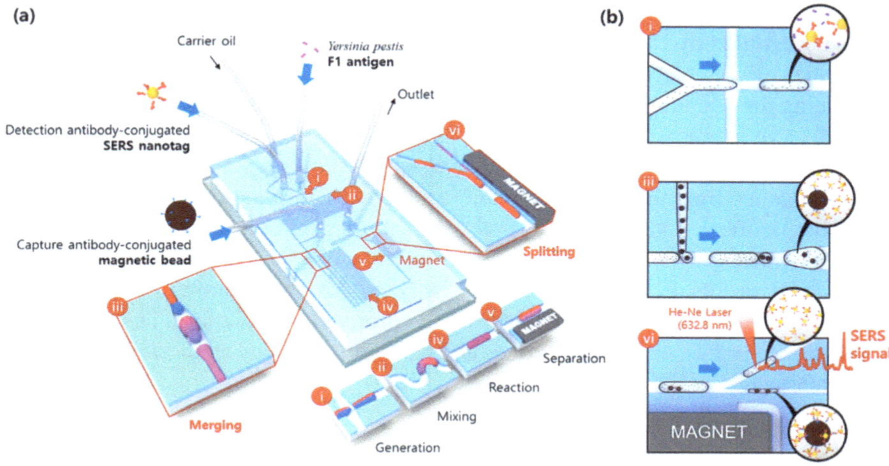

Fig. 3.10 (**a**) Schematic design of the integrated SERS-based microfluidic channel composed of six microdroplet compartments: (*i*) droplet generation from the shear force at the interface between the aqueous and oil phases, (*ii*) droplet mixing for the first immunoreaction, (*iii*) droplet merging for the formation of magnetic immunocomplexes, (*iv*) droplet mixing for the second immunoreaction, (*v*) droplet splitting for the wash-free immunoassay, and (*vi*) Raman detection of unbound SERS nano tags in supernatant solution droplets. (**b**) Extended images of (*i*) droplet generation, (*iii*) droplet merging, and (*vi*) droplet splitting. (Reprinted with permission from Gao et al. [14]. Copyright (2017) American Chemical Society)

completely isolated it is possible to perform the immunoreaction of hazardous materials safely while avoiding contamination from the surrounding environment. In the third compartment, capture antibody-conjugated magnetic beads are introduced into the inlet positioned at the middle of the chip, generating a second droplet. Subsequently, the first droplet (containing the antigens and detection antibody-conjugated SERS nano tag mixtures) and the second droplet (containing capture antibody-conjugated magnetic beads) are sequentially merged through decompression in the wide merging channel (middle image in Fig. 3.10b). In this geometry, the two droplets are pushed together before being pulled apart to initiate merging. To ensure efficient and robust merging, the flow rates of the two droplets must be precisely controlled. In the fourth compartment, immunoreactions between the capture antibody-conjugated magnetic beads and the antigen-captured SERS nano tags occur during passage through a second set of winding channels, with magnetic immunocomplexes being formed in a droplet twice as large as the original droplets.

In the fifth compartment, a magnetic bar embedded in the channel initiates droplet splitting. As shown in the bottom image of Fig. 3.10b, magnetic immunocomplexes align on the bottom side of the droplet as each merged droplet passes close to the magnetic bar, while unreacted SERS nanotags and antigens in the supernatant solution remain on the other side of the droplet. Subsequently, each droplet is split into a pair of daughter droplets, which respectively contain primarily magnetic immunocomplexes and primarily unbound SERS nano tags. The critical parameter that controls droplet splitting under such conditions is the capillary number, which can be controlled by changing the droplet velocity since the surface tension and the viscosities of both phases are constant. Specifically, the widths of the microchannels containing the two daughter droplets were varied to control both the droplet velocity and effective magnetic field strength. For the defined process, removal of unbound SERS nano tags and antigens by washing was not necessary since they are automatically separated from the magnetic immunocomplexes during droplet splitting. In the final compartment, both daughter droplets pass through a set of winding channels for particle dispersion. Finally, Raman signals of the unbound SERS nano tags in the supernatant droplets are measured and analyzed for the quantitative analysis of antigen. Such an integrated SERS-based microdroplet assay platform has significant potential utility in the sensitive, rapid, and safe immunoanalysis of various hazardous materials.

3.5 Summary

During the past decade, there has been growing interest in the biomedical application of SERS-based microfluidics. In particular, SERS-based immunoassays are a potential clinical tool for highly sensitive and reproducible detection of multiple biomarkers. In this chapter we describe three different types of SERS-based assay platforms: (i) array-embedded, (ii) magnetic bead-based, and (iii) segment flow-based microfluidic devices. These SERS-based microfluidic platforms have many advantages, including low reagent consumption, short incubation time, wider concentration dynamic range, and cost savings compared with the conventional ELISA technique. The SERS-based immunoassay technique can be used for highly sensitive analysis of multiple biomarkers. With the use of appropriate antibody-functionalized SERS-coding nanoprobes, the optical signals for specific immunocomplexes can be amplified and analyzed using this novel technique. Consequently, this SERS-based immunoassay platform is expected to be a powerful clinical tool for early disease diagnosis.

Acknowledgements The National Research Foundation of Korea supported this work through grant numbers 2017M3D1A1039287 and 2018M3A7B4071203. This work was also supported by a grant of the Korea Health Technology R&D Project through the Korea Health Industry Development Institute (KHIDI), funded by the Ministry of Health & Welfare, Republic of Korea (grant number: HG18C0062) and by the Agency for Chemical & Biological Detection Research Center (CBDRC).

References

1. Brouzes E, Kruse T, Kimmerling R, Strey HH (2015) Rapid and continuous magnetic separation in droplet microfluidic devices. Lab Chip 15:908–919
2. Campbell FM, Ingram A, Monaghan P, Cooper J, Sattar N, Eckersall PD, Graham D (2008) SERRS immunoassay for quantitative human CRP analysis. Analyst 133:1355–1357
3. Chen L, Choo J (2008) Recent advances in surface-enhanced Raman scattering detection technology for microfluidic chips. Electrophoresis 29:1815–1828
4. Chen L, Wang G, Lim C, Seong GH, Choo J, Lee EK, Kang SH, Song JM (2009) Evaluation of passive mixing behaviors in a pillar obstruction poly(dimethylsiloxane) microfluidic mixer using fluorescence microscopy. Microfluid Nanofluid 7:267–273
5. Cheng Z, Choi N, Wang R, Lee S, Moon KC, Yoon SY, Chen L, Choo J (2017) Simultaneous detection of dual prostate specific antigens using surface-enhanced Raman scattering-based immunoassay for accurate diagnosis of prostate cancer. ACS Nano 11:4926–4933
6. Choi N, Lee K, Lim DW, Lee EK, Chang SI, Oh KW, Choo J (2012) Simultaneous detection of duplex DNA oligonucleotides using a SERS-based micro-network gradient chip. Lab Chip 12:5160–5167
7. Choi N, Lee J, Ko J, Jeon JH, Rhie G, deMello AJ, Choo J (2017) Integrated SERS-based microdroplet platform for the automated immunoassay of F1 antigens in Yersinia pestis. Anal Chem 89:8413–8420
8. Chon H, Lee S, Son SW, Oh CH, Choo J (2009) Highly sensitive immunoassay of lung cancer marker carcinoembryonic antigen using surface-enhanced Raman scattering of hollow gold nanospheres. Anal Chem 81:3029–3034
9. Chon H, Lim C, Ha SM, Ahn Y, Lee EK, Chang SI, Seong GH, Choo J (2010) On-chip immunoassay using surface-enhanced Raman scattering of hollow gold nanospheres. Anal Chem 82:5290–5295
10. Chon H, Lee S, Yoon SY, Chang SI, Lim DW, Choo J (2011) Simultaneous immunoassay for the detection of two lung cancer markers using functionalized SERS nanoprobes. Chem Commun 47:12515–12517
11. Chon H, Lee S, Yoon SY, Lee EK, Chang SI, Choo J (2014) SERS-based competitive immunoassay of troponin i and CK-MB markers for early diagnosis of acute myocardial infarction. Chem Commun 50:1058–1060
12. Chung HJ, Castro CM, Im H, Lee H, Weissleder R (2013) A magneto-DNA nanoparticle system for rapid detection and phenotyping of bacteria. Nat Nanotechnol 8:369–375
13. Costas C, López-Puente V, Bodelón G, González-Bello C, Pérez-Juste J, Pastoriza-Santos I, Liz-Marzán LM (2015) Using surface enhanced Raman scattering to analyze the interactions of protein receptors with bacterial quorum sensing modulators. ACS Nano 9:5567–5576
14. Gao R, Choi N, Chang SI, Lee EK, Choo J (2014) Real-time analysis of diaquat dibromide monohydrate in water with a SERS-based integrated microdroplet sensor. Nanoscale 6:8781–8786
15. Gao R, Ko J, Cha K, Jeon JH, Rhie G, Choi J, deMello AJ, Choo J (2015) Fast and sensitive detection of an anthrax biomarker using SERS-based solenoid microfluidic sensor. Biosens Bioelectron 72:230–236
16. Gao R, Cheng Z, deMello AJ, Choo J (2016) Wash-free magnetic immunoassay of the PSA cancer marker using SERS and droplet microfluidics. Lab Chip 16:1022–1029
17. Grubisha DS, Lipert RJ, Park HY, Driskell J, Porter MD (2003) Femtomolar detection of prostate-specific antigen: an immunoassay based on surface-enhanced Raman scattering and immunogold labels. Anal Chem 75:5936–5943
18. Gu SQ, Zhang YX, Zhu Y, Du WB, Yao B, Fang Q (2011) Multifunctional picoliter droplet manipulation platform and its application in single cell analysis. Anal Chem 83:7570–7576
19. Han B, Choi N, Kim KH, Lim DW, Choo J (2011) Application of silver-coated magnetic microspheres to a SERS-based optofluidic sensor. J Phys Chem C 115:6290–6296

20. Han Z, Liu H, Wang B, Weng S, Yang L, Liu J (2015) Three-dimensional surface-enhanced Raman scattering hotspots in spherical colloidal superstructure for identification and detection of drugs in human urine. Anal Chem 87:4821–4828
21. He LL, Rodda T, Haynes CL, Deschains T, Strother T, Diez-Gonzalez F, Labuza TP (2011) Detection of a foreign protein in milk using surface-enhanced Raman spectroscopy coupled with antibody-modified silver dendrites. Anal Chem 83:1510–1513
22. Howes PD, Rana S, Stevens MM (2014) Plasmonic nanomaterials for biodiagnostics. Chem Soc Rev 43:3835–3853
23. Huang JA, Zhang YL, Ding H, Sun HB (2015) SERS-enabled lab-on-a-chip systems. Adv Opt Mater 3:618–633
24. Jahn IJ, Zukovskaja O, Zheng XS, Weber K, Bocklitz TW, Cialla-May D, Popp J (2017) Surface-enhanced Raman spectroscopy and microfluidic platforms: challenges, solutions and potential applications. Analyst 142:1022–1047
25. Jang J, Cho M, Lee HR, Cha K, Chun JH, Hong KJ, Park J, Rhie G (2013) Monoclonal antibody against the poly-γ-d-glutamic acid capsule of Bacillus anthracis protects mice from enhanced lethal toxin activity due to capsule and anthrax spore challenge. Biochim Biophys Acta 1830:2804–2812
26. Kumar S, Lodhi DK, Goel P, Neeti MP, Singh JP (2015) A facile method for fabrication of buckled PDMS silver nanorod arrays as active 3D SERS cages for bacterial sensing. Chem Commun 51:12411–12414
27. Lee K, Kim C, Ahn B, Panchapakesan R, Full AR, Nordee L, Kang JY, Oh KW (2009) Generalized serial dilution module for monotonic and arbitrary microfluidic gradient generators. Lab Chip 9:709–717
28. Lee K, Kim C, Kim Y, Ahn B, Panchapakesan R, Bang J, Kim J, Yoon YK, Kang JY, Oh KW (2011a) Micro- fluidic concentration-on-demand. Microfluid Nanofluid 11:75–86
29. Lee M, Lee S, Lee JH, Lim H, Seong GH, Lee EK, Chang SI, Oh CH, Choo J (2011b) Highly reproducible immunoassay of cancer markers on a gold-patterned microarray chip using surface-enhanced Raman scattering imaging. Biosens Bioelectron 26:2135–2141
30. Lee M, Lee K, Kim KH, Oh KW, Choo J (2012) SERS-based immunoassay using a gold array-embedded gradient microfluidic chip. Lab Chip 12:3720–3727
31. Li J, Skeete Z, Shan S, Yan S, Kurzatkowska K, Zhao W, Ngo QM, Holubovka P, Luo J, Hepel M, Zhong CJ (2015) Surface enhanced Raman scattering detection of cancer biomarkers with bifunctional nanocomposite probes. Anal Chem 87:10698–10702
32. Lim C, Hong J, Chung BG, deMello AJ, Choo J (2010) Optofluidic platforms based on surface-enhanced Raman scattering. Analyst 135:837–844
33. Lin L, Crew E, Yan H, Shan S, Skeete Z, Mott D, Krentsel T, Yin J, Chernova NA, Luo J, Engelhard MH, Wang C, Li Q, Zhong CJ (2013) Bifunctional nanoparticles for SERS monitoring and magnetic intervention of assembly and enzyme cutting of DNAs. J Mater Chem B 1:4320–4330
34. Mark D, Haeberle S, Roth G, von Stetten F, Zengerle R (2010) Microfluidic lab-on-a-chip platforms: requirements, characteristics and applications. Chem Soc Rev 39:1153–1182
35. Marre S, Jensen KF (2010) Synthesis of micro and nanostructures in microfluidic systems. Chem Soc Rev 39:1183–1202
36. Morozov VN, Grove S, Turell MJ, Bailey C (2007) Three minutes-long electrophoretically assisted zeptomolar microfluidic immunoassay with magnetic-beads detection. J Am Chem Soc 129:12628–12629
37. Nge PN, Rogers CI, Woolley AT (2013) Advances in microfluidic materials, functions, integration, and applications. Chem Rev 113:2550–2583
38. Niu X, Gulati S, Edel JB, deMello AJ (2008) Pillar-induced droplet merging in microfluidic circuits. Lab Chip 8:1837–1841
39. Oh KW, Lee K, Ahn B, Furlani EP (2012) Design of pressure-driven microfluidic networks using electric circuit analogy. Lab Chip 12:515–545

40. Paul AM, Fan Z, Sinha SS, Shi Y, Le L, Bai F, Ray PC (2015) Bio-conjugated gold nanoparticle based SERS probe for ultrasensitive identification of mosquito-borne viruses using Raman fingerprinting. J Phys Chem C 119:23669–23675
41. Reza KK, Wang J, Vaidyanathan R, Dey S, Wang Y, Trau M (2017) Electrohydrodynamic-induced SERS immunoassay for extensive multiplexed biomarker sensing. Small 13:1602902
42. Schneider C, Schöler HF, Schneider RJ (2004) A novel enzyme-linked immunosorbent assay for ethynylestradiol using a long-chain biotinylated EE2 derivative. Steroids 69:245–253
43. Shaban M, Hady AGA, Serry M (2014) A new sensor for heavy metals detection in aqueous media. IEEE Sensors J 14:436–441
44. Shao F, Lu Z, Liu C, Han H, Chen K, Li W, He Q, Peng H, Chen J (2014) Hierarchical nanogaps within bioscaffold arrays as a high-performance SERS substrate for animal virus biosensing. ACS Appl Mater Interfaces 6:6281–6289
45. Shiddiky MJA, Vaidyanathan R, Rauf S, Tay Z, Trau M (2014) Molecular nanoshearing: an innovative approach to shear off molecules with AC-induced nanoscopic fluid flow. Sci Rep 4:3716
46. Shirtcliff EA, Granger DA, Schwartz EB, Curran MJ, Booth A, Overman WH (2000) Assessing estradiol in biobehavioral studies using saliva and blood spots: simple radioimmunoassay protocols, reliability, and comparative validity. Horm Behav 38:137–147
47. Stanczyk FZ, Clarke NJ (2010) Advantages and challenges of mass spectrometry assays for steroid hormones. J Steroid Biochem Mol Biol 121:491–495
48. Stopforth A, Burger BV, Crouch AM, Sandra P (2007) The analysis of estrone and 17-β-estradiol by stir bar sorptive extraction-thermal desorption-gas chromatography/mass spectrometry: application to urine samples after oral administration of conjugated equine estrogens. J Chromatogr B 856:156–164
49. Stroock AD, Dertinger SK, Ajdari A, Mezic I, Stone HA, Whitesides GM (2002) Chaotic mixer for microchannels. Science 295:647–651
50. Tekin HC, Gijs MAM (2013) Ultrasensitive protein detection: a case for microfluidic magnetic bead-based assays. Lab Chip 13:4711–4739
51. Tsai SQ, Iafrate AJ, Joung JK (2014) Genome editing: a tool for research and therapy: towards a functional understanding of variants for molecular diagnostics using genome editing. Nat Med 20:1103–1104
52. Vaidyanathan R, Shiddiky MJA, Rauf S, Dray E, Tay Z, Trau M (2014) Tunable "Nano-shearing": a physical mechanism to displace nonspecific cell adhesion during rare cell detection. Anal Chem 86:2042–2049
53. Valentini F, Compagnone D, Gentili A, Palleschi G (2002) An electrochemical ELISA procedure for the screening of 17β-estradiol in urban waste waters. Analyst 127:1333–1337
54. Wang G, Chen Z, Chen L (2010) Aptamer-nanoparticle-based optical probes. Prog Chem 22:489–499
55. Wang LY, Luo J, Shan SY, Crew E, Yin J, Zhong CJ, Wallek B, Wong SSS (2011) Bacterial inactivation using silver-coated magnetic nanoparticles as functional antimicrobial agents. Anal Chem 83:8688–8695
56. Wang Y, Rauf S, Grewal YS, Spadafora LJ, Shiddiky MJA, Cangelosi GA, Schlucker S, Trau M (2014) Duplex microfluidic SERS detection of pathogen antigens with nanoyeast single-chain variable fragments. Anal Chem 86:9930–9938
57. Wang J, Wu X, Wang C, Shao N, Dong P, Xiao R, Wang S (2015) Magnetically assisted surface-enhanced Raman spectroscopy for the detection of Staphylococcus aureus based on Aptamer recognition. ACS Appl Mater Interfaces 7:20919–20929
58. Wang R, Chon H, Lee S, Cheng Z, Hong SH, Yoon YH, Choo J (2016) Highly sensitive detection of hormone Estradiol E2 using surface-enhanced Raman scattering based immunoassays for the clinical diagnosis of precocious puberty. ACS Appl Mater Interfaces 8:10665–10672
59. Wu L, Wang Z, Fan K, Zong S, Cui Y (2015) A SERS-assisted 3D barcode chip for high-throughput biosensing. Small 23:2798–2806

60. Xin TB, Chen H, Lin Z, Liang SX, Lin JM (2010) A secondary antibody format chemiluminescence immunoassay for the determination of estradiol in human serum. Talanta 82:1472–1477
61. Xu BB, Zhang YL, Wei S, Ding H, Sun HB (2013) On-chip catalytic microreactors for modern catalysis research. ChemCatChem 5:2091–2099
62. Yu J, Jeon J, Choi N, Lee JO, Kim YP, Choo J (2017) SERS-based genetic assay for amplification-free detection of prostate cancer specific PCA3 mimic DNA. Sensors Actuators B 251:302–309
63. Zhou H, Yang D, Ivleva NP, Mircescu NE, Schubert S, Niessner R, Wieser A, Haisch C (2015) Label-free in situ discrimination of live and dead bacteria by surface-enhanced Raman scattering. Anal Chem 87:6553–6561

Chapter 4
Miniaturized Electrochemical Sensors to Facilitate Liquid Biopsy for Detection of Circulating Tumor Markers

Yi-Ge Zhou, Leyla Kermansha, Libing Zhang, and Reza M. Mohamadi

Abstract Miniaturized sensors to facilitate liquid biopsy for early cancer detection and monitoring the disease progression have been unprecedentedly advanced in the recent years, among which electrochemical sensors, offer advantages with their features such as simplicity, fast response, low cost and capability for miniaturization. In this chapter, we will provide an overview of recent advances in using miniaturized electrochemical sensors for detection of circulating tumor markers, including circulating tumor cells (CTCs), circulating nucleic acids (cNAs), and extracellular vesicles (EVs). Representative examples will be given for each marker based on different electrochemical approaches, and combination of electrochemistry with other technologies and strategies will be shown.

Keywords Electrochemical sensors · Liquid biopsy · Circulating tumor cell · Circulating tumor DNA · Exosomes

---

Y.-G. Zhou (✉)
Institute of Chemical Biology and Nanomedicine, Hunan University, Changsha, People's Republic of China

State Key Laboratory of Chemo/Biosensing and Chemometrics, College of Chemistry and Chemical Engineering, Hunan University, Changsha, People's Republic of China
e-mail: yigezhou@hnu.edu.cn

L. Kermansha
Institute of Biomaterials and Biomedical Engineering, University of Toronto, Toronto, Canada

Department of Pharmaceutical Sciences, University of Toronto, Toronto, Canada

L. Zhang · R. M. Mohamadi (✉)
Department of Pharmaceutical Sciences, University of Toronto, Toronto, Canada
e-mail: reza.mohamadi@utoronto.ca

M. Tokeshi (ed.), *Applications of Microfluidic Systems in Biology and Medicine*, Bioanalysis 7, https://doi.org/10.1007/978-981-13-6229-3_4

4.1 Introduction

4.1.1 Circulating Tumor Markers

Cancer is a group of diseases associated with the growth and spread of abnormal cells. It is considered a localized disease in its early stage but becomes systemic once the abnormal cells invade or spread to other parts of the body, causing metastasis, which is responsible for 90% of cancer related deaths [7]. Few symptoms are evident at the early stage of cancer; by the time the symptoms appear and progress to an advanced stage, cancer is far more difficult to treat. Cancer survival rates, therefore, tend to be very low due to the late-stage diagnosis as well as the limited access to timely and effective treatment. Thus, there is an urgent need for early and accurate detection of cancer. In addition, during the course of treatment, it is necessary to monitor the progress of the disease. Ideally, this monitoring could be done using less invasive blood tests. Once the blood sample is taken, it can be tested for several biomarkers simultaneously.

A tumor biomarker refers to a biomarker found in serum, urine, or body tissues that is altered in the presence of cancer. Specifically, circulating tumor markers (CTMs) are known as biomarkers present in the blood circulation whose concentration is proposed as a diagnostic marker for quantitative real-time assessment of the tumor burden. CTMs include cancer cells and a variety of biomolecules such as proteins, enzymes, nucleic acids, and small vesicles [20, 76]. Detection of CTMs has been acknowledged as "liquid biopsy" by its ability in dynamic and non-invasive monitoring of cancer progression featuring convenience, high reproducibility and low cost [28].

Great advancements have been made to the detection of CTMs, including fluorescence methods [49], polymerase chain reaction (PCR) [34], enzyme-linked immunosorbent assay (ELISA) [2], surface plasmon resonance [31], surface enhanced Raman scattering (SERS) [39], electrochemical assay [14], colorimetric assay [77], microcantilevers [61], and quartz crystal measurement (QCM) [53]. However, there remain inevitable challenges in the following aspects [20]. The low concentration of biomarkers, the insufficient binding efficiency, and the low signal transduction all make it difficult to perform early-stage cancer detection with high sensitivity. Even when sensitivity is high, the detections may exhibit low specificity due to a huge background of non-cancerous factors, causing false-positive signals. Moreover, due to the cancer diversity and the biological variability, confirming cancer by using only a single biomarker is not reliable.

4.1.2 Why Electrochemistry

Electrochemical approaches provide quantitative or semi-quantitative analytical information by reporting the changes in the potential, the current or the impedance that is generated from the interaction at the analyte-receptor interface. Compared to other analytical methods, electrochemical sensors commonly show clinically relevant sensitivity and specificity; in additions they feature simplicity, fast response, low cost and the capability for miniaturization. Various types of voltammetry (such as cyclic voltammetry, linear sweep voltammetry, stripping voltammetry, differential pulse voltammetry (DVP), etc.), amperometry, and electrochemical impedance are the most widely used electrochemical detection methods. Electrochemical sensors are usually married to new technologies to enhance their performances. To further increase the sensitivity and specificity of the sensors, nanotechnology and nanomaterials are commonly involved to construct nanostructured interfaces [32, 85] or nanomaterial-based reporting system [42, 71]. Microfluidics that requires lower volumes of samples and speeds up the interaction between the target analyte and the recognition species, is often integrated with electrochemical sensors to achieve multiplexing, automation and high-throughput detection [26, 67]. The miniaturization and integration is promising for constructing portable devices and realizing on-site analysis. In addition, technologies such as magnetic field [32], ELISA [56] and surface plasmon resonance (SPR) [80] are also reported to be combined to electrochemical sensors.

In this chapter, we focus on the advances of electrochemical strategies for the detection of CTMs over the last 5 years. We will discuss examples of different electrochemical approaches according to the category of circulating tumor markers with special interests on CTCs, cNAs and EVs, aiming to demonstrate how the latest developments meet the aforementioned challenges and highlight the remaining problems that are urgent to be dealt with.

4.2 Circulating Tumor Cells

CTCs are shed from primary tumor and circulate in blood stream to seed metastasis which is responsible for a majority of cancer-related deaths. Research on CTCs has become a very active field in early cancer detection that is important for clinical diagnosis, prognosis and cancer treatment.

Due to the fact that CTCs are very rare in blood, i.e. 1–10 CTCs per mL of whole blood versus a few million white blood cells, isolation of CTCs from blood samples with high yield, sensitivity and specificity remains a major technical challenge. CellSearch is the only FDA-approved technology for CTC isolation taking advantage of immunomagnetic separation method. In this method, CTCs bound to anti-EpCAM-coated ferrofluid nanoparticles are captured by a magnetic field and later detected by fluorescence imaging. Although CellSearch can reproducibly detect and enumerate CTCs with an average recovery rate of no less than 80% [3], this technology shows limitations in its EpCAM antibody enrichment strategy and variable detection rates [23]. A great many techniques have been developed aiming to achieve higher CTCs capture efficiency based on two main principles: physical properties-based isolation [11, 13, 40] and immuno-based isolation [30, 51, 71]. The use of microfluidic technologies is very common in both cases. Compared to traditional methods, microfluidic technologies reduce the processing time to a large extent and show advantages including low sample requirement, high throughput, high sensitivity, large surface area-to-volume ratios, automatic operation and multiplexed integration [54]. Magnetic separation has also been combined with microfluidics to capture CTCs with higher purity, especially when a microfluidic chip with specially designed geometry is used [46]. Most recently, an *in vivo* capture of CTCs with a surface-modified vein indwelling needle [86] and a biomaterial implant [4] were carried out in animal models. This technique can address the problems such as small sampling volume and sampling error caused by the irregular shedding of CTCs.

As previously mentioned, electrochemical methods provide attractive solutions to CTC detection exhibiting simplicity, low cost, fast sensing, and clinically related sensitivity and specificity (Table 4.1). Electrochemistry was recently reported to release [82–84] and lyse [82] the captured CTCs for further analysis. A number of advances have also been made in the electrochemical detection of CTCs using various electrochemical techniques, including voltammetry techniques such as cyclic voltammetry [59], linear sweep voltammetry [71], stripping voltammetry [38, 42, 79], and DVP [88], electrochemical impedance [22, 44, 85], amperometry [10, 29], and electrochemiluminescent (ECL) [74, 78, 87].

Voltammetry, for example, is one of the most powerful techniques among electroanalytical methods in which information about an analyte can be obtained by measuring the current with varying potential. Cyclic voltammetry (CV) is the most common type of voltammetry measurement where the working electrode potential changes with time according to a triangular waveform, and is widely used to study redox processes reaction kinetics. A number of detections use CV as readout method to report on the current difference from the electroactive labels before and after CTCs are introduced to the electrode [59]. However, charging current accompanying the redox current will compromise the sensitivity of the sensors.

DPV is another kind of voltammetry that applies a series of regular voltage pulses on the potential linear sweep. Since the effect of the charging current can be minimized by sampling the current just before the potential is changed, DPV is often applied to bio-detection and bio-analysis. By reducing background current introduced by macromolecules or cells, DPV shows a much improved sensitivity over normal voltammetry methods [60]. Kelley laboratory constructed a chip-based assay with nanostructured microelectrodes (NMEs) to analyze and classify prostate tumor cells in cultured cells and ultimately in a pilot study involving blood samples from 16 prostate cancer patients. In this integrated chip, DPV was used to detect RNA from lysed CTCs isolated by immune-magnetic beads [32]. The electrochemical reporter system employed here was two different redox-active probes, $Ru(NH_3)_6^{3+}$ and $Fe(CN)_6^{4-}$. The Ru(III) species accumulated at the sensor surface in proportion with the hybridization of the negatively charged RNA targets to the NME-tethered PNA probes and was electrochemically reduced to Ru(II) by applying a DPV scan. The Fe(II) species was introduced to chemically regenerate Ru(III) from Ru(II) for further electrochemical cycles, amplifying the signal. In another study,

Table 4.1 An overview of common electrochemical approaches for CTC detection

Electrochemical approaches	Electrochemical labels	Biomarkers	Detection limit	References
Impedance	Label-free	Nucleolin	794 cells/mL	Feng et al. [22]
Impedance	Label-free	N/A	~5 cells/mL	Martinez-Cisneros et al. [44]
Impedance	Label-free	Glycan	12 cells/mL 580 cells/mL	Zhang et al. [85]
ECL	Tris (2,2-bipyridyl) ruthenium (TBR)	No information	89 cells/mL	Ding et al. [17]
ECL	C-dot@Ag	Folate receptor	10 cells/mL	Wu et al. [78]
ECL	NiS@CdS composite	MUC1	12 cells/mL	Wen and Yang [74]
DPV	$Ru(NH_3)_6^{3+}$/$Fe(CN)_6^{4-}$	EpCAM (CTC capture)	10–100 cells/sensor	Ivanov et al. [32]
DPV	$Fe(CN)_6^{4-/3-}$	EpCAM	125 cells/sensor	Moscovici et al. [47]
DPV	*p*-aminophenol	CK18	2 cells/mL	Safaei et al. [56]
DPV	H_2O_2	N-glycan	10 cells/mL	Chen et al. [12]
DPV	Thionine	EpCA	~34 cells/mL ~42 cells/mL	Zheng et al. [88]
Apmerometry	H_2O_2	EpCAM	10 cells/assay	Hong et al. [29]
Apmerometry	Proton	EpCAM	8340 cells/assay	Maltez-Da Costa et al. [43]
Stripping	CdTe and ZnSe QDs	EpCAM and GPC3	10 cells/mL	Wu et al. [79]
Stripping	CdTe QDs	No information	50 cells/mL	Liu et al. [42]
Stripping	CdTe QDs	MUC1	100 cells/mL	Li et al. [38]
LSV	AgNPs, CuNPs and PdNPs	EpCAM, MUC1, HER2, PSMA, CK18 and nucleolin	2 cells/sensor	Wan et al. [71]
Bipolar electrochemistry	N/A	N/A	320 cells/mL	Zhang et al. [87]

Moscovici et al. developed a novel microfabricated glass chip that is able to count prostate cancer cells by measuring the decreased DPV signals of the electroactive molecule $[Fe(CN)_6]^{4-/3-}$ with the increasing number of cells specifically bound to the electrode surface that hinders the electron transfer [47] (Fig. 4.1). Pulse amperometry was also used for genetic profiling of single cancer cells by detecting a HRP catalyzed electrochemical reaction where HRP was attached to a single stranded DNA reporter probe [1].

Stripping voltammetry can sensitively analyze trace amount of electroactive analytes in solution in two steps: a preconcentration/preaccumulation step to

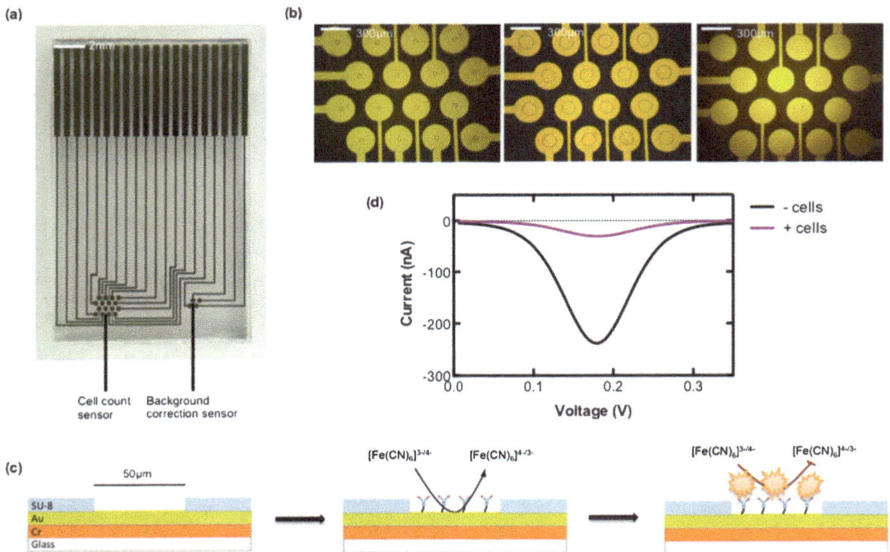

Fig. 4.1 Cancer cell sensing approach. (**a**) Photograph of a microfabricated glass chip used for cell detection. (**b**) Microscopic images of sensors with 50 μm, 150 μm, and 300 μm apertures. (**c**) Schematic of apertures created on a glass chip that serves as cell sensors. An anti-EpCAM antibody self-assembled monolayer is immobilized to gold surface as recognition layer. Upon binding of DU145 prostate cancer cells to the anti-EpCAM antibody, the interfacial electron transfer reaction of $[Fe(CN)_6]^{3/4-}$ is hindered due to the blocking of the gold surface, resulting in a decrease in the electrical signal. (**d**) DPV showing the decrease in electrical signal after target cancer cells are bound to the sensor. (Figure taken from Moscovici et al. [47])

concentrate the target analytes to the electrode surface and a stripping step to strip the preconcentrated analytes from the electrode and measure them in the form of current. Stripping voltammetry may have lowest detection limit among the commonly used electrochemical techniques due to the preconcetration/preaccumulation procedure [37]. Anodic stripping voltammetry, cathodic stripping voltammetry and adsorptive stripping voltammetry constitute the three commonly used stripping techniques. Among those, anodic stripping voltammetry is most frequently applied to biodetections, often with nanoparticles as labels, where nanoparticles are dissolved into metal ions before being analyzed using anodic stripping measurement to sensitively report the presence and level of target analytes [17]. The use of quantum dots (QDs), such as CdTe [38, 42, 79], and ZnSe [79], has been extensively reported as nanoparticle labels in surface marker based electrochemical sensing of CTCs. QDs can be functionalized with small molecules and biomacromolecules to achieve specific interactions with surface proteins of CTCs. After specifically bound to the surface of CTCs, QDs are dissolved in HNO_3 to form metal ions, such as Cd^{2+} and Zn^{2+}, which are determined using anodic stripping voltammetry to indicate the presence and the number of CTCs they bind to. To more reliably confirm cancer cells, in a recent study, Kang et al. used two kinds of QDs CdTe and ZnSe to

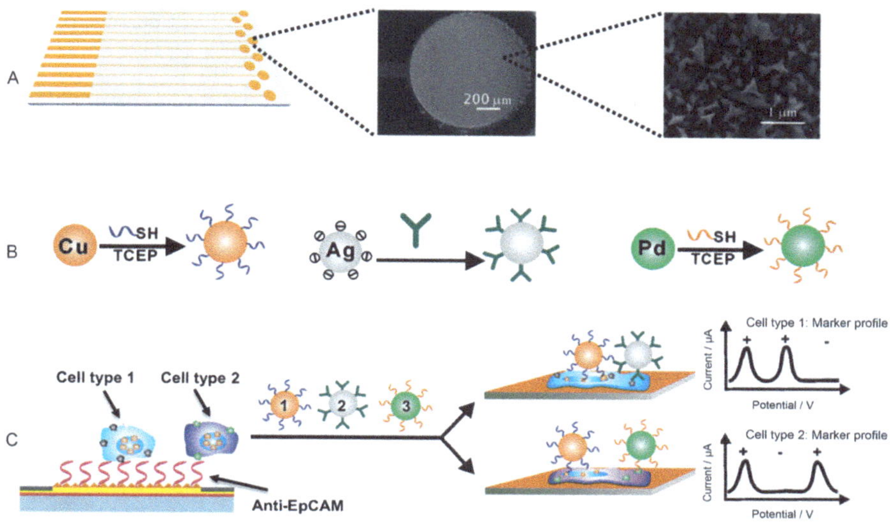

Fig. 4.2 Overview of the specific cancer cell detection using multi-nanoparticle approach. (**a**) Chip layout and SEM image of an electrode after plating. (**b**) MNP-based electrochemical labels are made specific to cell surface markers by conjugating MNPs to DNA aptamers or antibodies. (**c**) Working principle of the electrochemical sensor, including cancer cells capturing, MNP labels recognition, and electrochemical readout. (Figure taken from Wan et al. [71])

recognize two disease-specific biomarkers on the MCF-7 cell membrane [79]. By combining a simple, portable and low- cost paper-based microfluidic immunodevice with electrochemical-mediated signal amplification system, the present device enabled ultrasensitive detection and capture of low-abundant MCF-7 cells with a detection limit of 10 cells mL^{-1}.

Although anodic stripping technique had been successfully used in nanoparticle-labeled CTC detection with high sensitivity, the direct determination of the nano-particle labels without pre-dissolution and pre-deposition was never practiced in previous studies. Wan et al. later developed a new, chip-based strategy that exploited for the first time the direct electrochemical oxidation of metal nanoparticles (MNPs), including Ag, Cu and Pd, to report on the presence of specific surface markers [41] (Fig. 4.2). The electrochemical assay allowed simultaneous detection of multiple different biomarkers on the surface of cancer cells while discriminating between cancer cells and normal blood cells. Linear sweeping voltammetry (LSV) was used as the electrochemical readout method in the detection, however, without a pre-concentration step, this method may not be as sensitive as anodic stripping. By applying $NaBH_4$ as a signal amplification agent, the sensitivity of the electrochemical assay was successfully maintained or even elevated by achieving a detection limit of 2 cells per sensor.

Electrochemical impedance spectroscopy (EIS) is a highly sensitive electrochemical approach that measures electrode surface changes by determining the resistive and capacitive properties of materials bound to the electrode to probe complex

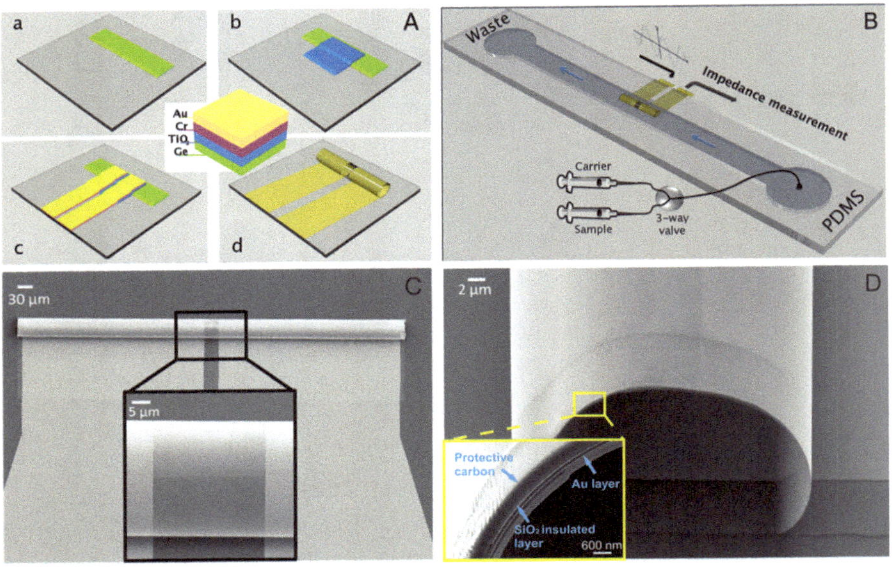

Fig. 4.3 Fabrication process for 3D tubular microsensors. (A) Sequence of the deposition layers: (a) 20 nm Ge sacrificial layer; (b) 60 nm TiO_2 layer; (c) 5 nm Cr layer and 10 nm Au electrodes on top of the planar structure; and (d) rolling up the nanomembranes into rolled-up sensor. (B) A 3D schematic representation of the experimental setup for in-flow sensing. (C) Scanning electron microscopy (SEM) image of the 3D tubular microsensor. (D) Closed-up lateral view of the tubular structure. Inset shows an FIB cut performed to the microtube to observe its cross section. (Figure taken from Martinez-Cisneros et al. [44])

bio-recognition events with a simple and label-free strategy. However, non-specific adsorptions may be very likely to limit the selectivity and sensitivity of this detection method, thus the use of blocking agents to eliminate non-specific adsorptions is important and common in electrochemical impedance detection [85]. EIS is widely used in CTC detection by measuring the change of resistance after CTCs are captured at the surface of an electrode [22]. Recently, Schmidt et al. reported an ultracompact three-dimensional tubular structures integrating Au-based electrodes as impedimetric microsensors for the direct in-flow determination of suspend HeLa cells [44] (Fig. 4.3). Particularly, the microsensor was free from surface functionalization, and showed a single cell resolution in a real-time manner. In another report, a simple and label-free impedimetric cytosensor was fabricated to detect MCF 7 cells, and a wide detection range from 30 to 1×10^6 cells per mL with a detection limit of 10 cells per mL was reached. By cleaving the deoxyuridines (dUs) of the capture aptamer, the cytosensor can be reused and the CTCs can be collected and contributed for further analysis [62].

ECL is a combination of chemiluminescence and electrochemistry, where light emission is initiated by a redox reaction occurring at an electrode surface. Due to its simplicity, high sensitivity and label-free operation, ECL has been used as a powerful technique for bioanalytical applications. Zhang *el al.* reported an

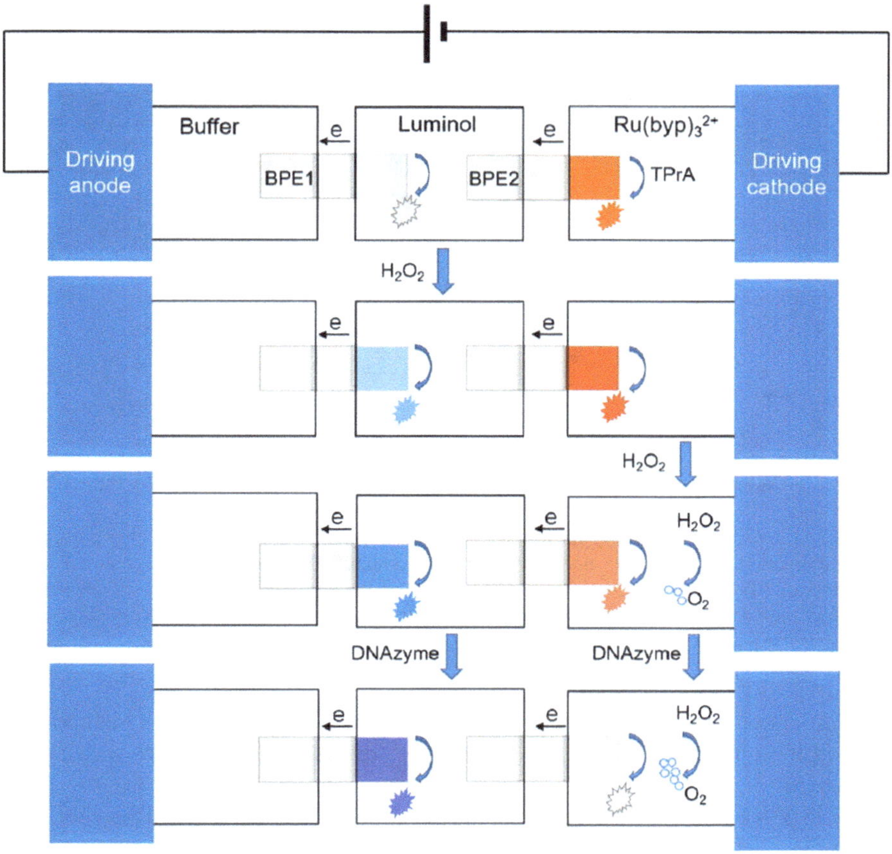

Fig. 4.4 Structure of D-BPE (dual-bipolar electrode) and working principle of BPE-ECL sensing platform. (Figure taken from Zhang et al. [87])

electrochemiluminescent method with a bare gold electrode platform and magnetic beads as isolation and purification tool that was able to detect Burkitt's lymphoma (Ramos) cells as low as 89 cells per mL [17]. Taking advantage of the nanostructured materials, a highly sensitive electrochemiluminescent cytosensing was developed using carbon nanodot@Ag hybrid material as probesand graphene nanosheets as signal amplification agents for cancer cell detection with high sensitivity and selectivity [78]. This ECL cytosensor showed superior cell-capture ability and exhibited a wide linear range and a much lower detection limit of 10 cells/mL. Most recently, Zhang et al. applied bipolar electrochemistry to ECL on a multichannel chip platform to realize visual color-switch biosensing of HL-60 cancer cell by detecting H_2O_2 produced by stimulating the cancer cells [87] (Fig. 4.4). The ECL intensity was much higher because of electron transfer facilitation through one bipolar electrode to the other, enabling quantitative detection of cancer cells and easy

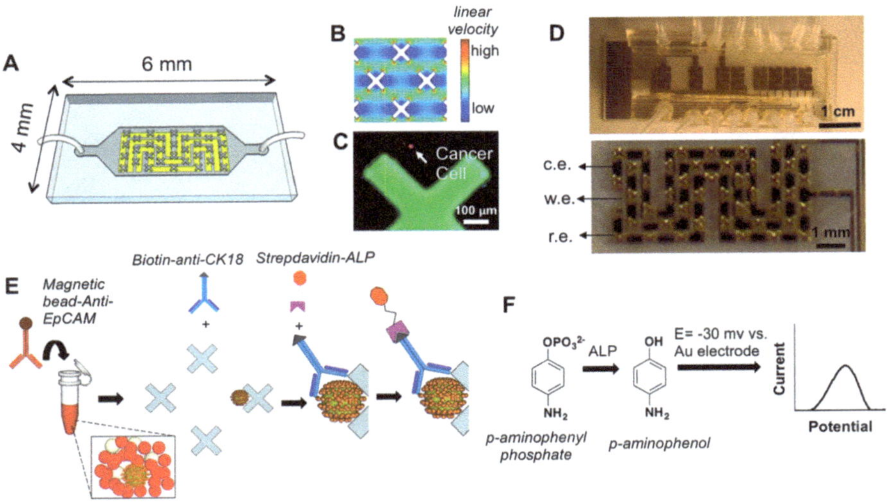

Fig. 4.5 Capture and electrochemical detection of cancer cells. (**a**) Microfluidic chip design. (**b**) COMSOL simulation of linear velocity gradient. (**c**) Immunofluorescent image of a captured cancer cell on chip. (**d**) Image of a chip with 8 independent sensors and a zoomed in view of one sensing chamber. (**e**) Cancer cell capture and ELISA construction. (**f**) Electrochemical readout. (Figure taken from Safaei et al. [56])

discrimination from normal cells. Meanwhile, by using ratiometric method that limited the interference factors, the possibility of false positive result was largely decreased.

ELISA has been extensively employed in clinical diagnostics. However, this technique usually requires relatively expensive test kits and bulky plate readers that limit its applications. Electrochemical ELISA (EC-ELISA) that takes advantage of the features of electrochemistry, such as simplicity, low cost and fast response, is able to largely overcome the limitations of the conventional ELISA. Safaei et al. recently reported a microfabricated system that integrated a sensitive EC-ELISA method within a microfluidic cell capture system, enabling the reliable detection of low levels of rare cancer cells in whole blood [56] (Fig. 4.5). Cancer cells were captured by immune-magnetic method based on the previously reported velocity valley (VV) chip with patterned microstructures [46]. Afterwards, the cancer cells were functionalized on-chip with biotinylated anti-CK18 and streptavidin-alkaline phosphatase (ALP) that converted *p*-aminophenyl phosphate to *p*-aminophenol, which was electrochemically detected using DPV to indicate the presence and the number of captured cells. In another report, Hong et al. presented an integrated multifunctional system constructed by conductive disulfide-biotin-doped polypyrrole nanowires (SS-biotin-Ppy NWs) for capture, release, and *in situ* quantification of CTCs using EC-ELISA [29]. The well-ordered 3-D nanowires not only significantly favored cell capture and release efficiency, but also showed high sensitivity and specificity towards the amperometric detection.

Most recently, Cao et al. reported a new technique that combines electrochemistry with a nanochannel-ion channel hybrid array for efficient CTC capture and sensitive detection [9]. In this work, CTCs were trapped in the ion cannel immobilized with aptamer that coupled with the protein overexpressed on the CTCs membrane, resulting in a dramatically blocked ionic flow through channels. Therefore, instead of detecting the electroactive labels that are routinely used in electrochemical detection of CTCs, this work monitored in real time the changed ionic transfer behaviors caused by the varied mass-transfer property using LSV.

Micro- or nano-structured electrodes and the construction of nano-structured electrode interface are very common in the electrochemical analysis of CTCs to achieve better assay performance with the effects of their surface morphology, size, optical and electronic properties, etc. Sage has applied the NMEs with the electrode materials of gold or palladium to many cancer-related detections and clinical diagnostics [57]. This electrochemical platform, with large surface area and very fine nanostructures, allows rapid and efficient analyte capture and more efficient mass transfer, enabling to achieve excellent specificity and sensitivity, and fast sensing in the detection of clinical samples at low concentrations. Taking advantages of their large surface area, excellent conductivity, biocompatibility and flexibility, nanostructured materials, such as carbon nanotubes [85], graphene [12], and soft nanowires [29] were used to construct biocomposites modified at the surface of the electrode with high stability and bioactivity for increasing analyte loading and enhancing the electrochemical signal. Apart from nanostructured electrodes or electrode surface modifications, nanoelectrocatalysts were designed and fabricated to catalyze the reactions of the electrochemical labels with high activity and efficiency for signal amplification of CTC sensors [43, 88]. Fe_3O_4@nanocage core (Ag-Pd) nanohybrid NPs, for example, were fabricated that showed superior electrocatalytic activities towards the nonenzymatic electrochemical reduction of thionine, the electrochemical label [88] (Fig. 4.6). These nanoelectrocatalysts showed signal-amplifying ability that could be used for ultrasensitive electrochemical cytosensing, with the detection limits of ~34 and ~42 cells mL^{-1} for MCF-7 and T47D cells, respectively.

Electrochemical processes can be easily adapted and combined with other useful technologies to achieve higher performance in biodetection by integrating the merits and advantages of each technology. As is reviewed in the previous text, electrochemistry has been married to microfluidics [79], magnetic interaction [32], ELISA [56], and nanotechnologies and nanomaterials [88]. Very recently, Wu et al. presented a new strategy of the combination of SPR and electrochemical method for real-time evaluation of live cancer cells [80] (Fig. 4.7). By monitoring the SPR signal changes from the morphology and mass changes of adsorbed cancer cells and the electrochemical measurement of the extracellular daunorubicin (DNR) residue after being treated with cancer cells, the present strategy showed great potential in evaluating therapeutic efficiency of bioactive agents to cells and monitoring relevant treatment processes in clinical applications.

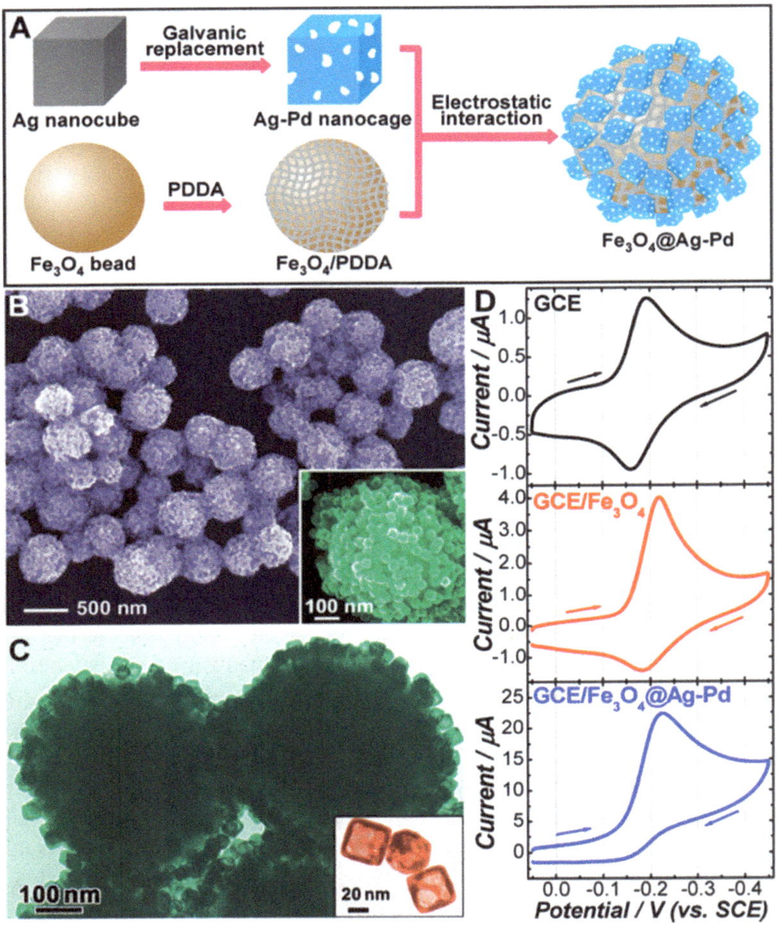

Fig. 4.6 (**a**) Scheme showing the fabrication of Fe_3O_4@Ag−Pd hybrid NPs. (**b**) SEM image of Fe_3O_4@Ag−Pd hybrid NPs. The inset shows one Fe3O4@Ag − Pd hybrid particle. (**c**) TEM image of Fe_3O_4@Ag−Pd hybrid NPs. The inset presents individual Ag−Pd nanocages. (**d**) CVs of a bare GCE, GCE/Fe_3O_4, and GCE/Fe_3O_4@Ag−Pd. (Figure taken Zheng et al. [88])

4.3 Circulgating Nucleic Acids

So far, most of molecular-based analyses for cancer diagnosis are performed using nucleic acids (NAs) extracted from nucleated cells by cell lysis. Unlike these nucleic acids, cNAs, or cell-free nucleic acids (cfNAs), are found outside cells, primarily in the serum and plasma with some in urine [24]. These nucleic acids such as DNA, mRNA and microRNA that are released from tumors and circulate in the blood of cancer patients have become a new generation of cancer biomarkers indicating the malignant progression of the disease [58].

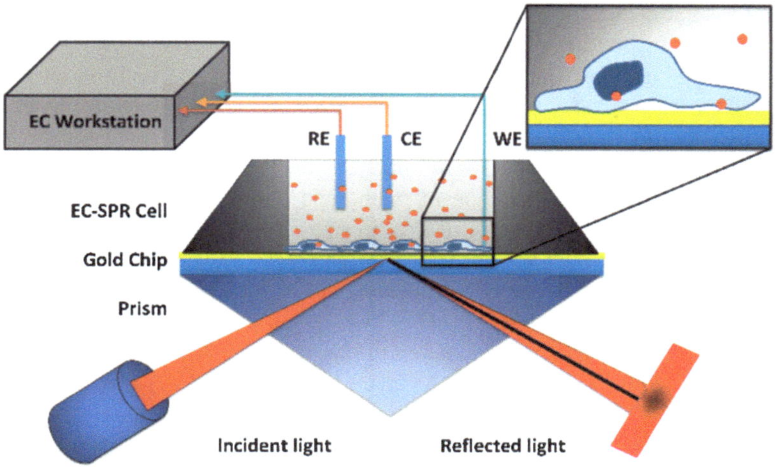

Fig. 4.7 Scheme of the combination of SPR and electrochemistry in real-time evaluation of live cancer cells. (Figure taken from Wu et al. [80])

The detection of cNAs could serve as a liquid biopsy, potentially replacing tumor-tissue biopsies in certain diagnostic applications, and has the potential to revolutionize the detection and monitoring of tumors in a non-invasive way [58]. However, the concentrations of cNAs are typically low and very variable. Thus, early sequencing technologies are often unable to detect cNAs and a number of efforts have been made to increase the detection. The conventional and common methods of detecting cNAs include PCR techniques, northern blotting, and fluorescent imaging [25, 64]. Unfortunately, these methods are normally expensive and time-consuming, and require large amount of samples.

Electrochemistry-based NAs sensors, a key area in the scope of biomolecular electrochemistry, are able to detect nucleic acid sequences or mutated genes associated with human disease by combining nucleic acid layers with electrochemical transducers. Basically all common electrochemical methods have been applied to NAs detection, in which amperometry [8], various voltammetry methods [19], and electrochemical impedance spectroscopy (EIS) [63] are reported the most. Most of electrochemical NAs sensors use label-based techniques that either covalently or non-covalently (such as electrostatic adsorption) binds electroactive reagents to NAs analytes. Label-free methods are also reported for this application, which are simpler but are generally not able to reach comparable level of sensitivity as the label-based methods. A number of literatures have reviewed the strategies of electrochemical NAs sensors based on these two methods [19, 37, 52]. In this part we will focus on new strategies employed in improving the performance of cNAs sensors.

Electrochemical NAs sensors are promising in constructing simple, accurate, and inexpensive platforms for cancer diagnostics with high sensitivity [19]. However, the crucial challenges still relate to further achieving high levels of sensitivity and specificity on clinical samples, especially when dealt with cNAs. Although there are currently not many studies concerning the electrochemical cNAs detections, smartly designed strategies have been presented in existing papers to improve the performance of the cNAs sensors, which often aim for signal amplification to enhance sensitivity. For example, Labib et al. developed a protein-facilitated sensor for the detection of miRNA signature of chronic lymphocytic leukemia (CLL) in human serum by electrochemical detection of methylene blue (MB) molecules using CV upon the hybridization of target miRNAs to two DNA adaptor strands bound to a universal interfacial probe [36]. Glucose oxidase was introduced as electrocatalyst to create multiple redox cycles for MB molecules in order to amplify the current from the electrochemical reduction of MB. The developed sensor was capable of distinguishing single base mismatches in the target miRNA and was employed for profiling of three endogenous miRMAs characteristic to CLL. Higher sensitivity of nucleic acid detection can also be achieved by the implementation of various nanomaterials, which act either as nanoelectrodes (immobilization interface for accumulation of increased amounts of nucleic acid probes), or as signal amplifiers of the hybridization event [52]. By using NMEs as a platform and amino acid/nucleic acid chimeras (ANAs) as probe molecules, Vasilyeva et al. reported a large-footprint electrochemical sensor that were able to capture large, slow-moving analytes [70] (Fig. 4.8). Using DPV, the sensor proved to be ultrasensitive and specific in the detection of mRNA derived from 10 cancer cells present in complex samples. Besides, Wen et al. demonstrated a three-dimensional (3D) DNA tetrahedral nanostructure-based interfacial engineering approach to enhance miRNAs binding recognition at the gold electrode surface and drastically improve the detection sensitivity by examining the electrochemical responses using CV and chronoamperometry [75]. By employing this DNA nanostructure, the resulting sensor can directly detect as few as attomolar miRNAs with high single-base discrimination ability. Recently, Fang et al. developed an ultrasensitive and highly specific electrochemical assay for miRNA-21 detection based on the selective binding of Just Another Zinc finger proteins (JAZ) between a ssDNA capture probe and a target miRNA [21]. In their work, the high signal amplification was obtained through enzymatic amplification by alkaline phosphatase (ALP) conjugated to JAZ coupled with electrochemical-chemical-chemical (ECC) redox cycling involving an ALP product (hydroquinone) mediated by $Ru(NH_3)_6^{3+}/Ru(NH_3)_6^{2+}$ redox system. The resulting detection limits for miRNA-21 in buffer and ten-fold diluted serum can be achieved as low as 2 and 30 fM, respectively.

Other strategies to improve cNAs detection performance include the invention of multi-mode detection [35] and the use of RNA binding viral protein as a selective biorecognition element [8]. To be more specific, Labib et al. developed a three-mode electrochemical sensor for miRNA analysis via direct hybridization with the target miRNA, p19 protein binding, and protein displacement, using both square wave voltammetry (SWV) and EIS measurement [35] (Fig. 4.9). The fabricated sensor showed a detection limit as low as 90 molecules per sample of miRNA and a wide dynamic range (from 10 aM to 1 mM). Very uniquely, the protein displacement allows the detection of any type of miRNA. Using an RNA binding viral protein as a

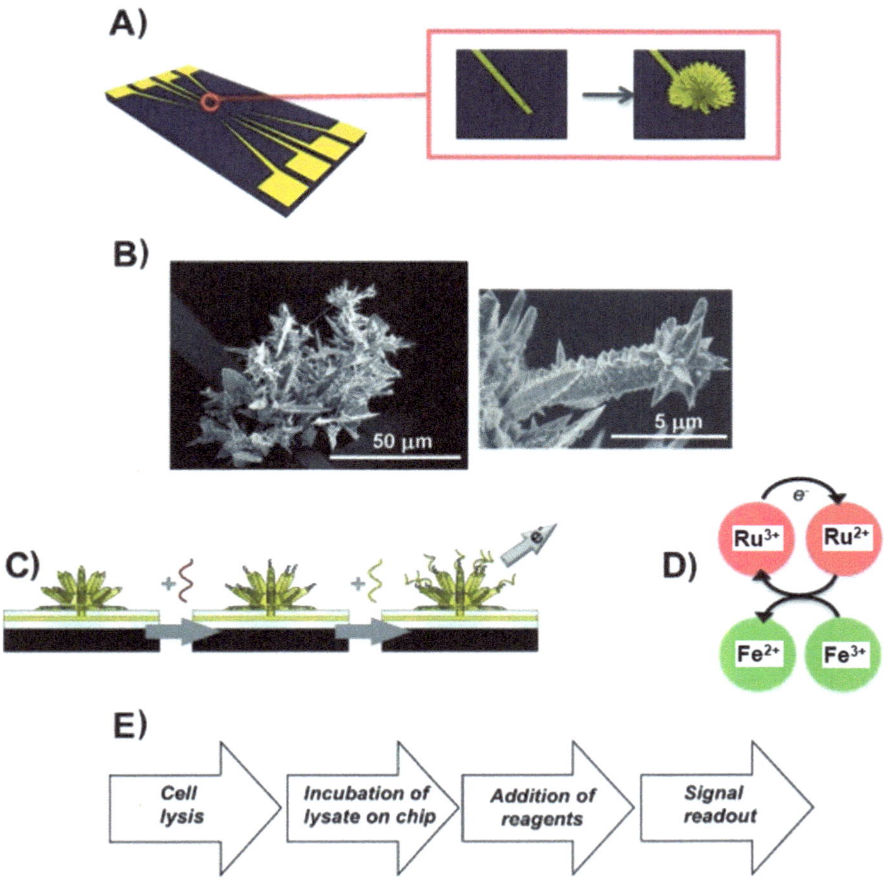

Fig. 4.8 Chip-based sensors for detection of chronic myeloid leukemia cells. (**a**) Chip layout. (**b**) SEM images of a 100 mm sensor formed by using gold electrodepositon on the surface of the chip. (**c**) Sequence of steps for nucleic acids analysis. (**d**) Ru^{III}/Fe^{III} reporter group permits hybridized nucleic acids to be detected. (**e**) Overall flow diagram of the analysis. (Figure taken from Vasilyeva et al. [70])

selective biorecognition element, Campuzano et al. developed an amperometric magnetobiosensor for miRNA quantification, which was able to detect 0.4 fmol of a synthetic target and endogenous miR-21 in only 2 h in total RNA extracted from cancer cells and human breast-tumor specimens [8].

The detection of cNAs for specific gene mutations, such as KRAS, is desirable because these gene mutations happen very frequently in many tumor types and contribute to tumor progression [58]. Recently, Das et al. reported an electrochemical chip-based clamp assay that detected mutated sequences KRAS and BRAF directly in patient serum, which was the first successful detection of cNAs without the need for enzymatic amplification [15] (Fig. 4.10). Using a collection of oligonucleotides that

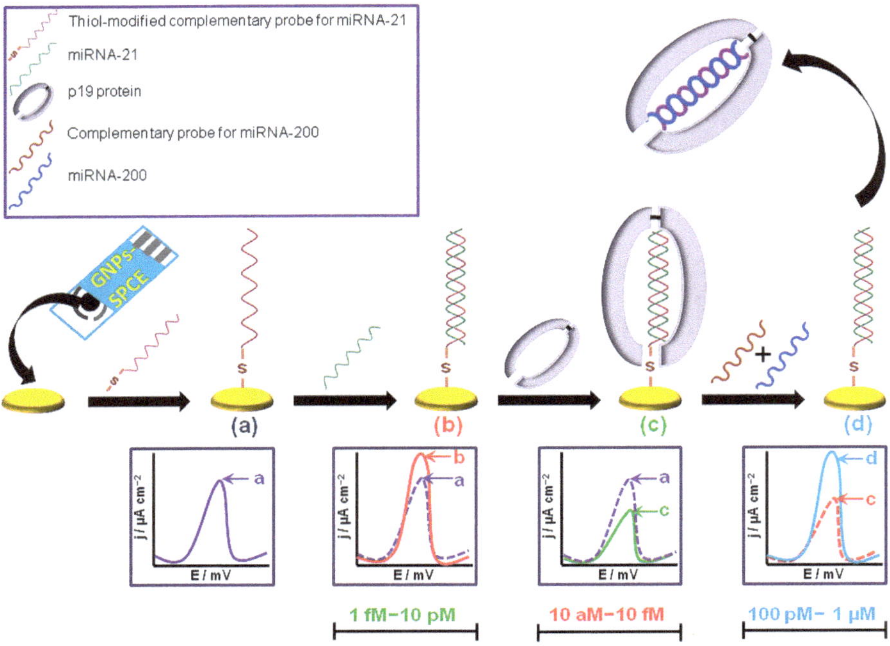

Fig. 4.9 Schematic representation of the 3-mode electrochemical sensor. (**a**) Modification of a thiol-modified complementary probe onto a gold nanoparticles-modified screen-printed carbon electrode for capturing miR-21. (**b**) A hybridization-based sensor showing an increase in the current intensity by the binding of the target miR-21. (**c**) A protein-based sensor showing a large decrease in current density by the binding of the p19 protein. (**d**) A displacement-based sensor showing a shift-back in the signal due to the dissociation of p19 protein forced by the hybridization product of miR-200 and its complementary probe. (Figure taken from Labib et al. [35])

sequester wild-type sequences in solution, the sensor allows only the mutated sequence to bind to its surface. Benefiting from this principle, the electrochemical clamp sensor exhibits excellent levels of sensitivity and specificity, and allows accurate detection of mutated sequences in a collection of samples taken from lung cancer and melanoma patients within 15 min readout time. On the basis of this work, the same group later reported the electrochemical detection of mutated circulating tumor DNA (ctDNA) in samples collected from cancer patients [16]. By employing DNA clutch probes (DCP) to prevent reassociation of the denatured DNA strands and ensure thereby that only mutated sequences associate with the detecting hybridization events, it was possible to readout the presence of mutated ctDNA. The assay exhibited excellent sensitivity and specificity of 1 fg/μL of a target mutation in the presence of 100 pg/μL of wild-type DNA, corresponding to detecting mutations at a level of 0.01% relative to wild-type.

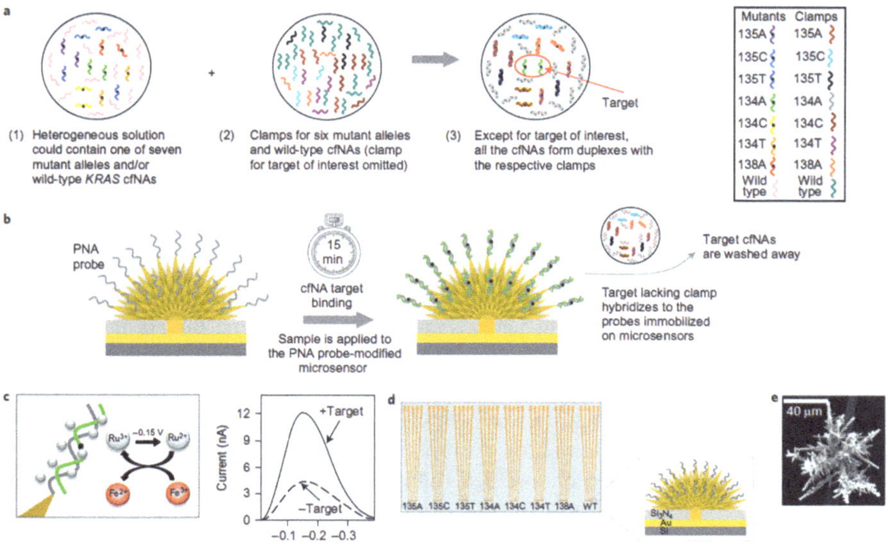

Fig. 4.10 Scheme of the clamp chip for the electrochemical analysis of mutated cfNAs. (**a**) Schematic representation of the clamp strategy. (**b**) Sequence of steps for sensor-based detection. (**c**) Electrochemical readout using DPV method. (**d**) Microchip layout. (**e**) SEM image of a NME sensor. (Figure taken from Das et al. [15])

4.4 Extracellular Vesicles

Many cell types, including tumor cells, release EVs. Exosomes (30–100 nm in diameter) and microsomes (100–1000 nm in diameter) are small EVs secreted by most mammalian cells which carry factors that facilitate intercellular communication. They are usually transported through biological fluids such as blood plasma [55], body fluids [31], and urine [50]. Likewise, exosomes and microsomes that are shed by tumor cells carry specific proteins, functional mRNAs, microRNAs and DNAs. These biomolecules have been acknowledged as potential biomarkers resource for non-invasive diagnosis of cancer without the need to access the tumor directly. Given this fact about exosomes and microsomes, they have attracted tremendous attention in the past decade. In the following section, we will focus on reviewing exosomes in terms of their isolation, characterization and electrochemical detection.

The isolation and characterization of exosomes and microsomes in clinical samples with the presence of billions of blood cells is technically challenging. Currently, the most commonly used isolation approaches for both exosomes and microsomes are based on differential ultracentrifugation steps (often accompanied by multi-step filtering). Since exosomes have much smaller sizes than microsomes, the centrifugation condition is even more critical. A high-speed centrifugation (up to 200,000 g) for more than 10 h is required which makes this approach tedious and time consuming [5]. Separation using affinity purification with specific antibodies is another option, however, it depends on the presence of binding-target proteins [65]. Since CD63 is one of the intrinsic molecular subtypes of exosomes and EpCAM is a ubiquitously expressed epithelial cancer marker, they are both commonly used as markers for capturing exosomes. Some techniques have emerged based on this separation principle, such as using affinity-binding beads and microfluidic immunocapture platforms [27, 45]. He et al. developed a magnetic-bead based microfluidic approach that allowed on-chip immunoisolation and *in situ* protein analysis of exosomes directly from patient plasma [27]. This approach enabled selective isolation of exosome subpopulations from ovarian and non-small-cell lung cancer patients by targeting various surface markers. The intervention with a dynamic force is also reported to combine with affinity purification method to enhance the specificity of capture. As an example, a multiplexed microfluidic device with high specificity for capture and detection of multiple exosome targets was designed using a tunable alternating current electrohydrodynamic (ac-EHD) methodology [67]. The generated nanoscaled fluid flow by electrical body forces within nanometers of an electrode surface enhanced the specific capture efficiency and also reduced nonspecific adsorption of weakly bound molecules from the electrode surface.

The analysis and characterization of exosomes have been mainly focused on their physical properties, protein expression and nucleic acids contents. Generally, physical properties such as size, concentration and morphology are obtained by optical and non-optical methods such as scattering and electron microscopy [68]. However, these methods, due to their intrinsic limitations in resolution, usually fail to quantitate the level of exosomes in patient samples. Conventional optical methods for cell analysis, such as fluorescence and dynamic light scattering (DLS), are not able to provide reliable information due to their high detection limit [68]. Exosomal proteins are typical markers for exosomes detection, and are mainly analyzed by western blot analysis and ELISA [50]. Recently, Vaidyanathan et al. analyzed exosomes based on their specific protein markers epidermal growth factor receptor 2 (HER2) and prostate specific antigen (PSA) after successfully capturing exosomes using an ac-EHD method on a microfluidic device [67]. By catalytically oxidizing peroxidase substrate 3,3′,5,5′-tetramethylbenzidine (TMB) with tagged horseradish peroxidase (HRP) on the surface of captured exosomes, this detection method provides a simple and rapid on-chip naked eye detection readouts and allows simultaneous detection of multiple exosome targets. Unfortunately, surface protein based detections, in general, have shortcomings such as poor sensitivity and highly manual workflows. Another approach for exosome characterization is to detect exosomal nucleic acids, such as miRNAs, mRNAs and DNAs by PCR and agarose gel electrophoresis [66]. However, exosomes have to be lysed before detection, which adds to the complexity of the testing. Very recently, Im et al. reported a label-free, high-throughput assay based on transmission surface plasmon resonance through periodic nanohole arrays functionalized with antibodies which enables profiling of exosome surface proteins directly from ascites samples from ovarian cancer patients [31] (Fig. 4.11). This approach offered improved sensitivity and allowed portable

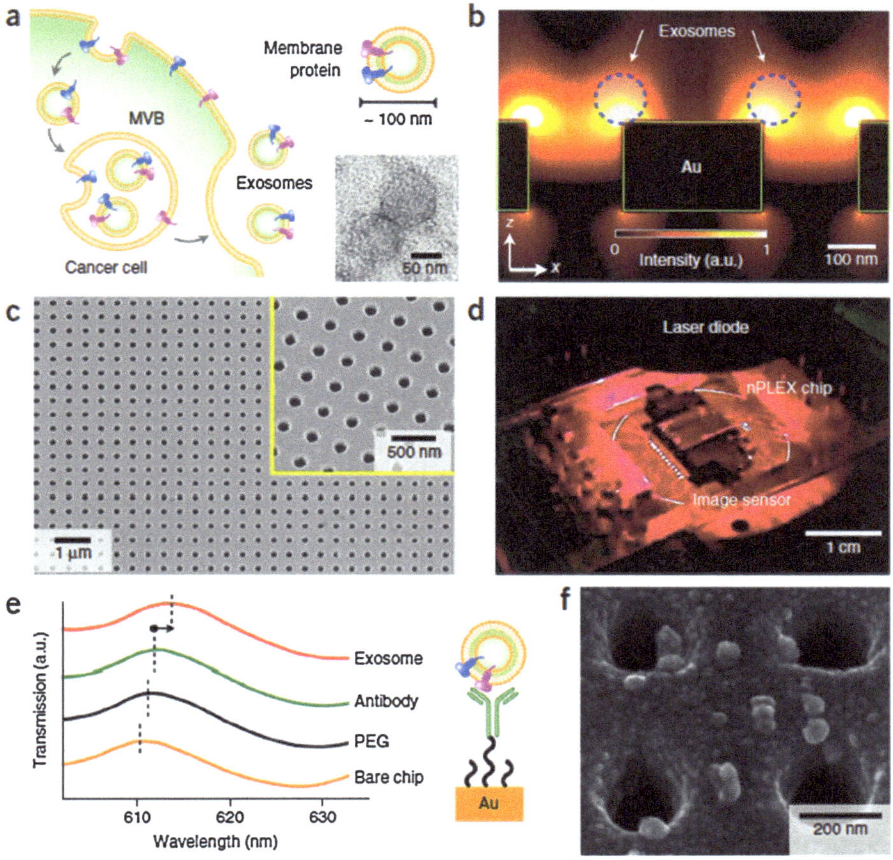

Fig. 4.11 Label-free detection of exosomes with nano-plasmonic exosome (nPLEX) sensor. (**a**) Exosomes secretion and high magnification transmission electron micrograph (TEM) (inset) showing that exosomes from CaOV3 culture have a diameter ~100 nm. (**b**) Finite-difference time-domain simulation showing the enhanced electromagnetic fields tightly confined near a periodic nanohole surface which overlaps with the size of exosomes captured. (**c**) SEM image of the periodic nanoholes in the nPLEX sensor. (**d**) A prototype miniaturized nPLEX imaging system for exosomes analyses. (**e**) A representative schematic of changes in transmission spectra showing exosome detection with nPLEX. (**f**) SEM image showing exosome capture by functionalized nPLEX. (Figure taken from Im et al. [31])

operation while requiring much smaller sample amount, but the detection relied on surface plasmon resonance instrument with a high level of cost and expensive maintenance.

Clinical analysis based on electrochemical biosensors is promising to overcome these limitations for their low cost, fast response and low material requirement. To date, electrochemical approaches have not been widely used for the detection of exosomes and other vesicles with clinically relevant levels of sensitivity, with only a few examples including amperometry [33, 73, 90], DPV [81], LSV [91], electrochemical impedance [41, 48] in the existing studies. The strategy of reporting the current decrease from the block of electron transfer by the introduction of target exosomes has been commonly used. For example, Yadav *et al.* reported the conventional immunoaffinity-based method to directly quantify the cancer-specific exosomes from cell culture media [81]. Using the $[Fe(CN)_6]^{4-/3-}$ redox system, a decrease in DPV current response was observed as the signal readout after the addition of exosomes to the electrode and block the electron transfer of $[Fe(CN)_6]^{4-/3-}$.

Aptamer-based biosensors (aptasensors) have drawn particular attention in the applications in cancerous exosome analysis for their advantages in detection sensitivity, speed, modest requirement of sample volumes and easy signal amplification [72, 90]. Zhou *et al.* developed an electrochemical aptasensor to detect exosomes from cell culture media [90]. In their study, aptamers specific to exosome transmembrane protein D63 were immobilized onto electrode surface, and MB labeled probing strands hybridized with the aptamers to emit an electrochemical signal. Upon the introduction of exosomes, the interaction of the aptamer-modified electrode with exosomes carrying D63 resulted in the displacement of the probing strands, causing the redox signal to decrease. Through a passivation procedure using the blocking agents, non-specific adsorption was successfully prevented. Furthermore, the assay was miniaturized by photolithography and integrated into microfluidic devices, showing the low requirement of sample volume and simple handling or processing steps. However, due to the flexibility of single-stranded DNA, aptamers on electrodes tend to undergo self-assembled monolayer aggregation or entanglement, largely impeding the accessibility of the target exosomes. In addition, the precise control of spatial orientation of single-stranded aptamers is also difficulty to achieve. Recently, Wang *et al.* constructed a portable electrochemical aptasensor that used DNA-based nanotetrahedron (NTH) for direct capture and detection of hepatocellular exosomes [72]. They also used $[Fe(CN)_6]^{4-/3-}$ redox system to signal the decrease of the electrochemical current upon the capturing of cancerous exosomes using square wave voltammetry (SWV) method. The NTH assisted structure allowed the oriented immobilization of aptamers that significantly improved the accessibility of an artificial aptamer to suspended exosomes with 100-fold higher sensitivity when compared to the single-stranded aptamer-functionalized aptasensor.

High-throughput measurement, such as flow cytometry, is of great importance and has becomes a trend in a wide range of topics in biology and medicine. However, flow cytometry tends to miss small vesicles (<200 nm) due to weak light scattering [69]. In order to realize high-throughput detection of exosomes, Jeong *et al.* constructed a portable sensor with multi-electrochemical channels for the detection of exosomes from plasma samples of ovarian cancer patients [33] (Fig. 4.12). The sensor was an EC-ELISA combining magnetic enrichment and enzymatic amplification. As a proof-of-concept, the authors implemented an eight-channel device that allowed for the simultaneous profiling of multiple protein markers within an hour, showing high sensitivity and speed over conventional assays. EC-ELISA was also

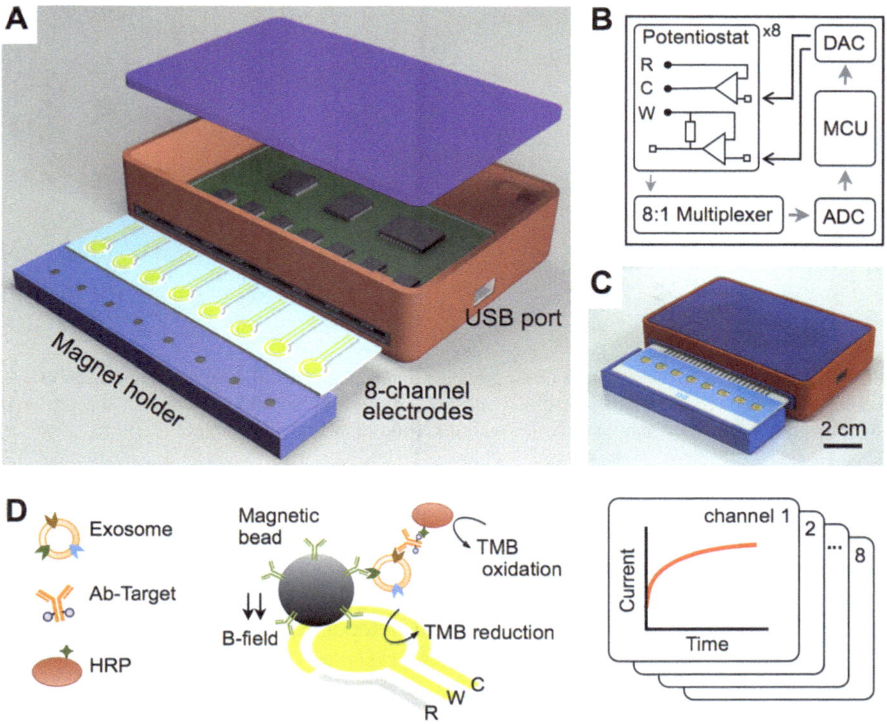

Fig. 4.12 Integrated magnetic–electrochemical exosome platform. (**a**) Sensor structure showing the simultaneous measurement of signals from eight electrochemical channels. (**b**) Circuit diagram. (**c**) Packaged device. (**d**) Schematic of assay. Exosomes are captured directly from plasma by magnetic beads coated with CD63 and labeled with HRP. W, C and R represent working, counter and reference electrode, respectively. HRP indicates horseradish peroxidase and TMB indicates 3,3′,5,5′-tetramethylbenzidine. (Figure taken from Jeong et al. [33])

reported for the determination of exosomes elsewhere with a detection limit as low as 200 exosomes per microliter [18].

Nanoparticle labeling was also used in exosome analysis. Zhou et al. extended the MNPs labeling method to the non-enzyme detection of exosomes and microsomes, which was based on direct electro-oxidation of MNPs to specifically recognize surface markers of exosomes and microsomes [91] (Fig. 4.13). They also used a simple centrifugation procedure to isolate exosomes and microsomes form cell culture supernatant or directly from blood serum with a maximum centrifugation of 16,100 g for 30 min. Compared to the aforementioned detection methods, this strategy is fast in response (within 10 s), simpler in instrumentation, and more cost effective. It also provides the simultaneous detection of multi-markers. By testing the levels of exosomal and microsomal EpCAM and PSMA derived from prostate cancer patients, it was shown that exosomes and microsomes had potential for cancer diagnosis at early stages. QDs were also reported as signal amplifiers in a stripping voltammetric immunoassay for the detection of disease-specific exosomes [6].

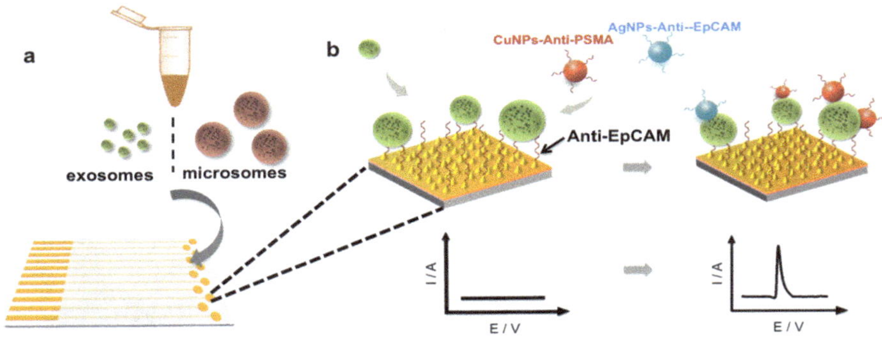

Fig. 4.13 Schematic representation of the isolation and analysis of exosomes and microsomes: (**a**) vesicles are captured to aptamer-modified sensors, (**b**) electrochemical detection of the captured vesicles with Cu and Ag nanoparticles as markers. (Figure taken from Zhou et al. [91])

4.5 Conclusions and Perspectives

Non-invasive early detection of cancer markers has been the main focus of cancer-control research over the past decades. In this chapter, we provide an overview of the recent developments in the electrochemical detections of new circulating CTMs, including CTCs, cNAs and EVs. We gave representative examples for each marker and showed the combination of electrochemical approaches with other remarkably progressed technologies and strategies, such as microfluidics, nanotechnology and bioconjugation techniques. Although the performances of electrochemical sensors have been much enhanced accordingly, there still exist obstacles that have to be overcome before they can be applied to point-of-care diagnostics.

Firstly, electrochemical transducers, except EIS that measures the change in electrochemical resistance, rely on the electron transfer from the electrochemical labels of cancer biomarkers to electrode surface in order to report the signal. Since most of the biomarkers, such as CTCs, proteins, and EVs are low in conductivity, the development of new signal amplification strategies [15, 71, 90, 91] is highly desired besides the common practice of constructing nanoscaled transducers with nanomaterials. Secondly, nanostructured electrodes with high roughness factors that aims for rapid and efficient analyte capture and more efficient mass transfer may sometimes result in opposite effects. For example, at very rough electrode surfaces the diffusion zones of the nanostructure may overlap or partially overlap, leading to linear to partially linear diffusion over the entire surface and producing less efficient mass transport [89]. In addition, the nanostructures with random sub-structures may lead to uncontrollable analyte orientation and configuration on the electrode surface, which may inhibit the transduction of the electrochemical signal. Therefore, a rational design in nanostructures should be carefully considered before the construction of the sensor. Thirdly, the efforts made to increase sensitivity also bring about selectivity issues from interferences of non-specific adsorption that usually occur when testing with complex biological matrices. Thus, the parameters that determine the specificity of the electrochemical devices have to be further optimized and refined to realize specific, reliable and simultaneous multi-marker detection in complex biological samples.

Furthermore, most of the electrochemical sensors are merely proof-of-concept demonstrations that are only practical to perform under optimized conditions in a lab. There is still a long way to go to achieve portable and user-friendly point-of-care diagnostics with complex real patient samples. High-throughput multiplexed sensing with highly automated systems for sample handling is a big trend to achieve a widespread use of point-of-care devices with fast response. Therefore, related technologies have to be further advanced, such as the design of a miniaturized multi-channel (hundreds and above) electrochemical circuit for high throughput and the invention of a robot for automatic sample handling. There are also some technologies that are well progressed but still exist constraints when integrated with electrochemistry. Microfluidics, for example, allows considerable portability and high capacity for device integration. However, it usually suffers from low reproducibility with electrochemical readout due to compact electrode configuration and small amount of liquid volume.

In summary, there are considerable issues to be dealt with in constructing efficient electrochemical platforms for portable, fast and cost-effective point-of-care diagnostic devices. New advances and improvements are expected from the cooperation of different communities to ultimately achieve this goal.

References

1. Acero Sánchez JL, Joda H, Henry OYF, Solnestam BW, Kvastad L, Akan PS, Lundeberg J, Laddach N, Ramakrishnan D, Riley I, Schwind C, Latta DC, O'Sullivan K (2017) Electrochemical genetic profiling of single cancer cells. Anal Chem 89:3378–3385
2. Alberti D, Van'T Erve M, Stefania R, Ruggiero MR, Tapparo M, Geninatti Crich S, Aime S (2014) A quantitative relaxometric version of the ELISA test for the measurement of cell surface biomarkers. Angew Chemie – Int Ed 53:3488–3491
3. Allard WJ, Matera J, Miller MC, Repollet M, Connelly MC, Rao C, Tibbe AGJ, Uhr JW, Terstappen LWMM (2004) Tumor cells circulate in the peripheral blood of all major carcinomas but not in healthy subjects or patients with nonmalignant diseases. Clin Cancer Res 10:6897–6904
4. Azarin SM, Yi J, Gower RM, Aguado BA, Sullivan ME, Goodman AG, Jiang EJ, Rao SS, Ren Y, Tucker SL, Backman V, Jeruss JS, Shea LD (2015) In vivo capture and label-free detection of early metastatic cells. Nat Commun 6:8094–8102
5. Berkman C (2014) Functional prostate-specific membrane antigen is enriched in exosomes from prostate cancer cells. Int J Oncol 44:918–922
6. Boriachek K, Islam MN, Gopalan V, Lam AK, Nguyen NT, Shiddiky MJA (2017) Quantum dot-based sensitive detection of disease specific exosome in serum. Analyst 142:2211–2219
7. Buck CB, Lowy DR (2011) Highly efficient capture and electrochemical release of circulating tumor cells by using aptamers modified gold nanowire arrays. J Clin Oncol 29:1506–1508
8. Campuzano S, Torrente-Rodrĩguez RM, Lõpez-Hernández E, Conzuelo F, Granados R, Sánchez-Puelles JM, Pingarrõn JM (2014) Magnetobiosensors based on viral protein p19 for microRNA determination in cancer cells and tissues. Angew Chem Int Ed 53:6168–6171

9. Cao J, Zhao XP, Younis MR, Li ZQ, Xia XH, Wang C (2017) Ultrasensitive capture, detection, and release of circulating tumor cells using a nanochannel-Ion channel hybrid coupled with electrochemical detection technique. Anal Chem 89:10957–10964
10. Chandra P, Noh HB, Pallela R, Shim YB (2015) Ultrasensitive detection of drug resistant cancer cells in biological matrixes using an amperometric nanobiosensor. Biosens Bioelectron 70:418–425
11. Chen W, Weng S, Zhang F, Allen S, Li X, Bao L, Al CET (2013) Nanoroughened surfaces for efficient capture of circulating tumor cells without using capture antibodies. ACS Nano 7:566–575
12. Chen X, Wang Y, Zhang Y, Chen Z, Liu Y, Li Z, Li J (2014) Sensitive electrochemical aptamer biosensor for dynamic cell surface N-glycan evaluation featuring multivalent recognition and signal amplification on a dendrimer–graphene electrode interface. Anal Chem 86:4278–4286
13. Cheng IF, Huang WL, Chen TY, Liu CW, Lin YD, Su WC (2015) Antibody-free isolation of rare cancer cells from blood based on 3D lateral dielectrophoresis. Lab Chip 15:2950–2959
14. Chikkaveeraiah BV, Bhirde AA, Morgan NY, Eden HS, Chen X (2012) Electrochemical immunosensors for detection of cancer protein biomarkers. ACS Nano 6:6546–6561
15. Das J, Ivanov I, Montermini L, Rak J, Sargent EH, Kelley SO (2015) An electrochemical clamp assay for direct, rapid analysis of circulating nucleic acids in serum. Nat Chem 7:569–575
16. Das J, Ivanov I, Sargent EH, Kelley SO (2016) DNA clutch probes for circulating tumor DNA analysis. J Am Chem Soc 138:11009–11016
17. Ding C, Ge Y, Zhang S (2010) Electrochemical and electrochemiluminescence determination of cancer cells based on aptamers and magnetic beads. Chem – A Eur J 16:10707–10714
18. Doldán X, Fagúndez P, Cayota A, Laíz J, Tosar JP (2016) Electrochemical sandwich immunosensor for determination of exosomes based on surface marker-mediated signal amplification. Anal Chem 88:10466–10473
19. Drummond TG, Hill MG, Barton JK (2003) Electrochemical DNA sensors. Nat Biotechnol 21:1192–1199
20. Duijvesz D, Luider T, Bangma CH, Jenster G (2011) Exosomes as biomarker treasure chests for prostate cancer. Eur Urol 59:823–831
21. Fang CS, Kim K, Yu B, Jon S, Kim MS, Yang H (2017) Ultrasensitive electrochemical detection of miRNA-21 using a zinc finger protein specific to DNA–RNA hybrids. Anal Chem 89:2024–2031
22. Feng L, Chen Y, Ren J, Qu X (2011) A graphene functionalized electrochemical aptasensor for selective label-free detection of cancer cells. Biomaterials 32:2930–2937
23. Fischer JC, Niederacher D, Topp S, Honisch E, Schumacher S, Schmitz N, Zacarias Föhrding L, Vay C, Hoffmann I, Kasprowicz NS, Hepp PG, Mohrmann S, Nitz U, Stresemann A, Krahn T, Henze T, Griebsch E, Raba K, Rox JM, Wenzel F, Sproll C, Janni W, Fehm T, Klein C, Knoefel WT, Stoecklein NH (2013) Diagnostic leukapheresis enables reliable detection of circulating tumor cells of nonmetastatic cancer patients. Proc Natl Acad Sci 110:16580–16585
24. Goessl C, Krause H, Müller M, Mu M, Schrader M, Sachsinger J (2000) Fluorescent methylation-specific polymerase chain reaction for DNA-based detection of prostate cancer in bodily fluids advances in brief fluorescent methylation-specific polymerase chain reaction for DNA-based detection of prostate cancer in bodily fluids. Cancer Res 1:5941–5945
25. Gueudin M, Plantier JC, Damond F, Roques P, Mauclere P, Simon F (2003) Plasma viral RNA assay in HIV-1 group O infection by real-time PCR. J Virol Methods 113:43–49
26. Han SI, Han KH (2015) Electrical detection method for circulating tumor cells using graphene nanoplates. Anal Chem 87:10585–10592
27. He M, Crow J, Roth M, Zeng Y, Godwin AK (2014) Integrated immunoisolation and protein analysis of circulating exosomes using microfluidic technology. Lab Chip 14:3773–3780
28. Henry NL, Hayes DF (2012) Cancer biomarkers. Mol Oncol 6:140–146
29. Hong WY, Jeon SH, Lee ES, Cho Y (2014) An integrated multifunctional platform based on biotin-doped conducting polymer nanowires for cell capture, release, and electrochemical sensing. Biomaterials 35:9573–9580

30. Hu X, Wei CW, Xia J, Pelivanov I, O'Donnell M, Gao X (2013) Tapping and photoacoustic detection of CTCs at the single cell per milliliter level with magneto-optical coupled nanoparticles. Small 9:2046–2052
31. Im H, Shao H, Park Y, Peterson VM, Castro CM, Weissleder R, Lee H (2014) Label-free detection and molecular profiling of exosomes with a nano-plasmonic sensor. Nat Biotechnol 32:490–495
32. Ivanov I, Stojcic J, Stanimirovic A, Sargent E, Nam RK, Kelley SO (2013) Chip-based nanostructured sensors enable accurate identification and classification of circulating tumor cells in prostate cancer patient blood samples. Anal Chem 85:398–403
33. Jeong S, Park J, Pathania D, Castro CM, Weissleder R, Lee H (2016) Integrated magneto-electrochemical sensor for exosome analysis. ACS Nano 10:1802–1809
34. Kazane SA, Sok D, Cho EH, Uson ML, Kuhn P, Schultz PG, Smider VV (2012) Site-specific DNA-antibody conjugates for specific and sensitive immuno-PCR. Proc Natl Acad Sci 109:3731–3736
35. Labib M, Khan N, Ghobadloo SM, Cheng J, Pezacki JP, Berezovski MV (2013) Three-mode electrochemical sensing of ultralow MicroRNA levels. J Am Chem Soc 135:3027–3038
36. Labib M, Khan N, Berezovski MV (2015) Protein electrocatalysis for direct sensing of circulating microRNAs. Anal Chem 87:1395–1403
37. Labib M, Sargent EH, Kelley SO (2016) Electrochemical methods for the analysis of clinically relevant biomolecules. Chem Rev 116:9001–9090
38. Li J, Xu M, Huang H, Zhou J, Abdel-Halimb ES, Zhang JR, Zhu JJ (2011) Aptamer-quantum dots conjugates-based ultrasensitive competitive electrochemical cytosensor for the detection of tumor cell. Talanta 85:2113–2120
39. Li M, Cushing SK, Zhang J, Suri S, Evans R, Petros WP, Gibson LF, Ma D, Liu Y, Wu N (2013) Three-dimensional hierarchical plasmonic nano-architecture enhanced surface-enhanced Raman scattering immunosensor for cancer biomarker detection in blood plasma. ACS Nano 7:4967–4976
40. Li P, Mao Z, Peng Z, Zhou L, Chen Y, Huang PH, Truica CI, Drabick JJ, El-Deiry WS, Dao M, Suresh S, Huang TJ (2015) Acoustic separation of circulating tumor cells. Proc Natl Acad Sci 112:4970–4975
41. Li Q, Tofaris GK, Davis JJ (2017) Concentration-normalized electroanalytical assaying of exosomal markers. Anal Chem 89:3184–3190
42. Liu H, Xu S, He Z, Deng A, Zhu JJ (2013) Supersandwich cytosensor for selective and ultrasensitive detection of cancer cells using aptamer-DNA concatamer-quantum dots probes. Anal Chem 85:3385–3392
43. Maltez-Da Costa M, De La Escosura-Muñiz A, Nogués C, Barrios C, Ibáñez E, Merkoçi A (2012) Simple monitoring of cancer cells using nanoparticles. Nano Lett 12:4164–4171
44. Martinez-Cisneros CS, Sanchez S, Xi W, Schmidt OG (2014) Ultracompact three-dimensional tubular conductivity microsensors for ionic and biosensing applications. Nano Lett 14:2219–2224
45. Mizutani K, Terazawa R, Kameyama K, Kato T, Horie K, Tsuchiya T, Seike K, Ehara H, Fujita Y, Kawakami K, Ito M, Deguchi T (2014) Isolation of prostate cancer-related exosomes. Anticancer Res 34:3419–3423
46. Mohamadi RM, Besant JD, Mepham A, Green B, Mahmoudian L, Gibbs T, Ivanov I, Malvea A, Stojcic J, Allan AL, Lowes LE, Sargent EH, Nam RK, Kelley SO (2015) Nanoparticle-mediated binning and profiling of heterogeneous circulating tumor cell subpopulations. Angew Chem Int Ed 54:139–143
47. Moscovici M, Bhimji A, Kelley SO (2013) Rapid and specific electrochemical detection of prostate cancer cells using an aperture sensor array. Lab Chip 13:940–946
48. Müller G (2012) Novel tools for the study of cell type-specific exosomes and microvesicles. J Bioanal Biomed 4:46–60
49. Nagrath S, Sequist LV, Maheswaran S, Bell DW, Irimia D, Ulkus L, Smith MR, Kwak EL, Digumarthy S, Muzikansky A, Ryan P, Balis UJ, Tompkins RG, Haber DA, Toner M (2007)

Isolation of rare circulating tumour cells in cancer patients by microchip technology. Nature 450:1235–1239
50. Nilsson J, Skog J, Nordstrand A, Baranov V, Mincheva-Nilsson L, Breakefield XO, Widmark A (2009) Prostate cancer-derived urine exosomes: a novel approach to biomarkers for prostate cancer. Br J Cancer 100:1603–1607
51. Ozkumur E, Shah AM, Ciciliano JC, Emmink BL, Miyamoto DT, Brachtel E, Yu M, Chen P, Morgan B, Trautwein J, Kimura A, Sengupta S, Stott SL, Karabacak NM, Barber T, Walsh JR, Smith K, Spuhler PS, Sullivan JP, Lee RJ, Ting DT, Luo X, Shaw AT, Bardia A, Sequist LV, Louis DN, Maheswaran S, Kapur R, Haber D, Toner M (2013) Inertial focusing for tumor antigen–dependent and –independent sorting of rare circulating tumor cells. Sci Transl Med 5:179ra47
52. Palecek E, Bartosik M (2012) Electrochemistry of nucleic acids. Chem Rev 112:3427–3481
53. Pan Y, Guo M, Nie Z, Huang Y, Pan C, Zeng K, Zhang Y, Yao S (2010) Selective collection and detection of leukemia cells on a magnet-quartz crystal microbalance system using aptamer-conjugated magnetic beads. Biosens Bioelectron 25:1609–1614
54. Qian W, Zhang Y, Chen W (2015) Capturing cancer: emerging microfluidic technologies for the capture and characterization of circulating tumor cells. Small 11:3850–3872
55. Rupp AK, Rupp C, Keller S, Brase JC, Ehehalt R, Fogel M, Moldenhauer G, Marmé F, Sültmann H, Altevogt P (2011) Loss of EpCAM expression in breast cancer derived serum exosomes: role of proteolytic cleavage. Gynecol Oncol 122:437–446
56. Safaei TS, Mohamadi RM, Sargent EH, Kelley SO (2015) In situ electrochemical ELISA for specific identification of captured cancer cells. ACS Appl Mater Interfaces 7:14165–14169
57. Sage AT, Besant JD, Lam B, Sargent EH, Kelley SO (2014) Ultrasensitive electrochemical biomolecular detection using nanostructured microelectrodes. Acc Chem Res 47:2417–2425
58. Schwarzenbach H, Hoon DSB, Pantel K (2011) Cell-free nucleic acids as biomarkers in cancer patients. Nat Rev Cancer 11:426–437
59. Seenivasan R, Maddodi N, Setaluri V, Gunasekaran S (2015) An electrochemical immunosensing method for detecting melanoma cells. Biosens Bioelectron 68:508–515
60. Selvaraju T, Das J, Han SW, Yang H (2008) Ultrasensitive electrochemical immunosensing using magnetic beads and gold nanocatalysts. Biosens Bioelectron 23:932–938
61. Shekhawat GS, Dravid VP (2013) Nanomechanical sensors: bent on detecting cancer. Nat Nanotechnol 8:77–78
62. Shen H, Yang J, Chen Z, Chen X, Wang L, Hu J, Ji F, Xie G, Feng W (2016) A novel label-free and reusable electrochemical cytosensor for highly sensitive detection and specific collection of CTCs. Biosens Bioelectron 81:495–502
63. Suni II (2008) Impedance methods for electrochemical sensors using nanomaterials. Trends Anal Chem 27:604–611
64. Swarup V, Rajeswari MR (2007) Circulating (cell-free) nucleic acids – a promising, non-invasive tool for early detection of several human diseases. FEBS Lett 581:795–799
65. Tauro BJ, Greening DW, Mathias RA, Ji H, Mathivanan S, Scott AM, Simpson RJ (2012) Comparison of ultracentrifugation, density gradient separation, and immunoaffinity capture methods for isolating human colon cancer cell line LIM1863-derived exosomes. Methods 56:293–304
66. Thakur BK, Zhang H, Becker A, Matei I, Huang Y, Costa-Silva B, Zheng Y, Hoshino A, Brazier H, Xiang J, Williams C, Rodriguez-Barrueco R, Silva JM, Zhang W, Hearn S, Elemento O, Paknejad N, Manova-Todorova K, Welte K, Bromberg J, Peinado H, Lyden D (2014) Double-stranded DNA in exosomes: a novel biomarker in cancer detection. Cell Res 24:766–769
67. Vaidyanathan R, Naghibosadat M, Rauf S, Korbie D, Carrascosa LG, Shiddiky MJ, Trau M (2014) Detecting exosomes specifically: a multiplexed device based on alternating current electrohydrodynamic induced nanoshearing. Anal Chem 86:11125–11132

68. Van Der Pol E, Hoekstra AG, Sturk A, Otto C, Van Leeuwen TG, Nieuwland R (2010) Optical and non-optical methods for detection and characterization of microparticles and exosomes. J Thromb Haemost 8:2596–2607
69. Van Der Pol E, Van Gemert MJC, Sturk A, Nieuwland R, Van Leeuwen TG (2012) Single vs. swarm detection of microparticles and exosomes by flow cytometry. J Thromb Haemost 10:919–930
70. Vasilyeva E, Lam B, Fang Z, Minden MD, Sargent EH, Kelley SO (2011) Direct genetic analysis of ten cancer cells: tuning sensor structure and molecular probe design for efficient mRNA capture. Angew Chem Int Ed 50:4137–4141
71. Wan Y, Zhou YG, Poudineh M, Safaei TS, Mohamadi RM, Sargent EH, Kelley SO (2014) Highly specific electrochemical analysis of cancer cells using multi nanoparticle labeling. Angew Chem Int Ed 53:13145–13149
72. Wang S, Zhang L, Wan S, Cansiz S, Cui C, Liu Y, Cai R, Hong C, Teng IT, Shi M, Wu Y, Dong Y, Tan W (2017) Aptasensor with expanded nucleotide using DNA nanotetrahedra for electrochemical detection of cancerous exosomes. ACS Nano 11:3943–3949
73. Wei F, Yang J, Wong DTW (2013) Detection of exosomal biomarker by electric field-induced release and measurement (EFIRM). Biosens Bioelectron 44:115–121
74. Wen Q, Yang PH (2015) In situ electrochemical synthesis of Ni-capped electrochemiluminescence nanoprobe for ultrasensitive detection of cancer cells. Anal Methods 7:1438–1445
75. Wen Y, Pei H, Shen Y, Xi J, Lin M, Lu N, Shen X, Li J, Fan C (2012) DNA nanostructure-based interfacial engineering for PCR-free ultrasensitive electrochemical analysis of microRNA. Sci Rep 2:867–872
76. Wu L, Qu X (2015) Cancer biomarker detection: recent achievements and challenges. Chem Soc Rev 44:2963–2997
77. Wu CH, Lin SR, Yu FJ, Wu DC, Pan YS, Hsieh JS, Huang SY, Wang JY (2006) Development of a high-throughput membrane-array method for molecular diagnosis of circulating tumor cells in patients with gastric cancers. Int J Cancer 119:373–379
78. Wu L, Wang J, Ren J, Li W, Qu X (2013) Reduced graphene oxide upconversion nanoparticle hybrid for electrochemiluminescent sensing of a prognostic indicator in early-stage cancer. Chem Commun 49:5675–5677
79. Wu Y, Xue P, Hui KM, Kang Y (2014) Electrochemical- and fluorescent-mediated signal amplifications for rapid detection of low-abundance circulating tumor cells on a paper-based microfluidic immunodevice. ChemElectroChem 1:722–727
80. Wu C, Rehman F, Li J, Ye J, Zhang Y, Su M, Jiang H, Wang X (2015) Real-time evaluation of live cancer cells by *in situ* surface plasmon resonance and electrochemical study. ACS Appl Mater Interfaces 7:24848–24854
81. Yadav S, Boriachek K, Islam MN, Lobb R, Möller A, Hill MM, Al Hossain MS, Nguyen NT, Shiddiky MJA (2016) An electrochemical method for the detection of disease-specific exosomes. ChemElectroChem 4:1–6
82. Yan S, Chen P, Zeng X, Zhang X, Li Y, Xia Y, Wang J, Dai X, Feng X, Du W, Liu BF (2017) Integrated multifunctional electrochemistry microchip for highly efficient capture, release, lysis, and analysis of circulating tumor cells. Anal Chem 89:12039–12044
83. Yu CC, Ho BC, Juang RS, Hsiao YS, Naidu RVR, Kuo CW, You YW, Shyue JJ, Fang JT, Chen P (2017) Poly(3,4-ethylenedioxythiophene)-based nanofiber mats as an organic bioelectronic platform for programming multiple capture/release cycles of circulating tumor cells. ACS Appl Mater Interfaces 9:30329–30342
84. Zhai YY, Ye D, Zhang QW, Wu ZQ, Xia XH (2017) Highly efficient capture and electrochemical release of circulating tumor cells by using aptamers modified gold nanowire arrays. ACS Appl Mater Interfaces 9:34706–34714
85. Zhang X, Teng Y, Fu Y, Zhang S, Wang T, Wang C, Jin L, Zhang W (2011) Lectin-based biosensor strategy for electrochemical assay of glycan expression on living cancer cells. Chem Sci 2:2353–2360

86. Zhang H, Jia Z, Wu C, Zang L, Yang G, Chen Z, Tang B (2015) In vivo capture of circulating tumor cells based on transfusion with a vein indwelling needle. ACS Appl Mater Interfaces 7:20477–20484
87. Zhang HR, Wang YZ, Zhao W, Xu JJ, Chen HY (2016) Visual color-switch electrochemiluminescence biosensing of cancer cell based on multichannel bipolar electrode chip. Anal Chem 88:2884–2890
88. Zheng T, Zhang Q, Feng S, Zhu JJ, Wang Q, Wang H (2014) Robust nonenzymatic hybrid nanoelectrocatalysts for signal amplification toward ultrasensitive electrochemical cytosensing. J Am Chem Soc 136:2288–2291
89. Zhou Y, Wan Y, Sage AT, Poudineh M, Kelley SO (2014) Effect of microelectrode structure on electrocatalysis at nucleic acid-modified sensors. Langmuir 30:14322–14328
90. Zhou Q, Rahimian A, Son K, Shin DS, Patel T, Revzin A (2016a) Development of an aptasensor for electrochemical detection of exosomes. Methods 97:88–93
91. Zhou YG, Mohamadi RM, Poudineh M, Kermanshah L, Ahmed S, Safaei TS, Stojcic J, Nam RK, Sargent EH, Kelley SO (2016b) Interrogating circulating microsomes and exosomes using metal nanoparticles. Small 12:727–732

Chapter 5
Spiral Inertial Microfluidics for Cell Separation and Biomedical Applications

Ning Liu, Chayakorn Petchakup, Hui Min Tay, King Ho Holden Li, and Han Wei Hou

Abstract The emergence of omics studies and single cell analysis in biomedicine has advocated a critical need to develop novel cell sorting technologies to process complex and heterogenous biological samples prior analysis. Spiral inertial microfluidics is an enabling membrane-free cell separation technique developed almost a decade ago for high throughput biophysical cell separation, and has since been widely exploited for different biomedical applications. In this chapter, we will provide a comprehensive review on spiral inertial microfluidics including (1) conventional and microfluidic cell sorting techniques, (2) introduction to inertial microfluidics and Dean-coupled inertial focusing, (3) classification of major spiral devices, (4) summary of different biomedical applications, (5) recent advances in next generation spiral cell sorters, and (6) highlight key challenges for future research. With increasing advancement in microfabrication and computational simulation, we envision that spiral inertial microfluidics will play a leading role in driving research and commercialization in clinical diagnostics, as well as other research areas in chemistry and material sciences.

Keywords Inertial microfluidics · Cell separation · Biomedical diagnostics

---

N. Liu · H. M. Tay
Lee Kong Chian School of Medicine, Nanyang Technological University, Singapore, Singapore

C. Petchakup · K. H. H. Li
School of Mechanical and Aerospace Engineering, Nanyang Technological University, Singapore, Singapore

H. W. Hou (✉)
Lee Kong Chian School of Medicine, Nanyang Technological University, Singapore, Singapore

School of Mechanical and Aerospace Engineering, Nanyang Technological University, Singapore, Singapore
e-mail: hwhou@ntu.edu.sg

M. Tokeshi (ed.), *Applications of Microfluidic Systems in Biology and Medicine*, Bioanalysis 7, https://doi.org/10.1007/978-981-13-6229-3_5

5.1 Introduction

Cell separation is an essential sample preparation step in biomedical research to purify target cell population and minimize cell-cell interactions prior analysis. For example, subpopulations of cells with unique biological signature and functions are often found in tissue and clinical samples that are highly heterogeneous. It is therefore necessary to isolate specific target cells to facilitate downstream biological assays. Liquid biopsy or blood diagnostic is another important application as blood is routinely obtained during medical checkup, and contains a myriad of information on the health status of an individual. Efficient isolation of blood cell components (white blood cells (WBCs), platelets) or rare diseased cells (cancer or bacteria (~1–100 cells/mL)) from a large cellular background (~5 billion red blood cells (RBCs)/mL) will hence greatly improve signal-to-noise ratio for unbiased and accurate clinical assessment [1].

5.1.1 Conventional Separation Techniques

Conventional "label-free" separation approaches include membrane filtration and centrifugation, which are widely available and simple to use. Membrane filtration works by size-exclusion through a porous matrix which retains cells above a certain size (larger than pore size) and allowing smaller cells to pass through. Target cells can be collected either on the membrane or in the filtrate. Although useful as a pre-enrichment step to remove cell aggregates, clogging issues persist in such systems and it is non-trivial to retrieve the larger target cells from the membrane. Variability in cell deformability can also affect size cut-off or optimal pressure conditions [3, 4]. Differential centrifugation is yet another popular method which separates cells based on size and density, but the isolation of rare cells (< 1000 cells) is not practical and risk substantial cell loss. Furthermore, fluid shear stresses from repeated centrifugation and resuspension can also damage or activate sensitive cells (e.g. neutrophils) [5].

With the growing repertoire of monoclonal antibodies, affinity-based cell separations methods such as fluorescence-activated cell sorting (FACS) and magnetic-activated cell sorting (MACS) have become increasing popular among biologists. Target cells are first immunolabelled with antibodies which bind to specific surface markers prior separation. FACS relies on fluorophore-conjugated antibodies for separation while MACS employs magnetic microbeads-conjugated antibodies. In FACS, fluorescently-labelled cells are hydrodynamically focused to a narrow stream and passed through a laser beam for signal interrogation. The resulting scatter and fluorescence signals are used to discriminate different cells types based on their cell size, granularity and surface markers. Target cells are subsequently identified based on these biophysical signatures and electrostatically deflected into separate reservoirs for collection [2]. Being a well-established method with high sensitivity and

Table 5.1 Comparison of conventional cell separation techniques

	Separation technique	Principle	Advantages	Disadvantages
Label-free	Filtration	Size	Simple to use	Low purity Low yield Non-specific Requires manual post-processing
	Centrifugation	Density	Relatively cheap	
Affinity-based	FACS	Antibody	High purity	High cost Additional labeling steps May affect cell function
	MACS	Antibody	High yield	
			Highly specific	

throughput, it is often considered as the "gold standard" for cell sorting [3]. A throughput of 2000–10,000 cells per second can be achieved with higher rates sacrificing purity [4]. On the other hand, MACS achieve immuno-magnetic separation by applying an external magnetic field to extract the magnetic bead-bound cells (positive selection) or eluting the non-labelled target cells (negative selection) [5]. In contrast to FACS which enables multiplexed cell sorting, MACS is a bulk and binary processing method, and does not provide individual cell analysis or multi-parametric outputs. Moreover, both methods are usually labor intensive, time-consuming, expensive, and the cell yield or recovery is highly dependent on user operations (Table 5.1).

5.1.2 Microfluidics Cell Separation

With advancement in microfabrication, the birth of microfluidics or lab-on-a-chip technologies since the 1990s has revolutionized chemical analysis and biological assays through unique physical phenomenon and flow control in the microscale [6]. The increasing demand for better and more sensitive assays have propelled the development of many novel microfluidics separation strategies integrated with single cell manipulation and analysis capabilities for point-of-care diagnostics [3, 7, 8]. These miniaturized systems offer numerous advantages including reduced sample and reagent consumption, faster processing time, high spatial resolution, low device cost and high portability [9]. Consequently, microfluidics has become an important toolbox for cell separation applications with the ability to achieve unprecedented size resolution and purity, and high throughput sample processing.

Generally, microfluidic cell separation techniques are classified into active and passive methods based on the involvement of external fields (*e.g.* electric, optical, acoustic, magnetic field) [8, 10, 11]. Active methods exploit external fields to impart different forces on cells to achieve separation, and common examples include magnetophoresis [12–14], dielectrophoresis [15], acoustophoresis [16, 17] and

Table 5.2 Comparison of various passive separation techniques

Separation technique	Mechanism/principle	Separation criteria	Throughput
Biomimetic	Hydrodynamic force/ Fahraeus effect	Size deformability	10 μL/h [19]
Hydrodynamic	Streamline manipulation	Size shape	20 μL/min [20] $>10^5$/min [21]
Hydrophoretic filtration	Pressure field gradient	Size	4×10^3/s [22]
Inertial	Lift force secondary flow	Size shape	~10^6/min [23]
Microstructure (Pillars and weirs)	Laminar flow/perturbation of flow	Size deformability	10^3 μm/s [24] 5 μL/min [25]
Surface affinity	Specific binding to surface markers	Size surface biomarkers	1–2 mL/h [26]
Pinched flow fractionation (PFF)	Laminar flow (Hydrodynamic force)	Size	~4×10^3/min [27] 20 μL/h [27]

optical sorting [18]. Passive separation methods rely on intrinsic hydrodynamic forces during fluid flow which are modulated by microchannel design and flow conditions (Table 5.2).

Numerous efforts have been focused on the development of passive separation methods including deterministic lateral displacement (DLD) [24, 28], pinched flow fractionation (PFF) [27, 29] and inertial focusing/microfluidics. Among these technologies, inertial microfluidics has emerged as a highly promising approach for size-based cell separation due to its ease of operation and high separation resolution. The first seminal work was reported by Di Carlo et al. in 2007 where they described particle inertial focusing effects in microfluidics and its application for high throughput size-based particles/cell separation [30]. This is shortly followed by the Papautsky's group who reported similar particle inertial focusing and Dean migration effects in spiral microchannels [23, 31]. With increasing understanding of the particle inertial focusing behavior [32], many researchers have started working in this exciting field to explore new frontiers and applications in fluid mechanics research and biomedical applications. Unsurprisingly, several review papers on inertial microfluidics have been recently published which give an excellent overview of inertial microfluidics in different channel geometries [33–36]. Spiral microchannel (hereafter termed as spiral inertial microfluidics) is one of the most widely used designs for cell separation as it exploits both size-dependent particle inertial focusing and secondary Dean-induced migration effects to achieve separation. In this chapter, we will first discuss the fundamental principles involved in spiral inertial microfluidics. This is followed by a review of the major types of spiral microfluidics devices and their separation principles. Next, we will provide an overview of spiral cell sorting applications for various bio-entities (cells and molecules) separation, and also highlight recent advances in spiral technologies with novel designs or multiplexing capabilities. Finally, we will conclude with current challenges and suggest future research directions in this area.

5.2 Theory

5.2.1 *Stokes Flow*

In fluid mechanics, the motion of viscous fluid can be described by the Navier-Stokes equation as shown below:

$$\rho(\partial \boldsymbol{u}/\partial t + \boldsymbol{u}\cdot\nabla\boldsymbol{u}) = -\nabla p + \mu\nabla^2\boldsymbol{u} + \boldsymbol{f}$$

where ρ represents the fluid density, μ is the fluid dynamic viscosity, $\boldsymbol{u}$ is the fluid velocity field, p represents fluid pressure field, and $\boldsymbol{f}$ is the vector field of external body forces imparting on fluid elements. $\rho(\partial \boldsymbol{u}/\partial t + \boldsymbol{u} \cdot \nabla\boldsymbol{u})$ corresponds to the inertial forces, $-\nabla p$ corresponds to pressures, and $\mu\nabla^2\boldsymbol{u}$ corresponds to viscous forces.

To characterize fluid flow, the channel Reynolds number (R_c) is proposed as a dimensionless quantity which describes the ratio of inertial to viscous forces:

$$R_c = \frac{\rho U_m D_h}{\mu}$$

where ρ represents the fluid density, μ is the dynamic viscosity of the fluid, D_h is defined as the hydraulic diameter where $D_h = d$ in a circular channel (d represents the diameter), or $D_h = 2wh/(w + h)$ in a rectangular channel (w and h denotes the width and height of the rectangular cross section, respectively), and U_m is the maximum velocity of the fluid flow. When considering finite-size particles in the channel flow, the particle Reynolds number (R_p) is helpful to describe the relationship between the particles and the channel dimensions:

$$R_p = R_c\frac{a^2}{D_h^2} = \frac{\rho U_m a^2}{\mu D_h}$$

where a is the particle size. Flow pattern in the channel will change when the size ratio of particle to channel varies. For $R_p \ll 1$, the viscous drag is dominant and the particles are deemed as "point-particle". As R_p increases, inertial effects become more apparent in the channel flow. Laminar flow occurs when R_c is below a critical value of approximately 2040 [37]. Due to the small channel dimensions (typically less than 1 mm) in microfluidics, R_c is usually less than 100 and fluid flow is completely laminar and the viscous forces of the fluid dominate the inertial forces (Stokes flow). Hence, the inertial portion in Navier-Stokes equation is neglected for most microfluidic systems by equating the left hand side of the equation to zero. Stokes flow lies within laminar regime, but the inverse is not true [33, 34]. Recently, the application potential of the long-ignored intermediate range flow ($\sim 1 < R_c < 100$) has received increasing attention and several inertial-based effects in microfluidics devices include improved mixing and precise particle control [30, 33, 34, 38].

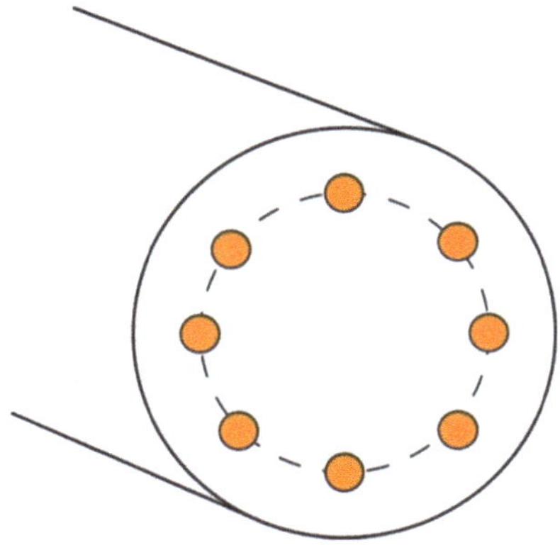

Fig. 5.1 Particle equilibrium positions in a circular straight channel as observed by Segre and Silberberg in 1960s

5.2.2 *Inertial Focusing*

Segre and Silberberg reported the first observation of particle inertial focusing effects and the earliest interpretation to explain this unintuitive phenomenon in early 1960s [39, 40]. In their experiment, the randomly dispersed particles (~1 mm diameter) were introduced into a cylindrical pipe (~1 cm diameter). After travelling a distance of 114 cm, the particles were distributed in an annulus between the center and the wall within the cross section of the pipe [40]. The mean radius of the annulus was measured to be ~0.6 times the pipe radius, as indicated in Fig. 5.1.

This unique particle lateral migration effect was later found to be the result of the interplay between two dominant inertial lift forces: the shear gradient induced lift force pushing particles in the medium away from the channel center, and the wall induced lift force repelling the particles away from the wall. Another important force to consider is the Stoke's drag force in secondary lateral flow. These forces will be explained in more details below.

5.2.2.1 Wall Induced Lift Force (F_{WL})

For a particle flowing near the channel wall, the interaction between the particle and the wall causes the particle to lag behind the fluid flow. In addition, the constricted flow space between the particle and channel wall will cause the fluid flow at the top side of the particle to be accelerated due to more streamlines diverted toward these side. This creates a relative lower pressure than the "wall side" of the particle and a lift force directed away from the wall is generated (Fig. 5.2a).

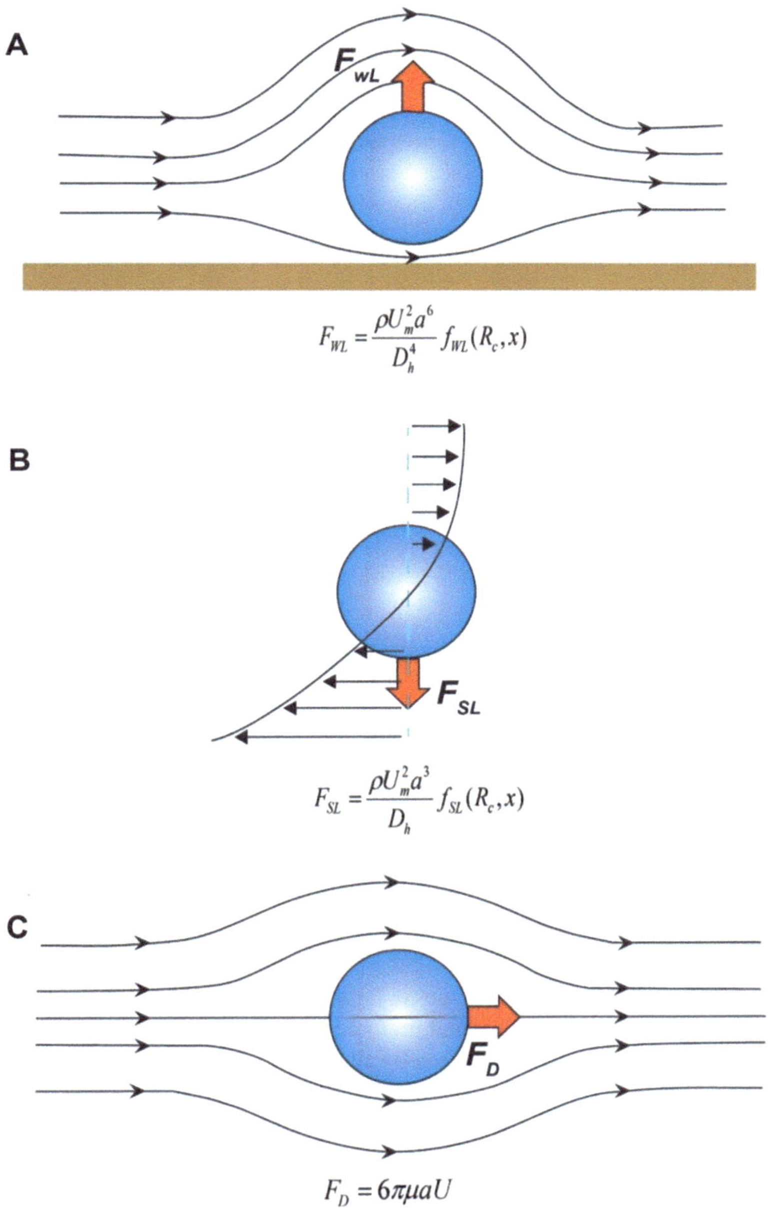

Fig. 5.2 Schematics illustration of the dominant forces experienced by particles in inertial microfluidics (**a**) Wall induced lift force (F_{WL}), (**b**) Shear gradient lift force (F_{SL}), (**c**) Secondary-flow drag force (F_D)

5.2.2.2 Shear Gradient Lift Force (F_{SL})

The shear gradient lift force arises from the curvature of the parabolic velocity profile. As shown in Fig. 5.2b, the velocity magnitudes on each side of the particle are different, leading to a pressure discrepancy between the top and bottom side of the particle. Due to the existence of the pressure difference, a shear gradient lift force is exerted on the particle which pushes it towards the channel wall (until it is balanced by the wall induced lift force).

5.2.2.3 Net Lift Force (F_L)

Following the discovery by Segre and Silberberg, matched asymptotic expansion methods were proposed to determine the lateral lift forces acting on particles during flow. Asmolov derived an analytical expression of the net lift force imparting on a rigid particle ($a/D_h \ll 1$) in a Poiseuille flow [41]:

$$F_L = \frac{\rho U_m{}^2 a^4}{D_h{}^2} f(R_c, x) = \frac{\mu^2}{\rho} R_p{}^2 f(R_c, x)$$

where $f(R_c, x)$ is the lift coefficient which depends on the particle position within the channel (x) and the channel Reynolds number (R_c).

Conventional theoretical predications are based on "point-particle" approximation which neglects the size effect of particles. However, when the particle size approaches the channel dimension, the disturbance to the flow will be affected by the particles. Di Carlo et al. experimentally demonstrated that the net lift force varies with the position in the microchannel [32]. For a particle of finite-size ($0.05 \le a/D_h \le 0.2$), the lift force scaling relationship is modified as: $F_{WL} = \frac{\rho U_m{}^2 a^6}{D_h{}^4} f_{WL}(R_c, x)$ near the channel wall, and as $F_{SL} = \frac{\rho U_m{}^2 a^3}{D_h} f_{SL}(R_c, x)$ near the channel center. The variation of lift forms is attributed to the disparate fluid dynamics in different positions of the channel. Near the center of the channel, the shear gradient lift force dominates, while the wall induced lift force is more significant near the channel wall.

5.2.2.4 Secondary-Flow Drag Force

Besides the wall induced lift force (F_{WL}) and the shear gradient induced lift force (F_{SL}), the secondary-flow drag force is the third major force responsible for particle inertial migration and focusing effects. In 1928, William Dean reported the presence of Dean vortices in curved channels due to the mismatch of fluid momentum within the channel cross section as a result of the centrifugal acceleration acting on the fluid flow [42]. Briefly, when fluid flows through a curved channel, the fluid velocity at

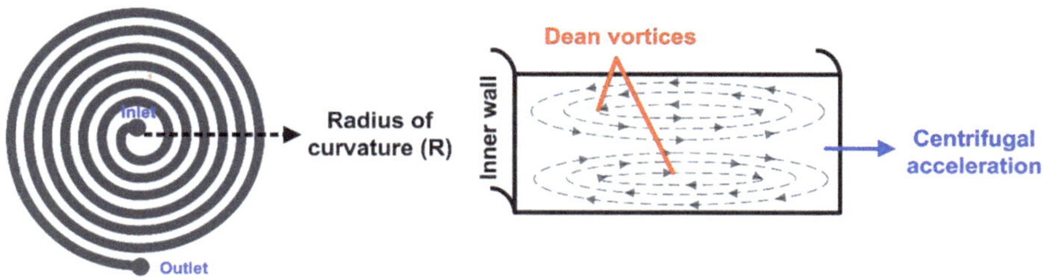

Fig. 5.3 Schematic of a pair of Dean vortices within a spiral channel with a rectangular cross section

the center of the channel is higher than the sides due to the parabolic flow profile. This introduces an additional momentum to the faster-moving fluid in the channel center, and is pushed toward the outer wall (the concave wall) of the channel curvature along the channel midline due to centrifugal acceleration. By conservation of mass, secondary counter-rotating flows are induced to compensate the fluid shifting. Hence, two symmetrical counter-rotating vortices are formed at the top and the bottom of the cross-sectional plane (Fig. 5.3). Hereafter, the terms Dean flow and secondary flow are used interchangeably [35].

The strength of the secondary flow can be characterized by a non-dimensional Dean number (*De*):

$$De = R_c\sqrt{\frac{D_h}{2R}}$$

where R_c is the channel Reynolds number defined as $R_c = \rho U_m D_h/\mu$, here U_m is the maximum channel velocity, μ and ρ is the dynamic viscosity and density of the fluid, respectively. The curvature ratio ($\delta = D_h/2R$, where R is the average radius of curvature of the channel) implies a faster turn in the channel (*i.e.* smaller radius of curvature R) and results in a stronger Dean flow.

For a given *De*, the expression of average transverse Dean velocity can be determined by [43]:

$$U_{De} = 1.84 \times 10^{-4} De^{1.63}\ (\mathrm{m/s})$$

Accordingly, the Dean drag force experienced by a particle located in this flow can be derived by assuming Stokes drag (Fig. 5.2c) and is expressed as:

$$F_D = 3\pi\mu U_{De} a = 5.4 \times 10^{-4} \pi\mu De^{1.63} a\ (\mathrm{N})$$

In spiral or curvilinear channels, the interplay of the net inertial lift force (F_L) and Dean drag force (F_D) gives rise to the Dean coupled inertial migration of particles. To characterize particle inertial focusing in curved channel, Di Carlo et al. proposed a key parameter R_f, the ratio of shear gradient lift force to Dean drag force, to describe the behavior [35, 44]:

$$R_f = \frac{F_{SL}}{F_D} \propto \frac{1}{\delta} \frac{a^2}{D_h^3}$$

It is well accepted that $R_f > 0.04$ (or $a/D_h > 0.07$ [23]) to achieve particle inertial focusing in microchannels. As R_f has a strong dependence on particle size ($\sim a^2$), this forms the basis for size-based particle separation in spiral devices since the particle equilibrium separation can be modulated by tuning these forces. This is explained in more details in the following section.

5.2.2.5 Dynamics of Particle Lateral Migration in Spiral Microchannel

Figure 5.4 depicts the schematic illustration of particle focusing dynamics in a low-aspect ratio spiral microchannel. Due to the asymmetrical parabolic flow profile,

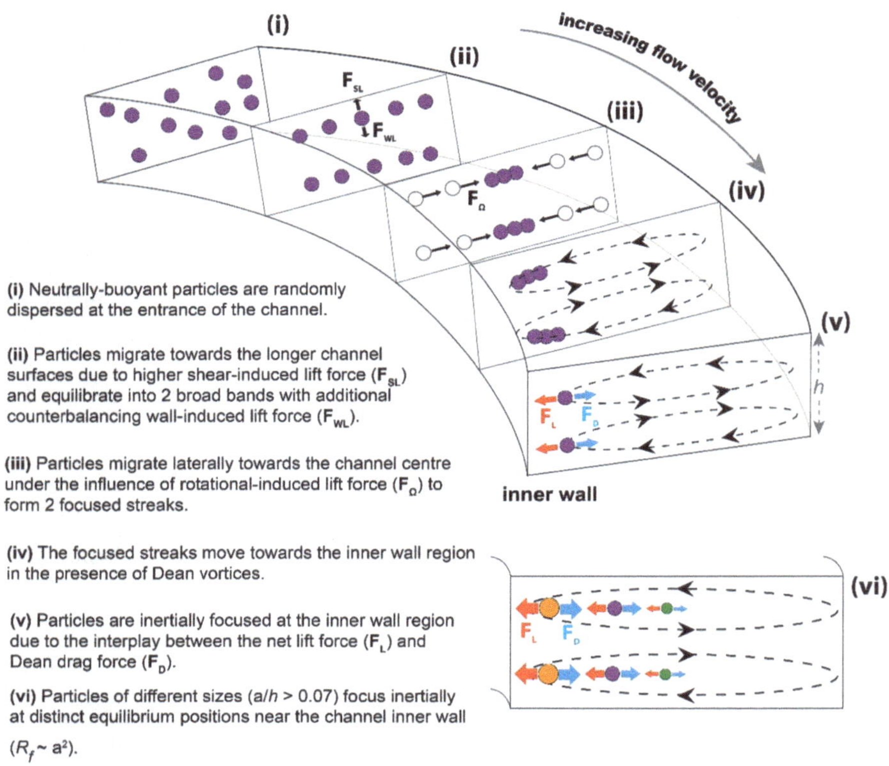

Fig. 5.4 Schematic illustration of particle migration dynamics in spiral microchannels

a steep shear gradient is generated along the vertical direction (height) of the channel. The randomly distributed particles (1) are first pushed by dominant shear gradient lift force (F_{SL}) across streamlines towards the channel top and bottom surfaces. Near the wall, the particles experience opposing wall-induced lift force (F_{WL}) and equilibrate into two broad bands at ~20% of channel height from surface (2). Next, a rotational-induced lift force (F_{Ω}), first proposed by Saffman [45], will act on the particles and they begin to migrate laterally towards the centre face of the channel wall to form two focused streaks (3). This lift force F_{Ω} is usually considered negligible compared with other lift forces (an order of magnitude lower than F_{WL} and F_{SL}), but becomes important in radially-asymmetric (e.g. rectangular) channels as particles exhibit spinning behavior in the presence of high (localized) shear rate close to the channel wall [46]. Other groups have reported the particle size dependency of F_{Ω} ($F_{\Omega} \sim a^3$) [47], and this effect is also recently used by Zhou et al. for particle separation in straight channels [48, 49]. With increasing flow velocity, the strength of the secondary Dean drag force becomes more significant and particles begin to migrate towards the inner wall region in the direction of the Dean vortices (4). Finally, a single focusing point is achieved near the inner wall region in each Dean vortice as a result of the balance of net inertial lift force (F_L) and Dean drag force (F_D) (5). As this equilibrium position is strongly dependent on particle size ($R_f \sim a^2$), larger particles ($a/h > 0.07$, as $D_h \approx h$ in a low aspect ratio channel[50]) will focus closer to the inner wall due to stronger inertial lift ($F_L \sim a^3$ vs. $F_D \sim a$) while smaller particles are positioned further away from inner wall [23] (6). For all particle sizes, a further increase in flow velocity will shift the focused particle streams back towards the outer wall due to increasing F_D [51]. Although an increase in flow velocity results in a greater lift force ($F_L \sim U_m^{\ 2}$) as compared to the Dean drag ($F_D \sim U_m^{\ 1.63}$), the particle movement away from the inner wall can be explained by a decrease in the lift coefficient $f(R_c, x)$ which is dependent on particle position within the channel [23].

5.3 Classification of Spiral Devices

Fluid mixing at the microscale poses a variety of challenges due to the dominant viscous drag forces (low *Re*), and molecular diffusion remains the main transport mechanism in this laminar flow regime. By taking advantage of the transverse Dean vortices in curvilinear channels, spiral microdevices have been used as micromixer to enhance fluid mixing [52, 53], and rotate cells in electroporation systems for efficient gene delivery [54]. Since flow conditions and channel geometries can affect the shape and magnitude of Dean vortices, these parameters have been extensively investigated in spiral inertial microfluidics to enhance cell sorting capabilities. In this section, we will describe four major types of spiral devices with different cells/ particles focusing mechanisms and features: (1) rectangular spiral microfluidics, (2) trapezoidal spiral microfluidics, (3) double-inlet spiral termed as Dean Flow Fractionation (DFF), and (4) High-resolution Dean Flow Fractionation (HiDFF).

5.3.1 Rectangular Spiral Microfluidics

Spiral microchannels with rectangular cross-section are one of the most widely used geometry, due to the well-established microfabrication techniques (photolithography and deep reactive ion etching (DRIE)) which can generate uniform channel/feature height and high aspect ratio vertical sidewall. The small feature size resolution (~1–10 μm) in microfabrication also enables the design of complicated outlet bifurcations (8–15 outlets) for multiplexed separation. In low-aspect ratio rectangular spiral channels, the regular cross-section geometry will lead to the formation of symmetrical Dean vortices along the channel midline. Inertial forces are weaker across the longer dimension (due to the blunting of the velocity profile) and particles will first migrate along the shorter channel dimensions (channel height) to the top and bottom surfaces due to higher shear rate [50]. Once particles have reached the z-direction equilibrium positions, they will then migrate along the channel width to the final equilibrium position near the inner wall region. Martel and Toner have performed a systematic characterization of inertial focusing dynamics in spiral microchannels of varying widths (Fig. 5.5a). Generally, as channel dimensions (D_h) increase, particles experience less shear-induced inertial forces (blunting of velocity profile) and more Dean drag force ($\sim D_h^{1.5}$). They proposed slight modifications to R_f (ratio of F_L/F_D) by using $U_{Dean,\ ave}$ in the equation to reduce variability of R_f to ~1 for quality focusing in different spiral designs [55]. A straightened composite image was also generated from empirically-determined particle focusing images to more clearly visualize the particle focusing streak width and position along the entire channel length (Fig. 5.5b).

As described previously, the interplay between inertial lift (F_L) and Dean drag forces (F_D) is important to focus particles of different sizes at distinct equilibrium positions. Early work by Kuntaegowdanahalli et al. [23] and Russom et al. [56] clearly demonstrated the capability of spiral microchannels for continuous high throughput size-based particle separation into different outlets. As shown on Fig. 5.6a, large particles ($a/h > 0.07$) focused inertially close to the channel inner wall due to dominant F_L. At high flow conditions (De~10–15, ~3 mL/min), significant Dean drag force would move these focused streams farther away from the channel inner wall based on particle size, with the largest particles being closest to the inner channel wall. This phenomenon was exploited to separate closely-spaced microparticles (10 μm, 15 μm, 20 μm) in a 500 μm wide Archimedean spiral device (130 μm height) [23]. Xiang et al. also developed a smaller spiral device (160 μm width, 50 μm height) for binary separation of 4.8 μm and 2.1 μm particles Fig. 5.6b [57]. In addition, they also described the non-focusing behavior of smaller 2.1 μm particles ($a/D_h \ll 0.07$) and the resultant particle-free regions at low flow conditions (De~1–5). Besides particle sorting and filtration applications, inertially-focused particle stream in rectangular spiral channels is used for cell self-ordering for deterministic single-cell droplet encapsulation [58]. Noteworthy, the above mentioned Dean-coupled particle inertial focusing effects can also be applied in other curvilinear channels with symmetrical cross section geometry including double

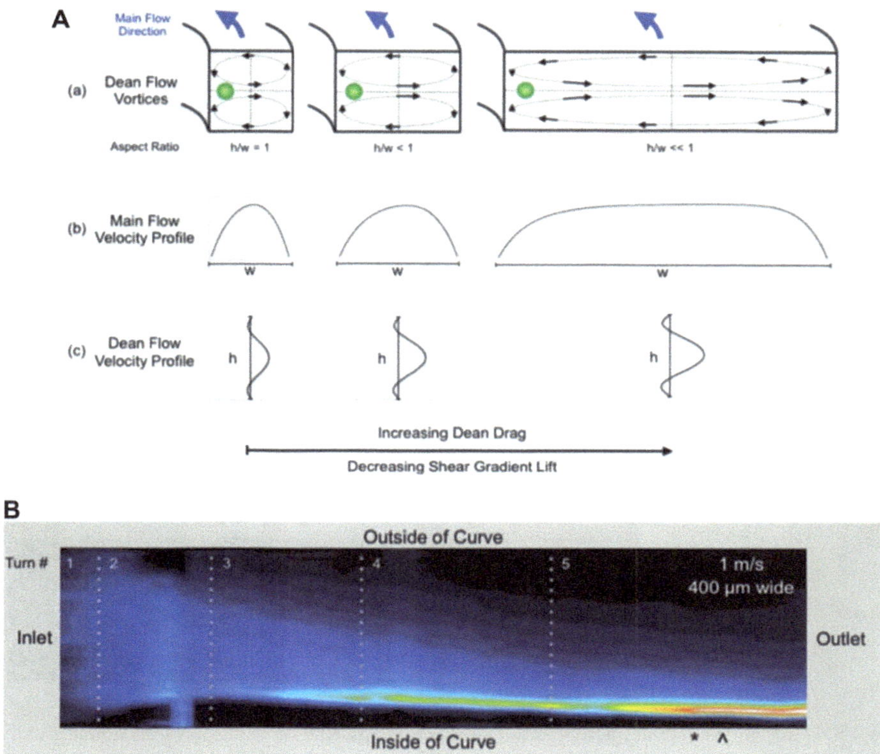

Fig. 5.5 Rectangular spiral microfluidics (**a**) Schematic illustrations of different flow profiles and associated forces in curved channels of different widths, (**b**) Straightened image of the inertial focusing behavior of 15 µm beads in a spiral channel. *Dotted lines represent channel curvature changes*. (Reproduced with permission from Ref. [55])

spiral [59–61], serpentine [30], as well as soft microtubes (circular cross section) coiled in planar or 3D spiral (helical) [62].

Recently, Nivedita et al. performed experimental and numerical simulation studies on fluid flow dynamics in spiral microchannels under high flow conditions. They reported the presence of multiple pairs of secondary flow vortices at high *Re* (>100) and *De* (~20–40), and defined a non-dimensional parameter termed as critical Dean number (De_c) to describe this novel flow observations [63]. According to them, the formation of additional Dean vortices was due to the large pressure gradient between the high velocity area and the channel outer wall. Above De_c, the primary Dean vortices were unable to maintain the pressure across the channel width. In order to balance the pressure, the primary vortices would thus split to recirculate the fluid near the outer wall region. This led to the formation of secondary Dean vortices which can entrain particles or cells at higher flow rates (Fig. 5.6c).

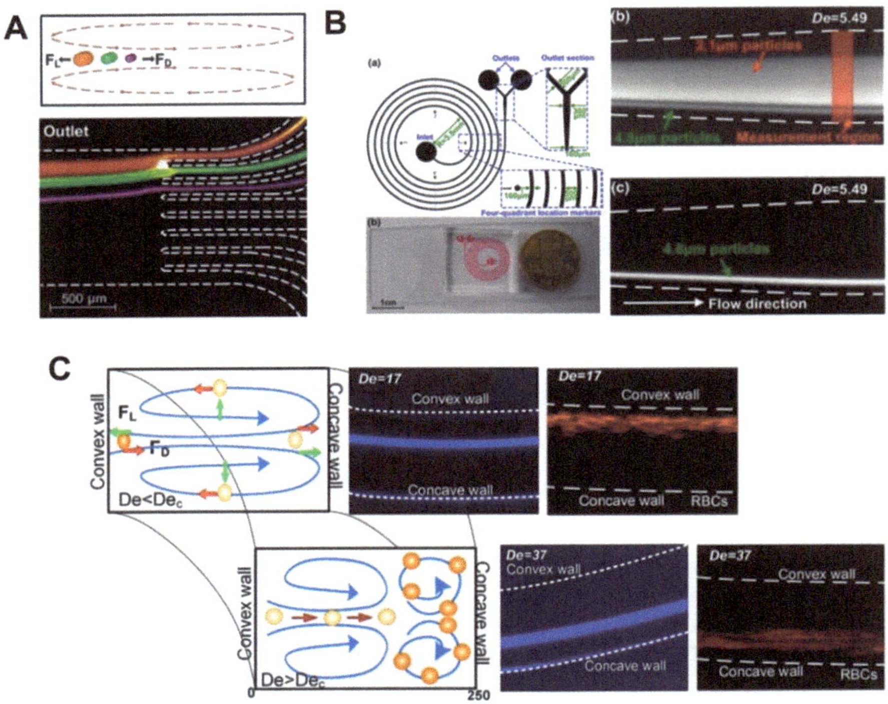

Fig. 5.6 Rectangular spiral microfluidics (**a**) Schematic illustration and fluorescence image indicating multiplexed separation of 10 µm, 15 µm, 20 µm microparticles into different outlets in a 500 µm wide and 130 µm tall Archimedean spiral device. (Reproduced from Ref. [23] with permission from The Royal Society of Chemistry) (**b**) CAD design and optical image of a 5-loop spiral device *(filled with red dye)* fabricated in polydimethylsiloxane (PDMS). Fluorescent images indicating tight focusing of 4.8 µm particles near the inner wall and the wider 2.1 µm particles band at the channel centre. (Reproduced with permission from Ref. [57].) **c** Schematic and fluorescent images of 10 µm particles *(blue)* and RBCs *(red)* entrapment in additional Dean vortices at high *De (~37)*. (Reproduced from Ref. [63] under Creative Commons)

5.3.2 *Trapezoidal Spiral Microfluidics*

Intuitively, one can modulate the Dean vortices to increase the separation distance between particles of different sizes and enhance the separation resolution. Trapezoidal cross section is an interesting design as the change in channel height from low (inner wall) to high (outer wall) results in a Dean flow velocity gradient across the channel width. Guan et al. first reported this behavior, and found that the skewed Dean vortices were beneficial for particles separation as the gradient in F_D across the channel would led to a sharp transition of size-based focusing behavior beyond a certain threshold flow rate [64]. Unlike in rectangular spiral channels where focused particles streams gradually migrate towards the outer wall with increasing flow rates,

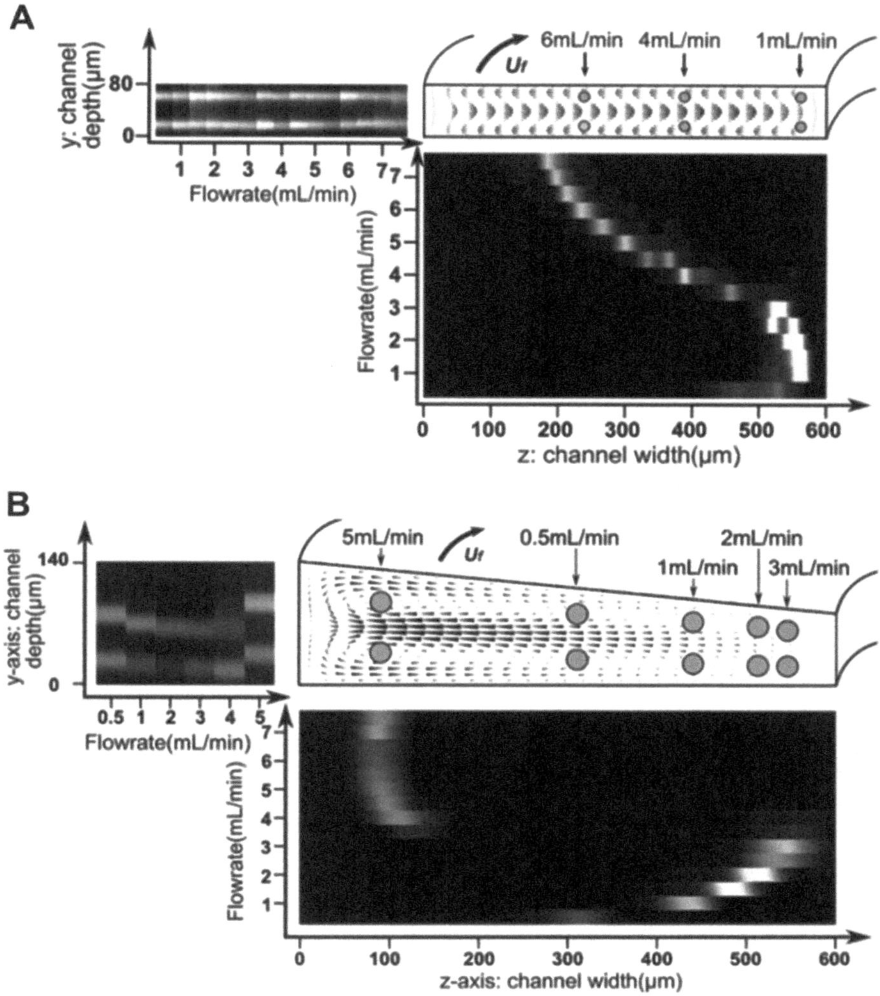

Fig. 5.7 Comparison of rectangular and trapezoidal spiral microchannels. CFD simulation and experimental results *(top and side view)* of (**a**) 15.5 μm particles focusing behavior in rectangular spiral microchannel (600 μm width, 80 μm height) and (**b**) 26.25 μm particles focusing behavior in trapezoidal spiral microchannels *(600 μm width, 80–140 μm height).* (Reproduced with permission from Ref. [64])

focused particles in trapezoidal spiral channels would "switch" to an equilibrium position located at the outer half (deeper side) of the channel as flow rate increases (Fig. 5.7). Side view analysis also revealed that these particles were trapped within the Dean vortices at the outer wall region [64].

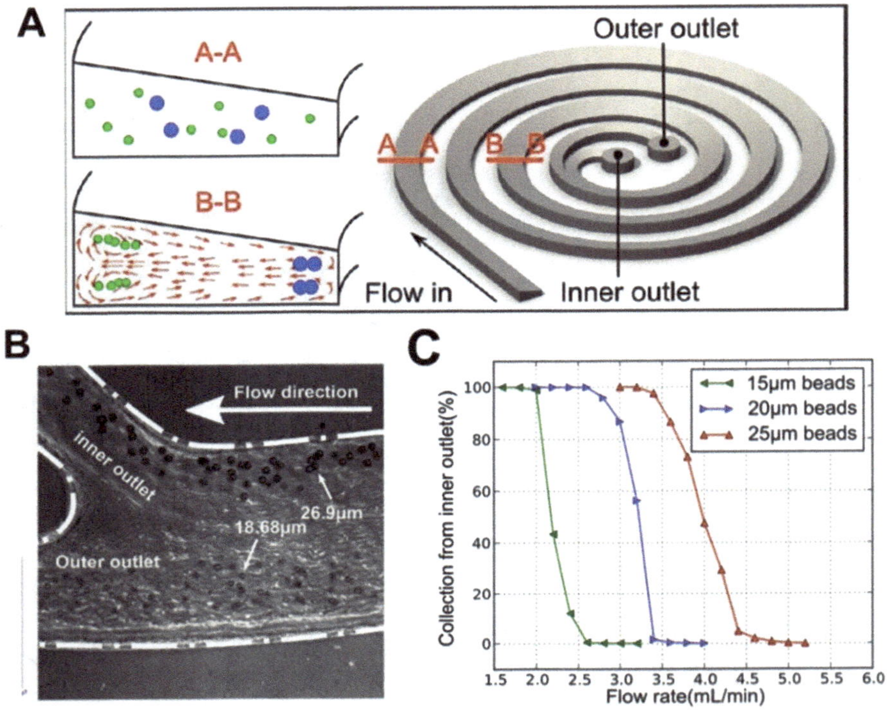

Fig. 5.8 Binary particle separation in trapezoidal spiral microchannels (**a**) Schematic illustration of particle focusing and trapping within the skewed Dean vortices in trapezoidal cross-section spiral microchannel (**b**) High speed image at the outlet bifurcation showing separation of 18.68 μm and 26.9 μm particles *(channel dimensions: 600 μm width, 80–140 μm height)* (**c**) Flow rate characterization of size-based "threshold" flow rate to switch particles focusing to outer wall focusing. (Reproduced with permission from Ref. [64])

As the threshold flow rate is a function of particle size, this novel focusing behavior was used to achieve efficient binary separation (~92–96%) of particle mixtures in a 2-outlet trapezoidal spiral channel (Fig. 5.8). This is ideal as one can process samples at high throughput (~3–4 mL/min) and higher particle concentrations (~10^7/mL) by minimizing the interactions between particles of different sizes. This method was applied for separation of leukocytes (~10–15 μm) [65] and circulating tumor cells (CTCs) (~15–20 μm) [66] from blood samples (RBCs ~6–8 μm), as well as macroscale filtration (~500 mL/min) of Chinese hamster ovary (CHO) and yeast cells in bioprocessing [67]. However, a major limitation of trapezoidal spiral microchannels is the use of micromilling [64] or 3D printing [68] techniques to fabricate the channel molds with uneven channel heights. Due to the relatively poor feature size resolution as compared to conventional microfabrication, the devices are limited to a 2-outlet bifurcation and thus only used for binary separation.

5.3.3 Double-Inlet Spiral/ Dean Flow Fractionation (DFF)

Rectangular and trapezoidal spiral microchannels are highly effective for size-based particle and cell separation with working sample concentrations of ~10^{5-7}/mL, but their use in blood-related applications is greatly limited by the large RBCs background (~45% *v/v*, ~10^9 RBCs/mL) as RBCs-RBCs interactions can severely affect the cell focusing behavior and hence deteriorate separation efficiency [69]. Typically, whole blood samples have to be diluted significantly (50–100 ×, ~0.1–2% hematocrit) which increases the processing time, making them unsuitable to process large volumes of blood samples required in clinical settings.

In a landmark work by Bhagat et al.[31], they developed a 2-inlet spiral channel to selectively introduce particle samples on the inner side of the channel with an additional sheath flow. As the particles flow through the spiral device (100 × 50 μm (w × h)), the Dean vortices would transpose the smaller particles ($a/D_h < 0.07$) towards the outer wall, and both inertial lift and Dean drag forces would equilibrate the larger particles ($a/D_h > 0.07$) near the inner wall. This was used for complete separation of 1.9 μm and 7.32 μm particles at $De = 0.47$ (Fig. 5.9a). It should be noted that unlike in 1-inlet rectangular spiral channels where all particles are inertially focused near the inner wall and their final equilibrium positions differ slightly based on size differences, the 1.9 μm particles did not undergo inertial focusing effects in the dual-inlet spiral device and solely migrated to the outer wall under the influence of Dean vortices. This was the primary reason for the significantly wider 1.9 μm particles focusing band as compared to 7.32 μm particles (Fig. 5.9a).

Inspired by this work, Hou et al. subsequently developed a 2-inlet 2-outlet spiral device (500 μm (*w*) × 160 μm (*h*)) for isolation of CTCs from whole blood, aptly termed as Dean Flow Fractionation (DFF) [70] (Fig. 5.9b). Based on the size difference between CTCs and blood cells [71, 72], the developed technique enables inertial focusing of larger CTCs (15–20 μm) near the inner wall while smaller blood components (RBC ~8 μm discoid; leukocytes ~7–12 μm) are solely affected by the Dean drag and transposed towards the outer wall, thus achieving separation. The authors further define this Dean-induced lateral migration in terms of 'Dean cycle' (DC) which can be modeled by COMSOL simulation (Fig. 5.9b). For instance, a particle which is initially positioned near the microchannel outer wall and migrates to the inner wall is said to have completed ½ Dean cycle (DC 0.5), and returning back to the original position near the channel outer wall completes a full Dean cycle (DC 1). The length for a complete Dean cycle migration (L_{DC}) can be approximated as $L_{DC} \sim 2w + h$ (where w is the microchannel width and h is the channel height). For a given microchannel length, the particles can thus undergo multiple DC migration with increasing flow rate conditions (Fig. 5.9b).

Notably, this separation principle is particularly useful for blood cell separation as the additional sheath buffer in DFF device facilitates the Dean migration of large volume of RBCs in a well-controlled manner. The authors reported that RBCs band broadened with increasing hematocrit due to cell-cell interaction induced dispersion

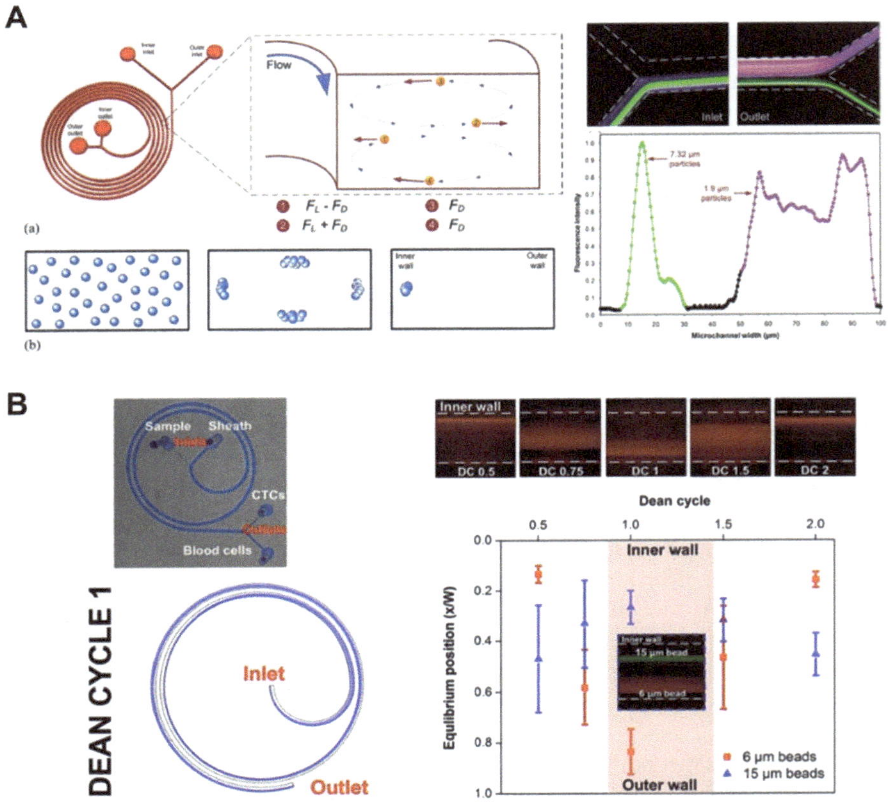

Fig. 5.9 Dean Flow Fractionation (DFF) using a 2-inlet spiral design (**a**) (*left*) Schematic illustration of the Dean-coupled inertial focusing separation principle in the 2-inlet spiral device. (*right*) Composite fluorescent images and intensity linescans illustrating complete separation of 1.9 µm (*purple*) and 7.32 µm (*green*) particles at $De = 0.47$. (Reproduced from Ref. [31] with permission from The Royal Society of Chemistry) (**b**) (*left*) Optical image of the DFF spiral microchannel (*filled with blue dye for visualization*) used for CTCs isolation from whole blood. Fluid simulation and particle tracking (*blue streamlines*) in the DFF spiral device at *Re* 50 (*DC 1*) indicating complete recirculation of the fluid elements at the outer wall region (*inlet*) and back to the outer wall (*outlet*) again. (*right*) Average composite fluorescence images and plot indicating equilibrium position of 6 µm and 15 µm beads at different DC. (Reproduced with permission from Ref. [70])

and a final hematocrit of 20% was chosen as it resulted in negligible RBCs contamination in the inner CTCs outlet (Fig. 5.10a). Compared to other inertial-based microfluidic separation methods, 20% hematocrit implies ~2× dilution of whole blood (original hematocrit ~40–45%). This translates to an unprecedented processing time of ~20 minutes for 1 mL of whole blood (at 100 µL/min). Interestingly, inertial focusing positions of cancer cells (MCF-7) remained similar in saline solution and 20% hematocrit blood samples, indicating that the lateral migration of RBCs did not affect their inertial focusing (Fig. 5.10b) [70]. The same group later

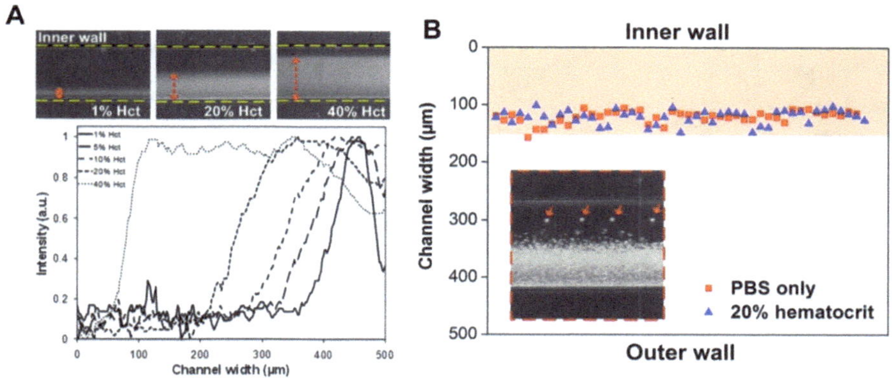

Fig. 5.10 High blood sample concentration processing in DFF (**a**) Averaged composite images and intensity plot illustrate broadening of RBCs occupied regions *(red dashed line)* for increasing hematocrit prior outlet bifurcation. *Yellow dotted lines* indicate position of channel walls, (**b**) Plot and high speed image captured at the channel outlet *(red dotted box)* indicate similar focusing positions of MCF-7 cancer cells suspended in PBS solution and 20% hematocrit blood samples at DC 1. *Shaded area (150 µm wide)* corresponds to the dimension of CTCs outlet. (Reproduced with permission from Ref. [70])

applied the DFF technology for label-free isolation of rare bacteria from whole blood by collecting the Dean-induced migrated bacteria at the channel outer wall [73].

Compared to other spiral technologies, DFF is clearly more versatile as it can achieve both size-dependent differential inertial focusing and multiplexed sorting of larger particles, as well as isolation of smaller micro or nanometer-sized elements through well-controlled Dean migration (a feat not possible with rectangular or trapezoidal spiral microchannels). Figure 5.11 illustrates the Dean migration profiles of 50 nm particles along the channel in DFF devices when introduced at either the inner or outer inlet. In both cases, the particles are solely affected by the Dean drag and migrated in the directions of the Dean vortices. By carefully tuning the sample to sheath flow rate ratio, DFF could thus serve as an efficient buffer exchange system to deplete smaller biological components including biomolecules (aptamers) [74] and nanoparticles [75] from target cells with high efficacy.

5.3.4 High-Resolution Dean Flow Fractionation (HiDFF)

In biomedical research, there exists a critical need to develop novel separation tools for sub-micron components in particle-based drug delivery system and purification of circulating microvesicles. While DFF enables well-controlled, Dean-induced migration of small microparticles (biomolecules, bacteria *etc.*) from inertially-focused larger target cells, a major drawback remains in its inability to further size-fractionate smaller microparticles (*e.g.* 1 µm vs. 2 µm) as they recirculate

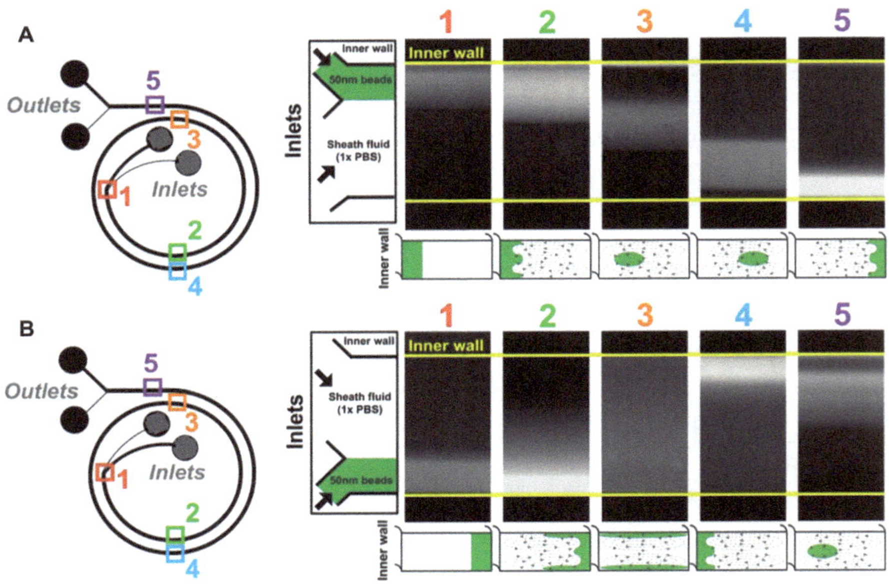

Fig. 5.11 Characterization of Dean-induced lateral migration of nanoparticles in DFF spiral devices. 50 nm fluorescent bead sample is introduced at the **a** inner wall or **b** outer wall at the inlet region. Average fluorescent stacked images indicate 50 nm bead positions along the channel. *Yellow lines* indicate positions of channel wall. Corresponding schematic images of channel cross section illustrate differences in bead migration pattern depending on initial position of beads. (Reproduced from Ref. [76] under Creative Commons)

continuously due to Dean vortices. Inertial focusing of small particles is highly challenging due to the requirement for smaller channel geometries (a/h > 0.07; $D_h \leq \sim 10$ μm for 1 μm bead focusing), and the large channel resistance would cause flow-induced channel deformation [77]. A spiral sorter was developed to inertially focus 2.1 and 3.2 μm microparticles, but the operating flow rate was low (10 μL/min) due to large pressure drop [78].

To address these limitations, the Hou group recently developed a novel 2-inlet, 2-outlet spiral sorting technology termed as High-resolution DFF (HiDFF) [76]. Contrary to current inertial microfluidics technologies, this was based on a novel phenomenon in spiral inertial microfluidics where particle transient innermost distance (D_{inner}) varied with size during Dean vortices-induced migration (Fig. 5.12a). Briefly, small microparticles ($a/D_h < 0.07$) introduced at channel outer wall experience Dean drag forces (F_D) due to Dean vortices and migrate laterally towards inner wall. As they migrate along the channel top or bottom, particles near the surface experience size-dependent wall-induced inertial lift forces ($F_{WL} \propto a^6$) that push particles away from the surface. Hence, particles flow at different fluid streamlines which leads to a differential transient innermost position (D_{inner}) at the inner wall before they recirculate back towards the outer wall (Fig. 5.12b). Of note, since

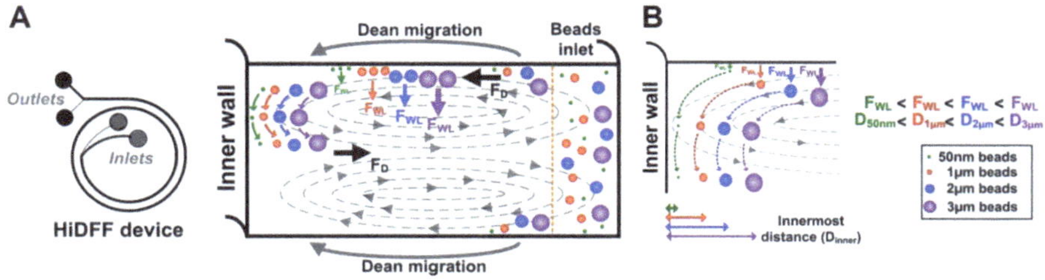

Fig. 5.12 High-resolution Dean Flow Fractionation (HiDFF) (a) Schematic illustration of HiDFF separation principle in a 2-inlet, 2-oulet spiral (300 μm (*w*) × 60 μm (*h*)) device. Particles introduced at the outer wall migrate laterally towards inner wall under the influence of Dean vortices. As the particles migrate along the channel top and bottom, they experience size-dependent wall-induced lift forces (F_{WL}) that push larger particles away from the surfaces, (b) Subtle differences in particle z-position (along height) lead to size-based transient innermost distance (D_{inner}) at the inner wall which can be exploited for small particle separation. (Reproduced from Ref. [76] under Creative Commons)

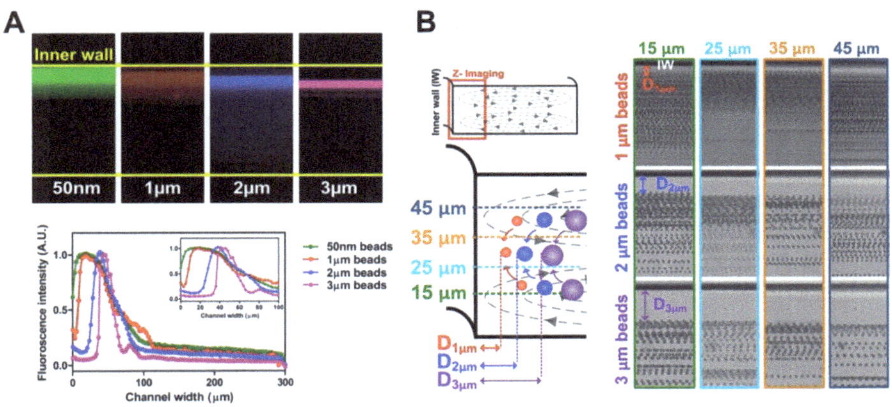

Fig. 5.13 Superior separation resolution of HiDFF (a) Fluorescence composite images and intensity linescans indicating distinct innermost distance (D_{inner}) for particles of different sizes (b) High speed, Z-imaging (40 × magnification) of inner wall region at different planes indicates similar D_{inner} along channel height. (Reproduced with permission from Ref. [76])

inertial focusing of particles is not necessary, channel dimensions are increased to minimize clogging while improving throughput significantly (~100 μL/min).

To characterize the dependence of D_{inner} with particle size, the authors tested beads of smaller diameters (50 nm, 1 μm, 2 μm and 3 μm) so that they would not undergo inertial focusing in the device (a/D_h <0.07). As shown on Fig. 5.13a, 50 nm beads migrated completely towards the inner wall while larger beads exhibited increasing D_{inner} at *Re* ~ 30–50. For all for bead sizes, there were negligible differences in their Dean migration towards outer wall at *Re* ~ 60–70. High speed imaging at the inner wall region was also performed to further understand the distinct differences in D_{inner} by visualizing the particle flow position at different planes along the channel height. Composite brightfield images clearly indicated increasing D_{inner} with particle size (~5.2 μm for 1 μm beads; ~17.4 μm for 2 μm beads; ~27.8 μm for 3 μm beads), a trend that was similar at different channel heights (Fig. 5.13b).

To determine if the subtle differences in D_{inner} during Dean migration can be exploited for separating particles with closely-spaced sizes, the authors characterized binary bead mixtures (2 μm and 3 μm beads; 1 μm and 2 μm beads) separation performance using HiDFF at different sample to sheath flow ratios. Smaller particles (with smaller D_{inner}) were positioned closer to inner wall and separated into the inner outlet (outlet 1), while larger particles were sorted into outer outlet (outlet 2). Separation efficiency improved significantly from 20% to 60% for 2 μm bead isolation (2 μm and 3 μm bead mixture), and from 5% to 40% for 1 μm bead isolation (1 μm and 2 μm bead mixture) at higher sheath flow, which translated to an enrichment of the smaller particles by ~1000 and ~100-fold, respectively (Fig. 5.14a, b). This was likely due to the smaller variation in particle initial y-position (along channel width) which enabled them to migrate laterally as a tight band towards the inner wall. As proof-of-concept for particle-based drug delivery applications, polydisperse poly(lactic-co-glycolic acid) (PLGA) microparticles were fabricated and fractionated into 3 different size groups (large, medium and small) in a 2-step HiDFF separation. Particle size and morphology were characterized using scanning electron microscope (SEM), which confirmed the distinct size differences in each group (unsorted, 3.3 ± 0.08 μm; large, 6.8 ± 0.13 μm; medium, 1.7 ± 0.03 μm; small, 0.89 ± 0.03 μm) (Fig. 5.14c).

5.3.5 Summary

So far, we have described four major types of spiral devices used for high throughput size-based particle separation. Each of these technologies has unique separation features in terms of particle size range, throughput and sample working concentration. These factors should be carefully considered when designing spiral devices for specific cells/particles sorting applications. Table 5.3 below provides a summary and comparison of key features in different spiral types.

5.4 Cell Applications

5.4.1 Cancer Cells

Cancer is the leading cause of death globally, and cancer metastasis (spreading from the primary tumor to secondary sites) is responsible for ~90% of cancer-related

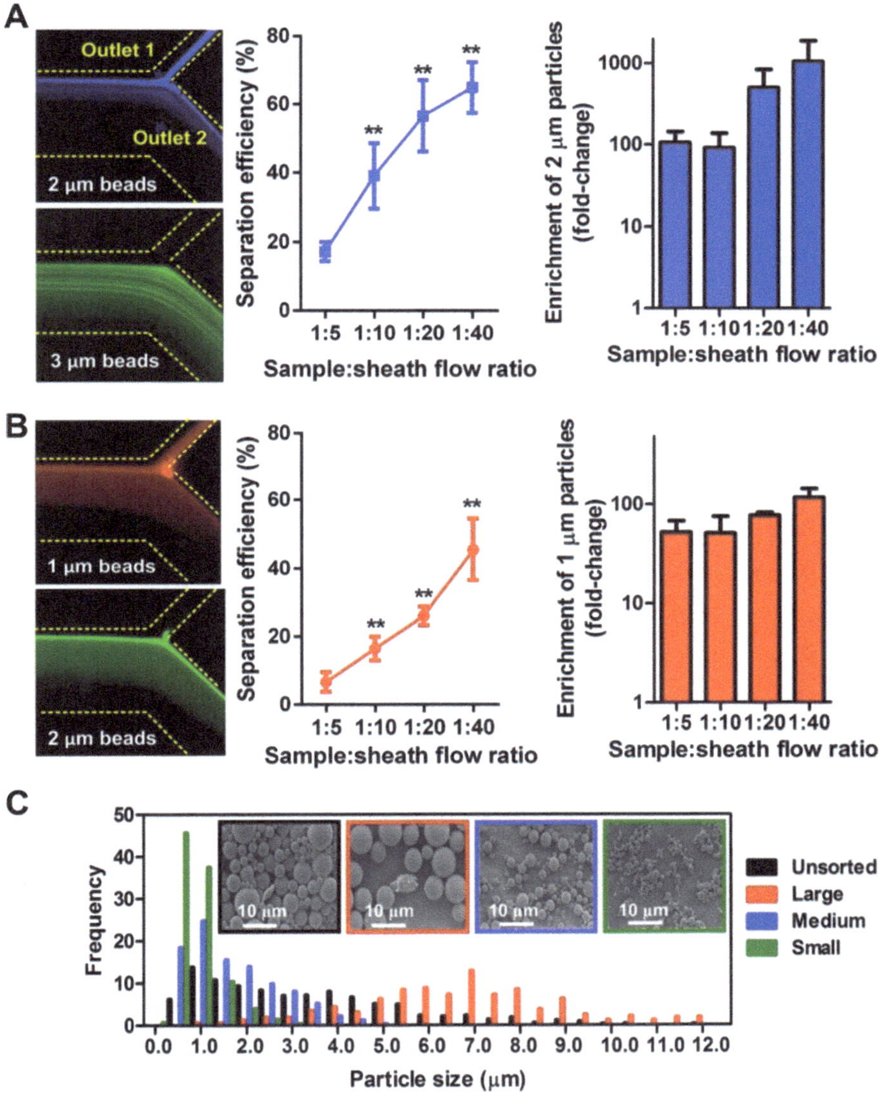

Fig. 5.14 Tunable size fractionation of microparticles using HiDFF. Average fluorescent composite images and separation performance of smaller particles at different sample to sheath flow ratio for (**a**) 2 and 3 μm, and (**b**) 1 and 2 μm binary bead mixture. *Yellow dotted lines* indicate positions of channel outlet bifurcation, (**c**) Size distribution plot of drug-loaded poly(lactic-co-glycolic acid) (PLGA) microparticles after a 2-step HiDFF separation into 3 sizes (large (>5 μm), medium (2–5 μm) and small (< 2 μm)). Inset SEM images highlight distinct size differences between sample (inlet) and different size groups. (Reproduced with permission from Ref. [76])

Table 5.3 Comparison of different spiral devices

Type	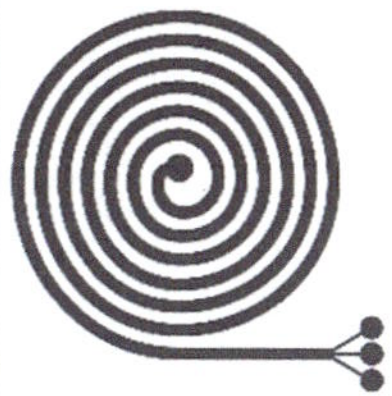 **Rectangular spiral microfluidics**	 **Trapezoid spiral microfluidics**	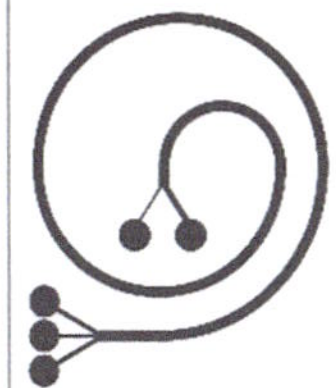 **Dean flow fractionation (DFF)**	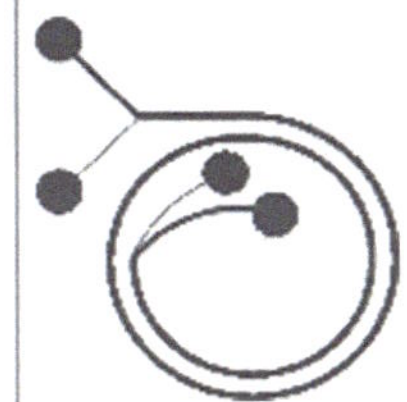 **High resolution DFF (HiDFF)**
Principle	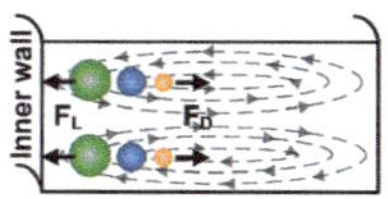	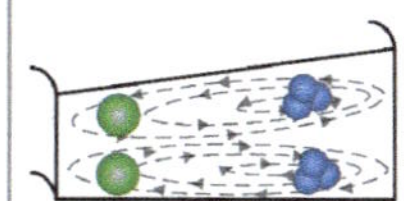	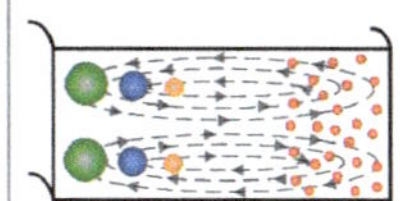	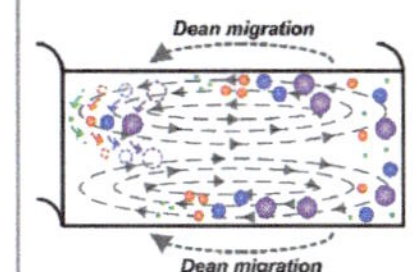
Size	 ~5–50 μm	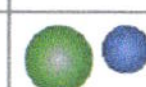 ~5–30 μm	 ~1–30 μm	 ~1–3 μm
Mode	Multiplexed sorting	Binary sorting	Multiplexed sorting	Binary sorting
Flow rate	~0.5–3 mL/min	~0.5–6 mL/min	~0.1–0.2 mL/min (sample flow)	~0.5–0.2 mL/min (sample flow)
[Cell]	~10^{5-6}/mL (~0.1–0.5% hematocrit)	~10^{6-7}/mL (~0.5–2% hematocrit)	~10^{6-8}/mL (~10–20% hematocrit)	~10^{6-8}/mL (~10–20% hematocrit)
Applications	Cell separation Particle concentrator Cell alignment and ordering	Rare cell isolation Microfiltration Plasma separation	Rare cell isolation Bacterial sorting Protein purification Buffer exchange	Microvesicle isolation Nanoparticles separation Bacterial sorting

fatalities [79]. During metastasis, cancer cells are shed from primary tumors into the peripheral blood and are known as circulating tumor cells (CTCs). Recent clinical studies have shown that CTCs frequency and their genetic information can be used as surrogate biomarkers to provide critical information for cancer diagnostic and monitoring [80, 81]. However, the technical challenge for CTCs isolation lies in the rarity of these cells (~1 to 10 CTCs/mL) in peripheral blood (~5 billion RBCs/mL) [82–85]. CELLSEARCH® is the only FDA approved CTC separation technology that uses antibodies to bind to epithelial cell adhesion molecule (EpCAM) on CTCs surfaces, but EpCAM expression is highly heterogeneous which can lead to considerable capture loss.

Microfluidic CTCs separation was first reported by Nagrath et al., where they captured CTCs from patient blood using anti-EpCAM antibody functionalized microposts (CTC-chip) [86]. Following this, the same group developed a herringbone device to increase the collision and capture frequency between CTCs and antibody-coated surfaces [87]. However, the same issue related to cell surface marker heterogeneity persists in microfluidic affinity-based cell sorting. Size-based separation is hence preferred as it can significantly reduce cell loss and preserve cell viability with its label-free sorting process. In most cancer types, the size of CTCs (~10–20 μm) is larger than blood cells (WBCs ~ 8–12 μm; RBCs ~ 8 μm; platelets ~ 2–3 μm), and this physical difference can be exploited in spiral inertial microfluidics for high throughout cell separation.

To date, many spiral microfluidic devices have been developed for cancer cells separation (Table 5.4) [23, 51, 60, 66, 70, 88–96]. Using Dean-coupled inertial migration, Kuntaegowdanahalli et al. first reported the use of spiral microfluidics for size-dependent cancer cell sorting (Fig. 5.15a) [23]. As proof-of-concept, they separated a mixture of neuroblastoma (~15 μm) and glioma cells (~8 μm), and achieved >80% separation efficiency at a high throughput of ~one million cells/min. These results were comparable to the performances obtained using commercial flow cytometry. To further improve the throughput, Warkiani et al. developed a multiplexed spiral device (three devices stacked together) to isolate spiked cancer cell lines from lysed blood samples. This high-throughput system can process 7.5 mL of lysed blood sample in 12.5 min and downstream fluorescence *in situ* hybridization (FISH) analysis was successfully performed on the eluted CTCs off-chip [94].

Recently, Guan et al. reported a slanted spiral device with trapezoid cross-section (80 and 130 μm in the inner and outer channel height, respectively) for cancer cell separation [64–66]. Due to the asymmetry of the channel cross-section, strong Dean vortices were generated at the outer half (deeper side) of the channel, which shifted smaller particles closer to the outer wall without affecting focusing of larger particles at the inner wall (Fig. 5.15b). They successfully isolated three cancer cell lines, MCF-7 (20–24 μm), T24 (16–17 μm) and MDA-MB-231 (10–15 μm) from whole blood with high recovery rate (>80%) and purity (400–600 WBCs/mL; ~4 log depletion of WBCs) [66]. In another study, Aya-Bonilla et al. designed a slanted spiral microfluidic device to isolate melanoma CTCs from lysed blood sample and obtained 80% recovery rate after one round of enrichment [98]. Kulasinghe et al.

Table 5.4 Applications of spiral inertial microfluidics for cancer cell separation

No.	Sample	Spiral type	Separation performance	References
1	SH-SY5Y and neuroblastoma cells	Rectangular spiral, single inlet	Throughput: ~2 mL/min	[97]
2	Neuroblastoma and glioma cells	Rectangular spiral, single inlet	Throughput: ~10^6 cells/min 90% recovery	[23]
3	MCF-7, MDA-MB-231, HeLa in 3 × diluted whole blood (15% hct)	DFF	Throughput: 100 μL/min >85% recovery	[70]
4	MCF-7, T24 and MDA-MB-231	Trapezoidal spiral	Throughput: 1.7 mL/min ≥80% recovery	[66]
5	1205Lu, A2058, SKMEL5, UACC62; Melanoma clinical samples	Trapezoidal spiral	Throughput: 1.7 mL/min >55% recovery	[98]
6	CAL27, RPMI2650, UD-SCC9, MDA-MB-486; HNC clinical samples	Trapezoidal spiral	Throughput: 1.7 mL/min 60–76% recovery	[95]
7	MCF-7 in 100 × diluted blood	Rectangular spiral, single inlet	Throughput: 400 μL/min 75.40% recovery	[93]
8	MCF-7 and HeLa	Rectangular double spiral, single inlet	Throughput: 3.33×10^7cells/min, 88.5% recovery	[60]
9	HeLa in 20 × diluted blood	Rectangular double spiral, single inlet	Throughput: 2.5×10^8 cells/min ~80% recovery	[88]
10	A549 (lung adenocarcinoma)	Rectangular double spiral, single inlet	Throughput: 25 mL/h. 74.4% recovery	[92]
11	DU-145 (prostrate)	Rectangular spiral, single inlet	Throughput: ~1 mL/min 67% recovery	[96]
12	Breast and lung cancer clinical sample (lysed blood)	Multiplexed DFF	Throughput: >1.5 mL/min	[90]
13	Breast and lung cancer clinical sample (lysed blood)	Multiplexed DFF	Throughput: >0.75 mL/min 20–135 CTCs/mL recovery	[91]
14	MCF-7 in leukocytes suspension	Rectangular spiral, single inlet	Throughput: 550 μL/min >86.8% recovery	[51]
15	MCF-7 (breast)	Rectangular spiral, single inlet	Throughput: >1 mL/min ~100% recovery	[89]

also performed CTCs isolation using trapezoidal spiral microfluidics in head and neck cancer (HNC) patients, and reported the presence of CTC clusters (large group with more than five CTCs) in a subset of these cancer patients [95].

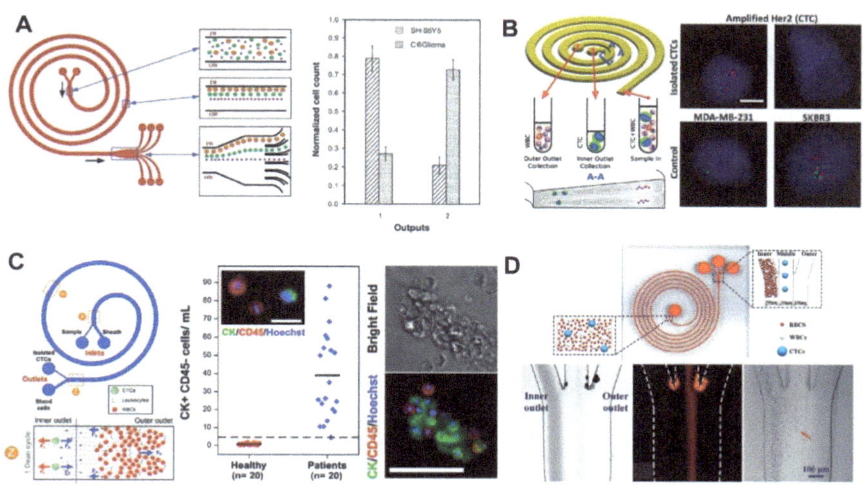

Fig. 5.15 Isolation of circulating tumour cells (CTCs) from whole blood using spiral inertial microfluidics (**a**) Distinct focusing and separation of microparticles *(~15–20 μm, similar size range to cancer cells)* into different outlets due to size-dependent inertial lift (F_L) and Dean drag (F_D) forces. Experimental results indicating efficient separation of SY5Y neuroblastoma cells from smaller C6 glioma cells. (Reproduced from Ref. [23] – published by The Royal Society of Chemistry) (**b**) Schematic illustration of CTCs isolation from blood using trapezoid spiral microchannels FISH analysis of HER2 expression in recovered CTCs after microfluidics isolation. (Reproduced from Ref. [66] – published by The Royal Society of Chemistry, under Creative Commons) **c** Schematic illustration of Dean Flow Fractionation (DFF) for CTCs isolation. Clinical validation of DFF in lung cancer patients. Fluorescence images and enumeration of isolated CTCs. Isolation of CTCs clusters using DFF. (Reproduced with permission from Ref. [70]) (**d**) Schematic illustration of a 1-inlet, 3-outlet spiral microfluidic device for cancer cell sorting. Due to the dominant F_D with smaller radius of curvature, cancer cells (MCF-7) focused inertially at the channel centre, while smaller RBCs equilibrated near the channel inner wall. (Reproduced from Ref. [93] – published by The Royal Society of Chemistry, under Creative Commons)

To increase blood processing throughput, Hou et al. developed a novel 2-inlet, 2-outlet spiral biochip for CTCs isolation based on DFF. Unlike other inertial focusing devices, the DFF device inertially focused the larger CTCs while the smaller hematologic cells (RBCs and leukocytes) were solely affected by Dean drag forces (Fig. 5.15c) [70]. This enabled accommodation of high RBCs content in the channel, which translated to a significant enhancement of RBCs processing (~20% sample hematocrit) and efficient cancer cell recovery of >85%.

Huang et al. also reported a simple 1-inlet, 3-outlet, 5-loop spiral cell sorter for CTCs isolation (Fig. 5.15d) [93]. Due to the smaller radius of curvature and dominant F_D, the larger cancer cells were recovered from the middle outlet, and the smaller blood cells were sorted into the inner outlet. To test the efficacy of the device, breast cancer cells (MCF-7) were spiked into diluted whole blood sample (1:100) and the device was able to remove ~99% of the hematologic cells after 2 rounds of separation at a throughput of 400 μL/min. The authors also reported that the processing capability can be further increased by using lysed blood samples.

Table 5.5 Applications of spiral inertial microfluidics for stem cell separation

No.	Samples	Spiral type	Separation performance	References
1	Human mesenchymal stem cells (hMSCs)	Rectangular spiral, single inlet	Throughput: 3 mL/min (~15 × 10^6 cells/h)	[104]
2	Neural stem cells (NSCs)	Rectangular spiral, single inlet	Throughput: 1 mL/min >80% recovery	[106]
3	Neural stem cells (NSCs)	DFF	Throughput: 3 mL/min ~93% recovery	[107]

5.4.2 *Stem Cells*

Stem cells are pluripotent cells that can be induced into other cell types using physical and biochemical cues. The capabilities of self-renewal and differentiation into other specialized cells have made stem cells highly important in regenerative medicine. Unsurprisingly, an unmet need for stem cell sorting is to identify novel biomarkers to isolate subpopulations that are more pluripotent ("more stemness"). Most of the existing techniques are based on cell surface expression and labeling, but this strategy is rather challenging due to the lack of well-established surface markers and cell heterogeneity.

Recently, the biophysical properties of stem cells such as cell size, stiffness and electrical properties have emerged as potential biomarkers for stem cell sorting [99–101] (Table 5.5). One of the key physical parameters is cell size, whereby the inherent size differences can be exploited for cell cycle synchronization [102]. This physical-based method has direct advantages over common chemical-based synchronization which can affect and possibly disrupt cell physiology and metabolism [103]. Lee et al. first proposed a spiral cell sorter for high-throughput cell cycle synchronization based on cell size differences (Fig. 5.16a) [104]. To satisfy the inertial focusing criteria ($a/h > 0.07$), the channel height was set between 130 to 150 μm depending on cell types. As proof-of-concept, they demonstrated the fractionation of human bone marrow-derived mesenchymal stem cells (hMSCs) into enriched subpopulations of G0/G1, S and G2/M phases. From outlet 1, 70.4% of the cell population collected were in S and G2/M phases (~24 μm) while 86.2% of the cells in outlet 4 were from G0/G1 phase (~15 μm). The device throughput (~15 × 10^6 cells/h) and cell viability (~95%) were much higher than those obtained by conventional methods. In a follow-up study, the group successfully identified a set of unique biophysical markers (small cell diameter, low cell stiffness and high nuclear membrane fluctuations) for the isolation of multipotent stem cells (Fig. 5.16a) [105].

Study of neural stem cells (NSCs) is pivotal for understanding disease progression and developing novel therapeutics for neurological diseases including Alzheimer's and Parkinson's diseases [108]. Neurospheres, which are clusters of hundreds to thousands of NSCs, are commonly used for *in vitro* study of neural precursor cells [109, 110]. To induce stem cell differentiation or conduct clonal analysis in this culture system, the original neurospheres are chemically and

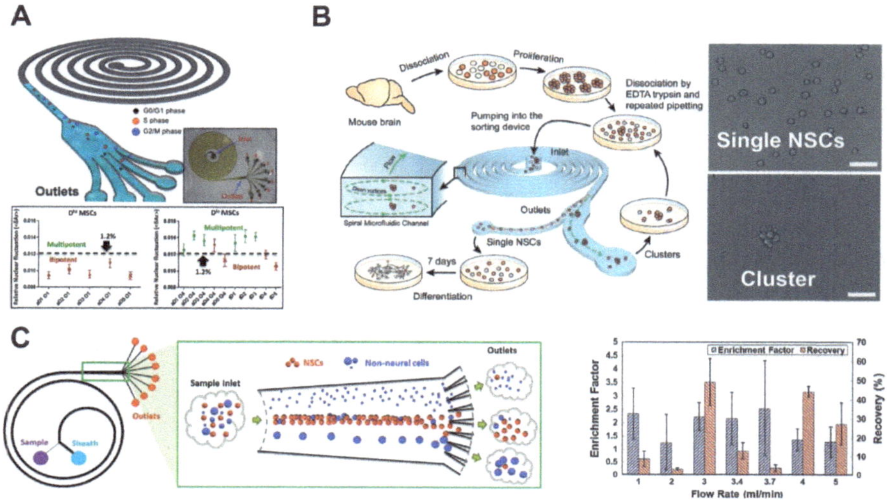

Fig. 5.16 Stem cell fractionation using spiral inertial microfluidics (**a**) Design of the 9-loop spiral microchannel for cell cycle synchronization (Adapted from Ref. [104] with permission from The Royal Society of Chemistry). Size-based sorting of mesenchymal stromal cells (MSCs) into different subpopulations for characterization of nuclear fluctuations *(NF).* (Reproduced with permission from Ref. [105]) (**b**) Overview of neural stem cells (NSCs) separation process from mice brain using spiral microfluidic device. Representative images of single NSCs and NSC clusters after separation at 1 mL/min *(scale bar: 50 μm).* (Reproduced from Ref. [106] with permission from The Royal Society of Chemistry) **c** Schematic illustration of iPSC-derived NSCs enrichment using a 2-loop spiral cell sorter. NSCs were collected in the middle outlets, while non-NSCs *(wide size range)* were collected in all outlets. Enrichment factor and recovery of NSCs at different flow rates. (Reproduced with permission from Ref. [107])

mechanically dissociated to produce single-cell suspension and plated under stringent conditions for growth. However, the dissociated single cells are often contaminated by a small population of stem cell clusters, which can affect subsequent cell identification and clonal analysis. An effective and rapid separation method is therefore highly desirable to separate these single cells from cell clusters. As neural stem cells (~8–14 μm) are smaller than cell clusters (~40–60 μm), Nathamgari et al. developed a 1-inlet, 2-outlet spiral microdevice for size-based isolation of single cells from chemically dissociated neurospheres (Fig. 5.16b) [106]. They showed that at low flow rates (e.g. 1 mL/min), large particles (e.g. 38 μm beads or cell clusters) were focused at the center of the channel while small particles (e.g. 7.7 μm beads or single cells) were focused near the inner wall. When the flow rate was increased to 3 mL/min, the beads focusing behavior were reversed. They eventually used a flow rate of 1 mL/min and reported that ~84% of single cells were isolated into outlet 1 and 2 (innermost outlets), while cell clusters equilibrated in the channel centre (sorted into outlet 3–5). In addition, as neural stem cells are sensitive to shear stress, a low working flow rate can help preserve the multipotency of the stem cells with high cell viability rate (>90%).

With the emergence of induced pluripotent stem cells (iPSC) technology, label-free cell purification methods are highly important for iPSC-derived cells enrichment. Song et al. designed a 2-inlet, 8-outlet DFF spiral device to enrich iPSC-derived NSCs from a heterogeneous cell mixture based on cell size differences [107]. They mixed NSCs (10–12 μm) and heterogeneously-sized non-neural cells (6–19 μm) in a ratio of 1:1 (final concentration of 2×10^6 cells/mL), and processed the sample at a flow rate of 3 mL/min. As expected, the NSCs focused into a tight band and were sorted into the middle outlets (outlet 4 and 5), while the non-neural cells were remained randomly distributed across the channel width and the larger cells were sorted into the inner wall outlets (outlet 7 and 8). They reported ~2.1 fold NSCs enrichment with a 93% recovery rate (Fig. 5.16c). A major limitation lies in the low purity, as the device is unable to deplete non-neural cells of similar sizes as NSCs.

5.4.3 Immune Cells

Neutrophils are the most abundant leukocytes in human blood and the key effector cells of the innate immunity. They are also implicated in major diseases including type 2 diabetes mellitus (T2DM) [111], cancer [112] and cardiovascular diseases [113]. In TD2M, numerous neutrophil dysfunctions such as cell stiffening [114, 115], impaired chemotaxis [116, 117] and phagocytosis [118] can lead to an increased susceptibility to bacterial infections. Traditionally, neutrophils are isolated using laborious methods such as density gradient centrifugation and RBC lysis, which not only require a large blood sample volume (>10 mL), but are also prone to induce neutrophil activation if not properly done. Commercial neutrophil isolation kits that utilize immunomagnetic labeling have been developed to negatively select untouched neutrophils (MACS xpress® (Miltenyi Biotec) and Easy SepTM (STEMCELL Technologies)), but these kits are expensive and not practical for large volume processing. Developing an efficient and cost-effective neutrophil sorting strategy is therefore necessary for accurate phenotyping in neutrophil studies and point-of-care testing. Table 5.6 shows a summary of spiral microdevices developed for this purpose.

Table 5.6 Applications of spiral inertial microfluidics for immune cell separation

No.	Samples	Spiral type	Separation performance	References
1	Monocytes in 10 × diluted blood	Rectangular spiral, single inlet	Throughput: 1.1 mL/min Recovery: not reported	[96]
2	Neutrophils in lysed blood	DFF	Throughput: 130 μL/min >90% purity	[119]
3	Leukocytes in 200 × diluted blood	Trapezoidal spiral	Throughput: 0.8 mL/min >80% recovery	[65]
4	Leukocytes in 500 × diluted blood	Rectangular spiral, single inlet	Throughput: 1.8 mL/min ~95% recovery	[120]

Capitalizing on the high separation resolution of DFF, Hou et al. developed a 4-outlet DFF device to purify neutrophils from whole blood without antibodies labelling. The separation is based on subtle cell size differences among leukocyte subtypes (neutrophils/monocytes (10–12 μm); lymphocytes (7–8 μm)) (Fig. 5.17a) [119]. This device only required small amount of blood (finger prick; ~100 μL) for neutrophil isolation, and sorted neutrophils also undergo simultaneous washing as they were eluted in fresh saline solution due to the buffer exchange capabilities of DFF. The group further characterized the rolling behavior of sorted neutrophils on E-selectin using microfluidics. In their clinical validation using healthy subjects and patients with T2DM, this developed microfluidic-based neutrophil sorting and phenotyping strategy revealed a significant difference in neutrophil rolling pattern between both groups, clearly suggesting neutrophil rolling speed as a potential functional biomarker for inflammatory profiling in T2DM patients.

In general, conventional blood cell separation methods (centrifugation, FACS, MACS) are laborious and highly dependent on user operation, and the phenotype of isolated leukocytes could be altered if not done carefully. To address these issues, Wu et al. developed a spiral microfluidic device with a trapezoid cross-section to isolate the larger leukocytes (~8–12 μm) from diluted whole blood (RBCs ~6–8 μm) [65]. The schematic in Fig. 5.17b illustrates the device design and working principle. At optimized working conditions, the device can separate polymorphonuclear leukocytes (PMNs) and mononuclear leukocytes (MNLs) from diluted human blood (1–2% hematocrit) with high efficiency (>80%). In addition, the activation in the device-sorted PMNs was negligible as compared to lysis method. Nivedita et al. also developed an Archimedean spiral device with <8 cm focusing length for leukocytes separation in diluted blood sample (1:500) at a flow rate of 1.8 mL/min, and achieved a high separation efficiency (~95%) and throughput (up to 1×10^6 cells/min) (Fig. 5.17c) [120].

5.4.4 Sperm Cells

Assisted reproductive techniques (ART) have benefited countless couples experiencing infertility, and one major aspect of ART is to select healthy spermatozoa *in vitro* fertilization (IVF). Traditionally, sperm cells are prepared using serial centrifugation or the swim-up methods, but repeated handling and centrifugation can cause damage to the sperm cells' DNA, or lead to the production of reactive oxygen species (ROS) [121, 122]. Conventional sperm cell sorting methods using microfluidics relies on sperm motility, but the approach is unable to identify viable but non-motile sperms for intracytoplasmic sperm injection, which is relevant for patients suffering from severe or complete asthenozoospermia [122]. Taking advantage of the size difference between sperm and blood cell, Son et al. developed a spiral microfluidics sperm cell sorter for non-motile sperms separation. (Fig. 5.18a) [123]. Sperm cells (1–2 million/mL) were mixed with RBCs (7–9 million/mL) and introduced into the spiral channel. At a flow rate of 0.52 mL/min, 81.2% of the sperm cells were sorted into the

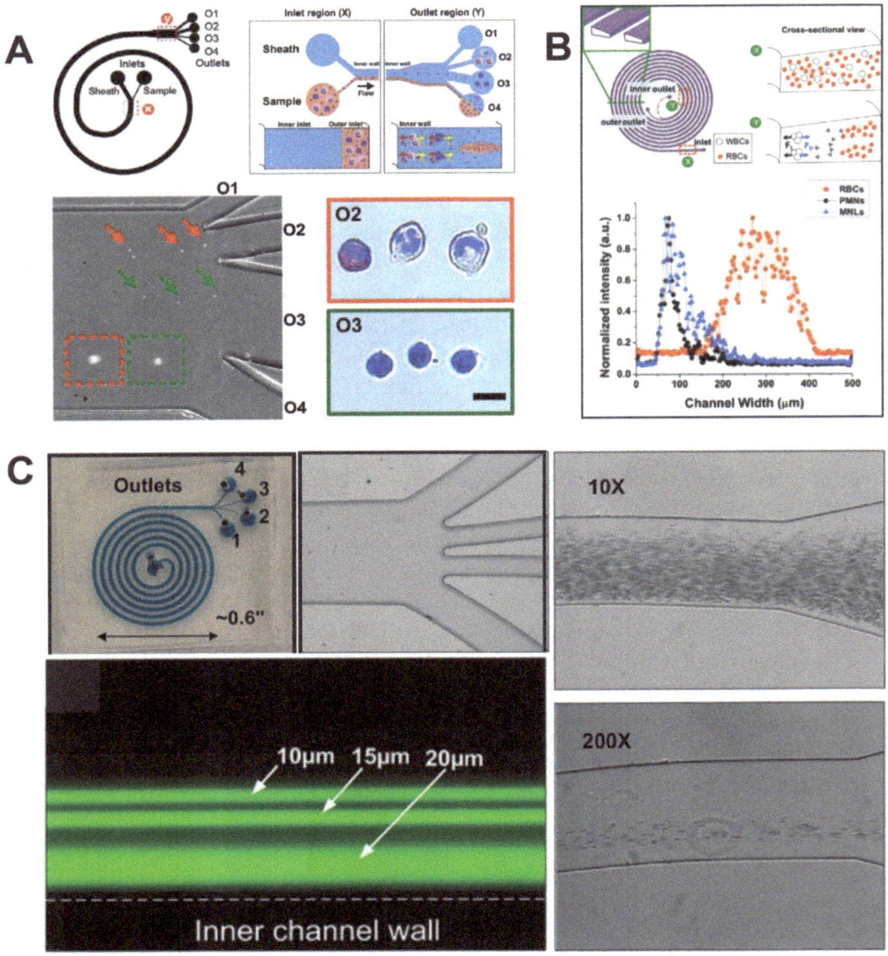

Fig. 5.17 Immune cells isolation using spiral inertial microfluidics (**a**) Rapid size-based neutrophil sorting and washing using DFF. High speed images indicating distinct focusing of larger neutrophils and smaller lymphocytes into different outlets. (Adapted from Ref. [119] under Creative Commons), (**b**) Immune cell isolation using trapezoid spiral device. Intensity line scans indicating distribution of polymorphonuclear leukocytes (PML), mononuclear leukocytes (MNL), and RBCs across channel width (Reprinted (adapted) with permission from Ref. [65]. Copyright (2014) American Chemical Society.) (**c**) Images of the 1-inlet, 4-outlet Archimedean spiral device (<1 in.^{2}). Focused streams of three particle populations (10 μm, 15 μm and 20 μm in diameter) at a flow rate of 2.2 mL/min. Images of focused RBCs at the outmost loop of spiral channel with 10 and 200-fold diluted whole blood. (Reproduced with permission from Ref. [120])

outer two outlets, while 99% RBCs were separated into the inner two outlets. As shown in Fig. 5.18b, the smaller sperm cells formed a broad focusing band near the channel centre, which can be attributed to their asymmetrical and irregular shape

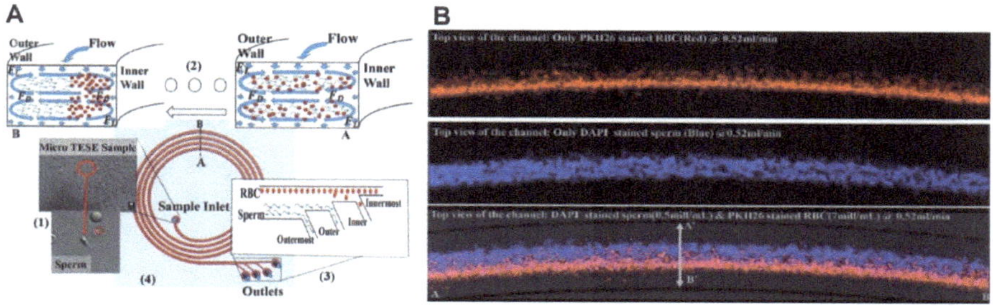

Fig. 5.18 Sperm cell isolation using spiral inertial microfluidics (**a**) Separation principle of sperm cells from RBCs using spiral microchannels (**b**) Fluorescent images of the focusing positions of sperm cells *(blue)* and RBCs *(red)* at a flow rate of 0.52 mL/min. (Adapted with permission from Ref. [123])

Table 5.7 Applications of spiral inertial microfluidics for microbes and biomolecules separation

No.	Samples/application	Spiral type	Separation performance	References
1	Non-motile sperm cell	Rectangular spiral, single inlet	Throughput: 0.52 mL/min 81% recovery	[123]
2	*E.coli, K. pneumoniae, P. aeruginosa, S. aureus, E. faecalis*	DFF	Throughput: 1.7 mL/min ~75% recovery	[73]
3	Algal cells	Rectangular spiral, single inlet	Throughput: 3.2 mL/min 77% recovery	[125]
4	*Phytophthora ramorum* sporangia	Rectangular spiral, single inlet	Throughput: 2 mL /min 95% recovery	[126]
5	Antibodies in serum	DFF	Throughput: 130 μL /min >80% recovery	[130]
6	Aptamer	DFF	Throughput: 160 μL /min (~2×10^6 cells/min) $\geq 10^6$ partitioning efficiency	[74]

(length of 4.79 ± 0.26 μm and width of 2.82 ± 0.23 μm). In contrast, larger RBCs (~7.5–8.7 μm) inertially focused into a tight stream at the inner wall, thus achieving separation (Table 5.6).

5.4.5 Microbes

Microorganism separation using spiral microfluidics (Table 5.7) is an area of considerable interest for bacterial diagnostics and environmental monitoring [124]. A major difference between microorganism and cell isolation is that microbes are smaller (~1–3 μm) as compared to mammalian cells (~10–20 μm), and thus will

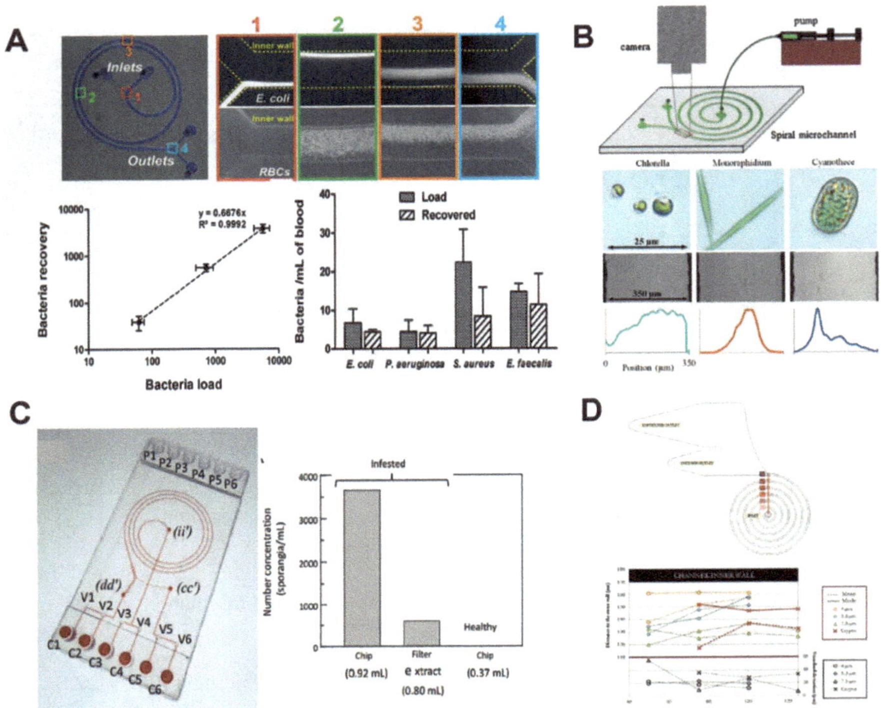

Fig. 5.19 Microbe isolation using spiral inertial microfluidics (**a**) Bacterial isolation using DFF. Plots indicating high bacterial recovery rate (>65%) at different bacterial loads and for low abundance bacteria (~10–50 CFU/mL). (Reproduced from Ref. [73] – published by The Royal Society of Chemistry.) (**b**) Experimental setup for bacteria separation using spiral microchannel. Images of different algae species (*Chlorella*, *Monoraphidium* and *Cyanothece*) and their focusing position across the channel width at 1.6 mL/min. (Reproduced from Ref. [125] under Creative Commons) (**c**) Photograph of a multi-layered microdevice consisting of pneumatic microvalves, spiral microchannel and fluid control channels. Significant enrichment of *P. ramorum* sporangia from infested *Rhododendron* leaves using the developed spiral sorter (Adapted with permission from Ref. [126]) (**d**) Spiral microfluidic channel for pathogens separation. Distribution of *C. parvum* as a function of distance to the outer wall. (Reproduced from Ref. [129] under Creative Commons)

experience less inertial (or drag) forces. Hence, channel dimensions have to be scaled down significantly to achieve similar inertial focusing effects.

To overcome this issue, Hou et al. utilized the DFF technique to isolate low abundance bacteria from whole blood based on cell size difference (Fig. 5.19a) [73]. By using a sheath flow to "pinch" the bacteria-containing blood sample at the inlet, they demonstrated well-controlled Dean migration of bacteria towards the outer wall while larger blood cells remained inertially focused near the inner wall to achieve separation. This approach enables continuous, species-independent isolation of clinical bacteria isolates spiked at low concentrations (~10–50 CFU/mL) from whole blood without affinity-based target labelling.

Besides bacteria, algae are also widely studied as they are often used as biosensors for monitoring and detecting environmental changes. Traditional manual algae identification method includes microscopy-based manual identification which is time-consuming and limits the sampling resolution. To automate the identification process, Schaap et al. developed a spiral inertial microfluidic device for algal cells separation based on cell size and shape [125]. Three morphologically different species of algae were used in the experiment: (i) *Chlorella* (spherical shaped, diameter of 6.0 ± 1.0 μm); (ii) *Cyanothece* (prolate spheroid shape, 15.6 ± 2.3 μm in long axis and 11.1 ± 1.0 μm in short axis); and (iii) *Monoraphildium* (cylindrical shape, 54.6 ± 14 μm in length, diameter of 3.14 ± 0.6 μm). The authors found that the shape of the algae can affect inertial focusing behavior in spiral channels. Even though *Cyanothece* and *Monoraphidium* possess equivalent spherical diameter, they can be separated based on their geometrical difference and a separation efficiency of 77% was achieved at a flow rate of 3.2 mL/min. The prolate spherical *Cyanothece* behaved similar to a 10 μm particle, while *Monoraphidium* has an effective diameter of 3.14 μm in the plane perpendicular to the flow, and thus experienced less lift forces as compared to *Cyanothece*. Since *Chlorella* did not fulfil the inertial focusing criteria, they remained randomly distributed spread across the channel. The experimental setup, algae images and distribution across the channel cross section are shown in Fig. 5.19b.

Phytophthora ramorum is a fungal plant pathogen that infects a large number of plant species and results in extensive damage to ecosystem. Hence, it is imperative to detect and prevent the spread of this fungus. The *P. ramorum* generally presents an ovoid shape with diameter ranging from 20 to 40 μm. To achieve efficient inertial focusing effects, Clime et al. developed a spiral microfluidic platform with a channel depth of 200 μm and width of 600 μm. The device was integrated with peristaltic microvalves for fluid operation and process control. Using samples derived from infested plant leaves, they were able to obtain 6.1-fold concentration of the fungi (Fig. 5.19c) [126].

Another crucial environmental application is the detection of waterborne pathogens in drinking water. It is a challenging task as pathogens are usually present in low numbers. For example, the presence of *Cryptosporidium* oocysts was reported less than 10 per 10 liters in the recreational lakes in Amsterdam, The Netherlands [127]. The standard waterborne pathogen monitoring process involves complex procedures including filtration, immune-magnetic separation, fluorescence staining and microscopy-based examination. These methods require long processing time (several days), expensive equipment and highly trained expertise [128]. In a work by Jimenez et al., they presented a 6-loop spiral focusing microchannel with depth of 30 μm and width of 170 μm to separate waterborne pathogens (Fig. 5.19d) [129]. The device has two wide outlets to enhance particle positions discrimination and separation resolution. To demonstrate the feasibility of this device, they separated waterborne pathogen *Cryptosporidium parvum*, (~4–5 μm) at a high flow rate of 500 μL/min with a separation efficiency of 100%.

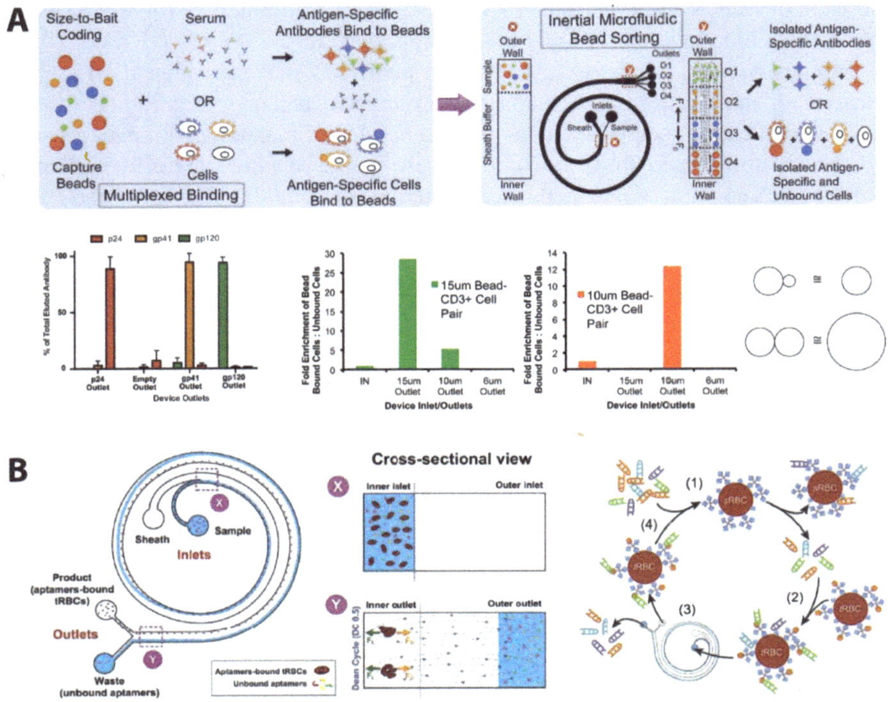

Fig. 5.20 Biomolecule separation using spiral inertial microfluidics (**a**) Workflow of multiplexed proteins or cells sorting using DFF coupled with affinity-based bead binding. Results indicating efficient separation of 3 major HIV antigen-specific antibodies, as well as the enrichment of bead-bound CD3+ lymphocytes. (Reproduced from Ref. [130] under Creative Commons) (**b**) Schematic illustration of the inertial microfluidic SELEX (iSELEX) for aptamer selection. Cells with bound aptamers are focused along the inner wall while the unbound aptamers migrate completely to the outer wall. (Adapted from Ref. [74] under Creative Commons)

5.4.6 *Biomolecules*

Purification of proteins or other biomolecules from complex background is essential in many biomedical applications and molecular assays. Unlike mammalian cells or bacteria, biomolecules are significantly smaller (nanometer scale) and hence challenging to establish inertial focusing effects. A possible strategy is to bind target biomolecules to functionalized microparticles to "artificially enhance" their sizes (Table 5.6). This unique method is presented in a work by Sarkar et al., where they reported a multiplexed affinity-based protein separation platform using DFF [130]. The working principle is shown in Fig. 5.20a. As proof-of-concept for HIV diagnostics, HIV antigens p24, gp41 and gp120, were coated on beads of three different sizes (10 μm, 4.5 μm and 1 μm) respectively to serve as capture agents. The coated beads were then incubated with serum obtained from HIV-infected patients and introduced into the DFF device. Based on inertial focusing effects, beads were sorted into different outlets based on their size. The captured antibodies were then eluted from the beads for analysis. Compared to traditional antibody purification methods, this high throughput sample processing platform (10^4–10^7 beads/s, milligram-of proteins) provided a ~ten-fold time reduction while enabling multiplexed protein purification.

Aptamers are short nucleic acid or peptide molecules that can selectively recognize distinct epitopes [131]. SELEX (systematic evolution of ligands by exponential enrichment) is a combinatorial chemistry technique used for selecting binding aptamers from a large random sequence pool *in vitro*, but this process is often iterative and time-consuming [132, 133]. To increase the selection efficacy, Birch et al. developed a novel microfluidic aptamer selection strategy based on the SELEX principle and DFF, termed as I-SELEX (Fig. 5.20b) [74]. Briefly, the pre-incubated cells and aptamer library mixtures were introduced into the spiral device. Due to distinct size differences between RBCs and nucleic acid molecules, unbound aptamers migrate along the Dean vortices towards the outer wall while the larger cells focus inertially near the inner wall. By using a wider spiral channel, the large separation distance between the unbound aptamers and aptamers-binding cells resulted in a high partition efficiency of ~10^6, which is comparable to the traditional "gold standard" capillary electrophoresis-based methods [134]. Besides high throughput processing (~2×10^6 cells/min), the developed technology was also used to identify novel high-affinity binding aptamer targets for malaria-infected RBCs.

5.5 Recent Advances

5.5.1 Novel Spiral Designs and Microstructures

Inspired by trapezoidal spiral devices, one can modulate the Dean vortices to manipulate particle focusing behavior by varying the channel geometries. For example, Sonmez et al. utilized an asymmetric serpentine channel design in a spiral device [135]. Due to the superposition of two different secondary flows induced by serpentine and spiral geometries, they demonstrated an increase in focusing quality of 9.9 μm beads as compared to conventional spiral and serpentine channels. The device can achieve ~99.5% purity of sorted 9.9 μm beads at flow rate of 2.5 mL/min.

In recent studies, circular channels are used in spiral devices without relying on conventional microfabrication techniques [62, 136]. This can be achieved by using Tygon® tubing [136] or circular cross section PDMS tube [62] wrapped around 3D-printed barrel to form curvilinear or helical structures (Fig. 5.21a). A key advantage of this strategy is that it is cheap and highly customizable as one can easily change barrel size and tubing diameter to modify the radius of curvature and channel dimension, respectively. Another attractive feature is that the tubings can be connected in a "plug-and-play" mode to other PDMS based microfluidic devices,

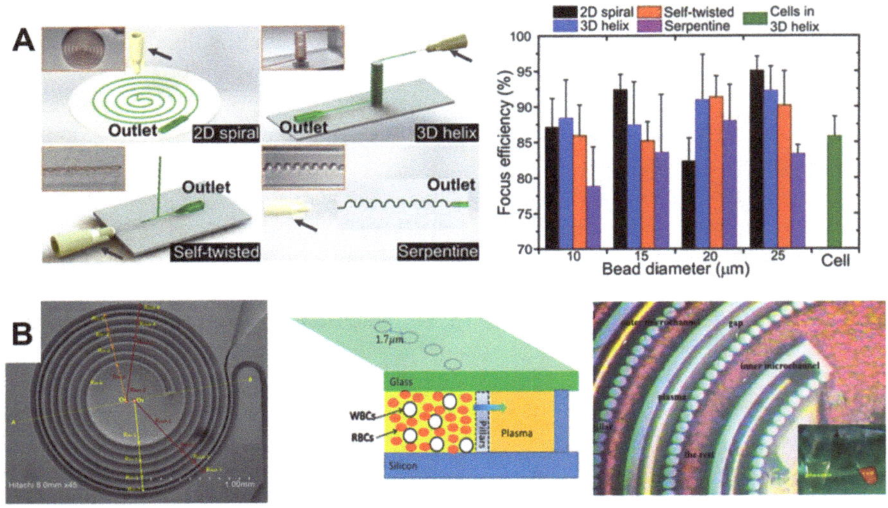

Fig. 5.21 Novel spiral designs and microstructures: (**a**) (*left*) Various designs of spiral inertial microfluidics created using circular PDMS microtubes. (*right*) Characterization of particle focusing efficiencies in different device designs. (Reproduced with permission from Ref. [62]) (**b**) (*left*) Spiral channel with micropillar array for plasma extraction. (*center*) Schematic illustration of separation principle. (*right*) Microscopic image showing efficient plasma extraction and RBCs retention. (Reproduced with permission from Ref. [137])

thus enabling in-line integration with different modalities. Hahn et al. utilized Tygon® tubing (inner diameter 190 μm, outer diameter 2 mm) in a helical spiral device and demonstrated separation of 15 and 25 μm particles into different outlets with recovery rates of 50.9 ± 5.3% and 99.5 ± 0.9%, respectively [136]. Xi et al also fabricated various designs of serpentine and spiral devices using circular PDMS tube (100 μm for inner diameter), and achieved high focusing efficiencies (>75%) for particle size ranging from 10–25 μm (Fig. 5.21a) [62]. For novel designs such as helical spiral and self-twisted spiral, an interesting feature is that Dean number (*De*) and radius of curvature remain constant for each loop as opposed to the standard planar Archimedean spiral devices [62]. The self-twisted spiral device also exhibits a stronger Dean flow profile due to the smaller radius of curvature.

Several groups have also introduced patterned microstructures such as micropillars or confined regions in their spiral channels [137, 138]. The presence of these obstacles induces additional secondary flow which can be exploited to manipulate particle position together with Dean flow to enhance separation efficiency [34]. Geng et al. designed a micropillar array (1.7 μm gap) in spiral channel for blood plasma extraction from diluted whole blood (Fig. 5.21b) [137]. Due to the combinatorial effects of physical filtration in pillar array and lateral Dean flow effects, "cell-free" plasma effectively filters through the micropillars towards the channel outer wall while the blood cells are retained in inner wall. Another important feature is the decreasing distance between pillars and the inner wall along the

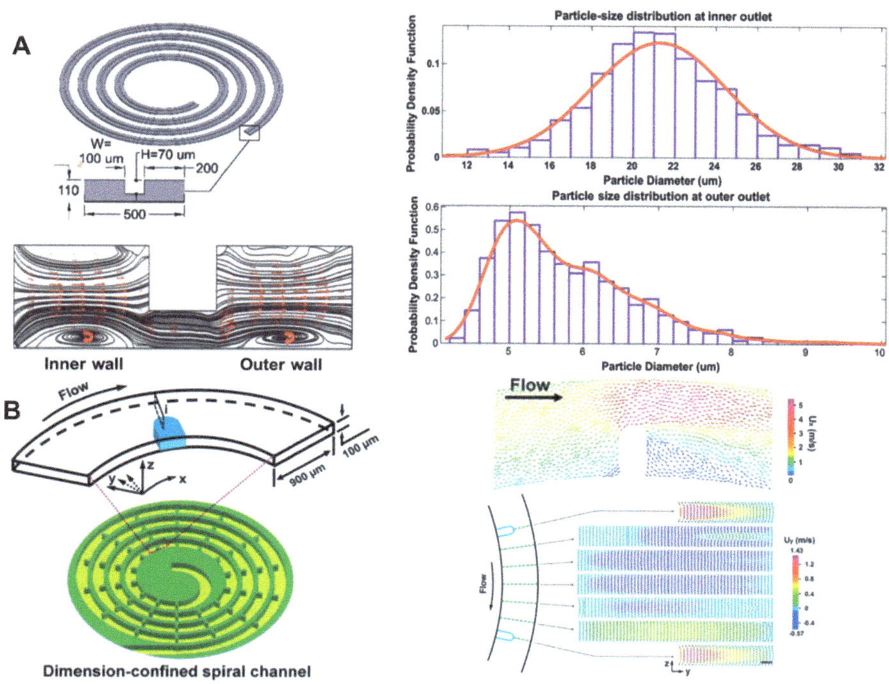

Fig. 5.22 Novel spiral designs and microstructures 2: (**a**) (*left*) Schematic illustration of the spiral device with stair-like cross-section. Cross-sectional view of the Dean vortices generated in the device. (*right*) Particle size distribution from the inner outlet and outlets. (Reproduced with permission from Ref. [138]) (**b**) (*left*) Spiral channel design with micro-obstacles. Velocity profiles at the confined region in xy plane (*right top*) and zy plane (*right bottom*) using fluid simulation. (Reproduced from Ref. [139] – published by The Royal Society of Chemistry)

microchannel to enhance the volume of extracted plasma. At an optimal flow rate of 10 μL/min, their device achieved 49.6% separation ratio (plasma volume to initial sample volume) from 250 μL of 20× diluted blood. Ghadami et al. also developed a novel spiral device with stair-like cross section in which the channel was divided into 3 sections namely inner, intermediate and outer wall [138] (Fig. 5.22a). Instead of having two counter-rotating Dean vortices in conventional spiral devices, the Dean vortices are positioned adjacent to each other with one located at the inner wall section and another at the outer wall section. Similar to DFF, larger particles would migrate towards the inner wall section while smaller particles remain in the outer wall section. They reported that the device can achieve a large separation distance of 260 μm between 7 μm particles and 20 μm particles. Recently, Shen et al. investigated various spiral designs with ordered narrow regions (Fig. 5.22b) [139]. As the fluid flows into the narrow regions, the increasing fluid velocity provides additional secondary flow acceleration which results in enhancement of particle focusing. To demonstrate the separation performance of the developed devices, they successfully sorted MCF-7 (97.5%) and HeLa (92.3%) cancer cells from diluted whole blood

(~2.5% hct) at 6.5 mL/min. Furthermore, they also performed blood plasma extraction at a lower flow rate of 3 mL/min whereby the blood cells were focusing close to the outer wall (removed through outer outlet) and plasma was extracted from other outlets. They reported a high blood cells rejection efficiency (99.96%) and plasma recovery (67.6%) from diluted blood (~2.5% hct) samples.

5.5.2 Integrating Multiplexing Spiral Devices

Unlike straight channel designs, a major limitation for spiral or curvilinear devices is the difficulty for massive planar parallelization to achieve higher throughput. To facilitate high volume processing, the spiral stacking strategy has been reported by several groups [67, 91, 140–143]. Khoo et al. developed a multiplexed spiral microfluidic device for label-free enrichment of CTCs at ultra-high throughput [140]. The stacked device was fabricated by stacking three spiral channels vertically with shared common inlets and outlets (Fig. 5.23a). To demonstrate the feasibility of this device, the authors processed blood samples (7.5 mL) from 10 healthy donors and 58 patients with metastatic breast or non-small cell lung cancer. The throughput was significantly enhanced (20-fold higher) which translated to processing 7.5 mL in less than 5 min with 100% detection sensitivity and high selectivity. Based on similar concept, Warkiani et al. utilized a stack of 40 trapezoidal spiral devices for CHO cell and yeast cells separation at 240 mL/min. [67]. Rafeie et al. later demonstrated multiplexing of 16 trapezoidal spiral devices on one layer for blood plasma separation from diluted blood [142]. They managed to achieve ~100% cell rejection ratio for ~0.5%–1% hematocrit at an optimal flow rate of 1.5 mL/min for a single device (24 mL/min for 16 devices). For stacking multiple devices, it should be noted that the top surface of the PDMS devices needs to be flat to prevent leakages between 2 consecutive layers. Precise alignment of the inlet and outlet ports is necessary to ensure that each device layer can receive equal flow distribution. In addition, stacking of multiple devices can suffer from slight pressure variance and might possibly affect individual device performance. To overcome these problems, several groups have introduced custom design manifold to distribute fluid flow equally. For example, Miller et al. developed a manifold to control pressure-driven flow in each spiral device [141]. As shown in Fig. 5.23b, they successfully incorporated 20 devices which can provide a large separation size range (2–300 μm) at a throughout of 1000 mL/min. Additionally, the system includes several unique sample processing features such as cascaded channels to enhance separation efficiency, and sample recirculation to increase purification and yield (Fig. 5.23c).

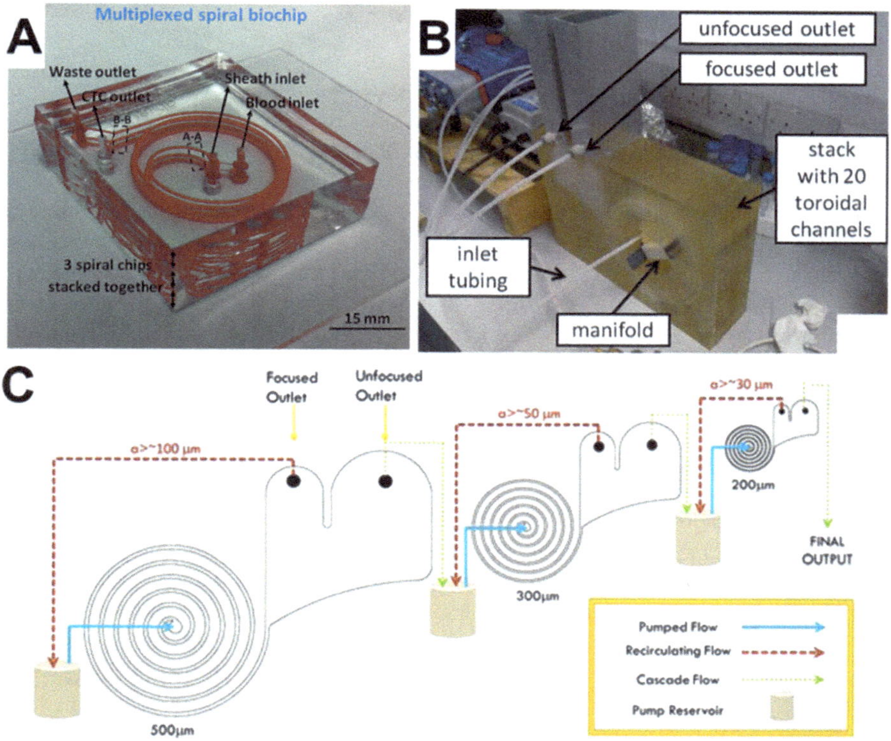

Fig. 5.23 Integrated multiplexing spiral devices: (**a**) Photo of a multiplexed spiral device by stacking 3 spiral devices together for high throughput CTCs isolation. (Adapted from Ref. [140] under Creative Commons) (**b**) Image of a stacked spiral system for large range particle separation (**c**) Schematic illustration of cascaded sample processing using different spiral devices. (Reproduced from Ref. [141] under Creative Commons)

5.5.3 Multiple Stage Spiral Device

Multiple stage spiral devices have been proposed to enhance sorting performance [141, 144]. Robinson et al. developed a two-stage spiral device for RBC depletion from diluted whole blood (2% hct). As a single spiral separation could not achieve high WBCs purity, they added another stage of separation at the end of first spiral device which has a bifurcation that equally splits into two smaller daughter spiral devices; one for subsequent RBCs exclusion and another one for transporting sorted RBCs to waste and balancing the fluidic resistances. They showed the device can effectively eliminate RBCs from diluted whole blood with 30-fold increase in RBCs depletion as compared to a single-stage spiral device [144].

5.5.4 Closed-Loop Sample Processing

Besides multiple stage sorting, recirculation strategy is another attractive alternative for processing of complex biofluid such as blood or high cell concentration samples [141, 143, 145]. Briefly, sample recirculation is achieved by using peristaltic pump to feed output eluent (containing target cells) back to the inlet sample reservoir so that the suspension can pass through the device again. The recirculation strategy provides several benefits. First, it continuously dilutes and removes the waste materials that can otherwise affect the separation performance. Secondly, it will enrich target cells into smaller volumes for subsequent processing without manual handling. In most cell sorting applications, the target cells in the samples are usually rare and do not contribute much to the sample concentration. Hence, the target cells recirculation will not compromise the separation performance. Fig. 5.24a shows a recirculation strategy developed by Ryu et al. [145]. In this work, they utilized spiral inertial microfluidics with recirculation strategy to extract polymorphonuclear leukocytes from patient-derived airway secretion or sputum. As sputum is highly complex and heterogeneous among patient sample, a single step sorting is often insufficient to achieve satisfactory separation performance. Recirculation can purify target cell suspending medium while continuously eliminates mucin aggregates from sample. The reported device retrieved ~95% PMNs from sputum from six patients and provided superior performance over traditional Sputalysin (DTT) protocol (Fig. 5.24a (*right*)). Kwon et al. recently described a closed-loop multiplexing spiral system for perfusion culture of Immunoglobulin G1 (IgG1) CHO cells with high IgG1 recovery rate (>99%), cell viability (>97%) and long term stability (18–25 days) [143].

5.5.5 Integrating with Other Separation Modalities

Coupling spiral microfluidics with other cell sorting modalities was recently proposed by Nivedita et al. [96]. In this work, they described a novel integrated platform which comprised of a spiral separator and a second-stage cell sorter based on lateral cavity acoustic transducer (LCAT). Briefly, angled lateral channel array was designed to trap air as depicted in Fig. 5.24b. Acoustic field was applied to the device which resulted in oscillation of the air/liquid interface and generation of microstreaming vortices. When larger particles approached these microvortices, they would be trapped into the circulating inner streamlines as smaller particles followed along the outer streamlines with the bulk flow [96]. By adjusting excitation voltage of acoustic actuator, the device can selectively capture different sized particles to further enrich target cells. As proof-of-concept, they applied the device to trap larger monocytes (>18 μm) from 10 × diluted blood. Smaller RBCs and WBCs, and intermediate sized cells (<18 μm) were eliminated to outlet 1 (spiral outlet) and outlet 2 (LCAT), while enriched larger monocytes remained trapped inside LCAT part. By flushing PBS through the channel, they were able to retrieve and concentrate target cells with an enrichment of 987-fold. They also demonstrated separation of DU-145 cells (>16 μm) from highly heterogeneous DU-145 population (7–28 μm) and achieved 91.7% purity and 67.5% recovery rate of target cells. Noteworthy, this work showed the unification of two separation techniques operating at different flow rates (1500 μL/min and 25 μL/min). This was achieved by adjusting the pressure drop/hydraulic resistance of the spiral outlet and transition region between 2 devices. Another simple modification that can assist in enrichment of cells was proposed by Wang et al. [92], in which they added a membrane filter at the outlet of the spiral device to further enrich CTCs for immunostaining.

For blood cell sorting applications, it remains challenging to use whole blood as a direct input for spiral devices due to the high RBCs concentration (~5 billion/mL) and secondary cell-cell interactions are known to have adverse effects on sorting performance [69, 146]. This makes off-chip sample pre-processing (dilution or RBC lysis) a prerequisite prior spiral separation. Recently, Ramachandraiah et al. proposed an integrated spiral device for leukocyte fractionation with on-chip RBC lysis (Fig. 5.24c) [146]. Whole blood and hypotonic solution (deionized water) were introduced into the device to selectively lyse the RBCs. The integrated lysis chamber with expansion chambers helped increase residence time for proper RBCs lysis. The lysed blood was then washed with sheath buffer stream (PBS) to prevent negative effects such as leukocytes swelling. Finally, the focused sample stream proceeded through the double spiral channel to fractionate out granulocytes, monocytes, and lymphocytes with a purity of 86%, 41% and 91% respectively. Such integrated device is favorable for point of care systems as it facilitates user operation and is less time-consuming process compared to sequential sample processing. Furthermore, it reduces cell loss that can occur during sample handling between each process. It should be noted that the abovementioned device required three syringe pumps to operate which can be a major drawback for point of care applications.

5.6 Conclusions and Future Outlook

Spiral inertial microfluidics has emerged as a superior separation technique for high throughput particle sorting in biomedical applications and clinical diagnostics. Since the introduction of inertial microfluidics by Di Carlo et al. [30] and the first demonstration of spiral microchannels for particle filtration [147] a decade ago, enormous efforts by us and other groups have focused on exploiting this technology for different cell sorting applications, achieving higher throughput by multiplexing, and enabling small micro and nanoparticles separation. Compared to other microfluidics systems, the key advantages of spiral devices include simplicity in microfabrication (3D printing is possible [68]), low clogging issues due to large channel dimensions, high separation resolution, and scalability for macro-scale volume processing.

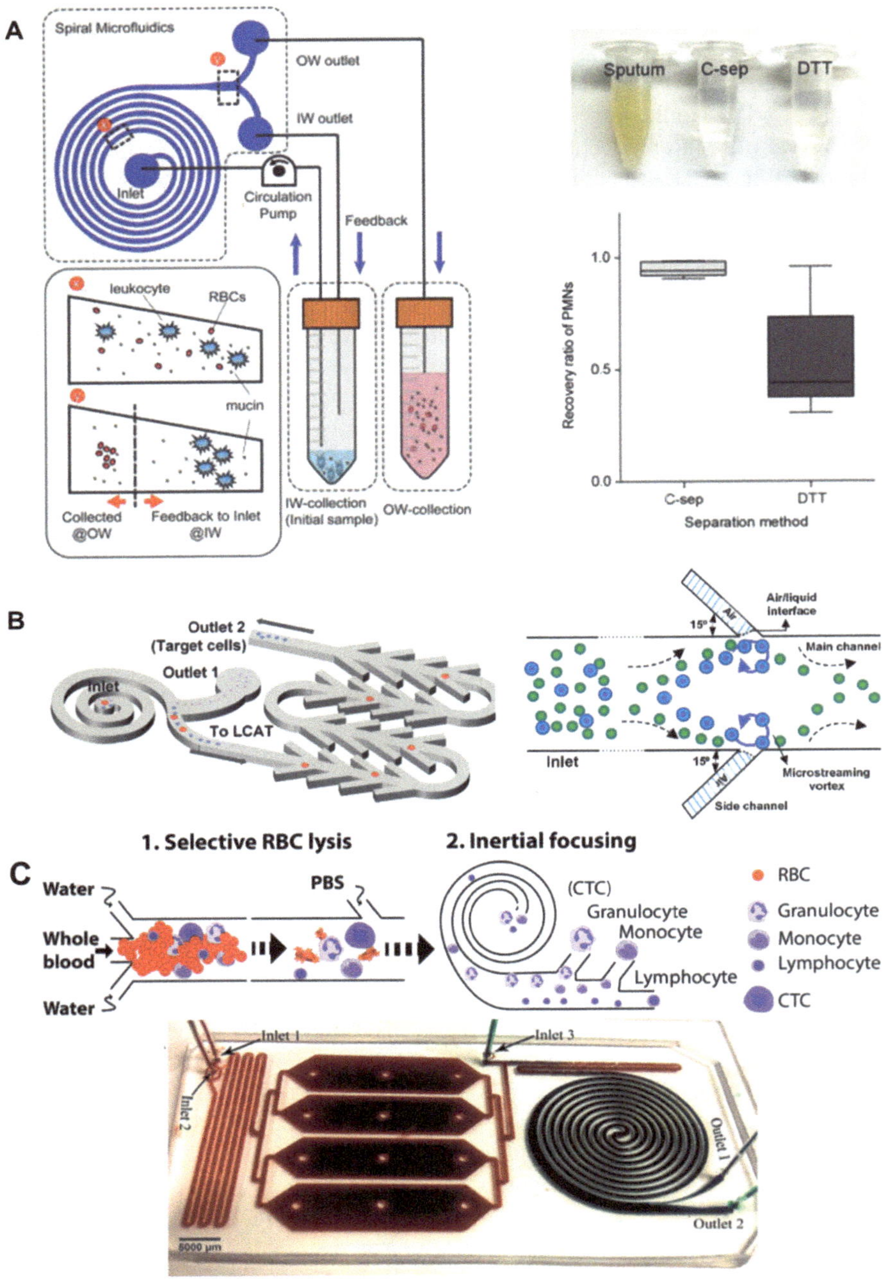

Fig. 5.24 (**a**) (*left*) Recirculation spiral microfluidics strategy (*right*) Images and plots showing sputum and purified samples using their system (C-sep) and DTT protocol. (Reprinted (adapted)

In this chapter, we have provided a comprehensive review on spiral inertial microfluidics and covered related topics including (1) conventional and microfluidic cell sorting techniques, (2) introductory theory in inertial microfluidics and Dean-coupled inertial focusing, (3) classification of major spiral devices, (4) sorting of cells and biomolecules using spiral technologies, and (5) recent advances in next generation spiral cell sorters. Noteworthy, we have also provided tables (in different sections) to highlight key principles and features in different spiral designs, as well as comparing separation performances in various cell-based applications. We believe this broad overview will be invaluable to readers who are new to the topic and keen to use spiral cell sorting technology.

As the spiral inertial microfluidics community continues to grow, there are several important areas that researchers can contribute to enable novel applications and improve understanding on the underlying physics. Being a size-based separation technology, its remains challenging to purify cells of closely-spaced sizes or with other biophysical differences (cell deformability, electrical properties *etc.*). To enhance sorting specificity, it is important to couple spiral inertial microfluidics with other active or passive sorting techniques which exploit other biophysical parameters such as electrical properties (dielectrophoresis), magnetic properties (magnetophoresis) and mechanical properties (acoustophoresis). The biggest challenge is the flowrate mismatch between spiral device and other techniques as high flow conditions commonly used in spiral channels may not be suitable for other sorting techniques. As cell heterogeneity is becoming increasingly appreciated, another major technological improvement is to integrate biosensing capabilities in spiral sorters to facilitate high throughput real-time single cell analysis. If successful, we are confident that these next-generation integrated spiral platforms will be of great significance for biomedical applications and commercialization opportunities.

Finally, improvement in imaging modalities to characterize particle equilibrium positions in 3D at high flowrates can help to elucidate the underlying physical mechanisms and inertial focusing dynamics [148]. A deeper quantitative analysis of particles inertial focusing in spiral channels can also validate simulation studies and machine learning models to optimize spiral designs and enhance separation capabilities. We envision that with increasing advancement in microfabrication and computational techniques, spiral inertial microfluidics will continue to play a leading role in driving biomedical research by enabling novel and tunable separation technologies with more precision and user-defined separation features.

Fig. 5.24 (continued) with permission from Ref. [145]. Copyright (2017) American Chemical Society.) (**b**) (*left*) Schematic of integrated spiral device with sequential lateral cavity acoustic transducer. (*right*) Microstreaming vortices in LCAT channel for particle trapping. (Adapted with permission from Ref. [96]) (**c**) (*top*) Workflow and image of the integrated spiral device with on-chip RBC lysis. (Reproduced from Ref. [146] – published by The Royal Society of Chemistry, under Creative Commons)

References

1. Antfolk M, Laurell T (2017) Continuous flow microfluidic separation and processing of rare cells and bioparticles found in blood – a review. Anal Chim Acta 965:9–35
2. Tomlinson MJ, Tomlinson S, Yang XB, Kirkham J (2013) Cell separation: terminology and practical considerations. J Tissue Eng 4:2041731412472690
3. Bhagat AAS, Bow H, Hou HW, Tan SJ, Han J, Lim CT (2010) Microfluidics for cell separation. Med Biol Eng Comput 48(10):999–1014
4. Faraghat SA, Hoettges KF, Steinbach MK, van der Veen DR, Brackenbury WJ, Henslee EA et al (2017) High-throughput, low-loss, low-cost, and label-free cell separation using electrophysiology-activated cell enrichment. Proc Natl Acad Sci 114(18):4591–4596
5. Gossett DR, Weaver WM, Mach AJ, Hur SC, Tse HT, Lee W et al (2010) Label-free cell separation and sorting in microfluidic systems. Anal Bioanal Chem 397(8):3249–3267
6. Manz A, Graber N, Widmer HM (1990) Miniaturized total chemical-analysis systems – a novel concept for chemical sensing. Sensors Actuators B-Chem 1(1–6):244–248
7. Lenshof A, Laurell T (2010) Continuous separation of cells and particles in microfluidic systems. Chem Soc Rev 39(3):1203–1217
8. Shields CW, Reyes CD, Lopez GP (2015) Microfluidic cell sorting: a review of the advances in the separation of cells from debulking to rare cell isolation. Lab Chip 15(5):1230–1249
9. Squires TM, Quake SR (2005) Microfluidics: fluid physics at the nanoliter scale. Rev Mod Phys 77(3):977–1026
10. Sajeesh P, Sen AK (2014) Particle separation and sorting in microfluidic devices: a review. Microfluid Nanofluid 17(1):1–52
11. Yan S, Zhang J, Yuan D, Li WH (2017) Hybrid microfluidics combined with active and passive approaches for continuous cell separation. Electrophoresis 38(2):238–249
12. Berger M, Castelino J, Huang R, Shah M, Austin RH (2001) Design of a microfabricated magnetic cell separator. Electrophoresis 22(18):3883–3892
13. Inglis DW, Riehn R, Austin RH, Sturm JC (2004) Continuous microfluidic immunomagnetic cell separation. Appl Phys Lett 85(21):5093–5095
14. Pamme N, Wilhelm C (2006) Continuous sorting of magnetic cells via on-chip free-flow magnetophoresis. Lab Chip 6(8):974–980
15. Cetin B, Li DQ (2011) Dielectrophoresis in microfluidics technology. Electrophoresis 32 (18):2410–2427
16. Lenshof A, Magnusson C, Laurell T (2012) Acoustofluidics 8: applications of acoustophoresis in continuous flow microsystems. Lab Chip 12(7):1210–1223
17. Ding XY, Peng ZL, Lin SCS, Geri M, Li SX, Li P et al (2014) Cell separation using tilted-angle standing surface acoustic waves. Proc Natl Acad Sci USA 111(36):12992–12997
18. MacDonald MP, Spalding GC, Dholakia K (2003) Microfluidic sorting in an optical lattice. Nature 426(6965):421–424
19. Yang S, Undar A, Zahn JD (2006) A microfluidic device for continuous, real time blood plasma separation. Lab Chip 6(7):871–880
20. Yamada M, Seki M (2005) Hydrodynamic filtration for on-chip particle concentration and classification utilizing microfluidics. Lab Chip 5(11):1233–1239
21. Yamada M, Kano K, Tsuda Y, Kobayashi J, Yamato M, Seki M et al (2007) Microfluidic devices for size-dependent separation of liver cells. Biomed Microdevices 9(5):637–645
22. Choi S, Song S, Choi C, Park JK (2007) Continuous blood cell separation by hydrophoretic filtration. Lab Chip 7(11):1532–1538
23. Kuntaegowdanahalli SS, Bhagat AAS, Kumar G, Papautsky I (2009) Inertial microfluidics for continuous particle separation in spiral microchannels. Lab Chip 9(20):2973–2980
24. Davis JA, Inglis DW, Morton KJ, Lawrence DA, Huang LR, Chou SY et al (2006) Deterministic hydrodynamics: taking blood apart. Proc Natl Acad Sci USA 103(40):14779–14784

25. Sethu P, Sin A, Toner M (2006) Microfluidic diffusive filter for apheresis (leukapheresis). Lab Chip 6(1):83–89
26. Nagrath S, Inglis DW, Morton KJ, Lawrence DA, Huang LR, Chou SY et al (2007) Isolation of rare circulating tumour cells in cancer patients by microchip technology. Nature 450 (7173):1235–1239
27. Takagi J, Yamada M, Yasuda M, Seki M (2005) Continuous particle separation in a microchannel having asymmetrically arranged multiple branches. Lab Chip 5(7):778–784
28. Huang LR, Cox EC, Austin RH, Sturm JC (2004) Continuous particle separation through deterministic lateral displacement. Science 304(5673):987–990
29. Yamada M, Nakashima M, Seki M (2004) Pinched flow fractionation: continuous size separation of particles utilizing a laminar flow profile in a pinched microchannel. Anal Chem 76(18):5465–5471
30. Di Carlo D, Irimia D, Tompkins RG, Toner M (2007) Continuous inertial focusing, ordering, and separation of particles in microchannels. Proc Natl Acad Sci USA 104(48):18892–18897
31. Bhagat AAS, Kuntaegowdanahalli SS, Papautsky I (2008) Continuous particle separation in spiral microchannels using dean flows and differential migration. Lab Chip 8(11):1906–1914. https://doi.org/10.1039/B807107A
32. Di Carlo D, Edd JF, Humphry KJ, Stone HA, Toner M (2009) Particle segregation and dynamics in confined flows. Phys Rev Lett 102(9)
33. Di Carlo D (2009) Inertial microfluidics. Lab Chip 9(21):3038–3046
34. Amini H, Lee W, Di Carlo D (2014) Inertial microfluidic physics. Lab Chip 14(15):2739–2761
35. Martel JM, Toner M (2014) Inertial focusing in microfluidics. Annu Rev Biomed Eng 16:371–396
36. Zhang J, Yan S, Yuan D, Alici G, Nguyen NT, Warkiani ME et al (2016) Fundamentals and applications of inertial microfluidics: a review. Lab Chip 16(1):10–34
37. Avila K, Moxey D, de Lozar A, Avila M, Barkley D, Hof B (2011) The onset of turbulence in pipe flow. Science 333(6039):192–196
38. Sudarsan AP, Ugaz VM (2006) Multivortex micromixing. Proc Natl Acad Sci USA 103 (19):7228–7233
39. Segre G, Silberberg A (1961) Radial particle displacements in Poiseuille flow of suspensions. Nature 189(476):209–210
40. Segré G, Silberberg A (1962) Behaviour of macroscopic rigid spheres in Poiseuille flow: part 2. Experimental results and interpretation. J Fluid Mech 14(1):136–157
41. Asmolov ES (1999) The inertial lift on a spherical particle in a plane Poiseuille flow at large channel Reynolds number. J Fluid Mech 381:63–87
42. Dean WR (1928) The stream-line motion of fluid in a curved pipe. (Second paper.). Philos Mag 5(30):673–695
43. Ookawara S, Higashi R, Street D, Ogawa K (2004) Feasibility study on concentration of slurry and classification of contained particles by microchannel. Chem Eng J 101(1–3):171–178
44. Gossett DR, Di Carlo D (2009) Particle focusing mechanisms in curving confined flows. Anal Chem 81(20):8459–8465
45. Saffman PG (1965) The lift on a small sphere in a slow shear flow. J Fluid Mech 22 (2):385–400
46. Cherukat P, McLaughlin JB (1994) The inertial lift on a rigid sphere in a linear shear flow field near a flat wall. J Fluid Mech 263:1–18
47. Loth E, Dorgan AJ (2009) An equation of motion for particles of finite Reynolds number and size. Environ Fluid Mech 9(2):187–206
48. Zhou J, Giridhar PV, Kasper S, Papautsky I (2013) Modulation of aspect ratio for complete separation in an inertial microfluidic channel. Lab Chip 13(10):1919–1929
49. Zhou J, Papautsky I (2013) Fundamentals of inertial focusing in microchannels. Lab Chip 13 (6):1121–1132
50. Bhagat AAS, Kuntaegowdanahalli SS, Papautsky I (2008) Enhanced particle filtration in straight microchannels using shear-modulated inertial migration. Phys Fluids 20(10):101702

51. Kim TH, Yoon HJ, Stella P, Nagrath S (2014) Cascaded spiral microfluidic device for deterministic and high purity continuous separation of circulating tumor cells. Biomicrofluidics 8(6):13. Art. no. 064117
52. Sudarsan AP, Ugaz VM (2006) Fluid mixing in planar spiral microchannels. Lab Chip 6 (1):74–82
53. Sudarsan AP, Ugaz VM (2006) Multivortex micromixing. Proc Natl Acad Sci 103 (19):7228–7233
54. Wang J, Zhan Y, Ugaz VM, Lu C (2010) Vortex-assisted DNA delivery. Lab Chip 10 (16):2057–2061
55. Martel JM, Toner M (2012) Inertial focusing dynamics in spiral microchannels. Phys Fluids 24 (3):032001
56. Russom A, Gupta AK, Nagrath S, Di Carlo D, Edd JF, Toner M (2009) Differential inertial focusing of particles in curved low-aspect-ratio microchannels. New J Phys 11:075025
57. Xiang N, Chen K, Sun D, Wang S, Yi H, Ni Z (2013) Quantitative characterization of the focusing process and dynamic behavior of differently sized microparticles in a spiral microchannel. Microfluid Nanofluid 14(1):89–99
58. Kemna EWM, Schoeman RM, Wolbers F, Vermes I, Weitz DA, van den Berg A (2012) High-yield cell ordering and deterministic cell-in-droplet encapsulation using Dean flow in a curved microchannel. Lab Chip 12(16):2881–2887
59. Seo J, Lean MH, Kole A (2007) Membrane-free microfiltration by asymmetric inertial migration. Appl Phys Lett 91(3):033901
60. Sun J, Li M, Liu C, Zhang Y, Liu D, Liu W et al (2012) Double spiral microchannel for label-free tumor cell separation and enrichment. Lab Chip 12(20):3952–3960
61. Sun J, Liu C, Li M, Wang J, Xianyu Y, Hu G et al (2013) Size-based hydrodynamic rare tumor cell separation in curved microfluidic channels. Biomicrofluidics 7(1):011802
62. Xi W, Kong F, Yeo JC, Yu L, Sonam S, Dao M et al (2017) Soft tubular microfluidics for 2D and 3D applications. Proc Natl Acad Sci 114(40):10590–10595
63. Nivedita N, Ligrani P, Papautsky I (2017) Dean flow dynamics in low-aspect ratio spiral microchannels. Sci Rep 7:44072
64. Guan G, Wu L, Bhagat AA, Li Z, Chen PCY, Chao S et al (2013) Spiral microchannel with rectangular and trapezoidal cross-sections for size based particle separation. Sci Rep 3:1475
65. Wu L, Guan G, Hou HW, Asgar A, Bhagat S, Han J (2012) Separation of leukocytes from blood using spiral channel with trapezoid cross-section. Anal Chem 84(21):9324–9331
66. Warkiani ME, Guan G, Luan KB, Lee WC, Bhagat AAS, Chaudhuri PK et al (2014) Slanted spiral microfluidics for the ultra-fast, label-free isolation of circulating tumor cells. Lab Chip 14(1):128–137
67. Warkiani ME, Tay AKP, Guan G, Han J (2015) Membrane-less microfiltration using inertial microfluidics. Sci Rep 5:11018
68. Lee W, Kwon D, Choi W, Jung GY, Au AK, Folch A et al (2015) 3D-printed microfluidic device for the detection of pathogenic bacteria using size-based separation in helical channel with trapezoid cross-section. Sci Rep 5:7717
69. Hou HW, Bhagat AAS, Lee WC, Huang S, Han J, Lim CT (2011) Microfluidic devices for blood fractionation. Micromachines 2(3):319–343
70. Hou HW, Warkiani ME, Khoo BL, Li ZR, Soo RA, Tan DS-W et al (2013) Isolation and retrieval of circulating tumor cells using centrifugal forces. *Sci Rep* 3:1259
71. Vona G, Sabile A, Louha M, Sitruk V, Romana S, Schutze K et al (2000) Isolation by size of epithelial tumor cells – a new method for the immunomorphological and molecular characterization of circulating tumor cells. Am J Pathol 156(1):57–63
72. Zabaglo L, Ormerod MG, Parton M, Ring A, Smith IE, Dowsett M (2003) Cell filtration-laser scanning cytometry for the characterisation of circulating breast cancer cells. Cytometry A 55 (2):102–108

73. Hou HW, Bhattacharyya RP, Hung DT, Han J (2015) Direct detection and drug-resistance profiling of bacteremias using inertial microfluidics. Lab Chip 15(10):2297–2307. https://doi.org/10.1039/C5LC00311C
74. Birch CM, Hou HW, Han J, Niles JC (2015) Identification of malaria parasite-infected red blood cell surface aptamers by inertial microfluidic SELEX (I-SELEX). Sci Rep 5:11347
75. Yeo DC, Wiraja C, Zhou Y, Tay HM, Xu C, Hou HW (2015) Interference-free micro/nanoparticle cell engineering by use of high-throughput microfluidic separation. ACS Appl Mater Interfaces 7(37):20855–20864
76. Tay HM, Kharel S, Dalan R, Chen ZJ, Tan KK, Boehm BO et al (2017) Rapid purification of sub-micrometer particles for enhanced drug release and microvesicles isolation. NPG Asia Mater 9:e434
77. Sollier E, Murray C, Maoddi P, Di Carlo D (2011) Rapid prototyping polymers for microfluidic devices and high pressure injections. Lab Chip 11(22):3752–3765
78. Johnston ID, McDonnell MB, Tan CKL, McCluskey DK, Davies MJ, Tracey MC (2014) Dean flow focusing and separation of small microspheres within a narrow size range. Microfluid Nanofluid 17(3):509–518
79. Dong Y, Skelley AM, Merdek KD, Sprott KM, Jiang CS, Pierceall WE et al (2013) Microfluidics and circulating tumor cells. J Mol Diagn 15(2):149–157
80. Cristofanilli M, Budd GT, Ellis MJ, Stopeck A, Matera J, Miller MC et al (2004) Circulating tumor cells, disease progression, and survival in metastatic breast cancer. N Engl J Med 351 (8):781–791
81. Hayes DF, Cristofanilli M, Budd GT, Ellis MJ, Stopeck A, Miller MC et al (2006) Circulating tumor cells at each follow-up time point during therapy of metastatic breast cancer patients predict progression-free and overall survival. Clin Cancer Res 12(14):4218–4224
82. Yu L, Ng SR, Xu Y, Dong H, Wang YJ, Li CM (2013) Advances of lab-on-a-chip in isolation, detection and post-processing of circulating tumour cells. Lab Chip 13(16):3163–3182
83. Chen P, Huang YY, Hoshino K, Zhang XJ (2014) Multiscale immunomagnetic enrichment of circulating tumor cells: from tubes to microchips. Lab Chip 14(3):446–458
84. Hajba L, Guttman A (2014) Circulating tumor-cell detection and capture using microfluidic devices. Trends Anal Chem 59:9–16
85. Murlidhar V, Rivera-Baez L, Nagrath S (2016) Affinity versus label-free isolation of circulating tumor cells: who wins? Small 12(33):4450–4463
86. Nagrath S, Sequist LV, Maheswaran S, Bell DW, Irimia D, Ulkus L et al (2007) Isolation of rare circulating tumour cells in cancer patients by microchip technology. Nature 450 (7173):1235–1239
87. Stott SL, Hsu CH, Tsukrov DI, Yu M, Miyamoto DT, Waltman BA et al (2010) Isolation of circulating tumor cells using a microvortex-generating herringbone-chip. Proc Natl Acad Sci USA 107(43):18392–18397
88. Sun JS, Liu C, Li MM, Wang JD, Xianyu YL, Hu GQ et al (2013) Size-based hydrodynamic rare tumor cell separation in curved microfluidic channels. Biomicrofluidics 7(1):11802
89. Burke JM, Zubajlo RE, Smela E, White IM (2014) High-throughput particle separation and concentration using spiral inertial filtration. Biomicrofluidics 8(2):17, Art. no. 024105
90. Khoo BL, Warkiani ME, Tan DSW, Bhagat AAS, Irwin D, Lau DP, et al. (2014) Clinical validation of an ultra high-throughput spiral microfluidics for the detection and enrichment of viable circulating tumor cells. PLoS One 9(7):7, Art. no. e99409
91. Warkiani ME, Khoo BL, Tan DS-W, Bhagat AAS, Lim W-T, Yap YS et al (2014) An ultra-high-throughput spiral microfluidic biochip for the enrichment of circulating tumor cells. Analyst 139(13):3245–3255. https://doi.org/10.1039/C4AN00355A
92. Wang JD, Lu WJ, Tang CH, Liu Y, Sun JS, Mu X et al (2015) Label-free isolation and mRNA detection of circulating tumor cells from patients with metastatic lung cancer for disease diagnosis and monitoring therapeutic efficacy. Anal Chem 87(23):11893–11900

93. Huang D, Shi X, Qian Y, Tang WL, Liu LB, Xiang N et al (2016) Rapid separation of human breast cancer cells from blood using a simple spiral channel device. Anal Methods 8 (30):5940–5948
94. Warkiani ME, Khoo BL, Wu LD, Tay AKP, Bhagat AAS, Han J et al (2016) Ultra-fast, label-free isolation of circulating tumor cells from blood using spiral microfluidics. Nat Protoc 11 (1):134–148
95. Kulasinghe A, Tran THP, Blick T, O'Byrne K, Thompson EW, Warkiani ME et al (2017) Enrichment of circulating head and neck tumour cells using spiral microfluidic technology. Sci Rep 7
96. Nivedita N, Garg N, Lee AP, Papautsky I (2017) A high throughput microfluidic platform for size-selective enrichment of cell populations in tissue and blood samples. Analyst 142 (14):2558–2569
97. Bhagat AAS, Kuntaegowdanahalli SS, Kaval N, Seliskar CJ, Papautsky I (2010) Inertial microfluidics for sheath-less high-throughput flow cytometry. Biomed Microdevices 12 (2):187–195
98. Aya-Bonilla CA, Marsavela G, Freeman JB, Lomma C, Frank MH, Khattak MA et al (2017) Isolation and detection of circulating tumour cells from metastatic melanoma patients using a slanted spiral microfluidic device. Oncotarget 8(40):67355–67368
99. Diogo MM, da Silva CL, Cabral JMS (2012) Separation technologies for stem cell bioprocessing. Biotechnol Bioeng 109(11):2699–2709
100. Zhu BL, Murthy SK (2013) Stem cell separation technologies. Curr Opin Chem Eng 2(1):3–7
101. Machado HL, Kittrell FS, Edwards D, White AN, Atkinson RL, Rosen JM et al (2013) Separation by cell size enriches for mammary stem cell repopulation activity. Stem Cells Transl Med 2(3):199–203
102. Davis PK, Ho A, Dowdy SF (2001) Biological methods for cell-cycle synchronization of mammalian cells. BioTechniques 30(6):1322−+
103. Choi S, Song S, Choi C, Park JK (1964-1968) Microfluidic self-sorting of mammalian cells to achieve cell cycle synchrony by hydrophoresis. Anal Chem 81(5):2009
104. Lee WC, Bhagat AAS, Huang S, Van Vliet KJ, Han J, Lim CT (2011) High-throughput cell cycle synchronization using inertial forces in spiral microchannels. Lab Chip 11 (7):1359–1367
105. Lee WC, Shi H, Poon ZY, Nyan LM, Kaushik T, Shivashankar GV et al (2014) Multivariate biophysical markers predictive of mesenchymal stromal cell multipotency. Proc Natl Acad Sci USA 111(42):E4409–E4418
106. Nathamgari SSP, Dong BQ, Zhou F, Kang WM, Giraldo-Vela JP, McGuire T et al (2015) Isolating single cells in a neurosphere assay using inertial microfluidics. Lab Chip 15 (24):4591–4597
107. Song HJ et al (2017) Spiral-shaped inertial stem cell device for high-throughput enrichment of iPSC-derived neural stem cells. Microfluid Nanofluid 21(4):1–9
108. Rossi F, Cattaneo E (2002) Opinion – neural stem cell therapy for neurological diseases: dreams and reality. Nat Rev Neurosci 3(5):401–409
109. Marshall GP, Reynolds BA, Laywell ED (2007) Using the neurosphere assay to quantify neural stem cells in vivo. Curr Pharm Biotechnol 8(3):141–145
110. Deleyrolle LP, Rietze RL, Reynolds BA (2008) The neurosphere assay, a method under scrutiny. Acta Neuropsychiatrica 20(1):2–8
111. Alba-Loureiro TC et al (2007) Neutrophil function and metabolism in individuals with diabetes mellitus. Braz J Med Biol Res 40(8):1037–1044
112. Gregory AD, Houghton AM (2011) Tumor-associated neutrophils: new targets for cancer therapy. Cancer Res 71(7):2411–2416
113. Papa A, Emdin M, Passino C, Michelassi C, Battaglia D, Cocci F (2008) Predictive value of elevated neutrophil-lymphocyte ratio on cardiac mortality in patients with stable coronary artery disease. Clin Chim Acta 395(1–2):27–31

114. Ernst E, Matrai A (1986) Altered red and white blood-cell rheology in type-ii diabetes. Diabetes 35(12):1412–1415
115. Pecsvarady Z, Fisher TC, Darwin CH, Fabok A, Maqueda TS, Saad MF et al (1994) Decreased polymorphonuclear leukocyte deformability in Niddm. Diabetes Care 17(1):57–63
116. Mowat AG, Baum J (1971) Chemotaxis of polymorphonuclear leukocytes from patients with diabetes mellitus. N Engl J Med 284(12):621–627
117. Delamaire M, Maugendre D, Moreno M, LeGoff MC, Allannic H, Genetet B (1997) Impaired leucocyte functions in diabetic patients. Diabet Med 14(1):29–34
118. Bagdade JD, Root RK, Bulger RJ (1974) Impaired leukocyte function in patients with poorly controlled diabetes. Diabetes 23(1):9–15
119. Hou HW, Petchakup C, Tay HM, Tam ZY, Dalan R, Chew DEK et al (2016) Rapid and label-free microfluidic neutrophil purification and phenotyping in diabetes mellitus. Sci Rep 6
120. Nivedita N, Papautsky I (2013) Continuous separation of blood cells in spiral microfluidic devices. Biomicrofluidics 7(5):54101
121. Agarwal A, Ikemoto I, Loughlin KR (1994) Effect of sperm washing on levels of reactive oxygen species in semen. Arch Androl 33(3):157–162
122. Swain JE, Lai D, Takayama S, Smith GD (2013) Thinking big by thinking small: application of microfluidic technology to improve ART. Lab Chip 13(7):1213–1224
123. Son JY, Murphy K, Samuel R, Gale BK, Carrell DT, Hotaling JM (2015) Non-motile sperm cell separation using a spiral channel. Anal Methods 7(19):8041–8047
124. Hou HW, Gan HY, Bhagat AAS, Li LD, Lim CT, Han J (2012) A microfluidics approach towards high-throughput pathogen removal from blood using margination. Biomicrofluidics 6 (2):24115
125. Schaap A, Dumon J, den Toonder J (2016) Sorting algal cells by morphology in spiral microchannels using inertial microfluidics. Microfluid Nanofluid 20(9):125
126. Clime L, Li K, Geissler M, Hoa XD, Robideau GP, Bilodeau GJ, et al (2017) Separation and concentration of *Phytophthora ramorum* sporangia by inertial focusing in curving microfluidic flows. Microfluid Nanofluid 21(1):13 Art. no. 5
127. Schets FA, van Wijnen JH, Schijven JF, Schoon A, Husmant A (2008) Monitoring of waterborne pathogens in surface waters in Amsterdam, The Netherlands, and the potential health risk associated with exposure to Cryptosporidium and Giardia in these waters. Appl Environ Microbiol 74(7):2069–2078
128. Bridle H, Kersaudy-Kerhoas M, Miller B, Gavriilidou D, Katzer F, Innes EA et al (2012) Detection of Cryptosporidium in miniaturised fluidic devices. Water Res 46(6):1641–1661
129. Jimenez M, Miller B, Bridle HL (Jan 2017) Efficient separation of small microparticles at high flowrates using spiral channels: application to waterborne pathogens. Chem Eng Sci 157:247–254
130. Sarkar A, Hou HW, Mahan AE, Han J, Alter G (2016) Multiplexed affinity-based separation of proteins and cells using inertial microfluidics. Sci Rep 6
131. Cheung YW, Dirkzwager RM, Wong WC, Cardoso J, D'Arc Neves Costa J, Tanner JA (2017) Aptamer-mediated Plasmodium-specific diagnosis of malaria. Biochimie. in press
132. Tuerk C, Gold L (1990) Systematic evolution of ligands by exponential enrichment: RNA ligands to bacteriophage T4 DNA polymerase. Science 249(4968):505–510
133. Ellington AD, Szostak JW (1990) In vitro selection of RNA molecules that bind specific ligands. Nature 346(6287):818–822
134. Berezovski M, Drabovich A, Krylova SM, Musheev M, Okhonin V, Petrov A et al (2005) Nonequilibrium capillary electrophoresis of equilibrium mixtures: a universal tool for development of aptamers. J Am Chem Soc 127(9):3165–3171
135. Sonmez U, Jaber S, Trabzon L (2017) Super-enhanced particle focusing in a novel microchannel geometry using inertial microfluidics. J Micromech Microeng 27(6):065003
136. Hahn Y, Hong D, Kang J, Choi S (2016) A reconfigurable microfluidics platform for microparticle separation and fluid mixing. Micromachines 7(8):139

137. Geng Z, Ju Y, Wang W, Li Z (2013) Continuous blood separation utilizing spiral filtration microchannel with gradually varied width and micro-pillar array. Sensors Actuators B Chem 180:122–129
138. Ghadami S, Kowsari-Esfahan R, Saidi MS, Firoozbakhsh K (2017) Spiral microchannel with stair-like cross section for size-based particle separation. Microfluid Nanofluid 21(7):115
139. Shen S, Tian C, Li T, Xu J, Chen S-W, Tu Q et al (2017) Spiral microchannel with ordered micro-obstacles for continuous and highly-efficient particle separation. Lab Chip. https://doi.org/10.1039/C7LC00691H
140. Khoo BL, Warkiani ME, Tan DS, Bhagat AA, Irwin D, Lau DP et al (2014) Clinical validation of an ultra high-throughput spiral microfluidics for the detection and enrichment of viable circulating tumor cells. PLoS One 9(7):e99409
141. Miller B, Jimenez M, Bridle H (2016) Cascading and parallelising curvilinear inertial focusing systems for high volume, wide size distribution, separation and concentration of particles. Sci Rep 6:36386
142. Rafeie M, Zhang J, Asadnia M, Li W, Warkiani ME (2016) Multiplexing slanted spiral microchannels for ultra-fast blood plasma separation. Lab Chip 16(15):2791–2802. https://doi.org/10.1039/C6LC00713A
143. Kwon T, Prentice H, Oliveira JD, Madziva N, Warkiani ME, Hamel J-FP et al (2017) Microfluidic cell retention device for perfusion of mammalian suspension culture. Sci Rep 7 (1):6703
144. Robinson M, Marks H, Hinsdale T, Maitland K, Cote G (2017) Rapid isolation of blood plasma using a cascaded inertial microfluidic device. Biomicrofluidics 11(2):024109
145. Ryu H, Choi K, Qu Y, Kwon T, Lee JS, Han J (2017) Patient-derived airway secretion dissociation technique to isolate and concentrate immune cells using closed-loop inertial microfluidics. Anal Chem 89(10):5549–5556
146. Ramachandraiah H, Svahn HA, Russom A (2017) Inertial microfluidics combined with selective cell lysis for high throughput separation of nucleated cells from whole blood. RSC Adv 7(47):29505–29514. https://doi.org/10.1039/C7RA02992F
147. Seo J, Lean MH, Kole A (2007) Membraneless microseparation by asymmetry in curvilinear laminar flows. J Chromatogr A 1162(2):126–131
148. Goda K, Ayazi A, Gossett DR, Sadasivam J, Lonappan CK, Sollier E et al (2012) High-throughput single-microparticle imaging flow analyzer. Proc Natl Acad Sci 109 (29):11630–11635

Chapter 6
Worms on a Chip

Han-Sheng Chuang, Wen-Hui Wang, and Chang-Shi Chen

Abstract Advancement in miniaturization in recent years has enabled high-throughput, in-parallel, rapid, and precise operations in modern medical and biological research. Although numerous biomimetic devices have been inspired by nature cues, the artificial gadgets still cannot be on a par with their natural counterparts. *Caenorhabditis elegans* (*C. elegans*), the smallest multi-cellular model animal, has become a popular platform for drug screening, biosensing, genetic engineering, neuroscience, developmental biology, and so forth since its first debut made by Sydney Brenner nearly five decades ago. The nematode *C. elegans* features small size, transparency body, fully sequenced genomes, high genetic similarity with humans, short life cycle, and simple neural network. The combination of *C. elegans* and microchip can prompt promising uses in some aspects. To cope with the new demands, the scientific community has endeavored great efforts to meet all sorts of worm maneuvers, such as sorting, immobilization, long-term imaging, confined culture, and biomechanics. The proposed manipulation repertoire then leads to realizations of a wide applications. Examples may include drug screening for pharmaceutics, point-of-care testing (POCT) for diseases, and fundamental research. Although worms-on-a-chip (WoC) appears to remain in its infancy stage of development, intensive research has gradually unveiled novel possibilities in many potential fields. This chapter aims to introduce the current development of WoCs and provides examples according to their categories. Pros and cons will be addressed in the end. Some practical uses will also be suggested for the future prospects.

H.-S. Chuang (✉)
Department of Biomedical Engineering, National Cheng Kung University, Tainan, Taiwan

Medical Device Innovation Center, National Cheng Kung University, Tainan, Taiwan
e-mail: oswaldchuang@mail.ncku.edu.tw

W.-H. Wang
State Key Laboratory of Precision Measurement Technology and Instrument, Department of Precision Instrument, Tsinghua University, Beijing, China

C.-S. Chen
Department of Biochemical and Molecular Biology, National Cheng Kung University, Tainan, Taiwan

M. Tokeshi (ed.), *Applications of Microfluidic Systems in Biology and Medicine*, Bioanalysis 7, https://doi.org/10.1007/978-981-13-6229-3_6

Keywords *C. elegans* · Microchip · Screening · Model animal · Microfluidics

6.1 Introduction

The nematode *Caenorhabditis elegans* is a well-known model animal in research fields, such as neuroscience, genetic engineering, and drug discovery. *C. elegans* was firstly studied and introduced to the world by Sydney Brenner in 1963 [1]. The tiny multi-cellular organism measures around 1 mm in length and about 4 pg in wet weight [2]. In addition, the worm possesses only 302 neurons, 959 somatic cells, a short life cycle (~3 days), and more than 60% genetic similarity with human being. In 1998, *C. elegans* was even the first animal to have its genome fully sequenced in a pilot project led by John Sulston and Bob Waterston. All of the above features make *C. elegans* a unique model animal in preliminary research. With such advantages, Sydney Brenner once praised *C. elegans* a nature's gift to science in his Nobel lecture in 2002. Until now, several Nobel prizes have been awarded to those scientists who devoted to the *C. elegans* research, including Sydney Brenner, Robert Horvitz, and John Sulston (the 2002 Nobel Prize in Physiology or Medicine), Andrew Fire and Craig Mello (the 2006 Nobel Prize in Physiology or Medicine), and Martin Chalfie shared the 2008 Nobel Prize in Chemistry. These facts make the tiny animal a real giant in science.

After more than five decades of intense development and support from the science community, *C. elegans* has become a great platform used in a broad spectrum of fields. To facilitate the research based on the worm, a worm bank, named Caenorhabditis Genetics Center (CGC), has been established at University of Minnesota to provide various mutants and genetically-engineered worms to all worm breeders around the world. Instead of conventionally tedious breeding, nowadays researchers only need to place their orders for specific worms and will receive them in a few days, thus saving time and effort in their work. Moreover, some online encyclopedias, such as WormBook, WormBase, and WormAtlas, can also guide beginners to start their work in a timely fashion. A clear trend shows that rapidly growing researchers have been brought into this area recently not only because of the significance of the worm but also the friendly environment. Of the researchers, the engagement from engineers especially throws unique impact to this community by introducing new tools and mindsets.

The prior studies [3–8] have revealed that *C. elegans* can respond to many stimuli with chemotaxis, thermotaxis, electrotaxis, magnetotaxis, phototaxis, and more. The capability of multiple responses implies that various sensory neurons are involved inside the worm's body. The basic communications between sensory neurons and motor neurons can form different pathways to activate relevant motions mentioned above. The locomotory gait of *C. elegans* is a sophisticated behavior in response to environmental changes [4–10]. Through image recognition of the gait changes, one can further understand the interactions between worms and their environmental stimuli. In anatomy, most worms are hermaphrodite, leaving only 5% of the total population to be male. Therefore, the self-reproduction results in slow process in genetic diversity. The highly genetic similarity between parents and offspring avoid inconsistency in experiments involving with several generations. Nevertheless, individual worms still can behave differently from each other even receiving same stimulus. The difference makes *C. elegans* an ideal model for investigating the correlation between decision making and the formation of neuronal network.

Although *C. elegans* is preferably cultivated on agar, they perform amphibious adaption in aqueous environments [11]. According to prior studies [12, 13], worms show no significant differences in lifespan, physiological conditions, and biochemical reactions between both situations. However, worms growing in an aqueous environment are more likely to result in vigorous undulating motion instead of crawling. Although the mechanism remains not very well understood yet, some studies suggest that viscosity change may form a mechanical stimulus to the worm's sensory neurons and then trigger the subsequent gait switch. Of the studies, some evidences show the gait change is rather a binary switch between swimming and crawling than a progressive shift [14]. In spite of the fact, aqueous media remain a popular culture environment for the study of *C. elegans*.

As *C. elegans* becomes the spotlight of model animals, many skills have been exploited to enhance the repertoire of the tiny little animal. In addition, their fast reproduction and easy maintenance provide a great statistical basis over other model animals. As compared with artificial intelligence, the natural-born primitive organism remains to outperform human technology in all aspects. For example, the neuronal network constructed by the 302 neuron cells imparts *C. elegans* sufficient ability to deal with various daily activities, such as foraging, mating, avoidance from adverse situations, sensing stimuli, decision making, and interactions with other worms or environments. As a result, the worm draws not only attention from biologists but also engineering scientists in recent years. By properly incorporating *C. elegans* into a microchip, the worm has potential to work as a multi-functional sensor, a drug carrier, a sophisticated animal platform, an actuator and more (Fig. 6.1). This chapter will begin with worm biology to introduce the uniqueness of the worm and then extend to some state-of-the-art WoCs. Next, we will look into some applications that require such kind of special worm chips and their niches. At last, we will finalize this chapter and envision some unmet needs and possible developments in the near future.

6.2 Worm Biology

6.2.1 Fundamental Features

The nematode *Caenorhabditis elegans* is a free-living and tiny multi-cellular organism that inhabits on rotted vegetables or fruits in nature. It was firstly studied by Sydney Brenner and introduced to the world as a great genetic model animal [15, 16]. Unlike its other parasitic cousins, such as threadworms, flukes, hookworms,

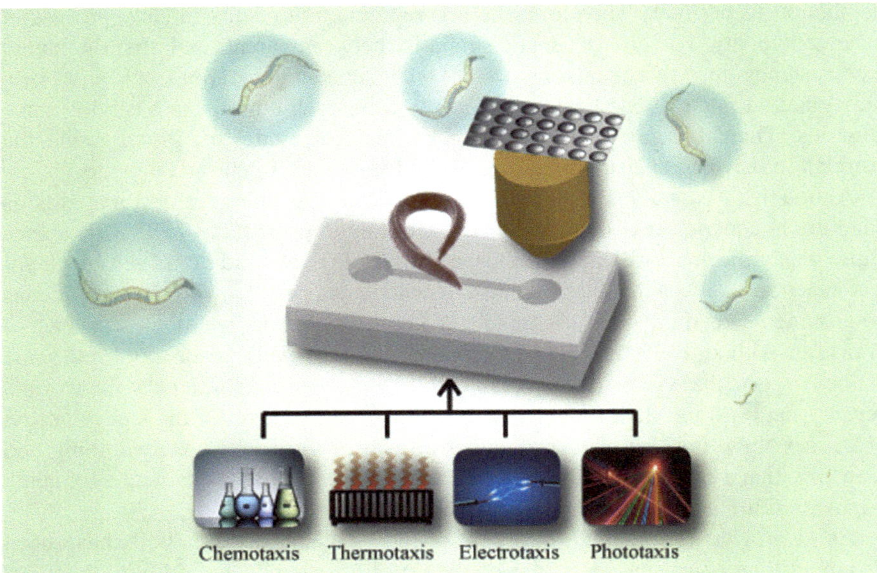

Fig. 6.1 Conceptual illustration of the universal WoC platform

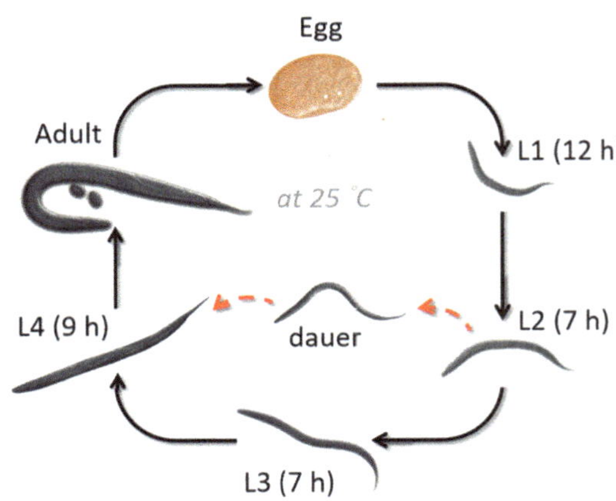

Fig. 6.2 The life cycle of *C. elegans* at 25 °C

and so forth, *C. elegans* has been proven to be non-hazardous, non-infectious, non-pathogenic, and thus safe in laboratory use. For research purposes, *C. elegans* is usually cultivated on agar and fed with non-pathogenic bacterium, *Escherichia coli* strain OP50. In general, the features that bring *C. elegans* on the spotlight are focused on their small size (~1 mm for adult worm and 959 somatic cells), transparent body, short life cycle (~3 days), a small number of neurons (302), and fully sequenced genome.

The distinct developmental stages of *C. elegans* include egg, larva, and adult (Fig. 6.2). In the larval stages, worms can be even divided into L1, L2, L3, and L4 according to their physiological phenotypes. At 25 °C, the growing periods for L1, L2, L3, and L4 stages are estimated to be 12, 7, 7, and 9 h, respectively. These predictable growth periods are therefore references used to synchronize worms. To this end, adult worms need to be bleached by strong basic solutions, such as alkaline hypochlorite, to remove their cuticles and leave eggs. Subsequently, all eggs will be synchronously hatched from the same time point and grow till the desired developmental stage. Notably, *C. elegans* may detour to a special larval stage, named dauer, when the worm experiences starvation before L2. The dauer worms are more durable to environmental stress, so they can survive for at least a month without food. This capability imparts them a high survival rate during critical conditions. As for the worm's appearance, the body lengths of L1, L2, L3, and L4 are estimated to be around <250, 360–380, 490–510, and 620–650 μm, respectively. In addition, the body length for a young adult worm is 900–940 μm and will become 1110 μm–1150 μm for a fully-grown adult worm. Especially, L4 stage worms can be explicitly recognized by a crescent mark loomed in the mid-body resulting from the intrusion of vulva.

6.2.2 Anatomy and Physiological Phenotypes

An adult *C. elegans* nematode comprises an alimentary system (mouth, pharynx, intestine, rectum, anus), a reproductive system (gonad, uterus, spermatheca, vulva in the hermaphrodite; gonad, seminal vesicle, vas deferens, cloaca in the male), a nervous system (302 neurons in the hermaphrodite, a cluster of synapses, and a nerve ring in the head), and an excretory system (a group of four cells believed to control osmolarity and the elimination of waste) (Fig. 6.3a and b). The cellular mitosis of somatic cells process only proceeds during the developmental stages. A total of 1030 somatic cells are generated in a hermaphrodite worm's time course of development. However, the number reduces to 959 after 131 cells are programmed to death in the later worm stage. Anatomically, the worm is simple in body constituent but sophisticated in functions. The animal is often described as a series of concentric tubes comprising cuticle and internal organs separated with a pseudocoelomic cavity (Fig. 6.3c). As a result, mass transfer in worm's body including metabolites and nutrients is primarily dependent on diffusion due to lack of a circulatory system. Three sets of muscles are responsible for controlling the foraging, locomotion, and egg laying. All the worm body movements and sensory activities are coordinated by the neuronal network out of the 302 neurons.

Apart from most organisms, wild-type *C. elegans* has two unique sexual forms, male and hermaphrodite. They can reproduce not only by mating but also self-fertilization. In nature, male worms only occur 0.1% of the overall population. As compared with the hermaphrodite in appearance, a male worm is usually distinguished by its fan-shaped tail. Since the male is not considered the majority of the

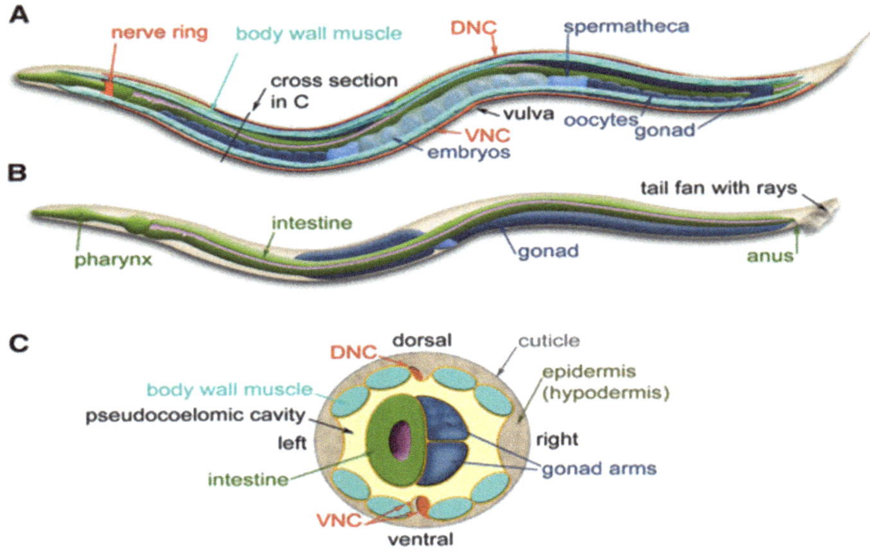

Fig. 6.3 Anatomy of *C. elegans*. (**a**) Anatomical features of hermaphrodite and (**b**) male worms. (**c**) Cross-sectional view of hermaphrodite in the anterior region. The schematic illustrates the four muscle quadrants surrounded by the epidermis and cuticle with the intestine and gonad housed within the pseudocoelomic cavity. (Reprint permission for all sub-figures from http://www.wormbook.org) [19]

worm population, unless mentioned otherwise, we will focus on hermaphrodite in the following content. Each hermaphrodite worm can produce around 300 eggs in its lifespan. The gonad of hermaphrodites forms an ovotestis that firstly produces haploid amoeboid sperm stored in the spermatheca in the L4 stage, and then the germline switches fate to produce much larger oocytes near adulthood. Young adult worms are mature in their germlines and start to lay eggs from adult day 1 to around adult day 6. Thereafter, the worm can keep living for another week or two. The short lifespan makes the animal easy for any life-long study.

Wild-type *C. elegans* performs undulating crawling to migrate on solid surfaces. Forward, backward and U-turn are three common locomotory gaits that worms generally maneuver. In a liquid phase, however, worms switch to thrashing-like swimming instead. The rate of body bends for N2 worms is around 1.5–2 Hz in average. In addition, a special behavior, termed quiescence/lethargy, can also be observed on *C. elegans* while they are in transition from one developmental stage to another. Although the mechanism remains not fully understood, lethargy is believed to be a sleep-like process during larval stages in *C. elegans* [12, 17]. Worms deprived of sleep can lead to growth defects or death.

Food uptake of *C. elegans* relies on constant pharyngeal pumping in the head. The pharyngeal pumping is usually associated with the healthy state of *C. elegans*. The pumping rate progressively declines after worms enter the late adulthood.

Egg-laying requires 16 muscle cells, including 4 vm1 and 4 vm2 vulval muscle cells and 8 uterine muscle cells [18]. Both vm1 and vm2 muscle cells are arranged in X-shape patterns around the vulva. The uterine muscle cells are responsible for pushing eggs out of the vulva by constricting the uterus. Of these cells, the 4 vm2 muscle cells are of particular importance to the vulval opening since only ablation of them causes an egg-laying defect. Nevertheless, loss of functions in these muscles increases the risk of eggs hatched internally in their parental worms, *a.k.a.* bag of worms (BOW). The occurrence of BOW is not rarely seen in nature. Age-related degeneration is a major cause of this phenotype. However, other factors from environmental toxins or genetic abnormality may trigger BOW as well.

6.2.3 Nervous System

Neurons in *C. elegans* differentiate in three developmental stages. The first stage is during the proliferation phase of embryogenesis, the second is at the late L1 stage, and the third is at the late L2 stage. According to the cells' loci in the hermaphrodite *C. elegans*, the nervous system is usually distributed in the anterior, the ventral cord, and the pharynx. Despite the simplicity of the worm's nervous system, its high degree of cellular diversity still astonishes researchers in many ways. In a general sense, all neurons in adult *C. elegans* can be classified into four functional categories: sensory neurons and motor neurons, interneurons, and polymodal neurons. Sensory neurons are specialized to sense a wide variety of environmental stimuli, including but not limited to mechanical, electrical, chemical, optical, and thermal variations. The perception of environmental cues is accomplished through 24 sensillar organs and various isolated sensory neurons (see Table 6.1). Each sensillum contains ciliated endings of one or more neurons. Most sensilla are located in the head, except for posterior deirids and phasmids. Motor neurons are responsible for locomotion, alimentary, and reproductive systems by making synaptic contacts onto muscle cells. The common behavioral phenotypes are forward, backward, and U-turn movements on agar or swimming/thrashing in liquid media. The motor neurons can also work together with sensory neurons in response to external inputs from gentle body touches or environmental vibration/tapping. Interneurons are bridges that connect different types of neurons. They function as information processors by receiving signals from one or more neurons and conveying decisions to other target neurons. They work as a primitive brain in modulating neuronal functions and process information. In addition to the three abovementioned neuron categories, polymodal neurons are some special neurons that can perform multiple circuit functions, such as ASH, M3, NSM, DVB, etc. An idealized network depicted in Fig. 6.4 is used to interpret the basic principle of a neuronal circuit for chemotaxis. Upon interactions and coordination between specialized neurons, *C. elegans* can maintain its life for all sorts of activities. To aim at the worm-based biosensors, the following paragraphs will focus on some common locomotory responses prompted by different sensations.

Table 6.1 Functions and input signals of some common sensory neurons

Neuron	Signal Sensed	Functions
ASE	Water-soluble chemicals	Chemotaxis (major)
AWC	Volatile odorants	Attractive chemotaxis, Lifespan, Navigation
AWA	Volatile odorants	Attractive chemotaxis, Lifespan (minor)
AWB	Volatile odorants	Avoidant chemotaxis
ASH	Touch, Chemicals, UV light, Osmolarity	Nociception: Osmotic avoidance, Nose touch avoidance, Chemical avoidance, Light avoidance
ASI	Water-soluble chemicals, Dauer pheromone	Dauer formation, Chemotaxis (minor), Navigation
ADF	Water-soluble chemicals, Dauer pheromone	Dauer formation, Chemotaxis (minor)
AFD	Temperature	Thermotaxis
ASG	Water-soluble attrantants (Na^+, Cl^-, cAMP, biotin), Hormonal signaling	Dauer formation (minor), Lifespan, Chemotaxis (minor)
ASJ	UV light, Electric field, Dauer pheromone), Hormonal signaling	Dauer formation and recovery, Chemotaxis (minor), Lifespan, Electrotaxis, Phototaxis
ASK	UV light), Hormonal signaling	Avoidance (minor), Chemotaxis (minor), Lifespan, Navigation, Phototaxis
ADL	Food	Avoidance (minor), Social feeding
URX, AQR, PQR	Oxygen, Food	Aerotaxis, Social feeding
PHA, PHB	Water-soluble chemicals	Avoidance (Antagonistic)

Modified from the WormBook [20]

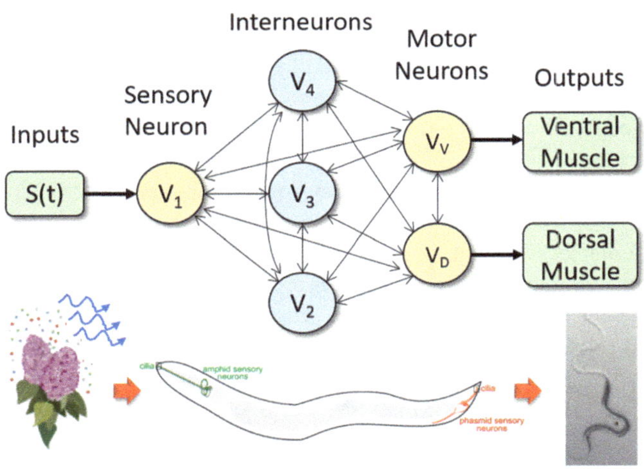

Fig. 6.4 An example of idealized neuronal network for chemotaxis in *C. elegans*

6.2.3.1 Phototaxis

Light sensation is one of the vital sensory organs in most animals. Phototaxis refers to a locomotion in response to optical stimuli. Unfortunately, *C. elegans* is a soil animal which lacks visionary capability to aid them navigating in the dark environment. However, this understanding was changed by Ward et al.'s study [10]. Ward and his colleagues investigated the negative phototaxis of *C. elegans* in 2008. Surprisely, they discovered that *C. elegans* can sense light via ciliated amphid neurons and maneuver repulsive reactions to it in a dose-dependent manner. Seven neurons, ASJ, AWB, ASK, ASH, ASI, AWC, and ADL, were verified to be involved with the avoidance response. Among them, ASJ, AWB, ASK, and ASH neurons seemed to show the strongest association with the head avoidance. In conclusion, Ward et al. indicated that *C. elegans* photoreceptor cells are primarily mediated by the CNG channels and the second messenger cGMP for the phototransduction. Notably, this phototaxis of *C. elegans* is associated with light wavelength and optical intensity. In another study, Edwards et al. [4] stated that *C. elegans* showed strong avoidance in relation to low wavelengths, for instance ultraviolet light. This particular relationship was inferred to the worm's physiological need of survival. Since sunlight exposure appears to be fatal to worms, sensing the right direction to migrate then becomes an essential ability to keep them safe from harm in the dark soil. By associating the noxious low wavelengths with the sunlight, the phototactic strategy can form immediately in worms to steer them away from the danger.

6.2.3.2 Chemotaxis

Chemosensation is vital to worms' survival since it can help worms find food, avoid noxious conditions, develop appropriately, and mate with male worms. The sensory neurons are mostly distributed in the cilia. There are roughly 16 pairs of amphid and phasmid neurons involved in chemosensation. The chemosensory neurons are interconnected with motor neurons to form so called "chemotaxis". To activate the reactions in *C. elegans*, target olfactory molecules must be firstly sensed by the neurons to trigger the subsequent signaling pathways. Therefore, *C. elegans* can be trained as a miniaturized robot to seek specific target molecules in an unknown environment. Neto et al. [21] attempted to use *C. elegans* as a biosensor to detect pulmonary tuberculosis (TB) through the exhaled breath from patients. Their study showed that TB-specific volatile organic compounds (VOCs), including methyl *p*-anisate, methyl nicotinate, methyl phenylacetate and *o*-phenylanisole, could be sensed by *C. elegans* and induced corresponding chemotactic responses. The approach improved the conventional diagnosis of TB which relies on a combined approach of clinical symptoms, a chest X-ray and sputum based laboratory testing including smear microscopy, bacterial culture and molecular methods. However, the observed olfactory response to the purified TB-specific VOCs demonstrated only preliminary feasibility. More work still needs to be done in order to verify the repeatability when the four VOCs are isolated from wild type *M. tuberculosis* and when spiked into human breath specimens.

6.2.3.3 Electrotaxis

In addition to odorants and light, *C. elegans* can sense electric fields and orient themselves to the negative poles as well. This tendency of movement toward the cathode in an electric field is then termed electrotaxis. Gabel and colleagues [6] firstly revealed the worm's electrosensory behavior and attributed the response to the amphid sensor neurons. Their in-depth investigation showed that ASJ and ASH were strongly stimulated when the worms' heads pointed to the anode in an electric field. ASJ and ASH apparently dominated the navigation and avoidance, respectively. As mentioned previously, these two neurons also play roles in mediating phototaxis and chemotaxis of *C. elegans* because ASJ typically responds to environmental cues during development and ASH is responsible for several nociceptive reflexes. According to their finding, a possible scenario as to how worms form their electrotactic orientation is by moving their heads randomly in an attempt to look for a direction that can lower the activity of these electrosensory neurons. Once the direction is found, they will orient their bodies to align with the electric field and move to the cathode. Although the physiological needs for such electrotaxis remains enigmatic, it may possibly be inferred to the inherent nature from foraging or mating lures the worms to move. Later, the electrotaxis of *C. elegans* was further applied in a microfluidic chip [22]. Evidences suggested that the effective electric field intensity changed with the worm's developmental stages. For worms younger than L3, they showed low response to the electric stimuli possibly because of immature sensory neurons. As worms aged (young adult), the required minimum intensity declined. The different responses to electric fields thus provide a way for size sorting [23–25].

6.2.3.4 Thermotaxis

Of all sorts of sensations, temperature is a unique factor that can regulate *C. elegans* growth rate and healthy states for the whole lifespan. Prior research [7, 26] has observed that *Caenorhabditis elegans* tends to migrate toward its cultivation temperature or isothermally tracks along that specific temperature in a spatial thermal gradient. The tendency of migrating up the thermal gradient is termed positive thermotaxis; otherwise, it is termed negative thermotaxis. It is a quite straightforward strategy to study the navigational skills in *C. elegans* to understand the plasticity and programming of sensor-motor circuits in a small nervous system [27]. In the thermotactic model, thermophilic neurons (AFD and AIY) and cryophilic neuron (AIZ) are involved in the behavior. Notably, worms were only tested between 15 °C and 25 °C and the gradient was as shallow as 0.2 °C/cm. It is obvious that temperature higher or lower than the thresholds that worms can bear will certainly harm the animals.

6.2.4 *Worm Cultures*

6.2.4.1 Growth on Agar

In general sense, *C. elegans* needs to be cultivated at a specific temperature and fed with sufficient food, *E. coli* OP50, on the nematode growth medium (NGM) agar dish. To this end, a drop of 0.1 mL of *E. coli* is firstly seeded on agar in a large petri dish with a pipette. Noted that the lawn of *E. coli* should be spread in the center of the dish to prevent worms from crawling to the edge. After seeding, the lawn of *E. coli* will grow overnight at room temperature or at 37 °C for 8 h [28]. Since the worm's growth is strongly regulated by its culture temperature, worms are grown in an incubator to ensure temperature is well controlled. Worm culture usually starts from eggs after bleach to achieve synchronization. The growth period in each stage can then be estimated according to the set temperature.

When worms reach adulthood, they will be able to lay eggs for the first 5~6 days. To avoid food shortage and interference from self-offspring, worms need to be transferred to a new agar dish daily to maintain the consistency. Several methods can be chosen for such purpose. A quick and convenient method is "chunking", which a chunk of agar from an old dish is cut and removed to a fresh dish. Another method is to pick single worms with a picker connected with a platinum wire (or eyelash) on one end. Between transferring, the platinum wire needs to be sterilized with fire to avoid contaminating the worm stocks. This method is performed when specific worms are selected. With sufficient food and space in a well-controlled environment, most worms can live up to 3 weeks.

6.2.4.2 Growth in Liquid Media

Since *C. elegans* is an amphibious animal, it can also be cultivated in liquid media. At macroscale, liquid cultures of *C. elegans* are usually grown on S medium using *E. coli* OP50 as a food source [29]. It is best to grow only one generation of worms in liquid before the worms are harvested. Whereas, growing worms for more than one generation, overcrowding can lead to dauer formation despite the presence of food. As a result, keeping food supply, *E. coli* OP50, sufficient in the liquid media is vital for a long-term worm culture. Worms usually consume food more and faster because of vigorous locomotion. Therefore, the amount of food needed should take into consideration the duration that worms are grown. As a reference, a 62.5 mL of liquid culture stock can support worms from a 100-mm petri dish to grow for 4~5 days [28].

At microscale, the maintenance is diverse due to various microchips and applications. Nevertheless, the amount of liquid determines the length of time that worms can survive. Without supplying fresh liquid with nutrient, single worms can live up to 4 days in a confined aqueous droplet as small as 125 nL. Otherwise, worms can live as long as those which are grown in an open or circulating environment. Notably, food appears to be associated with the worms' survival. Under proper condition, oxygen level and consumption can reach a safe equilibrium through gas exchange at the interface [12].

6.3 Basic Maneuvers for Worm Maintenance

6.3.1 Sorting

C. elegans worms at different developmental stages have distinct phenotypic behaviors and nerve responses [25]. Sorting *C. elegans* is of great significance in studying these stage-specific behaviors. Therefore, it is important to construct a rapid and accurate *C. elegans* sorting method. Manual picking is a preferred maneuver when the number of selected worms is small. However, the maneuver becomes time-consuming and labor-intensive as dealing with a large population. The commercially available automated sorter such as the BIOSORT/COPAS is massive and expensive, reducing its prevalence in the routine laboratory use [30]. Recently, microchip technology, relying on specially-designed micro-structure to sort *C. elegans* at different developmental stages, has emerged as an attractive and enabling platform to overcome these problems [31].

To date, sorting *C. elegans* on a microchip usually relies on the following three features [32]. (1) Fluorescence signal expression. For example, green fluorescent protein (GFP) expression in the URX neurons is consistent, but expression in AQR and PQR neurons is stage-specific, which can be used to classify *C. elegans* [33]. (2) Size difference. The life cycle of *C. elegans* consists of four developmental larvae stages (L1–L4) and an adult stage. In each stage, the nematode possesses different yet characteristic body sizes, morphological and anatomical features, which can be used to classify *C. elegans* [34]. (3) Behavioral phenotypes. It is known that *C. elegans* has distinct electrotactic responses to electrical fields. In the presence of an electric field, worms at different stages tend to crawl to the negative pole with different angles due to their deflecting electrotaxis [24].

Numerous sorter microchips using the electrotactic features have been developed. Chung et al. [35] fabricated a multivalve-integrated microfluidic chip that could identify the phenotypic features of the *C. elegans* under investigation and sort them one by one based on automatically classifying the fluorescent markers with customized software (Fig. 6.5). Worm screening throughput was achieved as high as several hundred worms per hour and the sorting accuracy reached over 95%. The microchip relies on a set of on-chip valves to control the suspension of *C. elegans* and differs from previous work in the integration of image-based feedback that enables robust

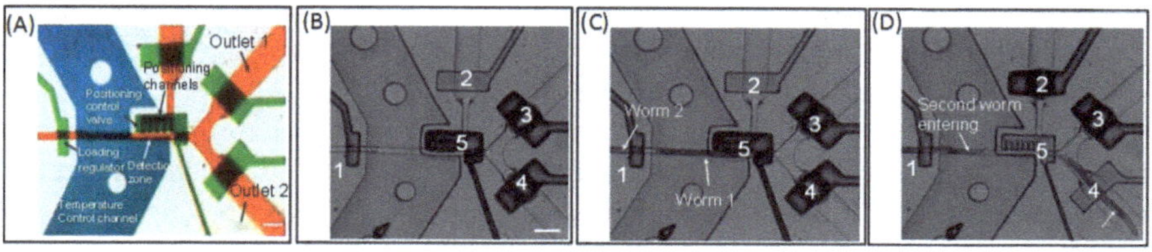

Fig. 6.5 Multivalve-integrated microfluidic microchip for *C. elegans* sorting. (**a**) Micrograph of the microchip full with different dyes showing active regions. (**b–d**) Video frames showing the chamber awaiting a worm to enter (**b**), with the worm loaded to prevent another from entering (**c**), and the worm exit from the chamber while a second worm automatically coming in (**d**). (Reprint with permission from [35]. Copyright 2008 Springer Nature)

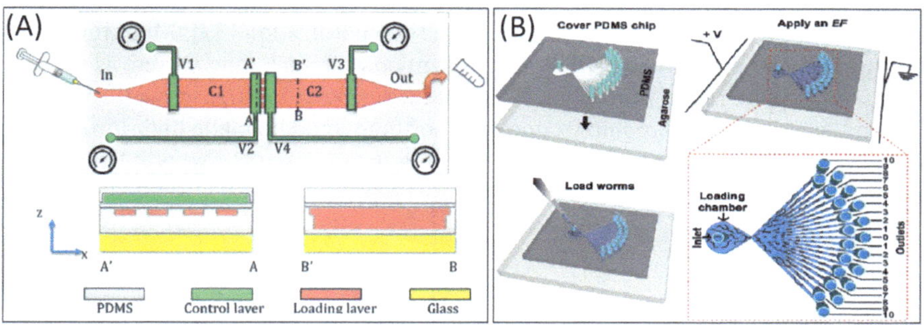

Fig. 6.6 Size- and electrotaxis-dependent sorting microchips. (**a**) Size-dependent sorting microchip. (Reprint with permission from [36]. Copyright 2015 Royal Society of Chemistry.) (**b**) Electrotaxis-dependent sorting microchip, with its schematic structure and working principle. (Reprint with permission from [24]. Copyright 2015 Royal Society of Chemistry)

operation, which is critical for high-throughput sorting. Particularly, the system is empowered with a customized control program to automate the entire processes of image acquisition and identification without human intervention. However, the microchip requires sucking *C. elegans* for still and high-magnification imaging, in which mechanical immobilization of the worm may cause traumatic damage and the sequential working mode limits the throughput.

Dong et al. [36] proposed a microfluidic chip for size-dependent sorting of *C. elegans* (Fig. 6.6a), making use of the external pressure-deformable cross-sectional shape of PDMS channels that run through two neighboring on-chip worm chambers. By setting the control pressure properly, the system separated *C. elegans* of specific developmental stages from a mixed worm population. The efficiency reached nearly 100% and throughput was up to 3.5 worms per second. The chip comprises two layers of PDMS structure, where one is for fluidic control and the other is for valve pressure control (green in Fig. 6.6a). Sandwiched between the fluidic and control layers is a thin layer of deformable PDMS membrane. In the fluidic layer, two isolated worm chambers C1 and C2 are connected by four parallel transfer channels. The effective cross-section and shape of the channels are regulated by a pneumatic control chamber orthogonally placed above. In this way, the transfer channels function as adjustable valves for size-dependent filtering. The microchip appears to be simple, highly efficient, and flexible.

Wang et al. [24] proposed a microchip (Fig. 6.6b) to realize the sorting of a population of *C. elegans* of different stages based on the stage-specific deflecting electrotaxis of *C. elegans*. Through investigation of the electrotactic angles of *C. elegans* at different electric field strength, the PDMS-agarose hybrid microfluidic chip was configured to have a number of circle-evenly-spaced sorting channels with a 5-degree interval in the range of $-50°$ to $50°$. The chip was assembled simply by covering the surface of an agarose gel with the PDMS chip treated with oxygen plasma to produce a hydrophilic surface, providing a highly efficient, cost-effective, and harmless method.

We envision that the microchips for sorting would follow the path towards integration, high-throughput, automation, and standardization [37]. More functional modules are expected to be integrated in one microchip including sorting as the very first sample preparation procedure. To meet the needs of biologists, who are potential end-users of the microchip technology, sorting would be made high-throughput, automated and standardized to use. In this way, microchips will find more applications.

6.3.2 *Immobilization*

Worm immobilization is an essential step in many applications, such as high-resolution imaging [38], sorting [39], monitoring neuronal responses after various stimuli [40], laser ablation of neurons [41], and exogenous gene microinjection [42]. Considering that the vulnerability of *C. elegans*, gluing and anesthesia [43] are often performed in conventional worm immobilization. However, labor-intensive, low-throughput, and irreversible nature limit their practical uses. Recently, quite a few high-efficiency and smart immobilization techniques have been proposed, including mechanical constriction [44, 45], valve-based aspiration [46], anesthetic gas (*e.g.*, carbon dioxide) [47]. These methods generally work well for particular application cases. However, there still exist unwanted limitations. Typically, mechanical constriction methods could immobilize *C. elegans* at certain places but without on-demand function. Valve-based aspiration methods need complex fabrication procedure and peripheral control. Moreover, these two methods easily injure *C. elegans*. Anesthetic gas methods could long-time completely immobilize worms, but they will influence synaptic transmission and have detrimental effects on sub-cellular neuronal transport. By contrast, thermosensitive hydrogel methods and dielectrophoresis (DEP) methods could eliminate issues of gluing and implement low-injury, rapid, reversible immobilizing. Here we introduce such two methods for immobilizing *C. elegans*.

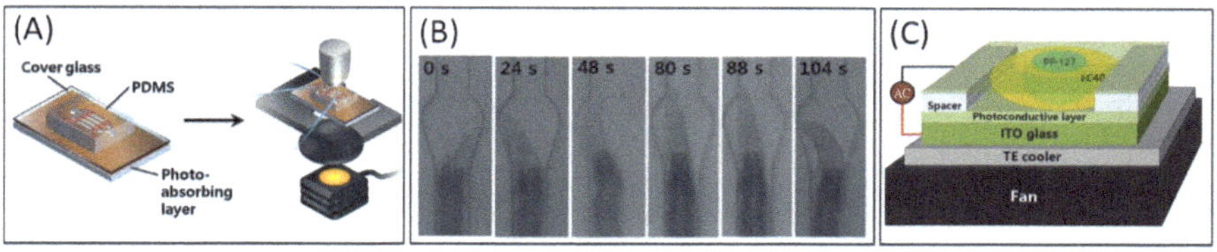

Fig. 6.7 Thermosensitive hydrogel methods for immobilizing *C. elegans*. (**a**) Schematic diagram of a protocol for immobilization of worms in a PDMS-glass microfluidic device. (**b**) Time-lapsed snapshots showing a worm trapped and immobilized in the microfluidic channel. Quantitative analysis of locomotion of the worm by tracking head position according to the time. (Reprint with permission from [49]. Copyright 2014 Royal Society of Chemistry) (**c**) Close-up view of the optoelectric device. (Reprint with permission from [50]. Copyright 2017 Elsevier)

6.3.2.1 Thermosensitive Hydrogels

Thermosensitive hydrogels can be used to immobilize worms by changing temperature to modulate the phase of the hydrogels. Pluronic F-127 is a thermo-reversible and biocompatible copolymer. By adjusting the ratio of PF-127 in a dissolving medium, its gelation temperature can be tuned to the worm's culture temperature (15–25 °C) [48]. Therefore, a slight temperature change can rapidly immobilize worms by constraining them in a gel. *C. elegans* appears to be unaffected and stay healthy. Hwang et al. applied a photo-absorbing layer to the substrate to raise temperature for gelation of the PF gel. This immobilization strategy requires only a normal white light source [49] (Fig. 6.7a). It can be configured to selectively immobilize a specific worm on demand by integrating a spatial light modulator. Fig. 6.7b shows an immobilized worm after 60 s of illumination.

Despite the abovementioned benefits, this method requires a long gelation time (45~60 s) and cannot immobilize/recover worms rapidly. Later, Chuang et al. [50] achieved faster worm immobilization based on bulk heating rather than surface heating by the optoelectric effect. The optoelectric device is composed of two indium-tin-oxide (ITO) glass plates separated by a spacer (Fig. 6.7c). One of the glass plates is coated with a photoconductive layer, which serves as an optical switch. Selected worms were fully or partially immobilized in 4 s and recovered in 1 s. This rapid and reversible immobilization technique provides insight to worm-based applications that require long-term and constant monitoring, such as drug assays.

The microchips with thermosensitive hydrogels feature simple, versatile, and biocompatible immobilization of *C. elegans*. Localized illumination allows selective immobilization of specific worms of all developmental stages at normal physiological conditions.

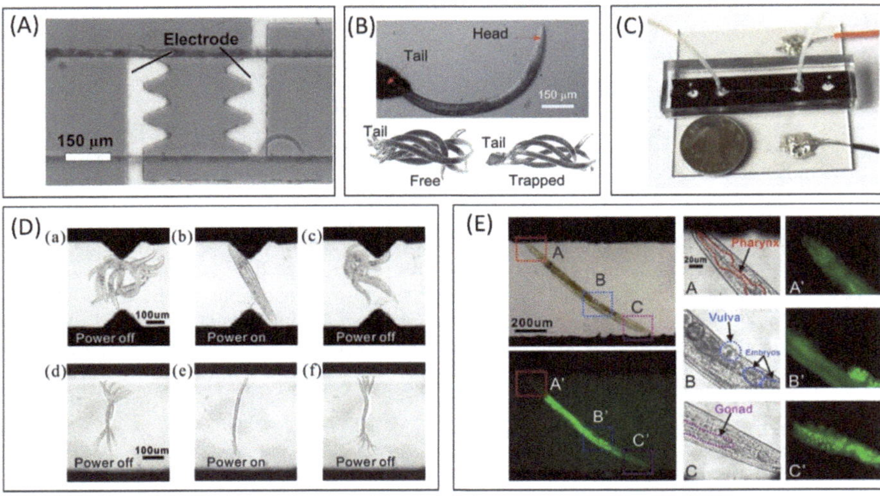

Fig. 6.8 Dielectrophoresis methods for immobilizing *C.elegans*. (**a**) Photograph of planar electrodes microfluidic device. (**b**) An adult worm trapped in the electric field. (Reprint with permission from [53]. Copyright 2011 Royal Society of Chemistry.) (**c**) Photograph of C-PDMS microfluidic device. (**d**) Immobilization results for *C. elegans* of different stages under different conditions. Multiple time-lapse images are superimposed in one picture for three states before/on/after immobilization. (**e**) Bright-field and corresponding fluorescent images of immobilized *C. elegans*, whole-body imaging at 10X and segmented high-resolution imaging at 60X. (Reprint with permission from [54]. Copyright 2018 Elsevier)

6.3.2.2 Dielectrophoresis

Numerous studies have demonstrated *C. elegans* immobilization with thermosensitive hydrogels. However, exposing to the hydrogel over a long period of time retards the growth of worms. To avoid the drawbacks, rapid, low-injury, reversible, and easy-to-implement immobilization techniques remain intensively pursued. Due to some notable merits such as little harm to organisms and fast response in action, DEP has been widely used to manipulate DNA, proteins and cells [51, 52]. *C. elegans* in a non-uniform alternating electric field can be polarized. The polarization DEP forces induce a torque to align the body with the electric field lines and thus immobilizes the worm. Chuang et al. [53] micropatterned a pair of planar gold electrodes on a glass slide (Fig. 6.8a). The worms were trapped at the location of the maximum electric field intensity. The magnitude of DEP forces can be readily controlled by adjusting the electric signals and worms can be trapped and released on demand (Fig. 6.8b).

Planar electrodes have showed great success to immobilizing worms. However, the electric field distribution is distance-dependent for planar electrodes. Accordingly, the field intensity significantly attenuates as worms stay away from the electrodes. This physical limitation impedes the immobilization of *C. elegans*. In order to alleviate the electric non-uniformity, Huang et al. [54] proposed an immobilization microchip with 200-μm-thick electrodes, which were made of carbon black PDMS (C-PDMS, carbon black nanoparticles mixed with PDMS) (Fig. 6.8c). The two parallel thick C-PDMS electrodes multiplex as the microfluidic channel sidewalls. When *C. elegans* swam inside the channel, switching on/off the electrodes could easily immobilize/release them. Through experiment, *C. elegans* of different stages were immobilized on demand in the microchannel, proving the efficiency and applicability of the method (Fig. 6.8d). High-resolution imaging of *C. elegans* was also demonstrated as a supportive evidence (Fig. 6.8e).

Both thermosensitive hydrogels and DEP technology provide simple, low-injury, versatile, and biocompatible capabilities to immobilize *C. elegans*. They can be integrated on microchips and have been demonstrated for immobilization of *C. elegans* of all developmental stages without physical damages. How to develop these immobilization platforms for more worm assays needs inputs from the *C. elegans* biologists.

6.3.3 *Confined Cultures of* C. elegans

Long-term cultures are crucial for investigations of *C. elegans*. The basic requirement for long-term survival of *C. elegans* is adequate nutrition supply, including oxygen and food. At present, *C. elegans* research is mainly carried out in liquid or on agar plates. Worms are bred in a large number of groups based on the traditional research method. As a result, plenty of reagents is needed and experiments are time-consuming, labor-intensive, and low throughput. Moreover, accurate transferring, controlling, and individual worm tracking are difficult. To tackle the problems, a wide variety of microchips for cultivating *C. elegans* were developed [31]. Polydimethylsiloxiane (PDMS) is commonly used in the microchip fabrication because of good permeability for gas exchange and therefore suitable for long-term cultures of worms [32]. There are two main ways to create a microenvironment for confined culture of *C. elegans*. One is to use channel-formed micro-chambers to trap, house and grow *C. elegans*. The other is to use micro-droplets to encapsulate *C. elegans* and transfer the droplets to a micro-chamber for confined cultures [31].

6.3.3.1 Microchannels

Channel microfluidics offers a continuous flow system, enabling long-term confined cultures for *C. elegans* because of the ability to exchange nutrients and excretions [55]. Krajniak et al. [56] proposed a channel-microchip that allowed long-term cultures of *C. elegans* starting at L1 stage (Fig. 6.9). The microchip enables to breed, individually track and repeatedly image *C. elegans* at high resolution while supporting normal development for *C. elegans* of any stage. The microchip is able to provide both nutrients and proper gas exchange to multiple worms bred in individual chambers, which are adopted to separate worms at all times, avoid cross-

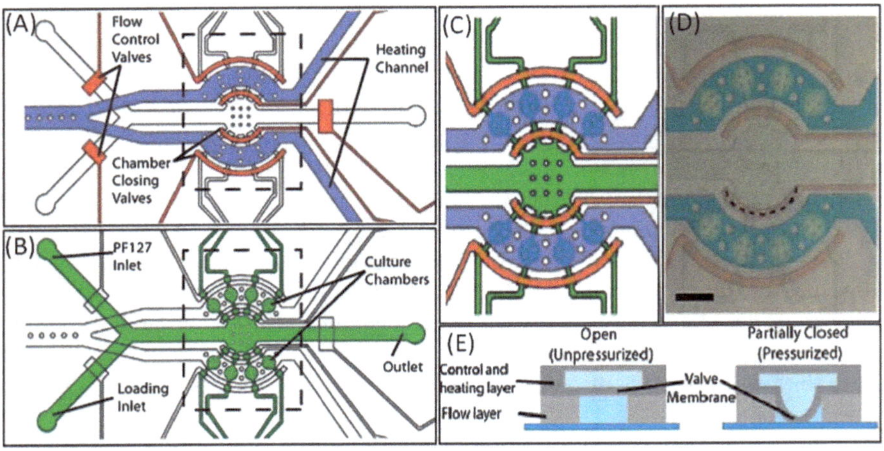

Fig. 6.9 Two-layer channel-microfluidic microchip for confined cultures of *C. elegans*. (**a**) The heating layer. (**b**) The fluidic layer. (**c–d**) Culture chambers. (**e**) The cross section of a valve. (Reprint with permission from [56]. Copyright 2010 Royal Society of Chemistry)

contamination and achieve single-worm analysis. Made of PDMS, the microfluidic chip has two layers including fluidic layer and heating layer. The fluidic layer has flow inlets, outlets and the individual culture chambers. The heating layer contains flow-control valves and temperature-control heating conduits. During loading process, *C. elegans* and culture medium are delivered to the chamber from the loading inlet, such that *C. elegans* enter and float in the individual culture chambers full with food and supplemental cholesterol. During analysis process, PF127 (thermoreversible and biocompatible copolymer, which can realize solution-to-gel transition when temperature is tuned) solution is flowed into the device from the PF127 inlet. The phase of PF127 can thus be tuned by controlling the flow rate and temperature of the fluids in heating-layer channels. At high temperature, PF127 is converted into gel which is capable of immobilizing *C. elegans*, so high-resolution image for analysis is accessible. At low temperature, PF127 is converted into solution, thus *C. elegans* can still be bred in the individual chambers.

The primary function of the device was designed to monitor specific developmental processes. However, the device seemed to be more excellent in maintaining culture conditions. In this channel-microfluidics device, early L1 stage *C. elegans* worms were loaded and maintained for up to 36 h until they became young adults. Moreover, the microchip can be scaled-up in terms of the array number to realize higher throughput.

6.3.3.2 Microdroplets

Operation processes in channel microfluidics require a longer time and thus consume large amounts of reagents. Moreover, channel microfluidics cannot control flow across channels independently [57]. As a result, droplet microfluidics becomes an

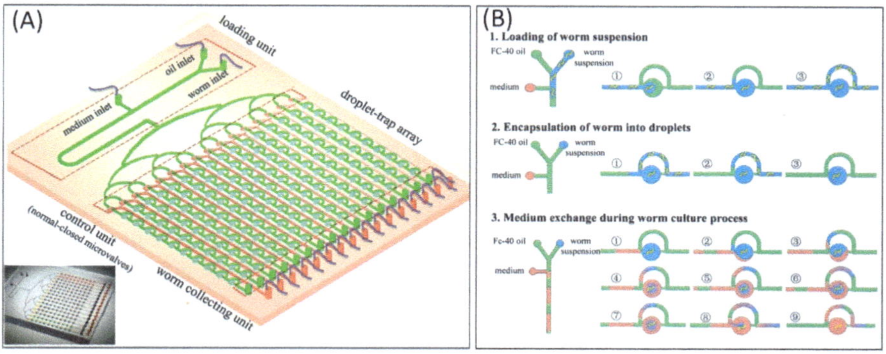

Fig. 6.10 Droplet-microfluidics microchip for *C. elegans* cultures. (**a**) Schematic diagram of the droplet chip. (**b**) Individual *C. elegans* loading, encapsulation and substance exchanges by using droplet operations. (Reprint with permission from [58]. Copyright 2015 Royal Society of Chemistry)

alternative method for confined cultures of *C. elegans* because of its discrete nature. Unlike the continuous channel microfluidics, droplet microfluidics creates discrete microdrops using immiscible phases. Moreover, droplet microfluidics can control each droplet independently, thus guaranteeing that each microchamber can be manipulated individually. As a result, droplet microfluidics can reduce the amount of reagents and processing time while offering greater potential for increased throughput and scalability than channel microfluidics [55]. Taking these advantages, Wen et al. [58] proposed a droplet-microfluidic microchip, which can not only handle individual worms flexibly through a controllable microenvironment, but also conduct high-throughput analysis for setting up bio-chemical conditions (Fig. 6.10). A series of droplets can be generated precisely with adjustable size. Then, they are used to encapsulate individual *C. elegans* and provide a collection of external cues to study the worm's behavioral response Integrating four different functional modules, including a sample-loading unit, a microvalve control unit, a droplet trapping array, and a worm collection unit, the droplet chip allows on-chip manipulations of worm sampling and long-term confined cultures, enabled by efficient nutrient exchange within the droplets. Parallel microfluidic channels are deployed and switched on/off to analyze multiple strains of individual *C. elegans* simultaneously. Sufficient substance exchange within the droplet is achieved by applying continuous coalescence and splitting steps between the interface of worm-in-droplet and multiple medium plugs, meeting the needs of long-term culture of *C. elegans*. This droplet-based microfluidic system is powerful for long-term study of developmental process of worms, in applications such as animal test in multiple drugs or environmental toxins.

To date, great progress has been made in the confined cultures of *C. elegans* on microchips even though it remains in the infancy stage. We envision integrating more analysis modules alongside cultures will be a challenge to face with. However, this step will boost the significance of confined culture modules in the real world.

6.3.4 Long-term Imaging

Long-term imaging is commonly applied in the environmental impact on worm development, nerve damage and regeneration, and the molecular pathways related to complex human dysfunctions [59, 60]. To achieve long-term imaging, specific worm individuals must be identified due to the need of repeated observation, thus it is important to culture and track animals individually [61]. Moreover, the microchip should have the capability of maintaining the normal development of *C. elegans* [56]. When immobilization is needed, for example, to observe specific small features such as cells/organs/nerves inside *C. elegans*, the immobilization module should be harmless to the animal.

Various microchips that can achieve long-term imaging have been proposed. They can be classified into two kinds based on whether an immobilization module is required. Normally, if imaging-based parameters may be obtained when the worm moves, like *C. elegans* locomotion metrics, forces, and body size, immobilization is unnecessary or prohibited. In this case, the key point is to guarantee that *C. elegans* can be cultured and tracked individually. If imaging-based parameters cannot be obtained unless *C. elegans* stays still, like *C. elegans* internal cells/organs/nerves, immobilization is required and the key thing is to ensure *C. elegans* maintains normal physiology that would not hamper imaging function under immobilization.

6.3.4.1 Microchips Without Immobilization

Churgin et al. [62] made a microchip called the WorMotel (WM). As shown in Fig. 6.11a, the WM is configured with hundreds of individual wells to house single worms. Each well is filled with necessary living materials including standard agar liquid media and bacterial food to support long-term cultivation and imaging of uniquely identified animals. Depending on the application, the WM can be imaged continuously or intermittently. Activity analysis could characterize aging and development of *C. elegans*. The worms could live more than 10 days in the WM. However, due to the liquid culture environment used in the WM, the operation of the microchip is challenging.

To avoid liquid handling, Pittman et al. [61] made a microchip called Worm Corrals that could culture 100~200 *C. elegans* individually, from hatching to death, on a solid gel medium supported by a microscope slide. The structure of the system is shown in Fig. 6.11b. Each animal occupies a single pad of food bacteria adsorbed to a PEG hydrogel, which is supported by a polyethylene frame attached to the glass slide. PDMS silicone cured atop the device binds covalently to the PEG gel wherever those materials are in direct contact (*i.e.* all locations except the positions of the food pads); thus each animal is confined while still movable on the two-dimensional surface of its food pad by covalent bonds in all directions. To implement hatching to death culture, one single egg is loaded first into each individual pad. This is done by dipping drops of bacteria medium onto the pad and then loading one egg into the medium by pipette.

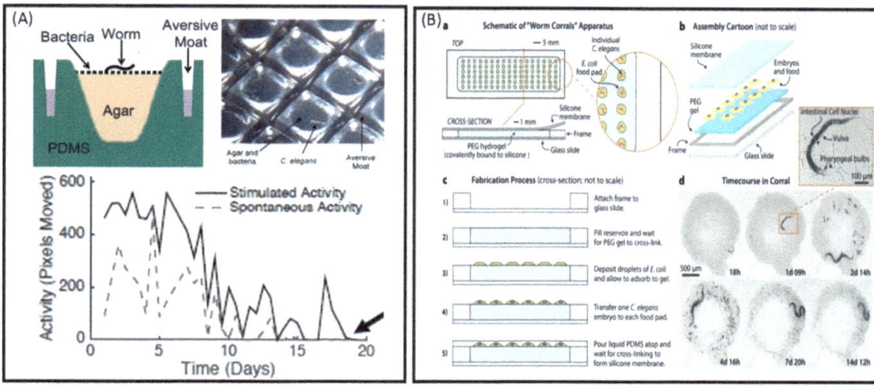

Fig. 6.11 Microchips with no immobilization module needed. (**a**) Design of the WM chip and *C.* elegans' activity data. (Reprint with permission from [62]. Copyright 2017 eLIFE) (**b**) Worm corrals, with schematic of the chip, components, fabrication process, and cultured *C. elegans*. (Reprint with permission from [61]. Copyright 2017 Royal Society of Chemistry)

6.3.4.2 Microchips with Immobilization

Reliable worm immobilization is highly demanded when specific parts or small features need to be monitored over a period of time. Hulme et al. [63] made a microchip that could not only culture worms individually in the chamber, but also immobilize worms when imaging. Using microfluidic "clamps" to immobilize worms periodically and temporarily, the microchip is useful to image individual worms from L4 stage until death in separate channels. The structure is shown in Fig. 6.12a. Only a worm in the early stage of L4 could get into the circular chamber due to the width of the right microchannel, then after 6 h at 24 °C, the worm becomes too large to exit the chamber from the right microchannel when the nutrient liquid is provided from the left. By reversing the direction of flow from right to left, the worm could be directed into the clamp for periodic, temporary immobilization. After imaging, the worm could be squeezed back to the chamber by directing flow from left to right.

Imaging *C. elegans* nervous system is of particular importance and interest, especially in investigations of the response of *C. elegans* to external stimuli, such as environment constraint and odor/chemical. Here we introduce two such classic microchips [38]. The first one is the behavior chip (Fig. 6.12b), which permits the correlation analysis of the worm locomotion in association with its neuronal activity. The behavior chip consists of a worm trap, which is a straight channel and differs from the above-mentioned tapered clamp. The trap dimensions were optimized to accommodate young adult worms. One design trick was that the width of the trap gradually decreases to nearly half at one end to prevent worm escape. Using the behavior chip, the group successfully imaged calcium transients in the AVA inter-neuron and used the data to explain how the neuron activity correlates with the direction of the traveling body wave. The second microfluidic device is called the olfactory chip (Fig. 6.12c), which integrates a worm trap channel with a microfluidic

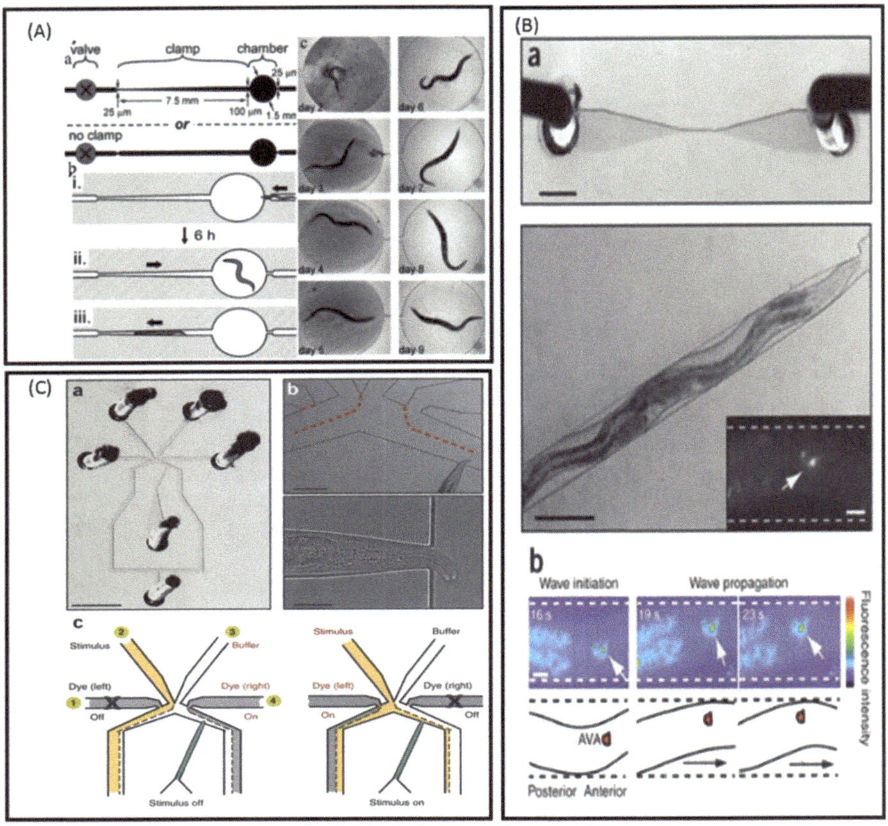

Fig. 6.12 Microchips with immobilization module for *C. elegans* imaging. (**a**) Clamp structure for *C. elegans* immobilization with culture chambers. (Reprint permission from [63]. Copyright 2009 Royal Society of Chemistry) (**b**) The behavior chip and its working for *C. elegans* neuron imaging associated with body wave propagation. (**c**) The olfactory chip and its working for *C. elegans* neuron imaging associated with stimulus solution. (Reprint permission from [38]. Copyright 2007 Springer Nature)

chemical delivery module. The trap channel for the young adult worm is the same as the behavior chip, but the constriction at one end is smaller. Therefore, when the worm is loaded, the very end of the nose protrudes out of the trap channel into the chemical delivery channel with flowing stream of the control or stimulus solution. The group applied high osmotic strength stimuli to the worm and simultaneously recorded calcium changes in ASH sensory neurons.

Microchips for long-term imaging often require chambers/wells/channels to culture worms, which may or may not be immobilized by compression or temperature-driven sol-gel transitions. It is crucial for the microchips to be able to maintain the normal physiological conditions for *C. elegans* during long-term imaging. With the aid of existing culture and immobilization techniques, long-term imaging of low-intensity fluorescent signals from neurons or rather strong bright-field signals for the whole animal has been made possible. We envision these microchips would find wide applications in drug testing and animal behavior screening.

6.3.5 Biomechanics

Like all animals, *C. elegans* is subject to continuous and a wide range of mechanical loadings during its daily life. The importance of *C. elegans* mechanics is highlighted by previous studies suggesting that the worm's locomotion patterns and foraging behaviors are strongly dependent on mechanical and osmotic stresses [64]. To date, studies of *C. elegans* mechanics focus on two aspects: one is to investigate its anatomy structure of cuticle, measuring its internal pressure or body stiffness; the other is to investigate *C. elegans* mechanical response to external conditions, measuring mainly its forces and locomotion parameters in the interplay with external environment. This chapter concentrates on the latter in which the worm is treated as an animal as a whole.

There are different microchips to measure forces, but all based on the displacement or bending of sensing elements, which get touch with the worm body [65–68]. Park et al. [65] fabricated a silicon piezoresistive cantilever which was utilized as a force-displacement measurement system (Fig. 6.13a). The worm is immobilized on the substrate and its body is touched by the cantilever tip, whose displacement is measured and converted to corresponding force. The system can measure forces between 100 nN and 1 mN for tip-worm distance of up to 100 μm with a resolution of 12 nN when the measuring frequency is between 0.1 Hz and 100 kHz. More recently, a microchip was developed with SU-8-based pillars. The pillars work as cantilever, and their tip gets touch with the worm body, but their bottom base is deployed with four 90°-spaced gold resistors as strain gauges (Figs. 6.13b and c) [66]. Measured by proper electronic instrumentation, the resistance change of the strain gauges gives the force applied at the tip of the pillar. Featuring sub-μN force resolution from 1 Hz to 1 kHz, >25μN range, kHz acquisition rate and biocompatibility, the device is smart in terms of using the electrical signal to map the force. However, the structure complexity, heat dissipation and optical occlusion from gold resistors cause unwanted interference to the locomotion pattern and instrumentation.

To overcome the drawbacks, a simpler force sensor was proposed by Ghanbari and colleagues [69]. The system still consists of an array of micropillars, but the pillars are made of transparent and elastic PDMS only (Fig. 6.14a). Correspondingly, the deflection of the cantilever-like pillar is measured from a vision-based algorithm, rather than from strain gauges. Subsequently, the force can be resolved from the deflection using the deflection-force model (Fig. 6.14b). The microchip provides a powerful platform for measuring continuous, multi-point forces of *C. elegans*, which can maintain its normal locomotion patterns. Experimental results showed that a resolution of 2.07 μN was achieved, enabled by the sub-pixel resolution for visual

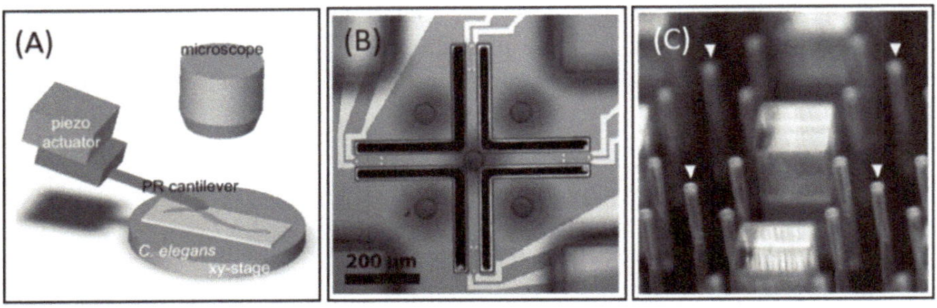

Fig. 6.13 Three microchips for measuring *C. elegans* forces. (**a**) Schematic of nematode measurement system with a piezoresistive cantilever. (Reprint with permission from [65]. Copyright 2007 PNAS) (**b**) SU-8-based pillar microchip viewed from the top. (**c**) The force sensing pillars, indicated by the arrows, are surrounded by passive spacer pillars and posts. (Reprint permission from [66]. Copyright 2009 Royal Society of Chemistry)

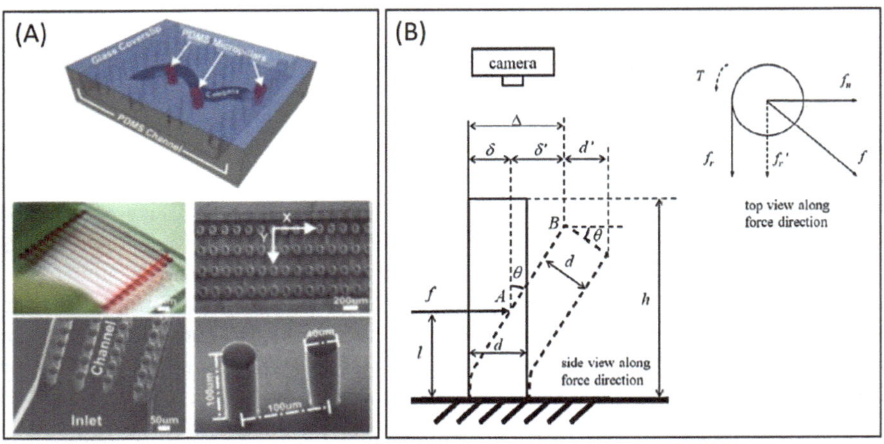

Fig. 6.14 PDMS pillar arrays for measuring multi-point forces. (**a**) The schematic of a worm deflecting PDMS micropillars while moving between them, and real pictures of the channel with pillars. (**b**) Schematic of the bending pillar for force analysis. (Reprint with permission from [69]. Copyright 2012 IOPScience)

tracking of the pillar deflection. 13 worm samples were experimented a maximum force of 61.94 μN was observed from 1571 data points.

Later, Wang and colleagues [70] extended the work to explore how the force pattern is correlated with locomotion patterns. To enable the research, they borrowed the ‘artificial dirt’ concept by using the pillars to mimic the worm’s soil-habitat environment. To incur different locomotion patterns, the microstructured pillar arrangement and spacing were adjusted. Two different pillar configurations were used in the microchip (Fig. 6.15a), including the ‘honeycomb’ (HC) design and the ‘lattice’ (LC) design. The spacing values were chosen to allow free movement of the

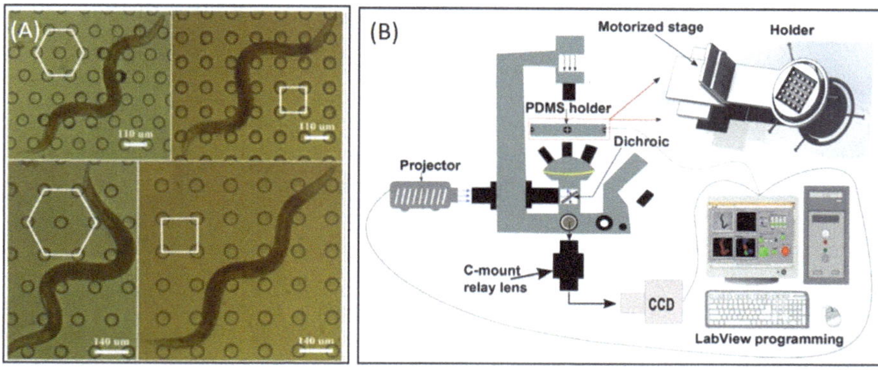

Fig. 6.15 PDMS pillar arrays configured for extended locomotion analysis and optigenetic manipulation. (**a**) PDMS device with two sets of micropillar configurations and two different spaces between the pillars. (Reprint with permission from [70]. Copyright 2013 Royal Society of Chemistry.) (**b**) Schematic view of the optogenetic platform. The microscope condenser provides bright-field light, the projector provides optogenetic light, the motorized stage actuates tracking of *C. elegans*, and the computer controls the hardware elements (*e.g.*, camera, stage, and projector) and relevant image processing software. (Reprint with permission from [71], Copyright 2015 AIP Publishing LLC)

worm inside the matrix of pillars. Interesting experimental results were obtained, and one is that the worm had smaller undulation frequency and locomotion speed for HC design than the LC counterpart, which was consistent with literature. They also provided evidence that the natural sinusoidal movement of *C. elegans* remains similar in the device, despite the presence of the PDMS micropillars.

Recently, Wang et al. [71] reported on the development of an optogenetic system (Fig. 6.15b) integrating optical illumination for neuron manipulation and a microfluidic chip with soil-environment-mimicking and force-measuring micropillar arrays, taking advantage of both optogenetics and microfluidics. The microchip is exactly the same structure as previously reported [70], but integrated with four control and instrumentation modules including illumination pattern generation, pattern projection, automatic visual tracking of the worm, and force measurement. They proposed a customized image-processing algorithm to isolate the worm's optical contour from the neighboring pillars, and track and extract its skeleton automatically. In return, the skeleton information is used to provide instantaneous position of the worm body such that specific illumination pattern can be projected onto the device through the lighting source. The system was demonstrated to work well and be useful for revealing subtle force and locomotion difference between *C. elegans* under normal or intensified lighting environments.

Overall, the biomechanics studies of nematodes may be promising though, more improvements still can be expected. One urgent point, for example, is to standardize the definition of behavioral parameters for nematode locomotion and forces. Therefore, data from different laboratories can be comparable. This will rely on more close collaborations between engineers and *C. elegans* biologists in the future.

6.4 WoC Applications in Environmental and Biomedical Fields

From the past to the present, humans have long been inspired by nature in all aspects. Despite the rapid development in technology, the diversity and complexity of natural species remain second to none. As compared with numerous biomimetic artifacts to date, the nature-born life may not be outstanding in a single domain. However, they are more robust and flexible to deal with multiplexed information in the real world. As a result, when lab-grown tissues still suffer from expressing full functions, the tiny nematode can certainly provide more comprehensive functionality than its counterparts. Unlike other higher animals, *C. elegans* is easy in maintenance, small in size, low in cost, but sophisticated in functions. They have been widely used in medical and biological applications. To take advantage of this multi-cellular organism, more researchers have attempted to incorporate *C. elegans* into different microchips to carry out a worm-based analysis system. Indeed, such kind of WoC can hence show us more possibilities in brand-new perspectives.

6.4.1 Drug Screening

Drug screening is usually a heavy-duty and lengthy process in the conventional pharmaceutics. In order to identify and optimize potential drugs for clinical trials, large libraries of chemicals must be screened beforehand. To this end, high throughput assays are always needed to improve the effects. Cells have long been used as targets to respond to the stimulations from different drugs [72–74]. However, there remains a gap between single cell analysis and animal tests. Although higher animals can provide more comprehensive information than single cells, high cost, long turnaround time, ethical issues, and sophisticated mechanisms hamper their prevalent use in reality. Instead, *C. elegans* is well known as a simplest multi-cellular organism in between single cells and higher animals. Therefore, the nematode can serve as an ideal candidate to provide biological cues in the selection of potential drugs. Putting the concerns into consideration, various worm-based drug screening studies have been emerging in recent years [75–77]. Chemotaxis is a common physical expression to show the worm's perceptions in the tested media. The visual responses may involve with developmental, physiological and behavioral changes. The olfactory system of the worm works as a major receptor to steer worms from an adverse environment to a favorable one. In general, numerous chemosensory neurons are interconnected by synapses in a compact neural circuit, allowing the worm to respond to sophisticated chemical stimuli. Of the 11 pairs of amphid chemosensory neurons (Fig. 6.16), the AWA, AWB, and AWC neurons are specialized for volatile odorants [78, 79], while ASE neurons are responsible for detecting salts and small soluble molecules [80]. Chemical and osmotic avoidance may be triggered by ASH, ASK, ADL, PHA, and PHB neurons. In particular, ASH

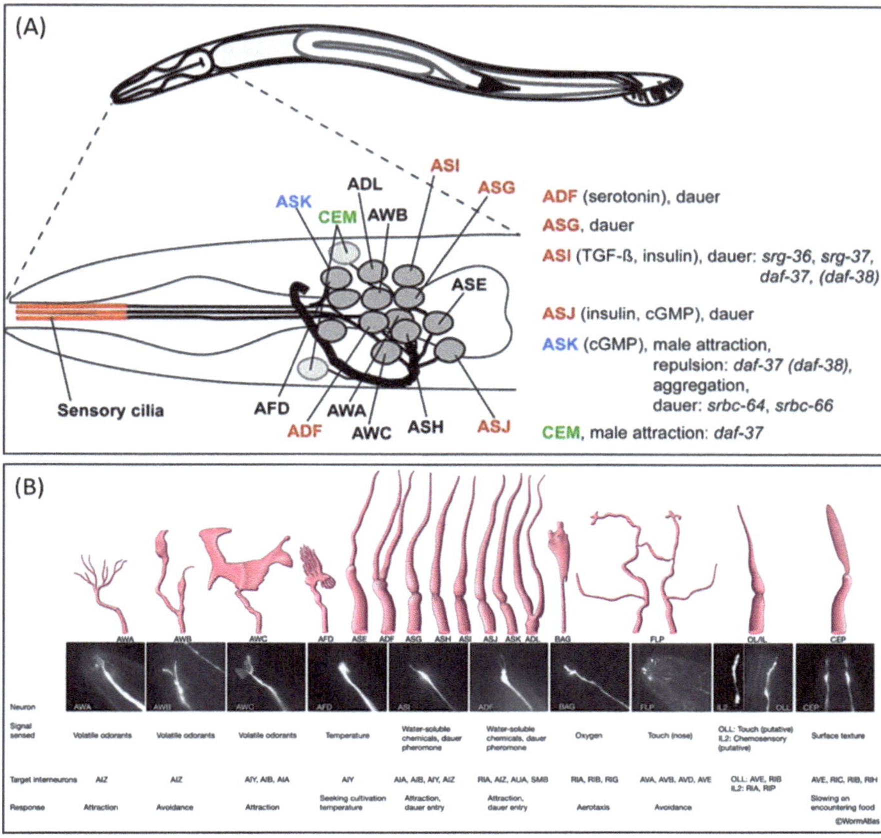

Fig. 6.16 (**a**) Head neurons in *C. elegans* and their roles for ascaroside-mediated phenotypes, downstream signaling pathways, and identified receptors (Reprint with permission from http://www.wormbook.org) [89]. (**b**) Cilia morphology of the sensory neurons in the head. (Reprint with permission from http://www.wormatlas.org) [90]

is a polymodal nociceptor analogous to pain-sensors in vertebrates [81–84]. It was found that high osmolarity, chemical repellants, or nose touch can all induce a sharp increase of the calcium level in ASH, activating a rapid response of avoidance. Repulsion from detergent and heavy metals, besides ASH, may also involve with additional ASK and AWE neurons. Parasitic environment and soil have vigorous metabolic activities, the ability of sensing oxygen and carbon dioxide becomes vital to the worm's survivorship in the nature. Prior studies [85–87] have shown that aerotaxis due to oxygen content can be associated with the URX, AQR, and PQR neurons. The neurons have a strong tendency to keep worms staying in an area of oxygen levels between 4% and 12% [87, 88]. According to Gray et al.'s investigation [87], shifting of the oxygen content can regulate the worm's social behaviors, such as clumping.

Luo et al. [91] once conducted a motility assay of *C. elegans* in an array of open wells to show their olfactory responses to odorants. Cycles of alternating streams of airborne isoamyl alcohol (IA)/benzaldehyde (BZ) and clean air were supplied to the wells. They found that a low dose (dilution ratio $D = 10^{-4}$) of IA and BZ can elicit attractive movement in wild-type worms. On the contrary, however, a relatively high dose (dilution ratio $D = 10^{-2}$) of IA and BZ elicits repulsive movement in wild-type worms. AWC and AWB neurons were identified capable of mediating the repulsion and attraction, respectively. In addition, the chemosensory neurons showed a clear dose-dependent trend as well.

Shi et al. [92] reported a microfluidic system composed of a droplet array (Fig. 6.17a). By trapping single worms in a droplet in the upstream and transport them to the droplet array in the downstream, they can achieve single-worm resolution while maintaining a high throughput. The innovation of this device is that no interference is incurred between worms, allowing single-worm tracking in a long timeframe. However, the encapsulation rate remains limited by the Poisson distribution. The microchip was used to investigate the neurotoxin, 1-methyl-4-phenylpyridinium (MPP+). The toxin causes the symptoms of Parkinson's disease in vertebrates by impairing their dopaminergic neurons. Since prior study has indicated that IC_{50} of MPP+ on worms was 11 mM, they attempted 3 mM and 5 mM of MPP+ on their microchip. The result revealed that 5 mM and 3 mM still exhibited significantly differences in stroke frequency and omega turn on worms. The slight changes in behaviors can be distinguished by the microchip. Later, similar design of static droplet arrays was adopted and modified by Aubry et al. [93] (Fig. 6.17b). With the flow chamber, they investigated worm's behavior while having them exposed to different concentrations of an anthelmintic drug, tetramisole. A dose-dependent and temporal drop in thrashing was also observed at concentration of the drug higher than 10 mM. The concentration showed even the same effect on the drug-resistant strain, CB211. In addition, the same group used the system to test a pheromone-enriched culture media. 77% of male worms were successfully evoked acute mating phenotypes (looped tail and curved tail) in such kind of man-made environment. After experiments, all worms could be safely unloaded from the droplet traps.

Nanoparticles have been extensively used in our consumer products and medical applications for a long while. However, their toxicity to animals remains controversial and debatable. *C. elegans* provides a good animal model to implement simple toxicity assays. Among so many nanomaterials, silver nanoparticles (AgNPs) have been widely used in our daily life, such as cosmetics, clothing, household items, medical devices, and food packaging [94, 95]. The broad prevalence of AgNPs thus draws the public attention to their environment impact to human health. Kim et al. [96] proposed a simple *C. elegans*-on-a-chip to rapidly monitor the toxicity (Fig. 6.17c). A mutant strain CL2122 which carries the *mtl-2::gfp* gene was used to report the existence of AgNPs. The *mtl-1* and *mtl-2* genes are two verified isoforms of metallothionein to detoxify heavy metals, such as Cd, Cu, Cr, Pb, Co, Ni, As, and so forth in *C. elegans* [97–100]. The worm's body size and green fluorescent intensity were measured and compared with the control. The

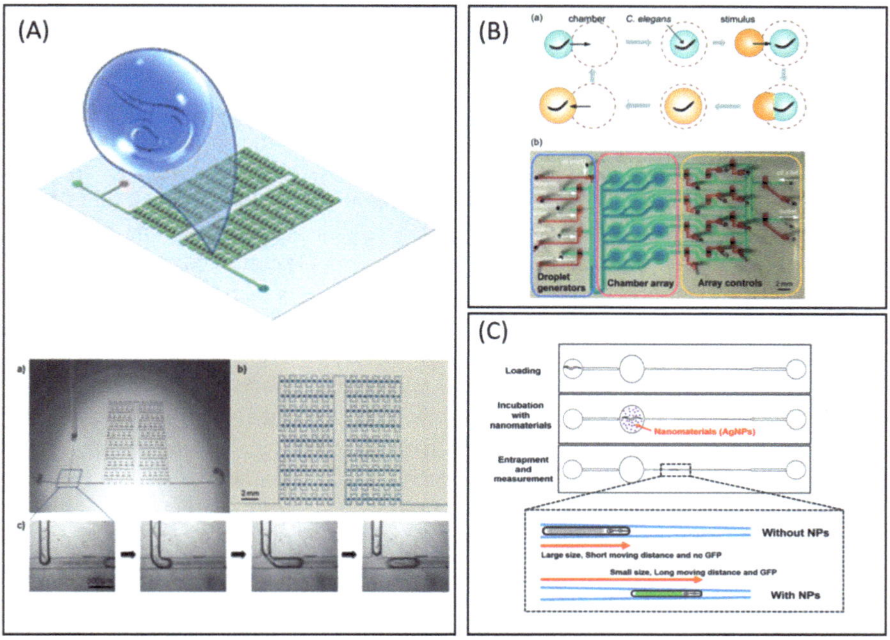

Fig. 6.17 (**a**) Top: Schematic of a droplet-based microfluidic system integrating a droplet generator and a droplet trap array. Bottom: Photograph of the microfluidic trap array immobilized with 180 droplets. Serial images showing an individual worm encapsulated in a droplet. (Reprint with permission from [92]. Copyright 2008 Royal Society of Chemistry.) (**b**) display of worms' body size, migrating distance, and specific gene expressions for Ag nanoparticle uptake and biotoxicity on the *C. elegans*-on-a-chip. (Reprint with permission from [93]. Copyright 2017 Royal Society of Chemistry.) (**c**) Droplet array platform for screening acute behavioral responses of *C. elegans* to chemical stimulation. (Reprint with permission from [96]. Copyright 2017 Springer Nature)

concentration of AgNPs as low as 5 μg/mL can be detected. The uptake of AgNPs in *C. elegans* led to reduction in body size and strong transcription of the *mtl-2* gene. However, higher concentration due to aggregation of nanoparticles impeded the worm's uptake, resulting in low adverse effects apart from the expectation. Nevertheless, the paper suggests high throughput and dose-dependent measurement can be done on such kind of droplet-based microfluidic system.

6.4.2 Biosensors for Diseases/Environmental Changes

C. elegans possesses 302 neurons, it can therefore sense a wide variety of stimuli, including electrical, chemical, optical, mechanical, magnetic, gravity, and so forth. Since the worm is small in size, it is particularly feasible for micro, precision, and versatile sensing applications. As compared with the cutting-edge microsensors

[101–107] and microactuators [108–111] developed so far in the research frontiers, *C. elegans* is obviously a perfect candidate. The use of animals to monitor environment or diseases is not whimsical. As a matter of fact, birds were commonly used by miners in mining sites to provide early warnings of noxious gases in the past; dogs are often trained to detect drug smuggling and explosives in airports nowadays; some dogs and cats can tell whether some people are ill by smelling special odors from humans [112, 113]. Therefore, it is convenient to take advantage of these super powers owned by animals to serve as sensors in fields of interest. The benefits of such worm biosensors may include, but not limited to, (1) consistent expressions due to rare genetic mutation via self-reproduction, (2) reliably statistical analysis due to ease of cultivating a large worm population, ensuring statistical correctness, (4) robustness, and (5) low failure rates. However, uncertainty and accuracy may still be a concern in such kind of animal-based detections.

Cancer detection by *C. elegans*, termed Nematode Scent Detection Test (NSDT), was firstly investigated by a Japanese group, Hirotsu and his colleagues [114] (Fig. 6.18a). To achieve non-invasive, economical, painless, and rapid cancer screening, urinary samples were collected from patients with various cancers (breast ca., gastric ca., colorectal ca., prostate ca., etc). High sensitivity (95.8%) and high specificity (95%) were obtained from all stages in the pilot runs. The measurement sensitivity could even reach 100% for those early stage samples. They observed that wild-type *C. elegans* was still attracted to 10^{-7} dilutions of the cancer-cell medium. Although the exact molecules that triggered the worm's chemotaxis remains unclear, the worm's corresponding neurons were found likely to rely on AWC and AWA neurons. The promising animal tests therefore lit up a hope for more disease prescreens based on *C. elegans*. Later, Tee et al. [115] developed another worm assay, termed *C. elegans* Sepsis Detection Assay (CESDA), to detect whether a person is under infection or sepsis though patient's urinary sample. To avoid interferences from cancer factors, all tested samples were diagnosed cancer-free at the measurement moment. In their study, sepsis could be detected in as early as 20 min while infection could be identified within 40 min among 45 subjects. However, as cancer and infection share many T-cell mediated processes, it is too early to conclude whether the both cancer and infection detections may lead to the same olfactory molecules. Further investigation will be necessary.

Metallic ions, pH level, temperature, and gases are part of the environmental constituents. Their fluctuations may alter the homeostasis of all animals. Therefore, *C. elegans* is equipped with all kinds of sensory neurons to deal with changes of those factors. Evidences have revealed that worms are attracted to oxygen (O_2) and repelled from carbon dioxide (CO_2). For instance, Bretscher et al. [116] reported that well-fed *C. elegans* avoids elevated CO_2 levels above 0.5%, which is ten-fold higher than the atmospheric CO_2 level. The sensation of CO_2 in the worm is promoted by the cGMP gated ion channel subunits TAX-2 and TAX-4. Notably, the CO_2 avoidance was found to be distinct from the avoidance of low pH. Wakabayashi et al. [117] stated that *C. elegans* employs two behavioral maneuvers, reversal and

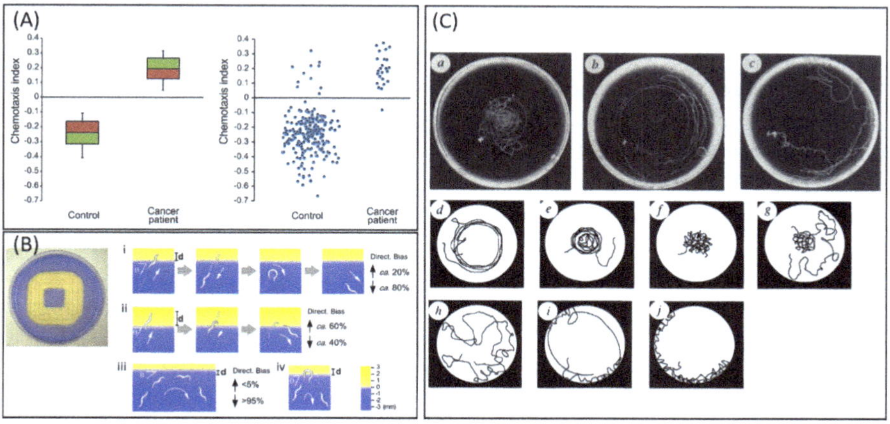

Fig. 6.18 (**a**) Box plots and dot plots of chemotactic responses of wild-type *C. elegans* to urine samples from control subjects (n = 218) or cancer patients (n = 24). (Reprint with permission from [114]. Copyright 2015 PLOS.) (**b**) Assay plate and acidic pH avoidance behavior. Schematic drawings of (**i**) long reversal avoidance, (**ii**) short reversal avoidance, (**iii**) gradual curve avoidance, and (**iv**) deep curve avoidance behavior. (Reprint with permission from [117]. Copyright 2015 BioMed Central.) (**c**) Tracks of wild-type and laser-operated worms on a thermal gradient ranging from 17 to 25 °C. Sub-figures (**a–c**) are responses of worms cultivated at 15 °C, 20 °C, and 25 °C, respectively. Other sub-figures (**d–j**) display tracks of different thermotactic phenotypes. (Reprint with permission from [120]. Copyright 1995 Springer Nature)

curve, to avoid acidic pH. The sensation of pH may be, in part, regulated by ASH and ASK neurons (Fig. 6.18b). The temperature responses of *C. elegans* was firstly discussed by Hedgecock et al. in 1975 [118]. Later, Jurado et al. [119] further delineated the thermal migration behavior of *C. elegans* in response to different thermal gradients. The work was based on their earlier finding [120] associated with the worm's thermosensation and neural plasticity (Fig. 6.18c). In Jurado et al.'s case, N2 worms were preconditioned at 17 °C and 23 °C before evaluations. When the worms were placed in the middle of a plate with a linear thermal gradient from 17 °C to 23 °C, they tend to move toward their preconditioned temperatures. However, they discovered that these responses were strongly affected by thermal gradients. In a shallow thermal gradient (0.5 °C/cm), thermophilic and cryophilic worms all migrated to their reconditioned temperatures. Whereas, only cryophilic worms returned to their preconditioned temperature in a steep thermal gradient (1.2 °C/cm). The difference was attributed to the natural protection mechanism encoded in the blind soil animals. In soil, steeper and warmer direction usually imply near the ground surface, which is very likely to endanger their lives by being heated by sun; conversely, cooler direction may imply deeper into the ground. Nevertheless, the appropriate temperature distribution may only be between 16 °C and 25 °C [118].

6.4.3 Neurosciences and Disease Modeling

Neurons are basic units to constitute a complex neural system in animals. The understanding of the roles of neurons can help scientists unveil the secrets of mind mapping, neuronal network, and neurodegenerative diseases. In this respect, *C. elegans* is apparently a good choice due to its manageable number of neurons. Laser ablation has been employed to study specific neurons in *C. elegans*. Considerable papers [6, 7, 10, 121–123] have revealed their findings on the roles of neurons. It is clear that neurons are not functioning alone in maintaining the worm's daily activities. By forming multiple neuronal networks, they are able to perform more complicated actions. This feature imparts *C. elegans* the ability to make more sophisticated decisions as encountering tasks. In order to understand such kind of intelligence, Qin and Wheeler [124] explored the learning capacity of *C. elegans* with three microfluidic mazes, including six T-maze, complex U-maze, and continuous T-maze. Two behavioral assays, group exploration and reward-related associative learning, were investigated. In their findings, worms' exploratory behavior was strongly affected by food (*E. coli*). Without food, however, the chance of exploration in each outlet was nearly equal. Surprisingly, this behavior could be acquired by learning. In the associative learning assay, worms' memory was progressively enhanced by first few trials with reward, leading to faster decision-making. Followed by similar trials without rewards, a high rate of the same worms could still make the right choice. Especially, the neurotransmitter dopamine was found to boost the learning. However, the memory seemed to lose rapidly as the no-reward trial continues. Nevertheless, this pilot study showed an insight to use the simple nematode for the future learning research.

Unlike single cell analysis, *C. elegans* is a multi-cellular model organism. As an alternative solution, the worm is suitable for studying neurodegenerative diseases. Transgenic worm models have been established for various neurodegenerative diseases, such as Alzheimer's disease (AD) [125], Parkingson's disease (PD) [126, 127], Huntington's disease (HD) [128], and Amyotropic Lateral Sclerosis (ALS) [129]. For such kind of research, long-term monitoring (>24 h) is usually needed to collect enough information for accurate time-resolved analyses. To this end, Cornaglia et al. [130] came up with a multi-functional microdevice to realize automated longitudinal monitoring of *in-vivo* neurodegenerative disease on *C. elegans* (Fig. 6.19a). The microdevice was composed of a heat sink, a ring-shaped thermoelectric module, a chip thermalization frame, a PDMS microfluidic chip, a glass coverslip, a thermally insulating plate, and a resistive temperature sensor. Their successfully-bred worms could grow from a desired developmental stage within the device up to 4 days. During the breeding, fresh food, *E coli* suspension, could be continuously supplied into the confinement chamber to maintain the worm's growth. Eggs were washed away to keep the observation free from unnecessary interferences. Pluronic F127 hydrogel was injected into the chamber to tentatively immobilize worms for acquiring high-resolution fluorescent images of protein aggregation inside the worms' bodies. The reversible gelation of PF127 was controlled by changing the chip temperature with the ring-shaped thermoelectric

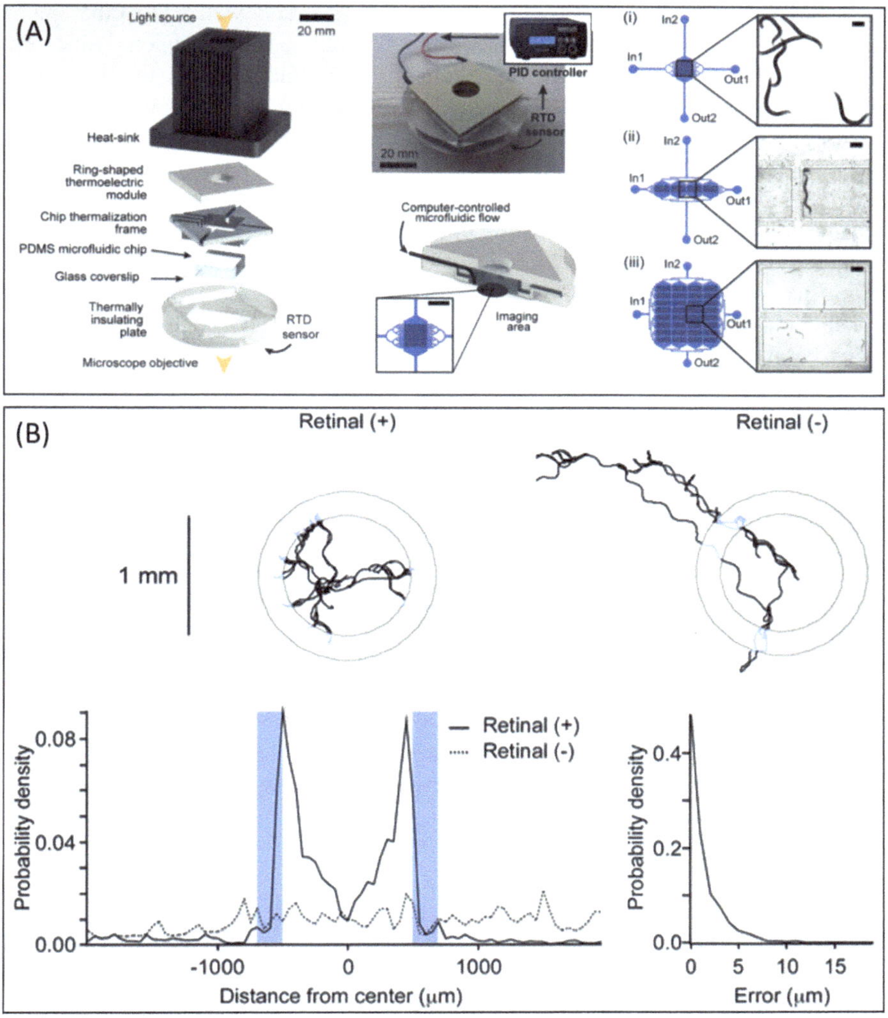

Fig. 6.19 (**a**) Schematic representation of the main constitutive components of the thermoelectric microfluidic chip for reversible worm immobilization. (Reprint with permission from [130]. Copyright 2016 BioMed Central.) (**b**) Channelrhodopsin-2 was specifically expressed in the osmo-sensitive neuron, ASH. Worms were tracked via a small cluster of neurons expressing a red fluorescent protein located in the head. ASH was photoactivated whenever the worm's head entered the annular region. (Reprint with permission from [140]. Copyright 2011 PLOS)

module. In the study, ALS transgenic *C. elegans* strain (AM725) was employed to monitor the dynamics of protein aggregation, SOD1-YFP, in the body wall muscle cells. Constant growing aggregates were found in all the disease models. However, it was verified that a treatment with an antibiotic, doxycycline, by activating UPR on the ALS transgenic worms effectively slowed down the progressive loss of their motility.

In addition to the neuron-based disease models, it is also well known that genes store biological traits and can be passed from generations to generations. Since the genome of *C. elegans* has been fully sequenced, specific genes in worms can be intentionally modified by genetic engineering to obtain anticipated strains. Currently, numerous transgenic strains, such as AD, PD, HD, ALS worms, have been established and stored in the CGC for order. Apart from that, another genetic engineering measure, optogenetics, has been emerging in recent years. Optogenetics is an innovative technology combining optics and genetics to manipulate the activity of individual neurons. Channelrhodopsin and Halorhodopsin are the two primary light-sensitive opsins used in the majority of optogenetic research. Nagel et al. [131] was the first group to demonstrate the activation of selected channels with blue light by expressing cRNA encoding Channelrhodopsin in oocytes of *Xenopus laevis*, in the presence of all-*trans* retinal. They found Channelrhodopsin can be a promising tool to mediate a large light-switched H^+ conductance to manipulate electrical and proton gradients across cell membranes simply by illumination. The idea was then rapidly prevailing in many research fields, such as PD [132, 133], AD [134], HD [135], and epilepsy [136–138]. As for *C. elegans*, the use of optogenetics can allow *in-vivo* observation of specific neurons without even removing them by microsurgery. Narayan et al. [139] once showed graded transmission between AFD and AIY neurons by expressing Channelrhodopsin-2 (ChR2) in AFD using *gcy-8* promoter. They found that the amplified signal in AFD was scaled in a quantitative manner. Faumont and colleagues [140] also studied locomotion of freely-moving *C. elegans* generated by distinct neuronal circuits using ChR2 (Fig. 6.19b). ChR2 was expressed specifically here in the osmo-sensitive neuron, ASH, to mimic the avoidant response from the worms. Their system can be used to create virtual environments by optogenetic activation of sensory neurons or to image activity in identified neurons at high magnification.

6.4.4 Parasitiology

Parasites, such as hookworms or whipworms, are rampant in undeveloped areas due to poor hygiene. Failing to control parasites may eventually cause life-threatening diseases in animals as well as humans. Unfortunately, like super bugs, some parasites also have gradually developed resistivity to drugs through natural evolution, weakening effectiveness of some anthelmintics. New anthelminitic treatments are urgently needed, but the current drug development pipeline remains very inadequate [141]. Although lab-bred *C. elegans* is recognized as hazard-free animals and can be safely used in non-bio-environments, it shares many biological similarities with its parasite relatives. The similarities make the worm a good candidate in the studies relevant to parasitiology. Among them, how anthelmintics can effectively affect parasites are extensively investigated in pharmacies. Chen et al. [142] characterized movement phenotypes of an animal parasite, *Oesophagostomum dentatum*, with a microfluidic device (Fig. 6.20a). They successfully found drug sensitivity

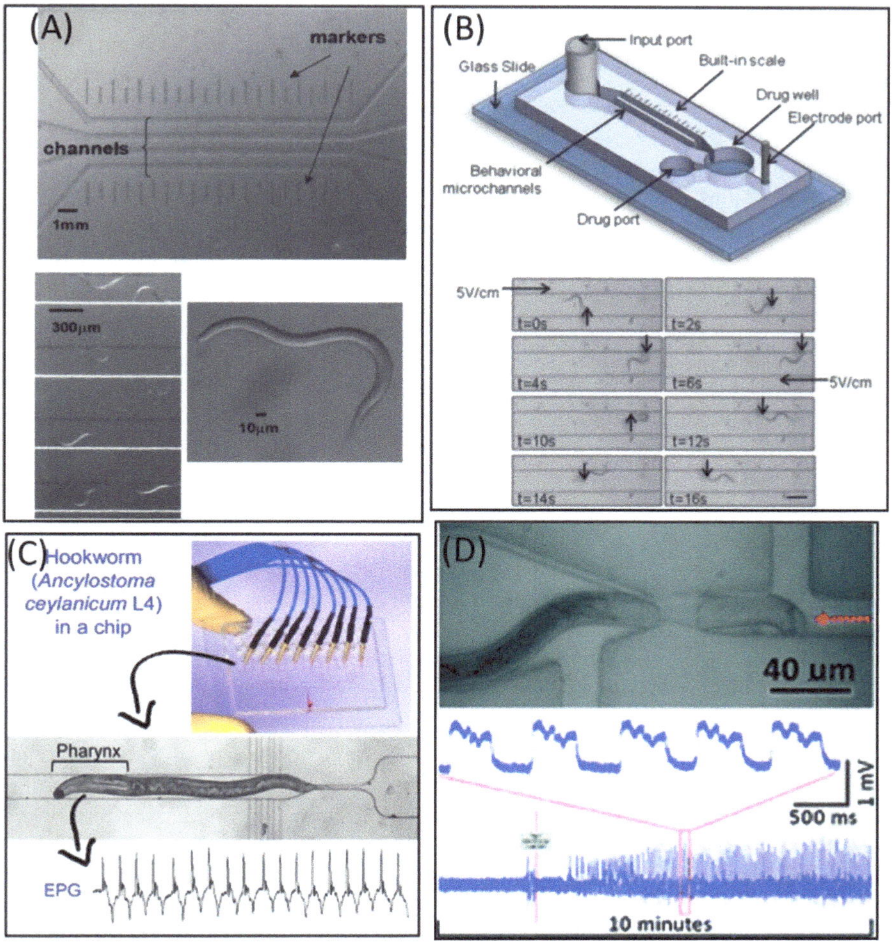

Fig. 6.20 (**a**) Microfluidic chamber and larvae. The four-channel microfluidic device was fabricated to study nematode locomotion. There are vertical markers to help calibrate the swimming velocity. (Reprint with permission from [142]. Copyright 2011 Cambridge University Press.) (**b**) Top: Illustration of the fabricated T-shaped microfluidic device showing the three electrode ports and sample electric fields at the respective ports. Bottom: Worms are inserted in the left port and their preference to move towards either cathode is tracked with the imaging unit. (Reprint with permission from [143]. Copyright 2011 Royal Society of Chemistry.) (**c**) Top: Microfluidic EPG recording device. Channels were filled with a dye to aid visualization in this image. Bottom: Enlarged view of a single recording module, with an *A. ceylanicum* L4 positioned tail-first in the worm trap. (Reprint with permission from [144]. Copyright 2016 Elsevier.) (**d**) Different trapping positions of J2 *G. pallida* generate electrical signals with different amplitude in the presence of 2 mM 5-HT. (Reprint with permission from [145]. Copyright 2014 Royal Society of Chemistry)

doses of levamisole for two drug-resistive *C. elegans* mutants, SENS and LEVR, within 45–60 min. They believed their microfluidic assay offered an improved and higher level of resolution. Carr et al. [143] also once reported a microfluidic platform for high-sensitivity and real-time drug screening on *C. elegans* (Fig. 6.20b). On this chip, a multi-parameter microfluidic bioassay has been developed to observe the innate locomotion of worms along with their transient and time-resolved responses to an anthelmintics, levamisole. Percentage of worms responsive, percentage of worms leaving the well, mean post-exposure velocity, and time until unresponsiveness were measured. To ensure worms can move in and out of the drug well as expected, electrotaxis was used as a guiding force in the research. With this new platform, the measurement duration of dose response was considerably improved from several hours in conventional worm motility assays to less than an hour. As compared to drug resistant strains, levamisole-sensitive N2 worms were paralyzed in a median duration of 0.4 min after exposing to 100 μM levamisole.

Many anthelmintic drugs target at proteins associated with neurotransmitter receptors and ion channels, so characterizing the electrophysiological functions can help the development of new drugs in parasitology. To understand electrophysiological signals and phenotypes of parasites, several microfluidic platforms were proposed. Weeks et al. [144] designed a microchannel to trap a single worm in a desired position and then inserted electro probes to measure their electropharyngeograms (EPGs) (Fig. 6.20c). EPGs reveal electrical activity of muscles and neurons of pharynx, which is important for feeding in worms. With the microchip, the worm's EPGs against environmental stimuli can be continuously recorded. Two parasite species, *A. ceylanicum* and *A. canium*, relevant to human health were removed from their hosts and measured *in vivo*. The study showed that such a new technology is capable of providing massive and real-time EPG information to assist evaluations of anthelmintic drugs on worms and understanding of the worm's electrophysiological statuses depending on their EPG readouts.

In addition to parasites in host animals, plants may be infested by invader worms as well. Plant parasitic nematodes (PPNs) are worms that infest the roots of their host plants impairing plant viability and reducing the yield of crops. For feeding, a stylet is protruded from a worm to stab into the plant root. However, the study of stylet activity remains technically challenging. To understand the underlying biology, electrophysiological recordings can commit some valuable information. A microfluidic device, named StyletChip, built up in such a purpose was developed by Hu et al. [145] (Fig. 6.20d). A PPN, *G. pallida*, was used to demonstrate the feasibility of the device. Recordings were made on the nematode within 2 days from hatching. Each worm was trapped at the test region with a positive pressure by arranged valve controls. 5-hydroxytryptamine (5-HT) is a drug used to stimulate the worm's stylet thrusting action. By 2 mM 5-HT was applied to the trapped worm, rhythmic electrical signals with a frequency of 1 Hz was recorded. The study showed that the proposed StyletChip was suitable for defining the actions of chemicals on stylet and median bulb activity with a view to developing new anthelmintics for crop protection.

6.5 Summary and Outlook

Since *C. elegns* was firstly introduced to the scientific community, it has been proven itself a quite successful and valuable model animal in all aspects over the past decades. Several scientific breakthroughs originated from *C. elegans* research have brought significant impact in our world. For instances, biocompatible green fluorescent proteins (GFP), insulin-like signaling pathway (ILS) for the secret of longevity, knocking down the genetic expressions with the biological tool, RNAi, and more. Nevertheless, challenges also arise when complex data, more information, and new tasks await processing in a shorter timeframe. Demands for new technologies to meet the new challenges are elevated. Fortunately, the advent of microfluidics has been progressively revolutionizing the worm research community. These novel microfluidic chips offer several advantages over conventional approaches: (1) maintaining well controllable micro-environments, (2) creating reproducible experimental conditions, (3) automating tedious experimental protocols, and (4) enabling high-throughput studies. The addition of the engineering elements extends advanced possibilities of the worm development. By taking advantaging of the worm's responses to different stimuli, strategies for size sorting, immobilization, long-term imaging, microsurgery, confined breeding have been well developed. The concept of WoC is even prevailing in broad applications, not only limited to the group of worm breeders nowadays. Drug screening, biosensors, disease modeling, parasitiology, neuroscience, and genetic engineering have long been benefited from *C. elegans* for decades. With WoC, however, all the applications appear to boost in a new era. Eventually, it can be expected to see the growing in WoCs as a promising branch of the *C. elegans* research in the near future.

6.5.1 Unmet Challenges

Despite the promising possibilities of the integration of microfluidics and *C. elegans*, there are still many hurdles laying ahead before the concept can really come to practice in the worm community. First, a general-purpose microfluidic platform remains undeveloped yet. Actually, most worm chips proposed so far in the papers are designed for specific goals, resulting in the use of microfluidics in *C. elegans* labor-intensive, time-consuming, costly, and high expertise needed. The consequence makes the technology very diverse and dependent on each lab, impeding its prevalence in the market. Second, worms are thought to be terrestrial animals although they can be amphibious. However, most microfluidic platforms require to culture worms in an aqueous environment, which may come to suggestions different from terrestrial cases. Third, how well can worm research be relevant to human health remains under debate. Although worms are well known to have more than 60% gene similarity with humans, it's not completely sure of whether the underlying genetic pathways preserved across species as diverse as humans and worms. Some of them have been proven to exist in both species [146, 147], but more are still waiting for confirmation. For a scientific standpoint, however, the uncertainty can be clarified faster when the microfluidics are incorporated.

6.5.2 Future Prospects

As many possibilities have demonstrated for WoCs to date, we envision more new gadgets to enrich the worm arsenal. Several routes are expected to receive focuses of the future prospects in WoCs. They are further addressed below:

(a) Automation of worm processing

Ease-to-use is always a concern in the worm research. Because of the nature of biological complexity, worm assays usually require plenty of expertise and training to complete worm processing. Automation accompanying with graphical user interfaces (GUIs) can be a great solution to ease researchers' life. Automation can also facilitate laymen to deal with the worm research. Despite the ideology, the major hurdle is attributed to the worm's diversity and uncertainty. Accordingly, a large number of data becomes key to obtain reasonable statistical analysis. The recent hotline concerning artificial intelligence (AI) perhaps provides an insight to the future development. With deep learning of the worm behavior, optimized automation can be tuned to increase the efficiency of the worm processing.

(b) High throughput, robustness, and high reliability for industrial applications

Unlike normal lab research, being eligible for industrial applications usually implies that a technique is sufficiently mature to proceed from theory to practice. In this phase, high throughput, robustness, and high reliability are particularly emphasized. Some potential fields, such as pharmaceutical and agriculture, have studied *C. elegans* as a model animal in drug tests for neurodegenerative diseases [148–150] and anthlmintic screening for parasites [151, 152]. However, the overall investigations still heavily relied on manual operations, limiting the yield and reliability. The introduction of powerful microdevices can certainly promote the throughput by several orders of number. Moreover, the reduction of labor can improve the reliability as well.

(c) An integrated platform to enable versatile goals

To this end, considerable methods have been conducted and discussed. Among them, the mostly novel proposal goes to a technique termed "LEGO microfluidics" [153]. Apart from the conventional way, module bricks of various microfluidic components were built in advance. When a specific assay is designated, the selected modules were assembled to form a complete platform for that task. Although such a highly integrated chips have not emerged yet, its appearance is for sure just a matter of time. Nevertheless, the conceptual work has been realized in many existing MEMS-based devices [154–158]. The pioneering idea of programmable lab on a chip was firstly attempted by Urbansiki in 2006 [159] and then extend to more areas [160, 161]. With all necessary components built on the same chip, general-purpose tasks can be done. The concept is to minimize the barriers from hardware, so that the core work can be executed with only software programming. Obviously, the use of such an integrated platform not only imparts WoCs versatility but also eliminates the gaps between labs, reducing the turnaround time in most worm studies.

Acknowledgements This work was supported by the Ministry of Science and Technology in Taiwan under the grants 107-2221-E-006-054-MY3 and 107-2622-E-006-022-CC2. W.-H.Wang is grateful to the financial support from the One-Thousand Young Talent Program of China.

References

1. Ankeny RA (2001) The natural history of *Caenorhabditis elegans* research. Nat Rev Genet 2:474
2. Johnson CD, Duckett JG, Culotti JG, Herman RK, Meneely PM, Russell RL (1981) An Acetylcholinesterase-defficient mutant of the nematode *Caenorhabditis elegans*. Genetics 97:261
3. Ward A, Liu J, Feng Z, Xu XZS (2008) Light-sensitive neurons and channels mediate phototaxis in *C. elegans*. Nat Neurosci 11:916
4. Edwards SL, Charlie NK, Milfort MC, Brown BS, Gravlin CN, Knecht JE, Miller KG (2008) A novel molecular solution for ultraviolet light detection in *Caenorhabditis elegans*. PLoS Biol 6:e198
5. Margie O, Palmer C, Chin-Sang I (2013) *C. elegans* Chemotaxis Assay. JoVE 74:50069
6. Gabel CV, Gabel H, Pavlichin D, Kao A, Clark DA, Samuel ADT (2007) Neural circuits mediate electrosensory behavior in *Caenorhabditis elegans*. J Neurosci 27:7586
7. Chung SH, Clark DA, Gabel CV, Mazur E, Samuel ADT (2006) The role of the AFD neuron in *C. elegans* thermotaxis analyzed using femtosecond laser ablation. BMC Neurosci 7:30
8. Vidal-Gadea A, Ward K, Beron C, Ghorashian N, Gokce S, Russell J, Truong N, Parikh A, Gadea O, Ben-Yakar A, Pierce-Shimomura J (2015) Magnetosensitive neurons mediate geomagnetic orientation in *Caenorhabditis elegans*. elife 4:e07493
9. Shen XN, Arratia PE (2011) Undulatory swimming in viscoelastic fluids. Phys Rev Lett 106:208101
10. Ward A, Liu J, Feng Z, Shawn Xu XZ (2008) Light-sensitive neurons and channels mediate phototaxis in *C. elegans*. Nat Neurosci 11:916–922
11. Van Voorhies WA, Ward S (1999) Genetic and environmental conditions that increase longevity in *Caenorhabditis elegans* decrease metabolic rate. Proc Natl Acad Sci U S A 96:11399–11403
12. Belfer SJ, Chuang H-S, Freedman BL, Yuan J, Norton M, Bau HH, Raizen DM (2013) Caenorhabditis-in-drop array for monitoring *C. elegans* quiescent behavior. Sleep 36:689–698
13. Ying D, Zhang K, Li N, Ai X, Liang Q, Wang Y, Luo G (2012) A droplet-based microfluidic device for long-term culture and longitudinal observation of *Caenorhabditis elegans*. Biochip J 6:197–205
14. Vidal-Gadea A, Topper S, Young L, Crisp A, Kressin L, Elbel E, Maples T, Brauner M, Erbguth K, Axelrod A, Gottschalk A, Siegel D, Pierce-Shimomura JT (2011) *Caenorhabditis elegans* selects distinct crawling and swimming gaits via dopamine and serotonin. Proc Natl Acad Sci U S A 108:17504–17509
15. Brenner S (1973) The genetics of behavior. Br Med Bull 29:269–271
16. Brenner S (1974) The genetics of *Caenorhabditis elegans*. Genetics 77:71–94

17. Raizen DM, Zimmerman JE, Maycock MH, Ta UD, You Y-j, Sundaram MV, Pack AI (2008) Lethargus is a *Caenorhabditis elegans* sleep-like state. Nature 451:569
18. White JG, Southgate E, Thomson JN, Brenner S (1986) The structure of the nervous system of the nematode *Caenorhabditis elegans*. Philos Trans R Soc Lond Ser B Biol Sci 314:1–340
19. Wightman B, Corsi AK, Chalfie M A transparent window into biology: a primer on *Caenorhabditis elegans* (June 18, 2015), WormBook (ed) The *C. elegans* research community, WormBook. https://doi.org/10.1895/wormbook.1.177.1.; http://www.wormbook.org
20. C.I.C.i.C.e.O. Bargmann (2006) WormBook (ed) The *C. elegans* research community, WormBook. https://doi.org/10.1895/wormbook.1.123.1.; http://www.wormbook.org.
21. Neto MF, Nguyen QH, Marsili J, McFall SM, Voisine C (2016) The nematode *Caenorhabditis elegans* displays a chemotaxis behavior to tuberculosis-specific odorants. J Clin Tuberc Other Mycobact Dis 4:44–49
22. Rezai P, Siddiqui A, Selvaganapathy PR, Gupta BP (2010) Electrotaxis of *Caenorhabditis elegans* in a microfluidic environment. Lab Chip 10:220–226
23. Manière X, Lebois F, Matic I, Ladoux B, Di Meglio J-M, Hersen P (2011) Running worms: *C. elegans* self-sorting by electrotaxis. PLoS One 6:e16637
24. Wang X, Hu R, Ge A, Hu L, Wang S, Feng X, Du W, Liu B-F (2015) Highly efficient microfluidic sorting device for synchronizing developmental stages of *C. elegans* based on deflecting electrotaxis. Lab Chip 15:2513–2521
25. Han B, Kim D, Hyun Ko U, Shin JH (2012) A sorting strategy for *C. elegans* based on size-dependent motility and electrotaxis in a micro-structured channel. Lab Chip 12:4128–4134
26. Luo L, Cook N, Venkatachalam V, Martinez-Velazquez LA, Zhang X, Calvo AC, Hawk J, MacInnis BL, Frank M, Ng JHR, Klein M, Gershow M, Hammarlund M, Goodman MB, Colón-Ramos DA, Zhang Y, Samuel ADT (2014) Bidirectional thermotaxis in *Caenorhabditis elegans* is mediated by distinct sensorimotor strategies driven by the AFD thermosensory neurons. Proc Natl Acad Sci U S A 111:2776–2781
27. Kimata T, Sasakura H, Ohnishi N, Nishio N, Mori I (2012) Thermotaxis of C. elegans as a model for temperature perception, neural information processing and neural plasticity. Worm 1:31–41
28. T.M.o.C.e.F. Stiernagle (2006) WormBook (ed) The *C. elegans* research community, WormBook. https://doi.org/10.1895/wormbook.1.101.1.; http://www.wormbook.org
29. J.A. Lewis and J.T. Fleming, Basic culture methods. Methods in cell biology, ed. H.F. Epstein and D.C. Shakes. Vol. 48. 1995, San Diego: Academic Press
30. Dupuy D, Bertin N, Hidalgo CA, Venkatesan K, Tu D, Lee D, Rosenberg J, Svrzikapa N, Blanc A, Carnec A (2007) Genome-scale analysis of in vivo spatiotemporal promoter activity in *Caenorhabditis elegans*. Nat Biotechnol 25:663
31. Shi W, Wen H, Lin B, Qin J (2011) Microfluidic platform for the study of *Caenorhabditis elegans*. In: Microfluidics. Springer, Berlin, pp 323–338
32. Wen H, Qin J (2012) Analysis of *Caenorhabditis elegans* in microfluidic devices. Science China Chem 55:484–493
33. Rohde CB, Zeng F, Gonzalez-Rubio R, Angel M, Yanik MF (2007) Microfluidic system for on-chip high-throughput whole-animal sorting and screening at subcellular resolution. Proc Natl Acad Sci 104:13891–13895
34. Ai X, Zhuo W, Liang Q, McGrath PT, Lu H (2014) A high-throughput device for size based separation of *C. elegans* developmental stages. Lab Chip 14:1746–1752
35. Chung K, Crane MM, Lu H (2008) Automated on-chip rapid microscopy, phenotyping and sorting of *C. elegans*. Nat Methods 5:637
36. Dong L, Cornaglia M, Lehnert T, Gijs MA (2016) Versatile size-dependent sorting of *C. elegans* nematodes and embryos using a tunable microfluidic filter structure. Lab Chip 16:574–585
37. Chronis N (2010) Worm chips: microtools for *C. elegans* biology. Lab Chip 10:432–437
38. Chronis N, Zimmer M, Bargmann CI (2007) Microfluidics for in vivo imaging of neuronal and behavioral activity in *Caenorhabditis elegans*. Nat Methods 4:727

39. Yang L, Hong T, Zhang Y, Arriola JGS, Nelms BL, Mu R, Li D (2017) A microfluidic diode for sorting and immobilization of *Caenorhabditis elegans*. Biomed Microdevices 19:38
40. Cho Y, Porto DA, Hwang H, Grundy LJ, Schafer WR, Lu H (2017) Automated and controlled mechanical stimulation and functional imaging in vivo in *C. elegans*. Lab Chip 17:2609–2618
41. Wu Z, Ghosh-Roy A, Yanik MF, Zhang JZ, Jin Y, Chisholm AD (2007) *Caenorhabditis elegans* neuronal regeneration is influenced by life stage, ephrin signaling, and synaptic branching. Proc Natl Acad Sci 104:15132–15137
42. Gilleland CL, Falls AT, Noraky J, Heiman MG, Yanik MF (2015) Computer-assisted transgenesis of *Caenorhabditis elegans* for deep phenotyping. Genetics 201:39–46
43. Kerr R, Lev-Ram V, Baird G, Vincent P, Tsien RY, Schafer WR (2000) Optical imaging of calcium transients in neurons and pharyngeal muscle of *C. elegans*. Neuron 26:583–594
44. Hulme SE, Shevkoplyas SS, Apfeld J, Fontana W, Whitesides GM (2007) A microfabricated array of clamps for immobilizing and imaging *C. elegans*. Lab Chip 7:1515–1523
45. Lee H, Kim SA, Coakley S, Mugno P, Hammarlund M, Hilliard MA, Lu H (2014) A multi-channel device for high-density target-selective stimulation and long-term monitoring of cells and subcellular features in *C. elegans*. Lab Chip 14:4513–4522
46. Aubry G, Zhan M, Lu H (2015) Hydrogel-droplet microfluidic platform for high-resolution imaging and sorting of early larval *Caenorhabditis elegans*. Lab Chip 15:1424–1431
47. Chokshi TV, Ben-Yakar A, Chronis N (2009) CO 2 and compressive immobilization of *C. elegans* on-chip. Lab Chip 9:151–157
48. Geng H, Song H, Qi J, Cui D (2011) Sustained release of VEGF from PLGA nanoparticles embedded thermo-sensitive hydrogel in full-thickness porcine bladder acellular matrix. Nanoscale Res Lett 6:312
49. Hwang H, Krajniak J, Matsunaga Y, Benian GM, Lu H (2014) On-demand optical immobilization of *Caenorhabditis elegans* for high-resolution imaging and microinjection. Lab Chip 14:3498–3501
50. Chuang H-S, Chuang W-Y (2017) Rapid, reversible and addressable immobilization of *Caenorhabditis elegans* in Pluronic F-127 using an optoelectric device. Sensors Actuators B Chem 253:376–383
51. Benhal P, Chase JG, Gaynor P, Oback B, Wang W (2014) AC electric field induced dipole-based on-chip 3D cell rotation. Lab Chip 14:2717–2727
52. Huang L, Tu L, Zeng X, Mi L, Li X, Wang W (2016) Study of a microfluidic chip integrating single cell trap and 3D stable rotation manipulation. Micromachines 7:141
53. Chuang H-S, Raizen DM, Lamb A, Dabbish N, Bau HH (2011) Dielectrophoresis of *Caenorhabditis elegans*. Lab Chip 11:599–604
54. Huang L, Zhao P, Wu J, Chuang H-S, Wang W (2018) On-demand dielectrophoretic immobilization and high-resolution imaging of *C. elegans* in microfluids. Sensors Actuators B Chem 259:703–708
55. Teh S-Y, Lin R, Hung L-H, Lee AP (2008) Droplet microfluidics. Lab Chip 8:198–220
56. Krajniak J, Lu H (2010) Long-term high-resolution imaging and culture of *C. elegans* in chip-gel hybrid microfluidic device for developmental studies. Lab Chip 10:1862–1868
57. Berthier J (2012) Micro-drops and digital microfluidics. William Andrew, Norwich
58. Wen H, Yu Y, Zhu G, Jiang L, Qin J (2015) A droplet microchip with substance exchange capability for the developmental study of *C. elegans*. Lab Chip 15:1905–1911
59. Kim E, Sun L, Gabel CV, Fang-Yen C (2013) Long-term imaging of *Caenorhabditis elegans* using nanoparticle-mediated immobilization. PLoS One 8:e53419
60. Zhuo W, Lu H, McGrath PT (2017) Microfluidic platform with spatiotemporally controlled micro-environment for studying long-term *C. elegans* developmental arrests. Lab Chip 17:1826–1833
61. Pittman WE, Sinha DB, Zhang WB, Kinser HE, Pincus Z (2017) A simple culture system for long-term imaging of individual *C. elegans*. Lab Chip 17:3909–3920
62. Churgin MA, Jung S-K, Yu C-C, Chen X, Raizen DM, Fang-Yen C (2017) Longitudinal imaging of *Caenorhabditis elegans* in a microfabricated device reveals variation in behavioral decline during aging. elife 6:e26652

63. Hulme SE, Shevkoplyas SS, McGuigan AP, Apfeld J, Fontana W, Whitesides GM (2010) Lifespan-on-a-chip: microfluidic chambers for performing lifelong observation of *C. elegans*. Lab Chip 10:589–597
64. Gilpin W, Uppaluri S, Brangwynne C (2015) Worms under pressure: bulk mechanical properties of *C. elegans* are independent of the cuticle. Biophys J 108:1887–1898
65. Park SJ, Goodman MB, Pruitt BL (2007) Analysis of nematode mechanics by piezoresistive displacement clamp. Proc Natl Acad Sci U S A 104:17376
66. Doll JC, Nahid H, Klejwa N, Kwon R, Coulthard SM, Petzold B, Goodman MB, Pruitt BL (2009) SU-8 force sensing pillar arrays for biological measurements. Lab Chip 9:1449–1454
67. Liu P, Mao D, Martin RJ, Liang D (2012) An integrated fiber-optic microfluidic device for detection of muscular force generation of microscopic nematodes. Lab Chip 12:3458–3466
68. Rabets Y, Backholm M, Dalnokiveress K, Ryu WS (2014) Direct measurements of drag forces in *C. elegans* crawling locomotion. Biophys J 107:1980–1987
69. Ghanbari A, Nock V, Johari S, Blaikie R, Chen XQ, Wang W (2012) A micropillar-based on-chip system for continuous force measurement of *C. elegans*. J Micromech Microeng 22:95009–95018
70. Johari S, Nock V, Alkaisi MM, Wang W (2013) On-chip analysis of *C. elegans* muscular forces and locomotion patterns in microstructured environments. Lab Chip 13:1699–1707
71. Qiu Z, Tu L, Huang L, Zhu T, Nock V, Yu E, Liu X, Wang W (2015) An integrated platform enabling optogenetic illumination of *Caenorhabditis elegans* neurons and muscular force measurement in microstructured environments. Biomicrofluidics 9:71–94
72. Ewa KZ-S, Wojciech W, Zbigniew M, Mariusz ZR (2012) Stem cells as a novel tool for drug screening and rreatment of degenerative diseases. Curr Pharm Des 18:2644–2656
73. Bithi SS, Vanapalli SA (2017) Microfluidic cell isolation technology for drug testing of single tumor cells and their clusters. Sci Rep 7:41707
74. Csöbönyeiová M, Polák Š, Danišovič LU (2016) Toxicity testing and drug screening using iPSC-derived hepatocytes, cardiomyocytes, and neural cells. Can J Physiol Pharmacol 94:687–694
75. O'Reilly LP, Luke CJ, Perlmutter DH, Silverman GA, Pak SC (2014) *C. elegans* in high-throughput drug discovery. Adv Drug Deliv Rev 69–70:247–253
76. Chen X, Barclay JW, Burgoyne RD, Morgan A (2015) Using *C. elegans* to discover therapeutic compounds for ageing-associated neurodegenerative diseases. Chem Cent J 9:65
77. Lockery SR, Hulme SE, Roberts WM, Robinson KJ, Laromaine A, Lindsay TH, Whitesides GM, Weeks JC (2012) A microfluidic device for whole-animal drug screening using electrophysiological measures in the nematode *C. elegans*. Lab Chip 12:2211–2220
78. Bargmann CI, Hartwieg E, Horvitz HR (1993) Odorant-selective genes and neurons mediate olfaction in *C. elegans*. Cell 74:515–527
79. Troemel ER, Kimmel BE, Bargmann CI (1997) Reprogramming chemotaxis responses: sensory neurons define olfactory preferences in *C. elegans*. Cell 91:161–169
80. Bargmann CI, Horvitz HR (1991) Chemosensory neurons with overlapping functions direct chemotaxis to multiple chemicals in *C. elegans*. Neuron 7:729–742
81. Hilliard MA, Bargmann CI, Bazzicalupo P (2002) *C. elegans* responds to chemical repellents by integrating sensory inputs from the head and the tail. Curr Biol 12:730–734
82. Hilliard MA, Bergamasco C, Arbucci S, Plasterk RHA, Bazzicalupo P (2004) Worms taste bitter: ASH neurons, QUI-1, GPA-3 and ODR-3 mediate quinine avoidance in *Caenorhabditis elegans*. EMBO J 23:1101
83. Colbert HA, Smith TL, Bargmann CI (1997) OSM-9, A novel protein with structural similarity to channels, is required for olfaction, mechanosensation, and olfactory adaptation in *Caenorhabditis elegans*. J Neurosci 17:8259
84. Culotti JG, Russell RL (1978) Osmotic avoidance defective mutants of the nematode *Caenorhabditis elegans*. Genetics 90:243–256
85. Cheung BHH, Cohen M, Rogers C, Albayram O, de Bono M (2005) Experience-dependent modulation of *C. elegans* behavior by ambient oxygen. Curr Biol 15:905–917

86. Chang AJ, Chronis N, Karow DS, Marletta MA, Bargmann CI (2006) A distributed chemosensory circuit for oxygen preference in *C. elegans*. PLoS Biol 4:e274
87. Gray JM, Karow DS, Lu H, Chang AJ, Chang JS, Ellis RE, Marletta MA, Bargmann CI (2004) Oxygen sensation and social feeding mediated by a C. elegans guanylate cyclase homologue. Nature 430:317–322
88. Dusenbery DB (1980) Appetitive response of the nematode *Caenorhabditis elegans* to oxygen. J Comp Pathol 136:333–336
89. Ludewig AH, Schroeder FC Ascaroside signaling in *C. elegans* (January 18, WormBook (ed) The *C. elegans* research community, WormBook. https://doi.org/10.1895/wormbook.1.155.1.; Available from: http://www.wormbook.org
90. Altun ZF, Hall DH (2010) Nervous system, neuronal support cells. In: WormAtlas. https://doi.org/10.3908/wormatlas.1.19. Edited for the web by LA Herndon. Last revision: June 6, 2012
91. Luo L, Gabel CV, Ha H-I, Zhang Y, Samuel ADT (2008) Olfactory behavior of swimming *C. elegans* analyzed by measuring motile responses to temporal variations of odorants. J Neurophysiol 99:2617–2625
92. Shi W, Qin J, Ye N, Lin B (2008) Droplet-based microfluidic system for individual *Caenorhabditis elegans* assay. Lab Chip 8:1432–1435
93. Aubry G, Lu H (2017) Droplet array for screening acute behaviour response to chemicals in *Caenorhabditis elegans*. Lab Chip 17:4303–4311
94. Fabrega J, Luoma SN, Tyler CR, Galloway TS, Lead JR (2011) Silver nanoparticles: behaviour and effects in the aquatic environment. Environ Int 37:517–531
95. Vance ME, Kuiken T, Vejerano EP, McGinnis SP, Hochella MF, Rejeski D, Hull MS (2015) Nanotechnology in the real world: redeveloping the nanomaterial consumer products inventory. Beilstein J Nanotechnol 6:1769–1780
96. Kim JH, Lee SH, Cha YJ, Hong SJ, Chung SK, Park TH, Choi SS (2017) *C. elegans*-on-a-chip for in situ and in vivo Ag nanoparticles' uptake and toxicity assay. Sci Rep 7:40225
97. Anderson GL, Cole RD, Williams PL (2004) Assessing behavioral toxicity with *Caenorhabditis elegans*. Environ Toxicol Chem 23:1235–1240
98. Boyd WA, Cole RD, Anderson GL, Williams PL (2003) The effects of metals and food availability on the behavior of *Caenorhabditis elegans*. Environ Toxicol Chem 22:3049–3055
99. Hilliard MA, Apicella AJ, Kerr R, Suzuki H, Bazzicalupo P, Schafer WR (2005) In vivo imaging of *C. elegans* ASH neurons: cellular response and adaptation to chemical repellents. EMBO J 24:63–72
100. Tvermoes BE, Boyd WA, Freedman JH (2010) Molecular characterization of numr-1 and numr-2: genes that increase both resistance to metal-induced stress and lifespan in *Caenorhabditis elegans*. J Cell Sci 123:2124–2134
101. Wang K-C, Kumar A, Williams SJ, Green NG, Kim KC, Chuang H-S (2014) An optoelectrokinetic technique for programmable particle manipulation and bead-based biosignal enhancement. Lab Chip 14:3958–3967
102. Wang J-C, Ku H-Y, Shieh D-B, Chuang H-S (2016) A bead-based fluorescence immunosensing technique enabled by the integration of Förster resonance energy transfer and optoelectrokinetic concentration. Biomicrofluidics 10:014113
103. Chuang H-S, Chen Y-J, Cheng H-P (2018) Enhanced diffusometric immunosensing with grafted gold nanoparticles for detection of diabetic retinopathy biomarker tumor necrosis factor-α. Biosens Bioelectron 101:75–83
104. Chung C-Y, Wang J-C, Chuang H-S (2017) Simultaneous and quantitative monitoring of co-cultured Pseudomonas aeruginosa and Staphylococcus aureus with antibiotics on a diffusometric platform. Sci Rep 7:46336
105. Jin W, Ho HL, Cao YC, Ju J, Qi LF (2013) Gas detection with micro- and nano-engineered optical fibers. Opt Fiber Technol 19:741–759
106. Wan Y, Su Y, Zhu X, Liu G, Fan C (2013) Development of electrochemical immunosensors towards point of care diagnostics. Biosens Bioelectron 47:1–11

107. Lafleur JP, Jönsson A, Senkbeil S, Kutter JP (2016) Recent advances in lab-on-a-chip for biosensing applications. Biosens Bioelectron 76:213–233
108. Conrad H, Schenk H, Kaiser B, Langa S, Gaudet M, Schimmanz K, Stolz M, Lenz M (2015) A small-gap electrostatic micro-actuator for large deflections. Nat Commun 6:10078
109. Norihan AH, Yunas J, Yeop Majlis B, Hamzah AA, Bais B (2015) Microfabrication of Si3N4-polyimide membrane for thermo-pneumatic actuator. Microelectron Int 32:18–24
110. Schneider U, Olofsson BR, Sörnmo O, Drust M, Robertsson A, Hägele M, Johansson R (2014) Integrated approach to robotic machining with macro/micro-actuation. Robot Comput Integr Manuf 30:636–647
111. Chuang H-S, Kumar A, Wereley ST (2008) Open optoelectrowetting droplet actuation. Appl Phys Lett 93:064104
112. Sonoda H, Kohnoe S, Yamazato T, Satoh Y, Morizono G, Shikata K, Morita M, Watanabe A, Morita M, Kakeji Y, Inoue F, Maehara Y (2011) Colorectal cancer screening with odour material by canine scent detection. Gut 60:814–819
113. György H, J. Gunvor af Klinteberg, Sven J, István H (2008) Human ovarian carcinomas detected by specific odor. Integr Cancer Ther 7:76–80
114. Hirotsu T, Sonoda H, Uozumi T, Shinden Y, Mimori K, Maehara Y, Ueda N, Hamakawa M (2015) A Highly Accurate Inclusive Cancer Screening Test Using *Caenorhabditis elegans* Scent Detection. PLoS One 10:e0118699
115. Tee LF, Tan TL, Neoh H-M, Jamal R (2017) Rapid detection of Sepsis using CESDA: the Caenorabditis elegans Sepsis detection assay. bioRxiv:144873
116. Bretscher AJ, Busch KE, de Bono M (2008) A carbon dioxide avoidance behavior is integrated with responses to ambient oxygen and food in *Caenorhabditis elegans*. Proc Natl Acad Sci 105:8044–8049
117. Wakabayashi T, Sakata K, Togashi T, Itoi H, Shinohe S, Watanabe M, Shingai R (2015) Navigational choice between reversal and curve during acidic pH avoidance behavior in *Caenorhabditis elegans*. BMC Neurosci 16:79
118. Hedgecock EM, Russell RL (1975) Normal and mutant thermotaxis in the nematode *Caenorhabditis elegans*. Proc Natl Acad Sci U S A 72:4061–4065
119. Jurado P, Kodama E, Tanizawa Y, Mori I (2010) Distinct thermal migration behaviors in response to different thermal gradients in *Caenorhabditis elegans*. Genes. Brain Behav 9:120–127
120. Mori I, Ohshima Y (1995) Neural regulation of thermotaxis in *Caenorhabditis elegans*. Nature 376:344
121. Zeng F, Rohde CB, Yanik MF (2008) Sub-cellular precision on-chip small-animal immobilization, multi-photon imaging and femtosecond-laser manipulation. Lab Chip 8:653–656
122. Li W, Feng Z, Sternberg PW, Shawn Xu XZ (2006) A *C. elegans* stretch receptor neuron revealed by a mechanosensitive TRP channel homologue. Nature 440:684–687
123. Gray JM, Hill JJ, Bargmann CI (2005) A circuit for navigation in *Caenorhabditis elegans*. Proc Natl Acad Sci U S A 102:3184–3191
124. Qin J, Wheeler AR (2007) Maze exploration and learning in *C. elegans*. Lab Chip 7:186–192
125. Link CD (1995) Expression of human beta-amyloid peptide in transgenic *Caenorhabditis elegans*. Proc Natl Acad Sci U S A 92:9368–9372
126. Kuwahara T, Koyama A, Gengyo-Ando K, Masuda M, Kowa H, Tsunoda M, Mitani S, Iwatsubo T (2006) Familial Parkinson mutant alpha-synuclein causes dopamine neuron dysfunction in transgenic *Caenorhabditis elegans*. J Biol Chem 281:334–340
127. Lakso M, Vartiainen S, Moilanen AM, Sirvio J, Thomas JH, Nass R, Blakely RD, Wong G (2003) Dopaminergic neuronal loss and motor deficits in *Caenorhabditis elegans* overexpressing human alpha-synuclein. J Neurochem 86:165–172
128. Faber PW, Voisine C, King DC, Bates EA, Hart AC (2002) Glutamine/proline-rich PQE-1 proteins protect *Caenorhabditis elegans* neurons from huntingtin polyglutamine neurotoxicity. Proc Natl Acad Sci U S A 99:17131–17136

129. Oeda T, Shimohama S, Kitagawa N, Kohno R, Imura T, Shibasaki H, Ishii N (2001) Oxidative stress causes abnormal accumulation of familial amyotrophic lateral sclerosis-related mutant SOD1 in transgenic *Caenorhabditis elegans*. Hum Mol Genet 10:2013–2023
130. Cornaglia M, Krishnamani G, Mouchiroud L, Sorrentino V, Lehnert T, Auwerx J, Gijs MAM (2016) Automated longitudinal monitoring of in vivo protein aggregation in neurodegenerative disease *C. elegans* models. Mol Neurodegener 11:17
131. Nagel G, Ollig D, Fuhrmann M, Kateriya S, Musti AM, Bamberg E, Hegemann P (2002) Channelrhodopsin-1: a light-gated proton channel in green algae. Science 296:2395
132. Gradinaru V, Mogri M, Thompson KR, Henderson JM, Deisseroth K (2009) Optical deconstruction of parkinsonian neural circuitry. Science 324:354
133. Steinbeck JA, Choi SJ, Mrejeru A, Ganat Y, Deisseroth K, Sulzer D, Mosharov EV, Studer L (2015) Optogenetics enables functional analysis of human embryonic stem cell-derived grafts in a Parkinson's disease model. Nat Biotechnol 33:204–209
134. Yamamoto K, Tanei Z-i, Hashimoto T, Wakabayashi T, Okuno H, Naka Y, Yizhar O, Fenno LE, Fukayama M, Bito H, Cirrito JR, Holtzman DM, Deisseroth K, Iwatsubo T (2015) Chronic optogenetic activation augments Aβ pathology in a mouse model of Alzheimer disease. Cell Rep 11:859–865
135. Cepeda C, Galvan L, Holley SM, Rao SP, André VM, Botelho EP, Chen JY, Watson JB, Deisseroth K, Levine MS (2013) Multiple sources of striatal inhibition are differentially affected in huntington's disease mouse models. J Neurosci Off J Soc Neurosci 33:7393–7406
136. Tønnesen J, Sørensen AT, Deisseroth K, Lundberg C, Kokaia M (2009) Optogenetic control of epileptiform activity. Proc Natl Acad Sci 106:12162–12167
137. Paz JT, Davidson TJ, Frechette ES, Delord B, Parada I, Peng K, Deisseroth K, Huguenard JR (2013) Closed-loop optogenetic control of thalamus as a new tool to interrupt seizures after cortical injury. Nat Neurosci 16:64–70
138. Ji Z-G, Wang H (2015) Optogenetic control of astrocytes: is it possible to treat astrocyte-related epilepsy? Brain Res Bull 110:20–25
139. Narayan A, Laurent G, Sternberg PW (2011) Transfer characteristics of a thermosensory synapse in *Caenorhabditis elegans*. Proc Natl Acad Sci 108:9667–9672
140. Faumont S, Rondeau G, Thiele TR, Lawton KJ, McCormick KE, Sottile M, Griesbeck O, Heckscher ES, Roberts WM, Doe CQ, Lockery SR (2011) An image-free opto-mechanical system for creating virtual environments and imaging neuronal activity in freely moving *Caenorhabditis elegans*. PLoS One 6:e24666
141. Geary TG, Sakanari JA, Caffrey CR (2015) Anthelmintic drug discovery: into the future. J Parasitol 101:125–133
142. Chen B, Deutmeyer A, Carr J, Robertson AP, Martin RJ, Pandey S (2011) Microfluidic bioassay to characterize parasitic nematode phenotype and anthelmintic resistance. Parasitology 138:80–88
143. Carr JA, Parashar A, Gibson R, Robertson AP, Martin RJ, Pandey S (2011) A microfluidic platform for high-sensitivity, real-time drug screening on C. elegans and parasitic nematodes. Lab Chip 11:2385–2396
144. Weeks JC, Roberts WM, Robinson KJ, Keaney M, Vermeire JJ, Urban JF, Lockery SR, Hawdon JM (2016) Microfluidic platform for electrophysiological recordings from host-stage hookworm and Ascaris suum larvae: A new tool for anthelmintic research. Int J Parasitol Drugs Drug Resist 6:314–328
145. Hu C, Kearn J, Urwin P, Lilley C, Connor VO', Holden-Dye L, Morgan H (2014) StyletChip: a microfluidic device for recording host invasion behaviour and feeding of plant parasitic nematodes. Lab Chip 14:2447–2455
146. Hengartner MO, Horvitz HR (1994) The ins and outs of programmed cell death during *C. elegans* development. Philos Trans R Soc Lond B Biol Sci 345:243
147. Kenyon C (2011) The first long-lived mutants: discovery of the insulin/IGF-1 pathway for ageing. Philos Trans R Soc B 366:9–16

148. Wu Y, Wu Z, Butko P, Christen Y, Lambert MP, Klein WL, Link CD, Luo Y (2006) Amyloid-β-induced pathological behaviors are suppressed by *ginkgo biloba* extract EGb 761 and ginkgolides in transgenic *Caenorhabditis elegans*. J Neurosci 26:13102
149. Braungart E, Gerlach M, Riederer P, Baumeister R, Hoener MC (2004) *Caenorhabditis elegans* MPP^+ model of parkinson's disease for high-throughput drug screenings. Neurodegener Dis 1:175–183
150. Gosai SJ, Kwak JH, Luke CJ, Long OS, King DE, Kovatch KJ, Johnston PA, Shun TY, Lazo JS, Perlmutter DH, Silverman GA, Pak SC (2010) Automated high-content live animal drug screening using *C. elegans* expressing the aggregation prone serpin α1-antitrypsin Z. PLoS One 5:e15460
151. Burns AR, Luciani GM, Musso G, Bagg R, Yeo M, Zhang Y, Rajendran L, Glavin J, Hunter R, Redman E, Stasiuk S, Schertzberg M, Angus McQuibban G, Caffrey CR, Cutler SR, Tyers M, Giaever G, Nislow C, Fraser AG, MacRae CA, Gilleard J, Roy PJ (2015) *Caenorhabditis elegans* is a useful model for anthelmintic discovery. Nat Commun 6:7485
152. Smout MJ, Kotze AC, McCarthy JS, Loukas A (2010) A novel high throughput assay for anthelmintic drug screening and resistance diagnosis by real-time monitoring of parasite motility. PLoS Negl Trop Dis 4:e885
153. Kevin V, Abraham Phillip L (2017) A truly Lego ® -like modular microfluidics platform. J Micromech Microeng 27:035004
154. Zhao W, Zhang L, Jing W, Liu S, Tachibana H, Cheng X, Sui G (2013) An integrated microfluidic device for rapid serodiagnosis of amebiasis. Biomicrofluidics 7:011101
155. Mellors JS, Jorabchi K, Smith LM, Ramsey JM (2010) Integrated microfluidic device for automated single cell analysis using electrophoretic separation and electrospray ionization mass spectrometry. Anal Chem 82:967–973
156. Jiang J, Zhao H, Shu W, Tian J, Huang Y, Song Y, Wang R, Li E, Slamon D, Hou D, Du X, Zhang L, Chen Y, Wang Q (2017) An integrated microfluidic device for rapid and high-sensitivity analysis of circulating tumor cells. Sci Rep 7:42612
157. Gómez-Sjöberg R, Leyrat AA, Pirone DM, Chen CS, Quake SR (2007) Versatile, fully automated, microfluidic cell culture system. Anal Chem 79:8557–8563
158. Lee C-C, Sui G, Elizarov A, Shu CJ, Shin Y-S, Dooley AN, Huang J, Daridon A, Wyatt P, Stout D, Kolb HC, Witte ON, Satyamurthy N, Heath JR, Phelps ME, Quake SR, Tseng H-R (2005) Multistep synthesis of a radiolabeled imaging probe using integrated microfluidics. Science 310:1793
159. Urbanski JP, Thies W, Rhodes C, Amarasinghe S, Thorsen T (2006) Digital microfluidics using soft lithography. Lab Chip 6:96–104
160. Fidalgo LM, Maerkl SJ (2011) A software-programmable microfluidic device for automated biology. Lab Chip 11:1612–1619
161. Amin AM, Thakur R, Madren S, Chuang H-S, Thottethodi M, Vijaykumar TN, Wereley ST, Jacobson SC (2013) Software-programmable continuous-flow multi-purpose lab-on-a-chip. Microfluid Nanofluid 15:647–659

Chapter 7
Microfluidic Devices for Gamete Processing and Analysis, Fertilization and Embryo Culture and Characterization

Séverine Le Gac, Verena Nordhoff, and Bastien Venzac

Abstract Assisted reproductive technologies (ART) include all techniques used to achieve pregnancies not only in case of human infertility, but also for the *in vitro* production of embryos in the livestock industry, and for the conservation of endangered species. Focusing on human ART only, the total number of babies born worldwide has been estimated to more than 7 million, and ART are more and more utilized as a consequence of an increase in human infertility. However, at the same time, even if ART are now used as routine clinical treatments, the success rate remains low, with less than 30% clinical pregnancies. Furthermore, the ART pregnancy rates are now stagnating, showing a need for new improvements. Finally, current protocols are lacking standardization and automation, and they are still dependent on the skills of highly trained personnel. In that context, microfluidics can offer a new paradigm in the ART field, by providing integrated and automated platforms. Furthermore, the use of microfluidics can introduce new approaches by performing some steps of the entire protocol. In this chapter, we summarize and discuss microfluidic developments in the field of ART, and specifically devices for gamete analysis, selection and processing, fertilization and embryo culture.

Keywords Assisted reproductive technology · Gamete processing and analysis · *In vitro* embryo manipulation · Integrated platforms · Microfluidics

S. Le Gac (✉) · B. Venzac
Applied Microfluidics for BioEngineering Research (AMBER), MESA+ Institute for Nanotechnology, TechMed Center, University of Twente, Enschede, The Netherlands
e-mail: s.legac@utwente.nl

V. Nordhoff
Centre of Reproductive Medicine and Andrology, University Hospital of Münster, Münster, Germany

M. Tokeshi (ed.), *Applications of Microfluidic Systems in Biology and Medicine*, Bioanalysis 7, https://doi.org/10.1007/978-981-13-6229-3_7

7.1 Introduction

Assisted reproductive techniques (ART), which comprise all techniques utilized to conceive offspring using (semi-)artificial ways have been utilized for 40 years to overcome human infertility. Next to this, ART are also widely used for the *in vitro* production of bovine and porcine embryos in the livestock industry as well as for the conservation of endangered species. Infertility affects 15% of the couples in the developed countries, and this figure continues to increase as a consequence of our modern lifestyle and repeated exposure to chemicals causing infertility such as endocrine disruptors. Currently, ART are well-established practices, and have become clinical routines in nearly all parts of the world. As a great honor for the ART field, the Nobel Prize in Medicine or Physiology was attributed to Sir Robert Edwards, one of the "medical fathers" of the first born IVF baby worldwide, Louise Brown. Since then, according to the European Society for Human Reproduction and Embryology (ESHRE, www.eshre.eu), more than 7 million babies have been conceived using ART worldwide. Furthermore, it has been estimated that during the years 2008–2010 over 1.1 million children have been born as the result of the various ART treatments [1], and, following the increasing infertility rate the number of treatment cycles has been continuously rising [2]. Equally important is the use of ART in the multi-billion dollar livestock industry with about half a million bovine embryos produced *in vitro* in 2013 [3]. The most commonly used techniques worldwide are intrauterine insemination (IUI), *in vitro* fertilization (IVF) and intracytoplasmic sperm injection (ICSI), as discussed later in this chapter.

In spite of this continuously growing number of ART cycles, the so-called "take-home baby rate" remains still low, with less than 30%. Furthermore, the year-to-year enhancement in the pregnancy rates reported in the last decades was not observed for the first time in 2008 [4], which may indicate that the techniques used in ART and for the selection of gametes have reached a saturation level regarding success rates. Altogether, these facts call for new technological developments and alternative approaches for the ART routines. Another main issue encountered with ART next to the low success rates is the higher chance for multiple pregnancies and multiple birth rates (www.eshre.eu; *ca.* 17.5% in 2014 in the EU), which can be accounted by the lack of reliable approaches to identify embryos with the highest developmental competence at the end of the pre-implantation development period. The implementation of the policy for elective single embryo transfer, or eSET, has helped decreasing the multiple pregnancy incidence, but it requires, more than ever, the availability of reliable means to grade and identify top-quality embryos before transfer. Next to this, almost all procedures in an ART protocol are entirely manual and differ from lab to lab. This lack of standardization together with the manual and subjective character of the processes explain the sometimes significant discrepancy in success rates between IVF centers.

In this chapter, we discuss how microfluidic technology could revolutionize the field of ART by providing not only new approaches to conduct the different steps of the process, but also integrated and automated platforms. We will first present in detail an entire ART protocol, using the most widely used insemination techniques, IUI, IVF and ICSI. Next, after a very brief introduction to microfluidic technology, we will highlight some of the key-features this technology presents, and which are of particular interest for the field of assisted reproduction. Following this, we will review selected platforms developed to conduct different individual steps of an ART routine; namely, the analysis, characterization and/or selection of the female and male gametes; the fertilization and/or cloning; the *in vitro* culture of the resulting embryos and, finally, their characterization in the view of identifying embryos with the highest developmental competence. In every section we will discuss the promises and identified shortcomings of microfluidic technology and the applicability of the so far reported platforms for true clinical routines. Finally, in the conclusion, we will introduce additional examples of microfluidic platforms either capable of performing an entire IVF process or developed for other promising applications related to ART.

7.2 A Standard ART Routine, as Performed in an ART Center

Although ART techniques have developed over the last decades and are differently managed and used in every ART clinic and lab, some basic methods are equal everywhere and form the basis for assisted reproduction the control of the ovarian cycle. In a regular ovulatory cycle of a woman (between 26 and 35 days) the growth of one follicle is stimulated by endogenous hormones (mainly Follicle-stimulating hormone, FSH) and, respectively, a single oocyte is ovulated 40–44 h after a Luteinizing hormone (LH) peak in the middle of the cycle. For assisted reproduction the female cycle is often supported or stimulated in a controlled manner by administration of exogenous FSH. In case of simple treatment options like timed intercourse or insemination (intrauterine insemination, IUI) only low doses of FSH are applied. Successful follicular growth is followed by ultrasound of the ovaries and if one or two follicles are confirmed ovulation is induced by a single high dose of human Choriogonadotropin (hCG), which mimics the natural LH peak. The couple can then either go for timed intercourse 24–36 h after ovulation induction or for insemination, with the placement of a prepared semen sample into the uterine cavity 36 h after application of the hCG dose.

If insemination is not the treatment of choice or has not resulted in a pregnancy, *in vitro* fertilization (IVF) and intra-cytoplasmic sperm injection (ICSI), the two major *in vitro* treatment options have to be used (Fig. 7.1). IVF is the treatment of choice if a female factor is the cause for the infertility, *e.g.*, obstruction of the oviducts or uncontrolled growth of endometrial cells in other body areas outside the uterus (endometriosis). ICSI is indicated if a male factor has been diagnosed. However, ICSI is also the treatment of choice if IVF has failed. For

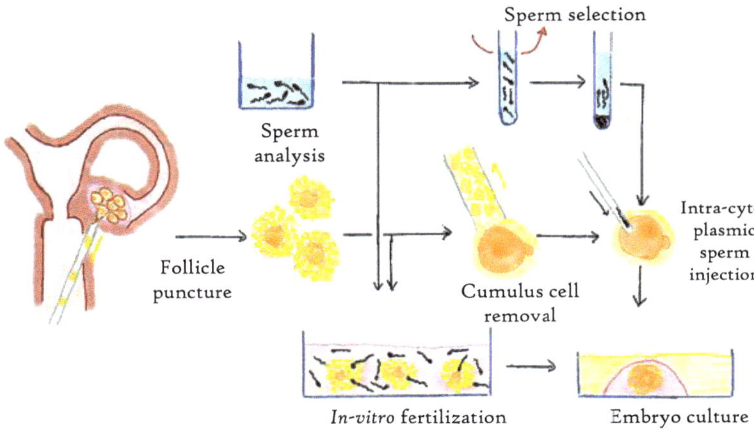

Fig. 7.1 Schematic representation of a conventional ART medical routine, using either IVF (*in vitro* fertilization) or ICSI (intra-cytoplasmic sperm injection) for the fertilization. Oocytes are retrieved from a woman's ovary, and incubated with processed sperm in a Petri dish for fertilization if the sperm count and motility are in the upper normal range. Otherwise, cumulus cells are removed around the oocyte, a single sperm is selected based on motility and ICSI is performed. The resulting embryos are kept a few days in culture until their intrauterine transfer for further development. Illustration by Dr. Venzac

IVF and ICSI women are treated with higher doses of hormones to achieve multi-follicular growth in both ovaries. This multi-follicular growth is necessary as not all oocytes may be of equal quality or possibly not mature enough to be used for IVF or ICSI. Additionally not all oocytes are fertilized or result in normally developing embryos or may although of good quality lead to a pregnancy.

Oocytes with their surrounding cumulus cells (cumulus-oocyte-complex, COC) are retrieved from the follicles 35–36 h after hCG application by an ultrasound-guided transvaginal puncture of both ovaries. In case of an IVF cycle the COCs are incubated with 25,000–100,000 motile spermatozoa from a prepared semen sample in droplets of 30–100 μL, often covered by light mineral oil to omit evaporation of fluid. For ICSI, the oocytes have to be separated from the cumulus cells by gentle denudation using an enzyme (hyaluronidase) and small pipettes. Mature oocytes (in Metaphase II stage) are then injected each with a single immobilized spermatozoon using a small injection needle. Fertilization success, meaning the appearance of a fertilized oocyte (pronucleus stage) is checked 15–18 h later. Further cultivation in 10–50 μL droplets [5–7] of special culture media for 2–5 days leads to the production of pre-implantation embryos, of which normally only one, sometimes also two or three (depending on quality and transfer policy of the respective country) is (or are) transferred back to the uterus of the patient. A biochemical pregnancy is confirmed by measuring β-hCG in blood, while a clinical pregnancy is checked via detection of the heartbeat of the fetus by ultrasound 1–2 weeks later.

For all ART treatments, meaning IUI, IVF and ICSI, the ejaculate has to be prepared. The ejaculate consists of two major fractions, a sperm-rich first fraction from the testes followed by a mostly seminal fluid fraction from the accessory glands [8]. Sperm have to be separated within 1 h from the seminal fluid as the latter contains components (*e.g.*, prostaglandins, zinc) and non-sperm cells that may damage intact spermatozoa. In the case of natural conception, these components and non-sperm cells cannot pass the cervical mucus at time of ovulation. However, in ART natural barriers are bypassed either by direct insertion of the sperm into the uterus of the women (IUI) or direct exposure of the oocytes to the sperm sample, so that these components become obstacles for the achievement of a pregnancy by possibly damaging sperm cells [9].

Sperm samples are typically prepared using different techniques, depending on the sperm concentration and motility in the native ejaculate. The choice of the method depends on which ART treatment is used [9]. For normozoospermic samples (meaning normal concentration (WHO reference value is 39 million/mL), normal motility (32% progressive motility) and normal morphology (4%)) the WHO recommends in their latest handbook [9] to use a simple washing step: the ejaculate is mixed with medium, centrifuged, the supernatant discarded and the pellet is simply resuspended with medium. In the swim-up procedure, which is ideal for samples in the normal range, the ejaculate is centrifuged, next overlaid with medium, and the motile sperm swim during an incubation step into the medium. If the semen sample presents a lower concentration or reduced sperm motility (oligoasthenozoospermia; oligo: reduction of concentration, astheno: reduction of motility) the WHO recommends the use of density gradients. Specifically, different layers of medium with various densities are prepared in a tube and the sperm sample is centrifuged through these different layers. Typically, spermatozoa accumulate in the pellet at the bottom of the tube, this pellet is re-suspended in medium for further use. Using the gradient approach high numbers of motile sperm are obtained, while the swim-up methodology provides lower number of yet highly motile sperm cells [10]. In very severe cases of oligoasthenoteratozoospermia (terato: reduction in morphology) with extremely low sperm numbers (<100,000 sperm in the complete sample) or in case of azoospermia (absence of spermatozoa in the ejaculate) it might even be necessary to retrieve sperm directly from the testes. During an operation the testis capsule is opened and the tubuli seminiferi, where spermatogenesis takes place, are screened using a microscope to detect areas with high probability of intact spermatogenesis. Small testicular biopsies are taken for testicular sperm extraction (TESE). Biopsies can be used either fresh or frozen to retrieve intact spermatozoa for ICSI. Normally, motility is the clear evidence of vitality and only vital spermatozoa will give reasonable fertilization rates. However in low quality TESE samples, it may happen that only immotile spermatozoa are retrieved. In this case additional methods for detection of vitality have to be used [11].

7.3 Why Microfluidics for ART?

Microfluidics, as pointed out by Georges Whitesides, can be defined as "the manipulation of fluids in channels with dimensions of tens of micrometers" [12]. For now more than two decades, this technology has been gaining more and more popularity, especially in the fields of biomolecule analysis and cell biology, and microfluidic devices are currently routinely used in healthcare and diagnosis, particularly for point-of-care applications. The success of microfluidics can be explained by the enhanced capabilities this technology brings about. Volumes in microfluidic devices are dramatically reduced down to the low nanoliter range. Therefore, analysis can be conducted on smaller-sized samples, and cell experimentation can easily be downscaled to the single cell level [13, 14]. Processes, including analyses, are much faster at the micrometer scale, and they are better controlled as a result of both the large surface-to-volume ratio and the unique properties of the flows [12]. For cellular studies, these characteristics associated with the high confinement level found in these devices enable to closely emulate conditions found *in vivo* [15]. At that scale, it is also straightforward to create dynamic conditions, diffusion-based delivery of substances [16, 17] or gradients [18]. Furthermore, microfluidic devices present a high level of integration. One device can for instance comprise a series of identical systems (horizontal integration), which is highly attractive for assay parallelization. Alternatively, a series of operations can be implemented in one single device (vertical integration), to perform a complete (analytical) process. Furthermore, microfluidic structures can be combined with add-on capabilities (smart integration), *e.g.*, for sensing of biochemical or physical parameters, to control the temperature, or for fluid actuation. Finally, these devices, which are realized using either conventional microfabrication techniques derived from the field of microelectronics or through replication approaches originating from the polymer industry, can be produced at a large scale, which drastically reduces their price and promotes their use as disposables.

Microfluidic technology is highly attractive for the field of ART, specifically to perform the different *in vitro* steps of the ART routine, namely from semen analysis and selection until embryo scoring and selection, including the fertilization step and the embryo pre-implantation culture. Some of the most promising features of microfluidic technology for ART are its high level of integration and its amenability to process automation, which would eventually allow implementing the whole ART routine in a fully automated single platform and additionally lead to standardization of the treatment across laboratories [19, 20]. Furthermore, using fully integrated platforms, the manipulation of the gametes and embryos would be minimized, together with human intervention, associated errors and risks of contamination, and, subsequently, the success rate of the treatment would be enhanced. Another main expected outcome of using microfluidic technology is a decrease in the treatment costs: not specifically due to the smaller volumes of medium required and the price of the devices, but also through an increase in the treatment success.

In the following sections, we will successively review microfluidic advances for the preparation of the gamete samples (oocytes and sperm), for the fertilization, and for the embryo culture and characterization. Finally, in the conclusion, new trends of fully integrated devices able to perform multiple steps of the ART protocol, or to support spermatogenesis *in vitro* will be briefly presented.

7.4 Gamete Preparation and Fertilization

7.4.1 Oocytes

In the frame of conventional *in vitro* fertilization routines, there is little or no preparation and selection of oocytes, as happens for the semen, before fertilization. As mentioned in Sect. 7.2, a commonly performed step consists of removing the cumulus cells surrounding the oocytes in the cumulus-oocyte-complex. However, there is evidence in the literature that pre-selecting the oocytes before fertilization does enhance the quality of the resulting embryos, and could be an alternative strategy to the currently used approach to select embryos after their pre-implantation culture and before their intra-uterine transfer. Furthermore, for ICSI-based fertilization, it may be advantageous to characterize the oocytes to ensure they are at the right maturation stage for fertilization (also before removal of the cumulus cells).

Various microfluidic platforms have been developed for the different purposes of oocyte denudation, oocyte characterization and/or selection, as well as oocyte maturation, as discussed in the following. Next to this, dedicated capabilities have been proposed for the controlled transportation and manipulation of oocytes in a microfluidic device, and for the fertilization step as well as cloning (discussed in sect. 7.4.3).

The first examples of microfluidic platforms for the manipulation of oocytes aimed at removing the cumulus cells around the oocytes, which is typically done *in vitro*, either before fertilization (in case of ICSI) or 15–18 h after IVF, when the oocyte is still at the zygote stage. Conventional approaches *in vitro* rely on an enzymatic treatment using hyaluronidase or a mechanical treatment, through pipetting or vortexing. The mechanical treatments in particular have been proven to be stressful for the zygotes. As an alternative approach, Zeringue *et al.* proposed a microfluidic protocol for the mechanical denudation of bovine oocytes, in a better controlled and gentler way, as illustrated in Fig. 7.2a. Their device comprised a long channel equipped with 2 constricted areas, followed by a 90° turn with smaller side channels [21]. While traveling through the constriction, the cumulus cells got loosened around the zygotes, and they could be entirely removed by applying suction from the side channels placed in the right angle turn. Rotation of the cumulus-oocyte-complex in the turn of the channel was found to be essential for efficient removal of all the cumulus cells. Comparison of this novel microfluidic approach with a vortex-based strategy revealed that embryos undergoing the

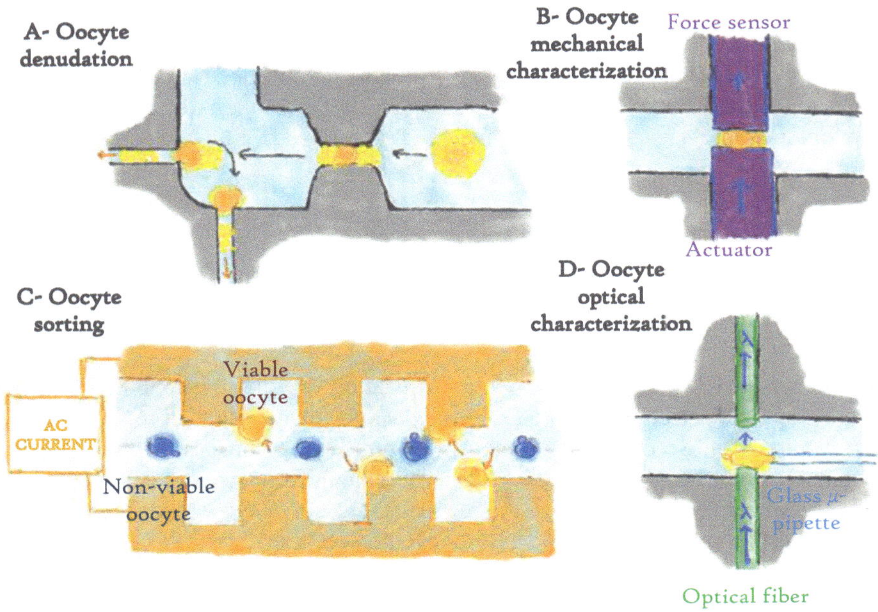

Fig. 7.2 Example of microfluidic platforms developed for the manipulation, characterization and sorting of oocytes. (**a**) Oocyte denudation in a microfluidic channel equipped by a constricted area and suction side-channels (illustration inspired by [21]); (**b**) Evaluation of the mechanical properties of oocytes using on-chip actuated beams (illustration inspired by [24]); (**c**) Dielectrophoretic-based sorting of oocytes (illustration inspired by [26]). (**d**) Optical characterization of oocytes using on-chip integrated optical fibers (illustration inspired by [27]). Illustrations by Dr. Venzac

microfluidic treatment were developing better, with higher cell counts than their counterparts having experienced vortexing [22]. Furthermore, the vortex group of embryos exhibited a peak in their transcriptomic activity just after removal of the cumulus cell, which may be indicative of their stressed state. Using a microfluidic format allows precisely controlling the shear exerted on the oocytes/zygotes by fine-tuning the dimensions of the microfluidic structures as well as the flow and negative pressure applied in the different channels to efficiently remove the cumulus cells, without compromising the viability of the oocytes/zygotes.

Developing novel integrated schemes for oocyte characterization has widely been explored, with three main applications in mind: (i) to determine the maturation stage of the oocyte before fertilization, (ii) to confirm fertilization after incubation with the semen sample, and (iii) to evaluate the oocyte viability. All microdevices reported for oocyte characterization rely on two approaches, using either mechanical sensors or optical means, which are successively discussed below.

The rationale behind the mechanical approach is that the properties of the zona pellucida, which is like a shell surrounding the oocyte, change during the different maturation stages through thickening [23], as well as after fertilization by "hardening". Furthermore, zona pellucida hardening has also been observed on "aged oocytes", which are of lower quality. In short, the mechanical properties of the zona pellucida directly correlate with the oocyte maturation stage and quality. A variety of mechanical sensors have been proposed and integrated in microfluidic devices to measure the mechanical properties of oocytes. In a first example, Wacogne et al. proposed a SU-8-based microtool consisting of a free-standing cantilever [24]. The oocyte was forced against a beam, and its mechanical properties derived by jointly measuring its deformation as well as the beam deflection (Fig. 7.2b). The system, which was tested on human oocytes, could be automated for successive characterization of a group of embryos. In an alternative and more straightforward approach, Luo et al. implemented a constriction in a microfluidic channel, through which oocytes could be forced and would deform [25]. In this device, mouse oocytes were exposed to well-defined flow and shear, by taking advantage of the microfluidic format. Here again, from its deformation, the mechanical properties of the oocyte could be determined.

Similarly, optical sensors have been developed to assess the oocyte maturation stage and the "fertility index" in a minimally invasive way. Optical fibers were integrated in a microfluidic device (Fig. 7.2d) to measure the transmission spectra of individual oocytes [27, 28]; the different maturation phases of the oocyte could successfully be distinguished by determining the minimum transmission intensity and the associated wavelength. Altogether, this approach allowed monitoring the oocyte maturation and deciding on the best moment for fertilization.

In the view of the selection of oocytes, dielectrophoresis (DEP) has been applied in a microfluidic format for the automated and non-biased identification and isolation of viable oocytes (Fig. 7.2c) [26, 29]. Viable and fertile oocytes were found to move faster. Oocytes with no or little dielectrophoretic motion gave rise, after fertilization, to embryos with a poor developmental competence, while in case of fast oocytes, reduced polyspermy was observed and higher blastocyst rates and cell numbers were found.

As recently reviewed by Yanez and Camarillo, key-advantages offered by microfluidic technology for the oocyte characterization and selection compared to better established approaches for the mechanical characterization of cells (*e.g.*, AFM and optical tweezers) [23] are its amenability for automated and high-throughput measurements, and the possibility to design integrated platforms for on-line fertilization of the oocytes. Next to this, in a microfluidic format, not only active elements like beams, for instance for mechanical measurements can be used, but also passive elements can be explored such as well-defined integrated constrictions and even more simply, a well-controlled flow. Furthermore, the fact that most of the materials used to fabricate microfluidic devices are transparent, they allow direct and continuous optical examination of the specimen during the measurements.

An essential aspect for this on-chip oocyte characterization is the capability of displacing the oocytes in the device in a gentle yet accurate manner. The Gharbi lab reported two different silicon-based strategies for, respectively, long-distance transportation and accurate capture of human oocytes at a characterization site [28, 30]. For long-range displacement, the oocyte was trapped in a translator consisting of a wet-etched cavity in a silicon free-standing beam, and motion of this translator was controlled using a dedicated stage. Accurate positioning of the oocytes was achieved by applying a negative pressure through micronozzles in the bottom substrate of the device. Nakahara et al. proposed another gentle protocol for controlled transportation of mouse oocytes in an open microfluidic device using a vibration-induced flow, towards an integrated on-chip robot for their mechanical characterization [31].

A last step of oocyte treatment, which has more recently been explored in a microfluidic format is the *in vitro* maturation of the oocytes. Normally, this step is performed *in vivo* through extensive hormonal treatment of the patient, as discussed in Sect. 7.2, with however a risk for significant complications such as the ovarian hyperstimulation syndrome. Alternatively, the oocytes can be exposed to a similar hormonal treatment *in vitro*, before fertilization, through exposure to gonadotropin [32]. Berenguel-Alonso et al. reported a COC-based platform (COC – cycloolefin copolymer), into which oocytes could be trapped with the help of a micromachined filter and exposed to maturation medium in the device [33]. Oocytes could easily be retrieved from the device, thanks to careful design of the microfluidic circuitry, for off-chip fertilization. A typical advantage of microfluidics for the oocyte maturation is the spatial and temporal control this technology provides for the perfusion of solutions. Furthermore, the mature oocytes could subsequently be fertilized *in situ* by replacing the maturation solution by prepared semen.

7.4.2 Semen Preparation, Analysis and Selection

In the frame of an ART routine, different operations are performed on the male gametes, including the semen preparation and selection for subsequent ART steps, as described in Sect. 7.2. Next to this, microfluidic systems have also been developed for fertility diagnosis, in the form of either lab-based or home-testing devices.

A male factor is involved in 40–50% of all infertility cases. This male factor can be caused by a variety of pathologies, and associated with different semen sample compositions, ranging from a total absence of sperm to a lower concentration of motile sperm in the ejaculate. Sperm analysis is therefore one of the first actions taken when a couple encounters fertility issues. The main assays focus on sperm characteristics, such as the concentration in the ejaculate, the motility of the sperm, their vitality and morphology, and their DNA integrity as detailed below. The threshold for a normal semen sample is considered to be 39 million cells per millilitre, as evaluated using a counting chamber or by automated treatment of microscopic images using a system called CASA (computer-assisted semen analysis). The motility is the second main parameter, which is also measured with a CASA system. Motility is a strong indicator for a successful fertilization. A semen with a good quality has to contain more than 32% of motile sperm [9]. Sperm cells can also be classified according to their motility, the latter being characterized as progressive (straight and fast movement), non-progressive and immotile. The vitality of the sperm is typically detected using classical dye exclusion or hypotonic swelling assays [9].

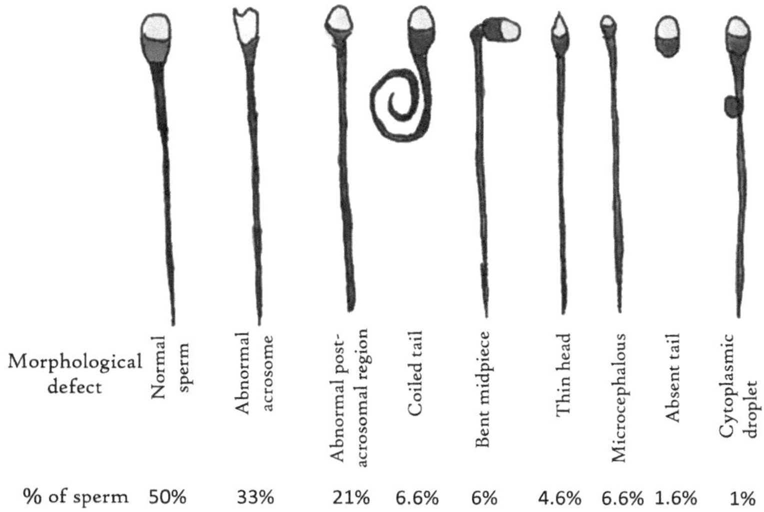

Fig. 7.3 Main morphological defects observed in sperm, with the average percentage of cells presenting this defect in samples from 1001 male partners of pregnant women. Illustration inspired by [34], and made by Dr. Venzac

Morphological defects (size of the head, double-headed, double-tailed, etc.) are easy to observe qualitatively (Fig. 7.3), but their quantification is more delicate, especially because a semen sample is still considered normal if only 4% of the sperm cells have a normal morphology. Moreover, the relationship between morphological defects and pregnancy rate is still not clear [35]. Finally, DNA integrity, which can be probed using a comet assay or a sperm chromatin structure assay, has been identified as a good predictor of embryo quality after IVF [36].

None of these analyses is currently automated; they all suffer from a lack of standardization, they use expensive equipment and they are time-consuming. Microfluidics is a mature technology, which has been proven to be attractive for diagnostic purposes, for other medical applications, as well as for the development of a variety of point-of-care devices [37]. In that frame, microfluidics not only can help to simplify, automate and even improve the diagnostic of male infertility, but also provides new opportunities to perform semen analysis at home. The so-far developed microfluidic systems for this purpose of semen analysis can be divided into lab-based systems and point-of-care devices.

First, microfluidic devices have been mainly developed to quantitatively evaluate the sperm concentration and/or motility in semen, in an automated way, in a specialized laboratory. A number of reviews has extensively discussed microfluidics-based analysis of sperm for interested readers [38, 39], and here, we will mainly focus on two different systems. Using impedance measurements of micrometre-sized objects flushed in a microchannel, Segerink et al. successfully distinguished sperm from other cell debris and white blood cells [40]. Furthermore, to derive the concentration of sperm cells in samples, a known amount of polystyrene beads was added to the raw semen and both the beads and sperm were counted. This approach eliminates the need to know the flow rate in the channel. The same system was next used to distinguish normal from morphologically impaired sperm that present a cytoplasmic droplet on the flagella [41]. Chu et al. proposed a simple passive system, in which a semen sample and medium are injected next to each other in a microfluidic chamber. After about 10 min, half of the motile sperm is expected to have migrated into the medium compartment [42]. The device was next subjected to a centrifugation step: each semen fraction formed a pellet in the device and the pellet areas were calculated to derive the concentration and ratio of motile sperm. The results with this device strongly correlated with those obtained with a CASA system. As for now, none of the microfluidic systems reported for semen analysis have been developed beyond the proof-of-concept stage. However, the performance and ease-of-use of these devices are promising for their future clinical translation as diagnostic tools.

Still, one of the main advantages of microfluidic devices remains their portability and their suitability for home-testing and as point-of-care devices. As a consequence, no medical appointment is needed for a first qualitative testing, which decreases the stress and anxiety related to testing in a specialized clinic as well as the price of the male fertility test. Point-of-care testing could also be of great interest for remote locations or developing countries with limited access to medical centres. Different technologies have been proposed for low-cost, easy-to-use and easy-to-readout assays for home-based semen analysis, as recently reviewed [43]. Lateral flow assays, which use the same principle as the pregnancy test, are already commercialized to determine at home the sperm concentration (*e.g.*, SpermCheck; https://www.spermcheck.com/). They only give a qualitative estimation of the sperm concentration, and must be therefore followed by a quantitative and complete analysis in a clinical laboratory in case of low sperm count. Furthermore, other main parameters (morphology, motility and viability) are not examined in this assay. Other systems based on the staining of sperm, which are to swim outside a sample semen drop (*e.g.*, SwimCount Sperm Quality Test, https://www.swimcount.com/), provide a qualitative concentration of motile sperm, with an accuracy of 95% as compared to the Gold standard (CASA).

Two other alternative microfluidic strategies offer multi-parametric, cheap and easy-to-use diagnostic. For instance, an automated CASA was implemented in a microfluidic chamber: sperm count and motility are being recorded and determined using the camera of a smartphone. Sample loading in this device is power-free and entirely manual. This system allows characterising the concentration, motility and velocity of the sperm with results similar to the traditional CASA [44]. Paper microfluidics, with its promises of ultra-low cost and ease-of-use, has also been applied for semen analysis. One device, reported by Nosrati et al. successfully determined the concentration of live and motile sperm in a sample [45]. Their assay relies on a colorimetric principle: an enzyme produced by live sperm converts a yellow tetrazolium dye into a purple compound. Next to this, the device includes a membrane with small pores, through which only motile sperm can pass. Therefore, only the motile sperm crossing the membrane could reach the dye and induce a colour change, reaching an accuracy similar to CASA. The sperm DNA integrity, which is acknowledged as the best predictor for a successful fertilization, was also measured in a paper microfluidic system using an electrokinetic separation of intact and damaged genomic DNA after on-chip sperm lysis [46]. In contrast to the aforementioned commercialized systems, and despite their promising performance, all these paper microfluidic systems are still at the proof-of-concept stage. Commercialization is also not foreseen, which is a difficulty shared with most of the paper microfluidic devices.

Even for perfectly fertile men, the ejaculated semen always contains a large amount of sperm with impaired characteristics. In a normal sample (according to the WHO definition), up to 68% of sperm can be immotile, and up to 96% can present morphological defects (See Fig. 7.3). As mentioned in sect. 7.2, during natural fertilization, sperm are subjected to a strong selection pressure in the female track (cervix/uterus/oviduct) [47, 48]. Specifically, sperm must be able (i) to swim independently, against a flow of oviductal fluid and in a complex environment in terms of topology and fluid composition, (ii) to escape the female immune system; (iii) to correctly reach the oocyte (long-range thermotaxis and short-range chemotaxis); (iv) to undergo hyperactivation and modification of their head composition (process also known as capacitation) after interactions with the oviduct epithelium.

From an initial quantity of several million cells, only a few hundred of them actually reach the oocyte, and all the immotile sperm presenting strong morphological defects are eliminated stepwise in the journey through the female reproductive track [48]. Arguably, similar selection has to be implemented *in vitro* to isolate the best sperm for subsequent IVF, and the unique optimal sperm for ICSI. Current selection techniques (*e.g.*, swim-up, density gradient), as discussed in sect. 7.2, merely sort sperm according to their motility and morphology, since morphologically abnormal sperm might not reach the swim-up medium. Furthermore, the centrifugation steps involved in the process are stressful to the sperm and may result in the production of reactive oxygen species (ROS) and, in turn, to DNA fragmentation [49]. Using microfluidics, the best sperm can be isolated using milder conditions and avoiding the stressful centrifugation steps, as illustrated by a number of devices proposed by several research labs.

A large number of microfluidic systems focus on passively selecting motile sperm in microchannels [50], by recovering the fastest swimmers and highly motile ones at the end of a microchannel. For instance, Nosrati et al. reported a circular high-throughput system with 500 microchannels, able to process 1 mL of raw semen in less than 20 min, and to isolate only motile sperm with a significant increase in DNA integrity (80%) compared to the raw sample, which is to be compared to the traditional 50% achieved using conventional methods involving centrifugation [51]. A similar device exploiting the swimming ability of motile sperm through a membrane having 8-μm pores provided an improvement in the DNA integrity, motility speed and morphology of the sperm compared to the standard swim-up technique [52]. Actually, sperm selected according to their motility always show better DNA integrity compared to the raw semen specimen, as most of the damaged sperm in the raw semen are not able to move efficiently anymore. Compared to more centrifugation-based traditional methods, the production of ROS and DNA damages are significantly decreased when using microfluidic devices, thanks to the use of gentler handling conditions. However, a large amount of the selected sperm is still morphologically deficient or damaged.

Using microfluidic devices, other selection criteria have been explored, which cannot easily be implemented using conventional dishware, to further improve the quality of the selected sample. *In vivo*, spermatozoa are able to direct themselves by following gradients of either factors secreted by the oocyte (short-range chemotaxis) [53], or temperature between the uterus and the fertilization location (long-range thermotaxis) [54], or even against weak flows of oviduct fluid (rheotaxis) [55].

Microfluidics, which provides a fine control over chemical and physical gradients (*e.g.*, temperature gradients) [18], as well as on flows, is the perfect format to design novel types of sperm selection devices, making use of different sorting strategies. For instance, chemotaxis has been combined with a motility screening assay, where swimming sperm in branching channels could choose to follow a gradient of factors either created by addition of chemical such as acetylcholine [56], or generated by cultured cumulus cells [57] or even directly by an oocyte [58]. The general finding of these systems is that the chemo-attracted sperm represents only a small fraction of the total initial cell numbers, *i.e.*, around 10%. Thermotaxis has also been explored as a means to quantify viable sperm, but not for selection [59]. Finally, Wu et al. proposed a high-throughput sorter (up to 200,000 sperm per minute) making use of the sperm cell ability to swim against a flow [60]. In a tapered channel, in which a gradient of fluid velocity is created, sperm that swims against the flow accumulate at the position corresponding to their maximum swimming speed, while non-motile cells and debris were flushed away by the flow (Fig. 7.4).

However, for all these examples, only proof-of-concept experiments have been reported. Furthermore, the DNA integrity, morphology and fertilization rate of the isolated sperm using these various taxis processes were not examined. Altogether, while showing great promises, these techniques still need further and more extensive clinical validation before they can effectively be utilized in the frame of ART routines.

In the case of ICSI-based fertilization, a single spermatozoon is required, and has to be chosen. For now, the choice is based on the cell motility and morphology, as evaluated by the embryologist under a microscope. However, this step is labor-intensive, time-consuming and dependent on the skills of highly trained personnel. Therefore, the availability of a fully automated system able to characterize and select individual sperm would be of great interest, especially if multiple parameters could be considered for this step (*e.g.*, cell vitality, DNA damage). To that end, de Wagenaar et al. [61] trapped individual sperm cells in geometrical constrictions in a microfluidic system, so as to evaluate the cell viability and chromosomal content using fluorescence microscopy. However, the sperm of interest could not be

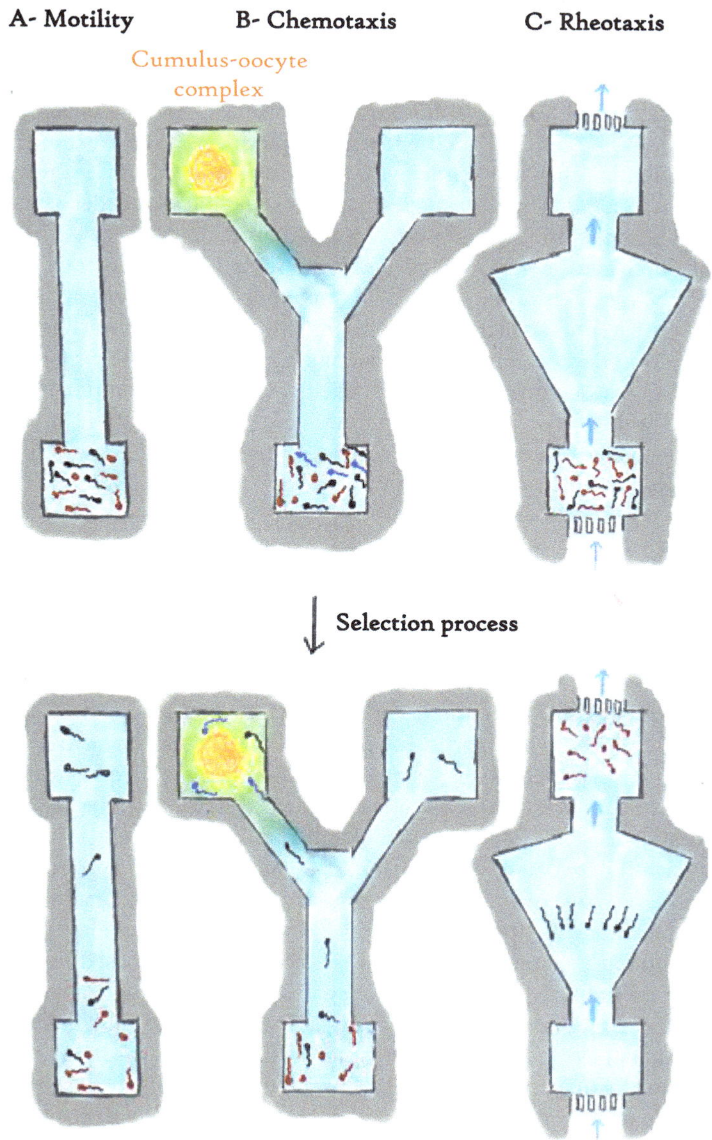

Fig. 7.4 Sperm selection in microfluidic systems based on different physical principles. (**a**) Motility (motile sperm in black, non-motile sperm in red); (**b**) Chemotaxis, where chemotactic-sensitive sperm (in blue) is attracted by a gradient of factors generated by a cumulus-oocyte-complex (illustration inspired by [57]); (**c**) Rheotaxis, where motile sperm swims against the flow while others cells are flushed away (illustration inspired by [60]). The top drawings correspond to the initial situation after loading of the semen and the bottom drawings to the final situation after the selection process. Illustrations by Dr. Venzac

retrieved from the device for subsequent ICSI. Alternatively, using opto-electronic tweezers immotile sperm were distinguished from dead cells in a microfluidic chamber for extreme cases of oligoasthenoteratozoospermia. Live sperm were successfully attracted to the opto-electronic tweezers, while dead cells were repelled [62]. When applied to ICSI, single sperm analysis can be a powerful strategy to identify the best gamete for subsequent fertilization. However, for extreme cases where a TESE (testicular sperm extraction) is needed, the resulting semen preparation is often composed of a small amount of often sperm in a large number of other germ cells, red and white blood cells, somatic cells and debris. Therefore, the initial sample should first be purified to yield a new sample containing only sperm, which could subsequently be analyzed at the single-cell level [63]. Microfluidics, which is an established technology for the isolation for rare cells [64] (as demonstrated for instance for circulating tumor cells [65] or bacteria), as well as for single cell analysis [10, 14] is without a doubt very attractive to provide integrated platforms for the isolation of the sperm from TESE samples, and their on-line characterization.

7.4.3 Fertilization and Cloning

After processing and analysis of the gametes, fertilization takes place using either IVF or ICSI. A few examples of microfluidic devices are found for this fertilization step. A first device was developed for porcine IVF; incubation of oocytes with a large number of sperm led to the penetration of multiple male gametes into the egg cell (polyspermic fertilization). Following this, two systems described by Clark et al. and Suh et al. consisted of a simple straight channel equipped with a constriction to trap the oocytes in the center of the channel. A gentle flow was applied to bring the sperm into the vicinity of the oocyte, decreasing thereby the contact time between the gametes and in turn polyspermy [66, 67]. Sperm sorting has directly been combined with fertilization in a handful of systems. Sano et al. used for fertilization sperm sorted according to their motility [68]. The developmental competence of the fertilized oocytes produced in their system was higher than using conventional IVF in a Petri dish and the monospermy rate improved.

Next to this, dedicated microfluidic approaches have been proposed for cloning purposes or genetic manipulation. For instance, Zeringue et al. exposed oocytes in a microfluidic device to a well-defined flow of lyzing acid solution to soften and remove the zona pellucida around the oocyte [69]. The Arai lab integrated magnetically actuated microrobots in a microfluidic platform for high-throughput oocyte enucleation and cloning [70]. The device specifically incorporated a microgripper for the capture of the oocyte and a microknife to rupture the zona pellucida. Mechanical squeezing of the oocyte thereafter allowed ejecting the nucleus out of the oocytes. Using this technique, enucleated oocytes were found to remain viable and gave rise to viable embryos after (off-chip) cloning.

7.5 Embryo Culture and Characterization

Next to semen analysis, embryo manipulation and *in vitro* culture is the second main aspect of an ART routine, which has been implemented in a miniaturized and/or microfluidic format. Interested readers are referred to recent reviews by us and others, which are dedicated to the steps of *in vitro* embryo culture and pre-transfer characterization [71–73], for more detailed information.

7.5.1 Embryo Culture

For the specific *in vitro* steps involving the embryos, namely their pre-implantation culture and characterization microfluidics exhibits a number of specific advantages. First, the single cell study capability offered by microfluidics is essential in the frame of the eSET policy since the use of microfluidics can allow culturing individual embryos in separate nanoliter structures and simultaneously monitoring *in situ* their development over time throughout the entire pre-implantation period. Embryo characterization would not be restricted to basic morphological criteria through simple imaging, but could include advanced morphokinetic studies and time-lapse imaging combined with metabolic measurements to yield a comprehensive picture of the embryo's developmental potential. Furthermore, from a culture perspective, using this single embryo approach issues associated with degenerating embryos that can have a bad influence on their healthy counterparts as found in group culture, would be entirely alleviated. Finally, the high control on the flow and on any physical and chemical parameter made possible at the micrometer-scale is ideally suited to finely tune the *in vitro* embryo microenvironment and to possibly refresh growth medium towards the possible design of an *in vivo*-like situation.

A variety of miniaturized platforms have been reported in the literature for the *in vitro* culture of pre-implantation mammalian embryos, although most of these platforms were tested on animal models. These platforms can be classified in different groups (Fig. 7.5): (i) microwell arrays, (ii) microfluidic devices, and (iii) semi-microfluidic platforms allying microfluidic structures with microliter volumes as conventionally used in an ART routine.

In microwell arrays, whose concept resembles that of the conventional well-of-the-well devices [74] embryos are individually captured in micrometer-sized wells, which are machined in the bottom of a plastic culture dish. While each embryo is isolated in one microwell, all embryos in the dish are in contact with each other through the culture medium. Furthermore, culture typically takes place in microliter volumes. The second category of platforms relies on the use of confined nanoliter volumes found in microfluidic structures, such as microchannels [77–79] and microchambers [75, 80, 81]. Finally, in the last category, embryo culture takes places in a conventional microliter droplet, while liquids are actuated and/or refreshed using microfluidic means in the form of an integrated peristaltic pump [82] or electrowetting-on-a-dielectric (EWOD) principle [76]. Most of these platforms are fabricated from PDMS, an elastomer material widely used in the field of

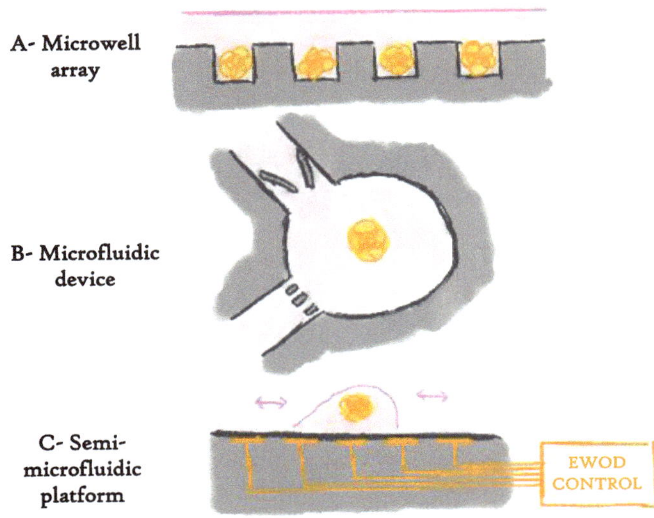

Fig. 7.5 Microfluidic and microfabricated platforms for the culture of mammalian embryos can be classified in 3 distinct categories: (**a**) microwell arrays (illustration inspired by [74]); (**b**) microfluidic structures (illustration inspired by [75]); and (**c**) semi-microfluidic platform, using here microliter-volumes and electrowetting-on-dielectric actuation for the dynamic culture of individual embryos (illustration inspired by [76]). Illustrations by Dr. Venzac

microfluidic technology, which has proven not to be toxic to the developing embryos. Alternatively, glass and silicon have been employed to produce embryo culture devices.

Typically, embryos are kept in culture for 3 days until reaching the 8-cell stage or up to 4–5 days, a time at which they are expected to reach the blastocyst stage. Thereafter, they are scored upon developmental rate and stage and, sometimes, cell counts. In two reports using microfluidic technology [75, 82] the full-term development of the *in vitro* cultured embryos was included in the embryo scoring to fully validate the potential of the proposed platforms and protocols. Overall, when compared to conventional droplet culture, microfluidics better supports embryo development: embryos are developing faster, they reach earlier the blastocyst stage and higher blastocyst rates are found. These higher developmental rates are generally attributed to the confinement found in these microfabricated structures that better emulate the *in vivo* conditions. Furthermore, the concentration in growth factors, which are secreted by the embryos and which are essential to their development is expected to be much higher in these small volumes. This last aspect has notably proven to be interesting to achieve single mouse embryo culture in a nanoliter chamber [75]. A conventional microliter-sized droplet format does not support single embryo development, in part because of the limited availability of these growth factors [83]. In contrast, in a nanoliter and highly confined environment growth factors produced by the embryo accumulate in its close vicinity, promoting thereby its development.

While one of the key-advantages of microfluidic technology would be to be able to downscale the pre-implantation culture to the single embryo level in a nanoscale environment, embryo culture has mostly been reported on groups of at least 10 mouse embryos [58, 77, 79, 82, 84] or, in one example, a pair of mouse embryos [81]. Only a few examples of microfluidic single embryo culture are found. For instance, Esteves et al. demonstrated viable single mouse embryo culture in nanoliter chambers in a first attempt of screening of *in vitro* culture parameters in a microfluidic format [75], and a similar device was next applied on donated human embryos [80]. In more recent work and using a completely different approach individual mouse embryos were successfully cultured in 1-μL droplet periodically actuated by EWOD [76].

Another promise of microfluidic culture compared to conventional droplet culture, is the possibility to implement dynamic culture conditions with tailored perfusion of fresh medium and/or growth factors. However, so far mostly static conditions have been tested, and embryos were cultured in microwells, microchambers or microchannels. Only a few reports actually describe dynamic culture, using either continuous flow, or (periodic) medium refreshment, or even physical shaking of the embryos. For instance, Hickman et al. implemented continuous pumping of liquid with a constant flow-rate [77]. Esteves et al. proposed one-time medium refreshment in the device using passive pumping [85] at a well-defined point [75]. Alternatively, Heo et al. developed a device with pulsatile delivery of fresh medium in a 10-μL volume, where embryos were cultured [82]. In another example, embryos were displaced in a channel presenting constrictions to emulate more closely the structure of the oviduct with the help of a tilting stage [78]. Finally, EWOD was used to regularly mechanically stimulate individual embryos in a 1-μL droplet of culture medium [76]. Comparison of static and dynamic culture shows contrasting results depending on the exact culture protocol. Continuous medium perfusion and the use of constrictions were associated with impaired development of the embryos [77, 78]. In contrast, periodic or one-time medium refreshment as well as regular embryo agitation resulted in enhanced blastocyst rates and cell numbers, and, in some cases better fetal rates [75, 82]. In those cases, the beneficial effect of the flow may be accounted for by the mild mechanical stimulation exerted on the embryos during medium refreshment, since this is expected to positively influence their development [86] while a too strong flow would result in embryo apoptosis [87].

7.5.2 Embryo Characterization

The last step of the ART routine which has been explored in miniaturized devices is the scoring of the pre-implantation embryos, either based on metabolic parameters or through continuous imaging and the use of morphokinetic parameters. Metabolic parameters provide information on the physiological status of the embryos, and they can be assessed in a non-invasive way using integrated sensors [88] and analytical modules, or on-line analysis by coupling the microdevice to spectroscopic techniques [89–91]. While individual embryos have a low metabolic activity, the use of small volumes as found in microfluidic devices will give rise to relatively large variations and detectable changes. Practically so far, metabolic measurements have focused on two metabolic parameters: the embryo oxygen consumption and their utilization of basic substrates (glucose, pyruvate and lactate), as discussed in the following.

The first ever introduced metabolic marker to evaluate embryo developmental competence is its oxygen consumption [92, 93]. Oxygen is an overall marker for the embryo metabolism, and based on previous studies a low oxygen consumption is linked to developmental arrest or embryo degeneration, while a too high consumption is an indicator for damages at the molecular level. Furthermore, an embryo oxygen utilization directly correlates with its developmental stage and presents a significant increase upon reaching the compaction stage. The embryo respiratory activity is typically characterized by measuring changes in the dissolved oxygen concentration in the culture medium, either electrochemically through oxygen reduction or optically using oxygen-sensitive probes (Fig. 7.6).

In the first case, the dissolved oxygen concentration is measured amperometrically in the vicinity of the embryo(s) with the help of (micro)electrodes, which are integrated in the embryo culture device. Two electrode configurations were tested: a linear series of 4 independent microelectrodes [94], or a ring-shaped sensor in the center on which a single embryo was placed [95]. In both cases, the oxygen concentration was mapped at different distances from the embryo and the embryo respiratory activity was derived from this measured oxygen depletion pattern using the spherical diffusion theory [96].

In the alternative strategy using an oxygen-sensitive light emitting probe, the dissolved oxygen concentration can be determined from variations in either the

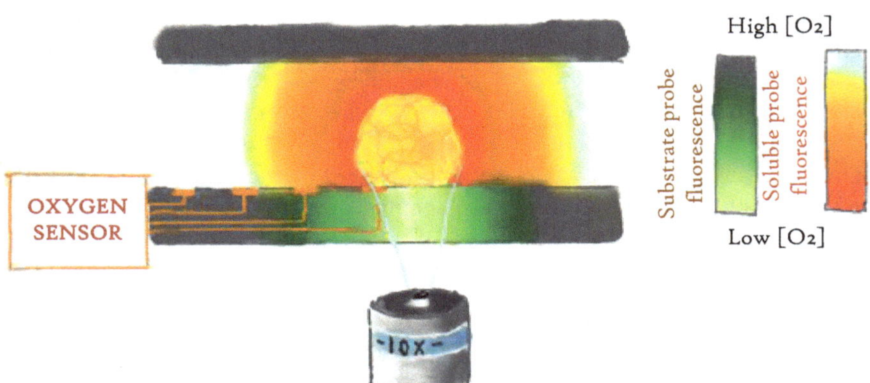

Fig. 7.6 Schematic representation of various embryo characterization schemes, which have been introduced in a microfluidic format: integrated electrodes to evaluate the embryo oxygen consumption, oxygen-sensitive probes, either added to the culture medium or embedded in the bottom of the device, and using time-lapse imaging *in situ* in the culture system. Illustration by Dr. Venzac

emitted light intensity or the fluorescence lifetime [97]. In a first report, O'Donovan et al. added a water-soluble oxygen-sensitive complex directly to the embryo culture medium and measured differences in oxygen consumption of embryos at the 2-cell and blastocyst stages in group culture [98]. Most of the time however, these complexes are not soluble in water but in organic solutions. Therefore, Komori et al. supplemented polystyrene material used as the bottom layer of their microwell array device with similar Pt-Porphyrin-based oxygen-sensitive dye [99] to follow the respiratory activity of individual mouse embryos.

Other metabolic parameters, which have been analyzed to evaluate the embryo quality also in a microfluidic format, are these three substrates: glucose, pyruvate and lactate. Embryos consume pyruvate and glucose, depending on their developmental stage and produce lactate, so that mapping the variations in the concentrations in the three substances provides information on the embryo metabolism and development.

These three substrates are typically quantified through enzymatic assays by analyzing products of an enzymatic reaction, either electrochemically or optically, depending on the set of enzymes used and the products formed. Using oxidase-type enzymes (*e.g.*, glucose oxidase for glucose analysis) hydrogen peroxide is produced, which can be quantified electrochemically. If dehydrogenases are used, in contrast, typically fluorescent or colored substances are generated. Using three different substrate-specific dehydrogenases, the substrate utilization of 10 individual mouse embryos was analyzed off-line from spent culture medium in an automated manner using parallelized nanoliter loop reactors in a microfluidic device [100]. In a second platform culture medium was automatically sampled from a funnel structure, where embryos were kept in culture [101]. Here again, spent culture medium was mixed with an enzymatic cocktail on-chip and the glucose consumption of groups of mouse embryos quantified through optical measurements.

Finally, embryo development can be monitored in microfluidic and miniaturized devices using continuous imaging or so-called time-lapse microscopy, by taking advantage of the transparent properties of the materials of which the devices are made. In particular, time-lapse imaging has been reported for microfabricated well array platforms [102–104].

7.6 Conclusion and Perspectives

Microfluidics undoubtedly presents high potentials for the field of ART. Namely, for individual steps of the ART procedure microfluidics allows automating processes while offering new opportunities. As reviewed in this chapter, these opportunities are in the form of new analysis and selection schemes for semen samples, and more *in vivo*-like conditions for the culture of pre-implantation embryos in well-defined and dynamic environments for instance. However, one key feature of microfluidics has not fully explored to date, which is the possibility to implement complex processes comprising multiple steps as found in a complete ART routine. Only few examples of integrated platforms combining different steps of the entire ART treatment are found [58, 84]. For instance, Ma et al. developed a device in which up to 16 individual

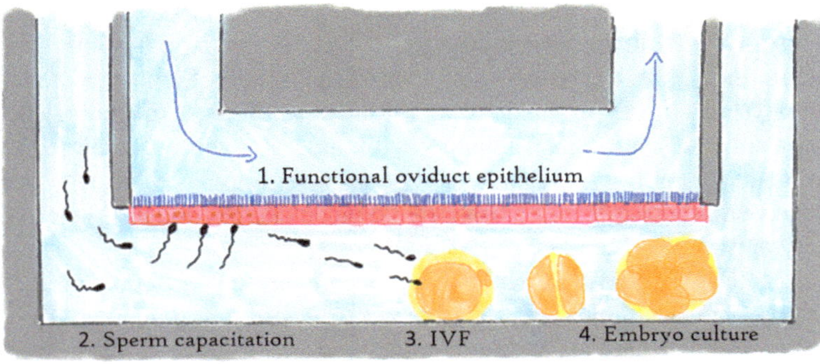

Fig. 7.7 Example of an integrated platform, in which the entire IVF process can be performed. Sperm interact with a mature epithelium grown on chip on a porous membrane (intermediate blue layer), to be capacitated, before they fertilize an oocyte. The resulting pre-implantation embryo is cultured on-chip in close vicinity of the oviduct epithelium. Illustration by Dr. Venzac

oocytes were individually trapped in octacolumn structures and fertilized *in situ* by sperm selected on-chip for their motility and ability to cross obstacles. The resulting embryos were cultured in the same device, after medium refreshment in the entire platform, during the pre-implantation period yet under static conditions, and could easily be retrieved one by one from the device for further development and/or analysis.

In a more recent example, and with the motivation to create a more biomimetic system to enhance the fertilization rates and the quality of the *in vitro* produced embryos, an oviduct was re-created in a compartmentalized microfluidic device and it was next successfully applied to perform an entire IVF process [105, 106]. First, a functional bovine oviduct epithelium was grown on a porous membrane, while being exposed to a continuous flow of growth medium, in both the apical and basolateral compartments (Fig. 7.7). Oocytes and sperm were separately injected on the apical side of the device for the process of fertilization, just after on-chip capacitation of the sperm by the oviduct epithelial cells. Embryos were kept in culture during the pre-implantation period in the close vicinity of the epithelium, and subjected to continuous perfusion of medium. Importantly, these embryos were found to present less molecular aberrations than their *in vitro* counterparts grown in a conventional Petri dish, which opens new avenues for the field of *in vitro* embryo production.

Along the same idea of exploiting the biomimetic environment provided by a microfluidic format, the process of spermatogenesis has recently successfully been recapitulated in a microfluidic device [107, 108]. In the frame of an ART procedure, the possibility to produce sperm *in vitro* from testis tissues is of particular interest for male patients that suffered from cancer at an early age of their life, and whose testicular stem cells were cryopreserved. Samples of mouse seminiferous tubules were successfully cultured for up to 6 months in a shallow chamber, while being exposed to a continuous flow of fresh nutrients. A porous membrane was placed between the tissues and the channel in which the flow was applied, to protect the tissues from any shear that would damage it. The on-chip generated sperm was functional and successfully used for fertilization through micro-insemination of mouse oocytes. Following this success one can expect similar development for the *in vitro* production of the female gametes and their maturation in a microfluidic device, starting from cryopreserved ovary tissues from female patients that were diagnosed with cancer at an early age.

References

1. Dyer S, Chambers GM, de Mouzon J, Nygren KG, Zegers-Hochschild F, Mansour R, Ishihara O, Banker M, Adamson GD (2016) International committee for monitoring assisted reproductive technologies world report: assisted reproductive technology 2008, 2009 and 2010. Hum Reprod 31(7):1588–1609. https://doi.org/10.1093/humrep/dew082
2. European IVFMCftESoHR, Embryology, Calhaz-Jorge C, de Geyter C, Kupka MS, de Mouzon J, Erb K, Mocanu E, Motrenko T, Scaravelli G, Wyns C, Goossens V (2016) Assisted reproductive technology in Europe, 2012: results generated from European registers by ESHRE. Hum Reprod 31(8):1638–1652. https://doi.org/10.1093/humrep/dew151
3. Blondin P (2015) Status of embryo production in the world. Anim Reprod 12(3):356–358
4. Ferraretti AP, Goossens V, de Mouzon J, Bhattacharya S, Castilla JA, Korsak V, Kupka M, Nygren KG, Nyboe Andersen A, European IVFm, Consortium for European Society of Human R, Embryology (2012) Assisted reproductive technology in Europe, 2008: results generated from European registers by ESHRE. Hum Reprod 27(9):2571–2584. https://doi.org/10.1093/humrep/des255
5. Balaban B, Sakkas D, Gardner DK (2014) Laboratory procedures for human in vitro fertilization. Semin Reprod Med 32(4):272–282. https://doi.org/10.1055/s-0034-1375179
6. Gardner DK, Reed L, Linck D, Sheehan C, Lane M (2005) Quality control in human in vitro fertilization. Semin Reprod Med 23(4):319–324. https://doi.org/10.1055/s-2005-923389
7. Swain JE (2015) Optimal human embryo culture. Semin Reprod Med 33(2):103–117. https://doi.org/10.1055/s-0035-1546423
8. Bjorndahl L, Kvist U (2003) Sequence of ejaculation affects the spermatozoon as a carrier and its message. Reprod Biomed Online 7(4):440–448
9. WHO (2010) WHO laboratory manual for the examination and processing of human semen. Cambridge University Press, Cambridge
10. Mortimer D (1994) Sperm recovery techniques to maximize fertilizing capacity. Reprod Fertil Dev 6(1):25–31
11. Nordhoff V (2015) How to select immotile but viable spermatozoa on the day of intracytoplasmic sperm injection? an embryologist's view. Andrology-US 3(2):156–162. https://doi.org/10.1111/andr.286
12. Whitesides GM (2006) The origins and the future of microfluidics. Nature 442 (7101):368–373. https://doi.org/10.1038/nature05058
13. Le Gac S, van den Berg A (2010) Single cells as experimentation units in lab-on-a-chip devices. Trends Biotechnol 28(2):55–62. https://doi.org/10.1016/j.tibtech.2009.10.005
14. Sims CE, Allbritton NL (2007) Analysis of single mammalian cells on-chip. Lab Chip 7 (4):423–440. https://doi.org/10.1039/b615235j
15. Young EWK, Beebe DJ (2010) Fundamentals of microfluidic cell culture in controlled microenvironments. Chem Soc Rev 39(3):1036–1048. https://doi.org/10.1039/b909900j
16. Lee PJ, Hung PJ, Rao VM, Lee LP (2006) Nanoliter scale microbioreactor array for quantitative cell biology. Biotechnol Bioeng 94(1):5–14. https://doi.org/10.1002/bit.20745
17. Tourovskaia A, Figueroa-Masot X, Folch A (2005) Differentiation-on-a-chip: a microfluidic platform for long-term cell culture studies. Lab Chip 5(1):14–19. https://doi.org/10.1039/b405719h

18. Keenan TM, Folch A (2008) Biomolecular gradients in cell culture systems. Lab Chip 8 (1):34–57. https://doi.org/10.1039/b711887b
19. Meseguer M, Kruhne U, Laursen S (2012) Full in vitro fertilization laboratory mechanization: toward robotic assisted reproduction? Fertil Steril 97(6):1277–1286. https://doi.org/10.1016/j.fertnstert.2012.03.013
20. Harper J, Magli MC, Lundin K, Barratt CL, Brison D (2012) When and how should new technology be introduced into the IVF laboratory? Hum Reprod 27(2):303–313. https://doi.org/10.1093/humrep/der414
21. Zerinque HC, Beebe DJ, Wheeler MB (2001) Removal of cumulus from mammalian zygotes using microfluidic techniques. Biomed Microdevices 3(3):219–224. https://doi.org/10.1023/A:1011463330597
22. Zeringue HC, Rutledge JJ, Beebe DJ (2005a) Early mammalian embryo development depends on cumulus removal technique. Lab Chip 5(1):86–90. https://doi.org/10.1039/b316494m
23. Yanez LZ, Camarillo DB (2017) Microfluidic analysis of oocyte and embryo biomechanical properties to improve outcomes in assisted reproductive technologies. Mol Hum Reprod 23 (4):235–247. https://doi.org/10.1093/molehr/gaw071
24. Wacogne B, Pieralli C, Roux C, Gharbi T (2008) Measuring the mechanical behaviour of human oocytes with a very simple SU-8 micro-tool. Biomed Microdevices 10(3):411–419. https://doi.org/10.1007/s10544-007-9150-7
25. Luo Z, Guven S, Gozen I, Chen P, Tasoglu S, Anchan RM, Bai B, Demirci U (2015) Deformation of a single mouse oocyte in a constricted microfluidic channel. Microfluid Nanofluid 19(4):883–890. https://doi.org/10.1007/s10404-015-1614-0
26. Choi W, Kim JS, Lee DH, Lee KK, Koo DB, Park JK (2008) Dielectrophoretic oocyte selection chip for in vitro fertilization. Biomed Microdevices 10(3):337–345. https://doi.org/10.1007/s10544-007-9141-8
27. Vidberg F, Zeggari R, Pieralli C, Amiot C, Roux C, Wacogne B (2011) Measurement of oocyte temporal maturation process by means of a simple optical micro-system. Sensor Actuat B-Chem 157(1):19–25. https://doi.org/10.1016/j.snb.2011.03.021
28. Zeggari R, Wacogne B, Pieralli C, Roux C, Gharbi T (2007) A full micro-fluidic system for single oocyte manipulation including an optical sensor for cell maturity estimation and fertilisation indication. Sensor Actuat B-Chem 125(2):664–671. https://doi.org/10.1016/j.snb.2007.02.045
29. Hwang H, Lee DH, Choi W, Park JK (2009) Enhanced discrimination of normal oocytes using optically induced pulling-up dielectrophoretic force. Biomicrofluidics 3(1):14103. https://doi.org/10.1063/1.3086600
30. Sadani Z, Wacogne B, Pieralli C, Roux C, Gharbi T (2005) Microsystems and microfluidic device for single oocyte transportation and trapping: toward the automation of in vitro fertilising. Sensor Actuat A-Phys 121(2):364–372. https://doi.org/10.1016/j.sna.2005.03.004
31. Nakahara K, Sakuma S, Hayakawa T, Arai F (2015) On-chip transportation and measurement of mechanical characteristics of oocytes in an open environment. Micromachines-Basel 6 (5):648–659. https://doi.org/10.3390/mi6050648
32. Jurema MW, Nogueira D (2006) In vitro maturation of human oocytes for assisted reproduction. Fertil Steril 86(5):1277–1291. https://doi.org/10.1016/j.fertnstert.2006.02.126
33. Berenguel-Alonso M, Sabes-Alsina M, Morato R, Ymbern O, Rodriguez-Vazquez L, Tallo-Parra O, Alonso-Chamarro J, Puyol M, Lopez-Bejar M (2017) Rapid prototyping of a cyclic olefin copolymer microfluidic device for automated oocyte culturing. SLAS Technol 22 (5):507–517. https://doi.org/10.1177/2472555216684625
34. Auger J, Eustache F, Andersen AG, Irvine DS, Jorgensen N, Skakkebaek NE, Suominen J, Toppari J, Vierula M, Jouannet P (2001) Sperm morphological defects related to environment, lifestyle and medical history of 1001 male partners of pregnant women from four European cities. Hum Reprod 16(12):2710–2717

35. Kovac JR, Smith RP, Cajipe M, Lamb DJ, Lipshultz LI (2017) Men with a complete absence of normal sperm morphology exhibit high rates of success without assisted reproduction. Asian J Androl 19(1):39–42. https://doi.org/10.4103/1008-682X.189211
36. Tandara M, Bajic A, Tandara L, Bilic-Zulle L, Sunj M, Kozina V, Goluza T, Jukic M (2014) Sperm DNA integrity testing: big halo is a good predictor of embryo quality and pregnancy after conventional IVF. Andrology-US 2(5):678–686. https://doi.org/10.1111/j.2047-2927.2014.00234.x
37. Pandey CM, Augustine S, Kumar S, Kumar S, Nara S, Srivastava S, Malhotra BD (2018) Microfluidics based point-of-care diagnostics. Biotechnol J 13(1). https://doi.org/10.1002/biot.201700047
38. Knowlton SM, Sadasivam M, Tasoglu S (2015) Microfluidics for sperm research. Trends Biotechnol 33(4):221–229. https://doi.org/10.1016/j.tibtech.2015.01.005
39. Nosrati R, Graham PJ, Zhang B, Riordon J, Lagunov A, Hannam TG, Escobedo C, Jarvi K, Sinton D (2017) Microfluidics for sperm analysis and selection. Nat Rev Urol 14 (12):707–730. https://doi.org/10.1038/nrurol.2017.175
40. Segerink LI, Sprenkels AJ, ter Braak PM, Vermes I, van den Berg A (2010) On-chip determination of spermatozoa concentration using electrical impedance measurements. Lab Chip 10(8):1018–1024. https://doi.org/10.1039/b923970g
41. de Wagenaar B, Dekker S, de Boer HL, Bomer JG, Olthuis W, van den Berg A, Segerink LI (2016) Towards microfluidic sperm refinement: impedance-based analysis and sorting of sperm cells. Lab Chip 16(8):1514–1522. https://doi.org/10.1039/c6lc00256k
42. Chen CY, Chiang TC, Lin CM, Lin SS, Jong DS, Tsai VF, Hsieh JT, Wo AM (2013) Sperm quality assessment via separation and sedimentation in a microfluidic device. Analyst 138 (17):4967–4974. https://doi.org/10.1039/c3an00900a
43. Yu S, Rubin M, Geevarughese S, Pino JS, Rodriguez HF, Asghar W (2018) Emerging technologies for home-based semen analysis. Andrology-US 6(1):10–19. https://doi.org/10.1111/andr.12441
44. Kanakasabapathy MK, Sadasivam M, Singh A, Preston C, Thirumalaraju P, Venkataraman M, Bormann CL, Draz MS, Petrozza JC, Shafiee H (2017) An automated smartphone-based diagnostic assay for point-of-care semen analysis. Sci Transl Med 9(382). https://doi.org/10.1126/scitranslmed.aai7863
45. Nosrati R, Gong MM, San Gabriel MC, Pedraza CE, Zini A, Sinton D (2016) Paper-based quantification of male fertility potential. Clin Chem 62(3):458–465. https://doi.org/10.1373/clinchem.2015.250282
46. Gong MM, Nosrati R, San Gabriel MC, Zini A, Sinton D (2015) Direct DNA analysis with paper-based ion concentration polarization. J Am Chem Soc 137(43):13913–13919. https://doi.org/10.1021/jacs.5b08523
47. Suarez SS, Pacey AA (2006) Sperm transport in the female reproductive tract. Hum Reprod Update 12(1):23–37. https://doi.org/10.1093/humupd/dmi047
48. Henkel R (2012) Sperm preparation: state-of-the-art--physiological aspects and application of advanced sperm preparation methods. Asian J Androl 14(2):260–269. https://doi.org/10.1038/aja.2011.133
49. Zini A, Mak V, Phang D, Jarvi K (1999) Potential adverse effect of semen processing on human sperm deoxyribonucleic acid integrity. Fertil Steril 72(3):496–499
50. Cho BS, Schuster TG, Zhu X, Chang D, Smith GD, Takayama S (2003) Passively driven integrated microfluidic system for separation of motile sperm. Anal Chem 75(7):1671–1675
51. Nosrati R, Vollmer M, Eamer L, San Gabriel MC, Zeidan K, Zini A, Sinton D (2014) Rapid selection of sperm with high DNA integrity. Lab Chip 14(6):1142–1150. https://doi.org/10.1039/c3lc51254a
52. Asghar W, Velasco V, Kingsley JL, Shoukat MS, Shafiee H, Anchan RM, Mutter GL, Tuzel E, Demirci U (2014) Selection of functional human sperm with higher DNA integrity and fewer reactive oxygen species. Adv Healthc Mater 3(10):1671–1679. https://doi.org/10.1002/adhm.201400058

53. Cohen-Dayag A, Tur-Kaspa I, Dor J, Mashiach S, Eisenbach M (1995) Sperm capacitation in humans is transient and correlates with chemotactic responsiveness to follicular factors. Proc Natl Acad Sci U S A 92(24):11039–11043
54. Bahat A, Tur-Kaspa I, Gakamsky A, Giojalas LC, Breitbart H, Eisenbach M (2003) Thermotaxis of mammalian sperm cells: a potential navigation mechanism in the female genital tract. Nat Med 9(2):149–150. https://doi.org/10.1038/nm0203-149
55. Miki K, Clapham DE (2013) Rheotaxis guides mammalian sperm. Curr Biol 23(6):443–452. https://doi.org/10.1016/j.cub.2013.02.007
56. Ko YJ, Maeng JH, Lee BC, Lee S, Hwang SY, Ahn Y (2012) Separation of progressive motile sperm from mouse semen using on-chip chemotaxis. Anal Sci Int J Jpn Soc Anal Chem 28 (1):27–32
57. Xie L, Ma R, Han C, Su K, Zhang Q, Qiu T, Wang L, Huang G, Qiao J, Wang J, Cheng J (2010) Integration of sperm motility and chemotaxis screening with a microchannel-based device. Clin Chem 56(8):1270–1278. https://doi.org/10.1373/clinchem.2010.146902
58. Ma R, Xie L, Han C, Su K, Qiu T, Wang L, Huang G, Xing W, Qiao J, Wang J, Cheng J (2011) In vitro fertilization on a single-oocyte positioning system integrated with motile sperm selection and early embryo development. Anal Chem 83(8):2964–2970. https://doi.org/10.1021/ac103063g
59. Li Z, Liu W, Qiu T, Xie L, Chen W, Liu R, Lu Y, Mitchelson K, Wang J, Qiao J, Cheng J (2014) The construction of an interfacial valve-based microfluidic chip for thermotaxis evaluation of human sperm. Biomicrofluidics 8(2):024102. https://doi.org/10.1063/1.4866851
60. Wu JK, Chen PC, Lin YN, Wang CW, Pan LC, Tseng FG (2017) High-throughput flowing upstream sperm sorting in a retarding flow field for human semen analysis. Analyst 142 (6):938–944. https://doi.org/10.1039/c6an02420c
61. de Wagenaar B, Berendsen JT, Bomer JG, Olthuis W, van den Berg A, Segerink LI (2015) Microfluidic single sperm entrapment and analysis. Lab Chip 15(5):1294–1301. https://doi.org/10.1039/c4lc01425a
62. Garcia MM, Ohta AT, Walsh TJ, Vittinghof E, Lin G, Wu MC, Lue TF (2010) A noninvasive, motility independent, sperm sorting method and technology to identify and retrieve individual viable nonmotile sperm for intracytoplasmic sperm injection. J Urol 184(6):2466–2472. https://doi.org/10.1016/j.juro.2010.08.026
63. Samuel R, Badamjav O, Murphy KE, Patel DP, Son J, Gale BK, Carrell DT, Hotaling JM (2016) Microfluidics: the future of microdissection TESE? Syst Biol Reprod Med 62 (3):161–170. https://doi.org/10.3109/19396368.2016.1159748
64. Shields CW, Reyes CD, Lopez GP (2015) Microfluidic cell sorting: a review of the advances in the separation of cells from debulking to rare cell isolation. Lab Chip 15(5):1230–1249. https://doi.org/10.1039/c4lc01246a
65. Saliba AE, Saias L, Psychari E, Minc N, Simon D, Bidard FC, Mathiot C, Pierga JY, Fraisier V, Salamero J, Saada V, Farace F, Vielh P, Malaquin L, Viovy JL (2010) Microfluidic sorting and multimodal typing of cancer cells in self-assembled magnetic arrays. Proc Natl Acad Sci U S A 107(33):14524–14529. https://doi.org/10.1073/pnas.1001515107
66. Clark SG, Haubert K, Beebe DJ, Ferguson CE, Wheeler MB (2005) Reduction of polyspermic penetration using biomimetic microfluidic technology during in vitro fertilization. Lab Chip 5 (11):1229–1232. https://doi.org/10.1039/b504397m
67. Suh RS, Zhu X, Phadke N, Ohl DA, Takayama S, Smith GD (2006) IVF within microfluidic channels requires lower total numbers and lower concentrations of sperm. Hum Reprod 21 (2):477–483. https://doi.org/10.1093/humrep/dei323
68. Sano H, Matsuura K, Naruse K, Funahashi H (2010) Application of a microfluidic sperm sorter to the in-vitro fertilization of porcine oocytes reduced the incidence of polyspermic penetration. Theriogenology 74(5):863–870. https://doi.org/10.1016/j.theriogenology.2010.04.011
69. Zeringue HC, Wheeler MB, Beebe DJ (2005b) A microfluidic method for removal of the zona pellucida from mammalian embryos. Lab Chip 5(1):108–110. https://doi.org/10.1039/b406703g

70. Ichikawa A, Sakuma S, Sugita M, Shoda T, Tamakoshi T, Akagi S, Arai F (2014) On-chip enucleation of an oocyte by untethered microrobots. J Micromech Microeng 24(9). https://doi.org/10.1088/0960-1317/24/9/095004
71. Le Gac S, Nordhoff V (2017) Microfluidics for mammalian embryo culture and selection: where do we stand now? Mol Hum Reprod 23(4):213–226. https://doi.org/10.1093/molehr/gaw061
72. Swain JE, Lai D, Takayama S, Smith GD (2013) Thinking big by thinking small: application of microfluidic technology to improve ART. Lab Chip 13(7):1213–1224. https://doi.org/10.1039/c3lc41290c
73. Swain JE, Smith GD (2011) Advances in embryo culture platforms: novel approaches to improve preimplantation embryo development through modifications of the microenvironment. Hum Reprod Update 17(4):541–557. https://doi.org/10.1093/humupd/dmr006
74. Vajta G, Korosi T, Du Y, Nakata K, Ieda S, Kuwayama M, Nagy ZP (2008) The well-of-the-well system: an efficient approach to improve embryo development. Reprod Biomed Online 17(1):73–81
75. Esteves TC, van Rossem F, Nordhoff V, Schlatt S, Boiani M, Le Gac S (2013) A microfluidic system supports single mouse embryo culture leading to full-term development. RSC Adv 3 (48):26451–26458. https://doi.org/10.1039/c3ra44453h
76. Huang HY, Shen HH, Tien CH, Li CJ, Fan SK, Liu CH, Hsu WS, Yao DJ (2015) Digital microfluidic dynamic culture of mammalian embryos on an electrowetting on dielectric (EWOD) chip. Plos One 10(5). https://doi.org/10.1371/journal.pone.0124196
77. Hickman DL, Beebe DJ, Rodriguez-Zas SL, Wheeler MB (2002) Comparison of static and dynamic medium environments for culturing of pre-implantation mouse embryos. Comp Med 52(2):122–126
78. Kim MS, Bae CY, Wee G, Han YM, Park JK (2009) A microfluidic in vitro cultivation system for mechanical stimulation of bovine embryos. Electrophoresis 30(18):3276–3282. https://doi.org/10.1002/elps.200900157
79. Raty S, Walters EM, Davis J, Zeringue H, Beebe DJ, Rodriguez-Zas SL, Wheeler MB (2004) Embryonic development in the mouse is enhanced via microchannel culture. Lab Chip 4 (3):186–190. https://doi.org/10.1039/b316437c
80. Kieslinger DC, Hao ZX, Vergouw CG, Kostelijk EH, Lambalk CB, Le Gac S (2015) In vitro development of donated frozen-thawed human embryos in a prototype static microfluidic device: a randomized controlled trial. Fertil Steril 103(3):680–U393. https://doi.org/10.1016/j.fertnstert.2014.12.089
81. Melin J, Lee A, Foygel K, Leong DE, Quake SR, Yao MWM (2009) In vitro embryo culture in defined, sub-microliter volumes. Dev Dynam 238(4):950–955. https://doi.org/10.1002/dvdy.21918
82. Heo YS, Cabrera LM, Bormann CL, Shah CT, Takayama S, Smith GD (2010) Dynamic microfunnel culture enhances mouse embryo development and pregnancy rates. Hum Reprod 25(3):613–622. https://doi.org/10.1093/humrep/dep449
83. Paria BC, Dey SK (1990) Preimplantation embryo development Invitro – cooperative interactions among embryos and role of growth-factors. Proc Natl Acad Sci USA 87 (12):4756–4760. https://doi.org/10.1073/pnas.87.12.4756
84. Han C, Zhang QF, Ma R, Xie L, Qiu TA, Wang L, Mitchelson K, Wang JD, Huang GL, Qiao J, Cheng J (2010) Integration of single oocyte trapping, in vitro fertilization and embryo culture in a microwell-structured microfluidic device. Lab Chip 10(21):2848–2854. https://doi.org/10.1039/c005296e
85. Walker GM, Beebe DJ (2002) A passive pumping method for microfluidic devices. Lab Chip 2 (3):131–134. https://doi.org/10.1039/b204381e
86. Fauci LJ, Dillon R (2006) Biofluidmechanics of reproduction. Annu Rev Fluid Mech 38:371–394. https://doi.org/10.1146/annurev.fluid.37.061903.175725
87. Xie YF, Wang FF, Zhong WJ, Puscheck E, Shen HL, Rappolee DA (2006) Shear stress induces preimplantation embryo death that is delayed by the zona pellucida and associated

with stress-activated protein kinase-mediated apoptosis. Biol Reprod 75(1):45–55. https://doi.org/10.1095/biolreprod.105.049791
88. Liu Q, Wu C, Cai H, Hu N, Zhou J, Wang P (2014) Cell-based biosensors and their application in biomedicine. Chem Rev 114(12):6423–6461. https://doi.org/10.1021/cr2003129
89. Gao D, Wei H, Guo GS, Lin JM (2010) Microfluidic cell culture and metabolism detection with electrospray ionization quadrupole time-of-flight mass spectrometer. Anal Chem 82 (13):5679–5685. https://doi.org/10.1021/ac101370p
90. Le Gac S, Rolando C, Arscott S (2006) An open design microfabricated nib-like nanoelectrospray emitter tip on a conducting silicon substrate for the application of the ionization voltage. J Am Soc Mass Spectrom 17(1):75–80. https://doi.org/10.1016/j.jasms.2005.09.003
91. Kalfe A, Telfah A, Lambert J, Hergenroder R (2015) Looking into living cell systems: planar waveguide microfluidic NMR detector for in vitro metabolomics of tumor spheroids. Anal Chem 87(14):7402–7410. https://doi.org/10.1021/acs.analchem.5b01603
92. Houghton FD, Thompson JG, Kennedy CJ, Leese HJ (1996) Oxygen consumption and energy metabolism of the early mouse embryo. Mol Reprod Dev 44(4):476–485
93. Mills RM, Brinster RL (1967) Oxygen consumption of preimplantation mouse embryos. Exp Cell Res 47(1–2):337–344. https://doi.org/10.1016/0014-4827(67)90236-4
94. Wu CC, Saito T, Yasukawa T, Shiku H, Abe H, Hoshi H, Matsue T (2007) Microfluidic chip integrated with amperometric detector array for in situ estimating oxygen consumption characteristics of single bovine embryos. Sensor Actuat B-Chem 125(2):680–687. https://doi.org/10.1016/j.snb.2007.03.017
95. Date Y, Takano S, Shiku H, Ino K, Ito-Sasaki T, Yokoo M, Abe H, Matsue T (2011) Monitoring oxygen consumption of single mouse embryos using an integrated electrochemical microdevice. Biosens Bioelectron 30(1):100–106. https://doi.org/10.1016/j.bios.2011.08.037
96. Shiku H, Shiraishi T, Aoyagi S, Utsumi Y, Matsudaira M, Abe H, Hoshi H, Kasai S, Ohya H, Matsue T (2004) Respiration activity of single bovine embryos entrapped in a cone-shaped microwell monitored by scanning electrochemical microscopy. Anal Chim Acta 522 (1):51–58. https://doi.org/10.1016/j.aca.2004.06.054
97. Ruggi A, van Leeuwen FWB, Velders AH (2011) Interaction of dioxygen with the electronic excited state of Ir(III) and Ru(II) complexes: principles and biomedical applications. Coordin Chem Rev 255(21–22):2542–2554. https://doi.org/10.1016/j.ccr.2011.05.012
98. O'Donovan C, Twomey E, Alderman J, Moore T, Papkovsky D (2006) Development of a respirometric biochip for embryo assessment. Lab Chip 6(11):1438–1444. https://doi.org/10.1039/b607622j
99. Komori K, Fujii S, Montagne K, Nakamura H, Kimura H, Otake K, Fujii T, Sakai Y (2012) Development of a well-of-the-well system-based embryo culture plate with an oxygen sensing photoluminescent probe. Sensor Actuat B-Chem 162(1):278–283. https://doi.org/10.1016/j.snb.2011.12.078
100. Urbanski JP, Johnson MT, Craig DD, Potter DL, Gardner DK, Thorsen T (2008) Noninvasive metabolic profiling using microfluidics for analysis of single preimplantation embryos. Anal Chem 80(17):6500–6507. https://doi.org/10.1021/ac8010473
101. Heo YS, Cabrera LM, Bormann CL, Smith GD, Takayama S (2012) Real time culture and analysis of embryo metabolism using a microfluidic device with deformation based actuation. Lab Chip 12(12):2240–2246. https://doi.org/10.1039/c2lc21050a
102. Chung YH, Hsiao YH, Kao WL, Hsu CH, Yao DJ, Chen CC (2015) Microwells support high-resolution time-lapse imaging and development of preimplanted mouse embryos. Biomicrofluidics 9(2). https://doi.org/10.1063/1.4918642
103. Hashimoto S, Kato N, Saeki K, Morimoto Y (2012) Selection of high-potential embryos by culture in poly(dimethylsiloxane) microwells and time-lapse imaging. Fertil Steril 97 (2):332–337. https://doi.org/10.1016/j.fertnstert.2011.11.042
104. Sugimura S, Akai T, Somfai T, Hirayama M, Aikawa Y, Ohtake M, Hattori H, Kobayashi S, Hashiyada Y, Konishi K, Imai K (2010) Time-lapse cinematography-compatible polystyrene-

based microwell culture system: a novel tool for tracking the development of individual bovine embryos. Biol Reprod 83(6):970–978. https://doi.org/10.1095/biolreprod.110.085522
105. Ferraz M, Rho HS, Delahaye J, Pinheiro N, Henning H, Stout T, Gadella B, Le Gac S (2017) Mimicking the bovine oviduct in a microfluidic device for advanced embryo in vitro culture systems. In: MicroTAS 2017, Savannah, GA, USA, 22–26 Oct. 2017
106. Ferraz MAMM, Rho HS, Hemerich D, Henning HHW, van Tol HTA, Hölker M, Besenfelder U, Mokry M, Vos PLAM, Stout TAE, Le Gac S, Gadella BM (2018) An oviduct-on-a-chip provides an enhanced in vitro environment for zygote genome reprogramming. Nat Commun 9 (1):4934. https://doi.org/10.1038/s41467-018-07119-8
107. Komeya M, Hayashi K, Nakamura H, Yamanaka H, Sanjo H, Kojima K, Sato T, Yao M, Kimura H, Fujii T, Ogawa T (2017) Pumpless microfluidic system driven by hydrostatic pressure induces and maintains mouse spermatogenesis in vitro. Sci Rep 7. https://doi.org/10.1038/S41598-017-15799-3
108. Komeya M, Kimura H, Nakamura H, Yokonishi T, Sato T, Kojima K, Hayashi K, Katagiri K, Yamanaka H, Sanjo H, Yao M, Kamimura S, Inoue K, Ogonuki N, Ogura A, Fujii T, Ogawa T (2016) Long-term ex vivo maintenance of testis tissues producing fertile sperm in a microfluidic device. Sci Rep 6. https://doi.org/10.1038/Srep21472

Chapter 8
Microfluidic Organs-on-Chips to Reconstitute Cellular Microenvironments

Yu-suke Torisawa

Abstract Recent advances in microsystems technology and tissue engineering have led to the development of biomimetic microdevices to model key functional units of human organs, known as organs-on-chips. By mimicking natural tissue architecture and microenvironmental chemical and physical cues within microfluidic devices, this technology realizes organ-level function *in vitro* that cannot be recapitulated with conventional culture methods. Since the physiological microenvironments in living systems are mostly microfluidic in nature, microfluidic systems facilitate engineering of cellular microenvironments. Microfluidic systems allow for control of local chemical gradients and dynamic mechanical forces, which play important roles in organ development and function. This organ-on-a-chip technology has great potential to facilitate drug discovery and development, to model human physiology and disease, and to replace animal models for efficacy and toxicity testing. This chapter shows an overview of the organ-on-a-chip technology to recapitulate cellular microenvironments and especially focuses on bone marrow-on-a-chip that enables culture of living bone marrow with a functional hematopoietic niche as a novel type of approach to develop organs-on-chips.

Keywords Organs-on-chips · Microfluidics · Tissue engineering · Cellular microenvironment

8.1 Introduction

8.1.1 Cellular Microenvironment

Cellular functions are precisely controlled by their specific microenvironment where they normally reside. The microenvironment contains a complex set of structural,

Y.-s. Torisawa (✉)
Hakubi Center for Advanced Research and Department of Micro Engineering, Kyoto University, Kyoto, Japan
e-mail: torisawa.yusuke.6z@kyoto-u.ac.jp

M. Tokeshi (ed.), *Applications of Microfluidic Systems in Biology and Medicine*, Bioanalysis 7, https://doi.org/10.1007/978-981-13-6229-3_8

chemical, and mechanical signals which are necessary to maintain cellular viability and function [1–3]. Since stem cells cannot maintain their stemness without specific stem cell niches [3–5], cellular microenvironments are crucial for maintaining cellular function, whereas current conventional culture methods do not contain these microenvironmental cues correctly. Because of this microenvironmental gap, cellular functions and responses *in vitro* are very different from those *in vivo*, and thus conventional 2D cultures cannot accurately predict cellular functions and responses inside the body [6–8]. It is necessary to reconstitute cellular microenvironments for developing reliable *in vitro* methods. Development of biomimetic microdevices that can recapitulate tissue structure and microenvironmental cues could be a useful platform to facilitate drug discovery and development and to develop predictive models of human physiology and diseases.

8.1.2 Cellular Microenvironments Within Microfluidic Systems

The physiological microenvironments inside the body contain largely microfluidic structures such as vascular networks and sinusoids. Thus, the use of microfluidic systems facilitates engineering of cellular microenvironments. Cell culture conditions in the microfluidic systems are very different from those in conventional culture dishes (Fig. 8.1); the volume of culture medium in the dish culture is much greater compared to cellular volume, whereas a majority of the volume in a tissue is taken up with cells inside the body [9]. This difference in fluid volume to cell volume ratio causes differences in the chemical microenvironment. The volume of culture medium in the microfluidic systems is nanoliter-scale which is similar to *in vivo*. Because of physiological ratios of cells to liquid volumes, microfluidic systems enable to maintain cellular interactions based on autocrine and paracrine signals, whereas dish cultures do not maintain these signals [9, 10]. This feature also facilitates the formation of local chemical gradients that may otherwise not be possible to be studied. Furthermore, microfluidic devices can apply fluidic shear stress, enabling reconstitution of the physiological mechanical microenvironment inside the body [11].

8.1.3 Control of Chemical Microenvironments Using Microfluidic Systems

Microfluidic cell culture systems are typically made from poly(dimethylsiloxane) (PDMS) using the soft lithography technique [12]. Microfluidic systems have been used to engineer chemical microenvironments such as concentration gradients of biochemicals. Gradients of biochemical signals (e.g. growth factors, hormones,

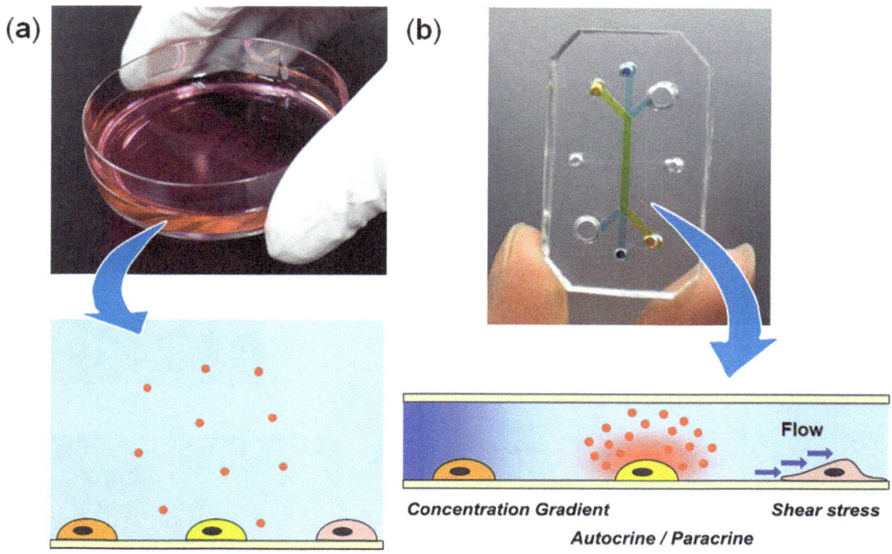

Fig. 8.1 Cellular microenvironment in a microfluidic culture and a conventional dish culture. (**a**) Cells cultured in a dish are maintained in a large volume of culture medium under static condition. (**b**) Cells cultured in a microfluidic device are maintained in a small volume of culture medium under static or dynamic conditions. Concentration gradients can be generated by flowing chemicals or cellular secretion and consumption and also fluid flow can generate shear stress within a microfluidic device

morphogens, and chemokines) play an important role in a wide range of biological processes including development, immune response, wound healing, and cancer metastasis [13, 14]. Although these biochemical gradients are very difficult to produce and maintain in conventional culture systems, microfluidic systems can generate arbitrary shapes of biochemical gradients [15, 16]. Thus, these microfluidic systems to generate concentration gradients have been used to study of immune responses, cancer metastasis, and stem cell biology [16–19]. Recently, development-on-chip has been developed by mimicking spatial and temporal chemical environments found *in vivo* during neural tube development using a microfluidic device [20]. This system was able to maintain simultaneous opposing and/or orthogonal gradients of developmental morphogens, resulting in neural tube patterning analogous to that observed *in vivo*.

Chemical gradients can be generated using cellular secreted factors by patterning different types of cells within microfluidic devices (Fig. 8.2). Cancer metastasis was modeled in a compartmentalized microfluidic device in which cancer cells were hydrodynamically patterned in a microchannel at spatially defined positions relative to source cells that secrete chemokines and sink cells that scavenge the chemokines (Fig. 8.2a) [21]. This system enabled to recreate a physiological cancer microenvironment, resulting in efficient chemotaxis under much shallower chemoattractant gradients than previously possible in other *in vitro* systems. These concentration

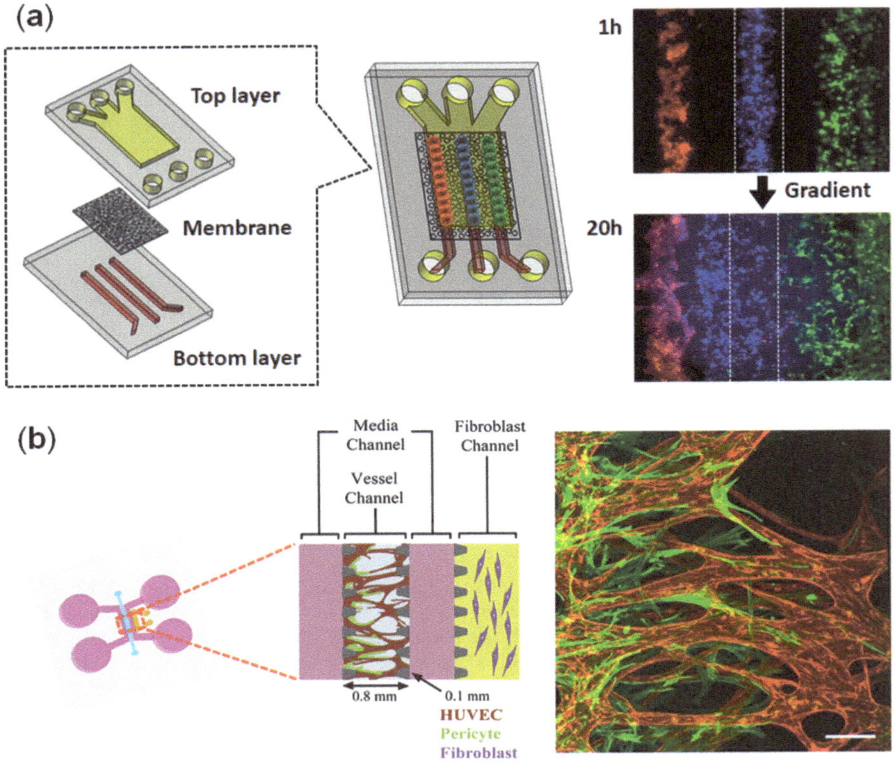

Fig. 8.2 Microfluidic co-culture systems to recreate chemical microenvironment. (**a**) A cancer metastasis model was engineered in a microfluidic device consisting of two PDMS microchannels separated by a porous membrane. Cancer cells (blue) were hydrodynamically patterned in the top channel at spatially defined positions relative to source (red) and sink cells (green), which generated chemoattractant gradients that induced cancer cell migration. (Reproduced from Ref. [21] with permission from the Royal Society of Chemistry). (**b**) A microfluidic system to form a 3D perfusable vascular network. The device consists of a central vessel channel, two adjoining media channels, and the outermost fibroblast channel. The vascular network (red) covered by pericytes (green) was formed in the central channel with assistance from the lateral fibroblasts. (Reproduced from Ref. [26]. Copyright 2015, Public Library of Science)

gradients of cellular secreted factors are very difficult to be maintained in conventional dish cultures. Using 3D cell culture techniques with microfluidic devices, 3D functional microvascular networks can be engineered [22–27]. Jeon's group has developed a microfluidic device which enables spatially patterned 3D co-culture of endothelial cells with stromal fibroblast cells [23] known to induce angiogenesis and support vascular formation (Fig. 8.2b) [25]. This system was able to model natural cellular programs found during development and angiogenesis to form perfusable networks of intact 3D microvessels. This system enabled the formation of 3D blood vessels with perfusable lumina that are similar in 3D architecture, intact barrier function, long term stability and salient biochemical markers to their *in vivo* counterparts. This model can be used to study interactions between pericytes and endothelial cells [26] and the role of interstitial flow during the formation of neovessels [27]. Since vascular networks are crucial to the maintenance of cellular viability and function in tissues and organs [28], engineering perfusable 3D vascular networks that can deliver nutrients and oxygen as well as cells could be a powerful platform to develop *in vitro* systems. This method has recently been applied to novel methods to vascularize 3D cell spheroids where a perfusable vascular network connected to microchannels is formed [29, 30]. These microfluidic devices may provide a novel approach to culture 3D tissues as well as organoids [31].

8.2 Organs-on-Chips to Mimic Cellular Microenvironmental Signals

Organs-on-chips are microfluidic cell culture devices made from elastomer, typically PDMS. By recapitulating tissue architectures and chemical and mechanical microenvironments, these microfluidic devices reconstitute organ-level cellular functionality not possible with conventional culture methods [32–35]. Especially, these devices can mimic biomechanical signals which cells normally experience inside the body.

8.2.1 Lung-on-a-Chip

A representative example of organs-on-chips is the lung-on-a-chip which reconstitutes alveolar function in the human lung by mimicking tissue-tissue interface and breathing mechanical environment (Fig. 8.3a) [36]. This system consists of a compartmentalized microfluidic device in which human alveolar epithelial cells are cultured in apposition with human pulmonary microvascular endothelial cells on a thin porous ECM-coated PDMS membrane that resembles the *in vivo* alveolar-capillary interface. This system is integrated with a mechanical actuation system to mimic breathing motions by applying cyclic suction to distort the PDMS microdevice which cyclically stretches the thin PDMS membrane. This biomimetic microdevice enables the reconstitution of organ-level cell responses not normally observed in conventional culture systems, such as recruitment of immune cells in responses to bacteria. Moreover, this system has revealed unexplained adverse effects of breathing-induced mechanical forces; the cyclic breathing motions increased cellular uptake of silica nanoparticles and translocation of nanoparticles from the alveolar airspace to the vascular compartment. This effect was confirmed in a mouse *ex vivo* model. This system also enables to model pulmonary edema induced by toxicity of the cancer drug interleukin-2 (IL-2) [37]. Clinical relevant concentration of IL-2 administration into the vascular channel caused continuous

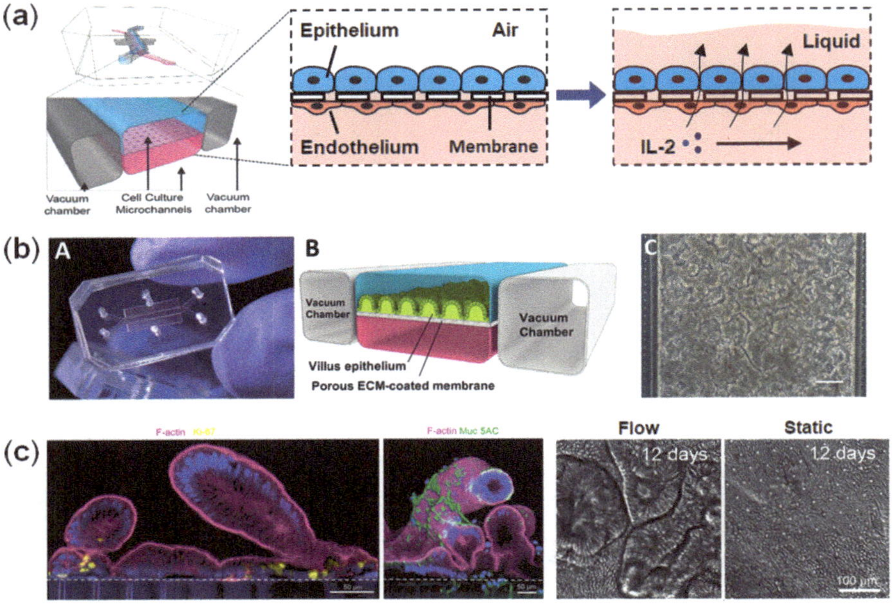

Fig. 8.3 Microfluidic organs-on-chips to recapitulate tissue architecture and mechanical microenvironment. (**a**) Lung-on-a-chip recapitulates the alveolar-capillary interface by culturing human alveolar epithelial cells on top of a thin porous PDMS membrane and human capillary cells on bottom. Breathing motions are recapitulated by applying cyclic suction to the side chambers, which deform the PDMS membrane to which the cell layers are attached. Administration of interleukin-2 (IL-2) into the vascular channel resulted in fluid leakage into the alveolar channel. (Reproduced from Ref. [44]. Copyright 2018, Nature Publishing Group). (**b**) Gut-on-a-chip. Photograph (A) and a schematic illustration (B) of the device in which human Caco-2 intestinal epithelial cells are cultured on a thin PDMS membrane to form 3D villi-like structures by applying peristalsis-like cyclic mechanical strain as well as fluid flow. (C) Micrograph of Caco-2 cells cultured for 6 days in the Gut-on-a-chip. Scale bar, 100 μm. (Reproduced from Ref. [43]. Copyright 2015, Public Library of Science). (**c**) Primary human small intestine-on-a-chip. A confocal image showing a cross-section of primary intestinal epithelium immunostrained (magenta, F-actin; yellow, Ki67; green, Muc5AC). Optical images of primary intestinal epithelium cultured on-chip for 12 days under continuous flow compared with a static culture condition. Formation of intestinal villi-like structures occurred only in the presence of flow. (Reproduced from Ref. [44]. Copyright 2018, Nature Publishing Group)

leakage of vascular fluid into the alveolar channel and complete flooding of the airspace (Fig. 8.3a). Importantly, this system revealed that the mechanical forces produced by breathing motions contribute to the development of increased vascular leakage and pulmonary edema. Thus, this system enabled the identification of novel therapeutics that inhibited mechanotransduction pathways. The lung-on-a-chip microdevice faithfully recapitulates organ-level cell function and predict physiological and pathological responses *in vivo*.

This system has been applied to a human lung small airway-on-a-chip which contains a differentiated, mucociliary bronchiolar epithelium exposed to air and an underlying microvascular endothelium that experiences fluid flow [38]. This system enabled active synchronized cilia beating and mucociliary transport whose velocity was nearly identical to that in healthy human lung airway. Exposure of the epithelium to IL-13 in this device reconstituted the goblet cell hyperplasia, cytokine hypersecretion, and decreased ciliary function of asthmatics. The small airway-on-a-chip generated with epithelial cells from individuals with chronic obstructive pulmonary disease reconstituted features of the disease such as selective cytokine hypersecretion, increased neutrophil recruitment, and clinical exacerbation by exposure to viral and bacterial infections. Thus, this device allows for analysis of organ-level lung pathophysiology *in vitro*. Furthermore, this device enabled modeling of smoke-induced lung injury by connecting to a smoking instrument that inhales and exhales whole smoke from burning cigarettes in and out of the epithelium-lined microchannel of the device [39]. This system was used to compare the effects of inhaled smoke on chips containing bronchiolar epithelium isolated from normal lungs or from lungs of chronic obstructive pulmonary disease (COPD) patients. This model led to identification of ciliary micropathologies, COPD-specific molecular signatures, and epithelial responses to smoke generated by electric cigarettes. This smoking airway-on-a-chip provides a tool to study normal and disease-specific responses of human lung to inhaled smoke across molecular, cellular, and tissue-level responses.

8.2.2 Gut-on-a-Chip

A similar compartmentalized microfluidic system was used to create a human gut-on-a-chip system that mimics physiological peristaltic-like motion and fluid flow of the intestine (Fig. 8.3b) [40–43]. The device consists of two layer of cell culture channels separated by a thin porous PDMS membrane in which human intestinal epithelial Caco-2 cells (colorectal carcinoma line) are cultured. Caco-2 cells cultured with fluid flow and cyclic mechanical distortion in the microfluidic device spontaneously formed 3D structures that resemble the architecture of the intestinal villus. This system enabled to reproduce the intestinal barrier function, high levels of mucus production, and intestinal absorptive characteristics, which will be useful for drug testing as well as development of intestinal disease models. This system also enabled long-term co-culture of various gut microbes with the epithelial cells by recapitulating the dynamic mechanical microenvironment, whereas conventional culture methods cannot maintain co-culture of microbes. Lack of mechanical signals was shown to trigger bacterial overgrowth similar to that observed in patients with ileus and inflammatory bowel disease. Thus, the human gut-on-a-chip can be used to analyze contributions of microbiome to intestinal pathophysiology and disease mechanisms in a controlled manner.

Caco-2 cells are widely used to estimate intestinal barrier function; however, these cells are cancer-derived cells and have some issues such as enzymatic activity. Recently, primary human small intestine-on-a-chip has been developed using intestinal biopsies [44]. Primary intestinal epithelial cells harvested from 3D organoids are cultured in a microfluidic device with human intestinal microvascular endothelium in a parallel microchannel under fluid flow. The intestinal epithelium formed villi-like structures lined by polarized epithelial cells that undergo multi-lineage differentiation (Fig. 8.3c). Transcriptomic analysis demonstrated that the intestine chip more closely mimicked whole human duodenum than duodenal organoids. This study demonstrated that recapitulation of the dynamic mechanical environment and tissue-tissue interface enabled the formation of 3D villi-like structures as well as the maintenance of intestinal functions. Since this system can maintain normal human intestinal function, it will be a powerful platform to study drug pharmacokinetics and to model human intestinal diseases.

8.2.3 Kidney-on-a-Chip

In vitro systems to model kidney functions are crucially needed to predict drug-induced kidney injury which is often observed in pharmacotherapy; however current available models do not recapitulate the biological functions of the kidney and poor predictions of drug-induced kidney injury [45]. A kidney proximal tubule-on-a-chip microdevice was developed that is lined by human proximal tubule epithelial cells exposed to physiologically relevant fluidic flow [46]. Mechanical forces caused by fluid flow have been recognized as a key determinant of cellular functions in the kidney. This microfluidic device consists of a luminal flow channel and an interstitial compartment separated by a thin porous membrane on which the kidney cells are cultured. This device demonstrated that low levels of fluid shear stress (0.2 dyn cm^{-2}) similar to that observed in the collecting ducts and proximal tubules of the kidney enhanced differentiation, increased molecular and drug transport functions, and produced more *in vivo*-like toxicity responses. Thus, this kidney proximal tubular model can mimic the structural, mechanical, transport, absorptive, and physiological properties of the human kidney. A 3D flow-directed kidney proximal tubule system has also been developed using a microfluidic device [47]. Human proximal tubular epithelial cells are seeded onto collagen extracellular matrix within a microchannel so that the cells self-assemble to form a 3D tubular structure. This system recapitulates the perfusion delivery and transport pathway of a solute which is perfused into a surrogate vascular channel, diffuses through the pseudo-interstitial space, and undergoes uptake and efflux across the epithelial barrier into the flowing perfusate in the tubular luminal channel. This model replicates the polarity of the proximal tubule, retains polarized expression and function of protein essential for reabsorptive and secretory transport, responds to physiological stimuli, and performs critical biochemical synthetic activities. This system provides a platform for modeling of renal drug clearance and drug-induced nephrotoxicity.

A kidney glomerulus model has recently been developed using a compartmentalized microfluidic device with human iPS (induced pluripotent stem) cell-derived podocytes which can produce glomerular basement-membrane collagen [48]. This device enabled to recapitulate the natural tissue-tissue interface of the glomerulus and the differential clearance of albumin and inulin by co-culturing podocytes and human glomerular endothelial cells under physiological fluid flow and cyclic mechanical strain. This device demonstrated that application of physiological relevant mechanical cues further augmented podocyte differentiation and maturation. Since existing immortalized podocyte cells and cultures poorly mimic glomerular function, this device could provide a better *in vitro* system for predicting nephrotoxicity and therapeutic development.

8.3 Bone Marrow-on-a-Chip to Reconstitute Hematopoietic Niche Physiology

Bone marrow is the only permanent hematopoietic organ in humans. Given the importance of bone marrow as the source of all blood cells, an *in vitro* culture system that can reconstitute function of living bone marrow will be a useful platform. The bone marrow microenvironment contains a complex set of cellular, chemical, structural, and physical cues necessary to maintain the viability and function of the hematopoietic system [4, 49]. This bone marrow microenvironment regulates the function of hematopoietic stem cells (HSCs), facilitating a balance between self-renewal and differentiation into progenitors that generate all mature blood cells. Current *in vitro* culture methods do not accurately model bone marrow physiology; hence, studies relevant to the hematologic system are usually conducted in animals [50–52]. Engineering an artificial bone marrow capable of reproducing the bone marrow microenvironment could be potentially used to study blood development and physiology, model diseases, and serve as a platform for drug development and toxicity studies.

8.3.1 Engineering of Bone Marrow In Vivo

Because of the complexity of the bone marrow microenvironment necessary to support HSC function as well as blood production, it has not been possible to recapitulate the complex bone marrow microenvironment as well as blood forming functions *in vitro* [52–55]. To overcome this challenge, we used a tissue engineering approach to first induce formation of new bone containing marrow *in vivo*, and then we surgically removed it whole and maintained it in a microfluidic device *in vitro* (Fig. 8.4a) [56]. New bone formation was induced in the subcutaneous tissue of a mouse by implanting a PDMS device containing cylindrical hole filled with a collagen gel and bone-inducing materials (demineralized bone powder and BMPs). This method produced a cylindrical shaped bone containing a bony cortex and an internal trabecular bone network that was filled with a normal appearing marrow

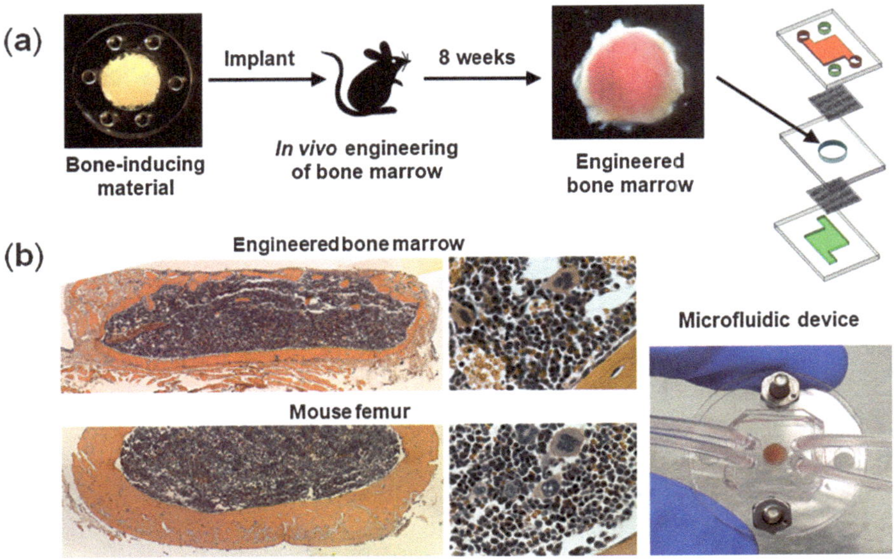

Fig. 8.4 Bone marrow-on-a-chip. (**a**) PDMS devices containing a cylindrical chamber filled with bone-inducing materials were implanted subcutaneously on the back of a mouse for 8 weeks and then surgically removed. The engineered bone marrow that formed within the PDMS device was placed into a similar shaped central chamber in a microfluidic system and then maintained in culture *in vitro*. (**b**) Low- (left) and high-magnification views (right) of histological hematoxylin-and-eosin-stained sections of the engineered bone marrow formed in the PDMS device at 8 weeks following implantation (top) compared with a cross section of bone marrow in the normal mouse femur (bottom). (Reproduced from Ref. [56] with permission from the Nature Publishing Group)

with a blood cell composition virtually identical to that of natural bone marrow (Fig. 8.4b). Previous reports showed that tissue engineering approaches with the bone-inducing materials resulted in the formation of marrow largely inhabited by adipocytes [57–59] which are known to inhibit hematopoiesis [60]. To reduce adipocyte content in the marrow, we used a PDMS device to restrict access of cells or soluble factors from the overlying adipocyte-rich hypodermis to the bone-inducing materials while maintaining accessibility to the underlying muscle tissue through the lower opening. Thus, the use of PDMS devices enabled to engineer bone marrow that resembles to natural bone marrow (Fig. 8.4b). We demonstrated that our method produced bone containing marrow with a hematopoietic cell composition nearly identical to that of natural bone marrow. Immunohistochemical analysis also confirmed that endothelial and perivascular cells as well as hematopoietic stem cells were located in their normal spatial positions in the engineered bone marrow. The presence of key cellular and molecular components of the hematopoietic niche indicated that the cellular content of the engineered bone marrow closely resembles the natural bone marrow microenvironment.

Several approaches have been reported that bone marrow can be engineered ectopically in mice. Human mesenchymal stem cells (MSCs) were seeded into collagen-based scaffolds and cultured *in vitro* to form a hypertrophic cartilage tissue [61]. When the engineered cartilage tissues were implanted into mice for 12 weeks, these tissues formed fully-fledged bone containing marrow with hematopoietic stem and progenitor cells which can reconstitute multilineage long-term hematopoiesis in lethally irradiated mice. Matrigel has also been used to engineer bone marrow ectopically [62]. When human MSCs were suspended in Matrigel and subcutaneously injected into immunodeficient (NSG) mice, MSCs spontaneously formed a bone marrow cavity through a vascularized cartilage that was progressively replaced by hematopoietic tissue and bone. This humanized MSC-derived microenvironment permitted homing and maintenance of long-term mouse hematopoietic stem cells (HSCs) as well as human HSCs after cord blood transplantation. This xenotransplantation model also enables robust engraftment of primary acute myeloid leukemia samples at levels much greater than those in unmanipulated mice [63]. Direct intraossicle transplantation accelerated engraftment and resulted in the detection of higher leukemia-initiating cell frequencies. These humanized ossicle xenotransplantation systems allow to model wide variety of human hematologic diseases. Therefore, human MSCs can induce the formation of bone marrow ectopically *in vivo* when they are implanted with bone-inducing materials. The *in vivo* engineering of bone marrow offers a new approach for analysis and study of hematopoiesis and hematologic disease.

8.3.2 In Vitro *Culture of Bone Marrow with a Functional Hematopoietic Niche*

Various *in vitro* culture systems have been developed to maintain and expand HSCs and hematopoietic progenitors because the expansion of HSCs would greatly improve bone marrow transplantation and facilitate the development of advanced cell therapies for many blood disorders and malignant diseases [64–68]. Bone marrow stromal cells are mostly used to mimic the bone marrow microenvironment *in vitro* [68]. Recently, 3D culture systems have been develop using scaffolds or hydrogels in which bone marrow stromal cells are cultured with hematopoietic cells [53, 69–72]. Bone-derived materials have also been used as a scaffold to recreate the bone marrow microenvironment [73]. Frozen human bones were decellularized, demineralized, and cut into pieces to make bone-derived scaffolds so that cells can be cultured with those scaffolds in conventional culture plates. Bone marrow stromal cells such as mesenchymal stromal cells (MSCs) and osteoblasts were cultured within the bone-derived scaffolds with hematopoietic stem and progenitor cells (HSPCs), resulting in enhancing expansion of HSPCs compared with conventional 2D culture systems. Microfluidic culture systems with scaffolds also have been developed to maintain HSPCs for a long time. A 3D co-culture model based on a hydroxyapatite coated zirconium oxide scaffold consisting of human MSCs and HSPCs enabled the culture of HSPCs for 4 weeks within a microfluidic device

[74]. Although HSCs rapidly expand after transplantation *in vivo*, *in vitro* studies indicate that control of HSPC self-renewal and differentiation in culture remains difficult. There is currently no method to recreate the entire bone marrow microenvironment and to expand hematopoietic stem cells *in vitro* because of the complexity of the bone marrow microenvironment.

As a proof of concept, we tested whether the engineered bone marrow could maintain a functional hematopoietic system *in vitro* [56]. The engineered bone marrow formed 8 weeks after implantation was surgically removed from the mouse, perforated with a surgical needle to permit fluid access, placed in a similar shaped chamber in a microfluidic device, and perfused continuously with culture medium for *in vitro* culture (Fig. 8.4). Flow cytometric analysis of cellular components of the cultured 'bone marrow-on-a-chip' revealed that the presence of hematopoietic stem and progenitor cells in similar proportions to those of freshly harvested bone marrow for up to 7 days on-chip. In contrast, bone marrow cultured on a stromal feeder cell layer in a culture dish, which is the current benchmark, exhibited a significant decrease of long-term HSCs and a concomitant increase in hematopoietic progenitor cells relative to freshly isolated from natural mouse bone marrow. Thus, the long-term HSCs, which are the only cells capable of self-renewal and multilineage potential, appeared to be differentiating into more specialized progenitor cells in the static stroma-supported culture system as previously reports noted [68–72]. Thus, the bone marrow-on-a-chip enabled maintenance of a significantly higher proportion of long-term HSCs while more effectively maintaining distribution of hematopoietic progenitors as well as mature blood cells. Moreover, because the engineered bone marrow autonomously produces factors necessary to maintain hematopoiesis, this system was able to maintain HSCs and hematopoietic progenitors in normal proportions in the cultured chip without the addition of exogenous cytokines. Importantly, the full functionality and robustness of this engineered bone marrow tissue was confirmed by demonstrating that bone marrow cultured on-chip could be used to fully reconstitute the entire blood system when transplanted into lethally irradiated mice. Cells harvested from the engineered bone marrow after *in vitro* culture successfully engrafted the mice at a similar rate to that of freshly isolated bone marrow and repopulated all differentiated blood cell lineages. Thus, the bone marrow-on-a-chip retains fully functional, self-renewing, multipotent HSCs as well as a functional hematopoietic niche *in vitro*. Furthermore, because of microfluidic perfusion culture, this bone marrow-on-a-chip can produce blood cells continuously and release them into the microfluidic circulation, while maintaining HSCs and hematopoietic progenitors in normal *in vivo*-like proportions inside the microfluidic device for at least 2 weeks in culture [75]. This system was able to induce red blood cell production by adding erythropoietin (EPO) which is known to stimulate erythrocyte formation. When EPO was administrated in the bone marrow chip, we detected a continuous increase in the number of erythrocytes released in the outflow. Thus, this bone marrow-on-a-chip can be used to study and test continuous blood cell production *in vitro*. This method enabled to maintain an intact 3D bone marrow microenvironment with functional hematopoietic cells *in vitro*.

8.3.3 Testing of Drugs Using Bone Marrow-on-a-Chip

To evaluate whether the maintenance of an intact bone marrow microenvironment could be effective for efficacy and toxicity testing, we tested whether the bone marrow-on-a-chip can be used to test radiation toxicity and pharmacological countermeasures that might help radiation injured bone marrow repair itself [56, 75]. Despite an active search for medical countermeasures to radiation toxicity, no effective drug has been approved by FDA that could help protect or reconstitute the hematopoietic system [76]. Current drug development studies rely on animal studies; however, these animal models rarely reflect human response [77]. Thus, there is a critical need for predictive *in vitro* models to test effects of radiation countermeasures. When the bone marrow-on-a-chip was exposed to varying doses of γ-radiation, it showed a significant radiation dose-dependent decrease in the proportion of HSCs and hematopoietic progenitors which closely mimics what is observed in the bone marrow of live irradiated mice. In contrast, the stroma-supported static culture exhibited suppressed responses and was significantly more resistant to the effects of radiation toxicity. We then evaluated radiation-protecting effects of two potential therapeutic proteins, granulocyte-colony stimulating factor (G-CSF) [78] and bactericidal/permeability-increasing protein (BPI) [79], which have been reported to accelerate recovery of hematopoiesis after radiation-induced bone marrow failure *in vivo*. When G-CSF was administrated to the bone marrow chip 1 day after exposure to γ-radiation, we detected a significant increase in the number of HSCs and progenitors within the bone marrow chip compared to untreated irradiated bone marrow chips, and furthermore we detected significant increases in the number of HSCs, progenitors, and myeloid cells in the outflow of the chips. When BPI was administrated to the bone marrow chip 1 day after exposure to γ-radiation, it also significantly increased the number of HSCs and myeloid cells in the outflow of the chips. Thus, these radiation countermeasure drugs were able to accelerate recovery of blood cell production after radiation injury in the bone marrow-on-a-chip *in vitro* as they have previously been reported *in vivo*. In contrast, when a conventional static culture was exposed to γ-radiation, it was not able to detect any effect of BPI. Therefore, the bone marrow-on-a-chip effectively mimics the *in vivo* response of living bone marrow to radiation countermeasure drugs, whereas the conventional static bone marrow cultures do not. These results suggest that the presence of a hematopoietic microenvironment is crucial for modeling radiation toxicity and testing effects of potential countermeasure drugs *in vitro*. Thus, it is because conventional culture methods do not accurately recapitulate the bone marrow microenvironment that drug development studies currently rely on animal models for efficacy and safety testing. Since the bone marrow-on-a-chip system contains a functional bone marrow microenvironment, this system can mimic physiological responses to radiation countermeasure drugs normally only observed *in vivo* and thus this system could offer a novel approach to replace animal testing. Therefore, recapitulation of cellular microenvironment is crucial for efficacy and toxicity testing and thus the organ-on-a-chip technology has great potential to generate reliable predictions of drug efficacy and toxicity in humans.

8.4 Culture of iPS Cells in Organs-on-Chips

The organ-on-a-chip technology has been used to culture human pluripotent stem cells including induced pluripotent stem cells (iPSCs) [80]. A microfluidic method has been reported to induce functional differentiation of human pluripotent stem cells on-chip [81]. This method demonstrated that extrinsic signal modulation though optimal frequency of medium delivery can be used as a parameter for improved germ layer specification and cell differentiation. This method enabled accurate spatiotemporal control of the soluble microenvironment around cells through regulation of periodic perfusion frequencies and achieved effective induction of hepatocytes on-chip with higher activity of hepatocyte specific functions under optimized perfusion conditions.

A disease model has been reported to reproduce disease pathophysiology *in vitro* using the organ-on-a-chip technology. A heart-on-a-chip was used to replicate contractile pathophysiology of Barth syndrome (BTHS) using engineered myocardial tissue assembled from BTHS or control iPSC-derived cardiomyocytes (iPSC-CMs) [82]. When iPSC-CMs were cultured onto thin elastomers micropatterned with fibronectin lines for 5 days, the iPSC-CMs self-organized into laminar anisotropic myocardium tissues. During electrical field stimulation, control iPSC-CM tissues contracted rhythmically, whereas BTHS iPSC-CM tissues developed significantly lower twitch and peak systolic stress compared to controls, demonstrating that BTHS engineered myocardial tissues replicate the BTHS myopathic phenotype. Importantly, this method enabled to model disease correction; treatment of BHTS iPSC-CMs with TAZ modRNA for 5 days restored contractile function of the cardiomyocyte tissues to levels comparable to those of controls. Patient-derived iPSCs have considerable potential to facilitate study of human disease and enable therapeutic screening [83]; however, currently there is the lack of *in vitro* models that reproduce disease pathophysiology. Thus, the use of organ-on-a-chip systems with the iPS cell technology could be a powerful platform to develop *in vitro* disease models [84].

Microfluidic systems have recently been utilized to engineer blood cells from pluripotent stem cells. To produce platelets, microfluidic bioreactors have been developed by mimicking the bone marrow and blood vessel microenvironments [85, 86] because platelets are generated from megakaryocytes in the bone marrow vasculature *in vivo* [87]. These bioreactors recapitulate extracellular matrix components, stiffness, endothelial cell contacts, and vascular shear forces. These systems demonstrated that physiological shear stress triggers proplatelet production and platelet release. By recapitulating physiological microenvironments, these bioreactors enabled the production of functional human platelets from megakaryocytes derived from adult hematopoietic progenitor cells as well as human iPSCs. The iPSC-based technology has recently enabled to produce stable immortalized megakaryocyte progenitor cell lines [88]. These iPSC-derived megakaryocyte progenitor cells can be expanded, cryopreserved, and differentiated into functional platelets. Producing platelets from iPSCs *in vitro* could offer donor-free blood products as well as genetically modified products and provide a way of autologous transfusion [89].

Microfluidic systems that can recapitulate biomechanical forces have also been used to engineer HSCs. It has been demonstrated that biomechanical forces, such as blood flow-induced shear stress, play an important role in hematopoietic development and the emergence of HSCs [90, 91]. The use of microfluidic systems enabled to apply physiological fluid shear stress (5 dyn cm^{-2}), demonstrating that fluid shear stress endows long-term multilineage engraftment potential on early hematopoietic tissues not previously shown to harbor HSCs [92]. This study revealed that effects on hematopoiesis are mediated in part by a cascade downstream of shear stress that involves stimulation of prostaglandin E_2 (PGE_2)-cyclic adenosine monophosphate (cAMP)-protein kinase A (PKA) signaling axis. Using mouse embryonic stem cells differentiated *in vitro*, it has been demonstrated that fluid shear stress increases hematopoietic colony-forming potential and expression of hematopoietic markers [93]. Thus, the use of the organ-on-a-chip technology that can recapitulate cellular microenvironment and biomechanical signals could facilitate engineering of HSCs [94]. Recently, production of HSCs from iPSCs has been achieved using transcription factors as well as *in vivo* transplantation [95]. Human iPSCs were differentiated into hemogenic endothelial cells *in vitro* and then treated with cocktails of transcription factors to endow the potential to engraft multi-lineage hematopoiesis *in vivo*. This method requires *in vivo* microenvironment to mature HSCs, suggesting signals from the microenvironment is necessary for HSC maturation. Therefore, the organ-on-a-chip technology may provide a way to produce HSCs *in vitro* by mimicking the microenvironment inside the body.

8.5 Future Prospects

Organ-on-a-chip microdevices produce organ-level function *in vitro* by recapitulating natural tissue arrangements and microenvironmental signals. Since these microdevices are created using human cells including primary human cells and cell lines as well as pluripotent stem cell-derived cells, these devices can predict human responses to medications, which provides reliable prediction of drug efficacy and safety in humans. These systems have also potential to serve as a platform to differentiate pluripotent stem cells into specific cell types. The use of iPS cells with organ-on-a-chip microdevices is especially attractive because iPS cells provide sources of patient-specific cells and enable genome editing. These systems enable to model human physiology and disease and further will provide opportunities to model organs that have not yet been possible to be studied *in vitro*.

As a novel approach to develop organs-on-chips, we have developed the bone marrow-on-a-chip in which an organ, bone marrow, is engineered inside a device *in vivo* and then maintained in culture. This system can sustain an intact bone marrow with a functional hematopoietic microenvironment *in vitro*, enabling to mimic complex tissue-level responses to radiation toxicity and to therapeutic agents. This system provides an interesting alternative to animal models because this offers the ability to manipulate individual hematopoietic cell populations or to insert other cell types such as cancer cells *in vitro* as well as *in vivo* before analyzing the response of the intact bone marrow to relevant clinical challenges. It is also possible to generate human bone marrow models by engineering bone marrow in immunocompromised mice and inserting human hematopoietic cells or human leukemia cells [63]. Therefore, this method offers a new approach for analysis of drug responses and toxicities and for study of hematopoiesis and hematologic diseases.

These organ-on-a-chip microdevices can be integrated via microfluidic linkage to model physiological interplay between different organs. Body-on-a-chip systems have been developed by integrating several organ models (typically 2~4 types) into a microfluidic device [96–100]. These systems realize to estimate the adsorption, distribution, metabolism, and elimination (ADME) of drugs *in vitro*. For example, when a drug was introduced into a microfluidic device, the drug converted in a liver compartment into its reactive metabolites and then circulated to other organ compartments such as a lung compartment or a cancer compartment to detect drug toxicity or effect. It might be possible to develop 'human body-on-a-chip' consisting of fluidically linked chambers representing different organs that could reliably predict drug toxicity and efficacy inside the human body in the future. Although the organs-on-chips have great potential, this technology still requires further validation and improvement. For integration of different organ chips, it requires a novel common culture medium, such as blood substitute, that can be perfused through the entire system to maintain viability and functions of all the cells on a chip. Furthermore, although PDMS devices are widely used, the use of PDMS could raise a problem for drug discovery applications because it absorbs small hydrophobic molecules. By improving materials, cell culture techniques, and tissue engineering techniques, organs-on-chips could provide novel approaches for drug discovery and development, which could reduce or replace animal testing in the future.

References

1. Shamir ER, Ewald AJ (2014) Three-dimensional organotypic culture: experimental models of mammalian biology and disease. Nat Rev Mol Cell Biol 15:647–664
2. Quail DF, Joyce JA (2013) Microenvironmental regulation of tumor progression and metastasis. Nat Med 19:1423–1437
3. Morrison SJ, Spradling AC (2008) Stem cells and niches: mechanisms that promote stem cell maintenance throughout life. Cell 132:598–611
4. Morrison SJ, Scadden DT (2014) The bone marrow niche for haematopoietic stem cells. Nature 505:327–334

5. Sugiyama T, Kohara H, Noda M, Nagasawa T (2006) Maintenance of the hematopoietic stem cell pool by CXCL12-CXCR4 chemokine signaling in bone marrow stromal cell niches. Immunity 25:977–988
6. Pampaloni F, Reynaud EG, Stelzer EH (2007) The third dimension bridges the gap between cell culture and live tissue. Nat Rev Mol Cell Biol 8:839–845
7. Griffith LG, Swartz MA (2006) Capturing complex 3D tissue physiology in vitro. Nat Rev Mol Cell Biol 7:211–224
8. Mazzoleni G, Di Lorenzo D, Steimberg N (2009) Modelling tissues in 3D: the next future of pharmaco-toxicology and food research? Genes Nutr 4:13–22
9. Paguirigan AL, Beebe DJ (2009) From the cellular perspective: exploring differences in the cellular baseline in macroscale and microfluidic cultures. Integr Biol 1:182–195
10. Helmlinger G, Endo M, Ferrara N et al (2000) Formation of endothelial cell networks. Nature 405:139–141
11. Shemesh J, Jalilian I, Shi A et al (2015) Flow-induced stress on adherent cells in microfluidic devices. Lab Chip 15:4114–4127
12. Whitesides GM, Ostuni E, Takayama S et al (2001) Soft lithography in biology and biochemistry. Annu Rev Biomed Eng 3:335–373
13. Gurdon JB, Bourillot PY (2001) Morphogen gradient interpretation. Nature 413:797–803
14. Keenan TM, Folch A (2008) Biomolecular gradients in cell culture systems. Lab Chip 8:34–57
15. Dertinger SKW, Chiu DT, Jeon NL, Whitesides GM (2001) Generation of gradients having complex shapes using microfluidic networks. Anal Chem 73:1240–1246
16. Kim S, Kim HJ, Jeon NL (2010) Biological applications of microfluidic gradient devices. Integr Biol 2:584–603
17. Jeon NL, Baskaran H, Dertinger SK et al (2002) Neutrophil chemotaxis in linear and complex gradients of interleukin-8 formed in a microfabricated device. Nat Biotechnol 20:826–830
18. Wang SJ, Saadi W, Lin F et al (2004) Differential effects of EGF gradient profiles on MDA-MB-231 breast cancer cell chemotaxis. Exp Cell Res 300:180–189
19. Park JY, Kim SK, Woo DH et al (2009) Differentiation of neural progenitor cells in a microfluidic chip-generated cytokine gradient. Stem Cells 27:2646–2654
20. Demers CJ, Soundararajan P, Chennampally P et al (2016) Development-on-chip: in vitro neural tube patterning with a microfluidic device. Development 143:1884–1892
21. Torisawa Y, Mosadegh B, Bersano-Begey T et al (2010) Microfluidic platform for chemotaxis in gradients formed by CXCL12 source-sink cells. Integr Biol 2:680–686
22. Chen X, Aledia AS, Popson SA et al (2010) Rapid anastomosis of endothelial progenitor cell-derived vessels with host vasculature is promoted by a high density of cotransplanted fibroblasts. Tissue Eng Part A 16:585–597
23. Newman AC, Nakatsu MN, Chou W et al (2011) The requirement for fibroblasts in angiogenesis: fibroblast-derived matrix proteins are essential for endothelial cell lumen formation. Mol Biol Cell 22:3791–3800
24. Moya ML, Hsu YH, Lee AP et al (2013) In vitro perfused human capillary networks. Tissue Eng Part C Methods 19:730–737
25. Kim S, Lee H, Chung M, Jeon NL (2013) Engineering of functional, perfusable 3D microvascular networks on a chip. Lab Chip 13:1489–1500
26. Kim J, Chung M, Kim S et al (2015) Engineering of a biomimetic pericyte-covered 3D microvascular network. PLoS One 10:e0133880
27. Kim S, Chung M, Ahn J et al (2016) Interstitial flow regulates the angiogenic response and phenotype of endothelial cells in a 3D culture model. Lab Chip 16:4189–4199
28. Potente M, Gerhardt H, Carmeliet P (2011) Basic and therapeutic aspects of angiogenesis. Cell 146:873–887
29. Nashimoto Y, Hayashi T, Kunita I et al (2017) Integrating perfusable vascular networks with a three-dimensional tissue in a microfluidic device. Integr Biol 9:506–518
30. Oh S, Ryu H, Tahk D et al (2017) "Open-top" microfluidic device for in vitro three-dimensional capillary beds. Lab Chip 17:3405–3414

31. Takebe T, Zhang B, Radisic M (2017) Synergistic engineering: organoids meet organs-on-a-chip. Cell Stem Cell 21:297–300
32. Huh D, Torisawa Y, Hamilton GA et al (2012) Microengineered physiological biomimicry: organs-on-chips. Lab Chip 12:2156–2164
33. Bhatia SN, Ingber DE (2014) Microfluidic organs-on-chips. Nat Biotechnol 32:760–772
34. Esch EW, Bahinski A, Huh D (2015) Organs-on-chips at the frontiers of drug discovery. Nat Rev Drug Discov 14:248–260
35. Zheng F, Fu F, Cheng Y et al (2016) Organ-on-a-chip systems: microengineering to biomimic living systems. Small 17:2253–2282
36. Huh D, Matthews BD, Mammoto A et al (2010) Reconstituting organ-level lung functions on a chip. Science 328:1662–1668
37. Huh D, Leslie DC, Matthews BD et al (2012) A human disease model of drug toxicity-induced pulmonary edema in a lung-on-a-chip microdevice. Sci Transl Med 4:159ra147
38. Benam KH, Villenave R, Lucchesi C et al (2016) Small airway-on-a-chip enables analysis of human lung inflammation and drug responses in vitro. Nat Methods 13:151–157
39. Benam KH, Novak R, Nawroth J et al (2016) Matched-comparative modeling of normal and diseased human airway responses using a microengineered breathing lung chip. Cell Syst 3:456–466
40. Kim HJ, Huh D, Hamilton G, Ingber DE (2012) Human gut-on-a-chip inhabited by microbial flora that experiences intestinal peristalsis-like motions and flow. Lab Chip 12:2165–2174
41. Kim HJ, Ingber DE (2013) Gut-on-a-chip microenvironment induces human intestinal cells to undergo villus differentiation. Integr Biol 5:1130–1140
42. Kim HJ, Li H, Collins JJ, Ingber DE (2016) Contributions of microbiome and mechanical deformation to intestinal bacterial overgrowth and inflammation in a human gut-on-a-chip. Proc Natl Acad Sci U S A 113:E7–E15
43. Villenave R, Wales SQ, Hamkins-Indik T et al (2017) Human Gut-on-a-chip supports polarized infection of coxsackie B1 virus in vitro. PLoS One 12:e0169412
44. Kasendra M, Tovaglieri A, Sontheimer-Phelps A et al (2018) Development of a primary human Small Intestine-on-a-chip using biopsy-derived organoids. Sci Rep 8:2871
45. Wilmer MJ, Ng CP, Lanz HL et al (2016) Kidney-on-a-chip technology for drug-induced nephrotoxicity screening. Trends Biotechnol 34:156–170
46. Jang KJ, Mehr AP, Hamilton GA et al (2013) Human kidney proximal tubule-on-a-chip for drug transport and nephrotoxicity assessment. Integr Biol 5:1119–1129
47. Weber EJ, Chapron A, Chapron BD et al (2016) Development of a microphysiological model of human kidney proximal tubule function. Kidney Int 90:627–637
48. Musah S, Mammoto A, Ferrante TC et al (2017) Mature induced-pluripotent-stem-cell-derived human podocytes reconstitute kidney glomerular-capillary-wall function on a chip. Nat Biomed Eng 1:0069
49. Asada N, Takeishi S, Frenette PS (2017) Complexity of bone marrow hematopoietic stem cell niche. Int J Hematol 106:45–54
50. Cao X, Wu X, Frassica D et al (2011) Irradiation induces bone injury by damaging bone marrow microenvironment for stem cells. Proc Natl Acad Sci U S A 108:1609–1614
51. Askmyr M, Quach J, Purton LE (2011) Effects of the bone marrow microenvironment on hematopoietic malignancy. Bone 48:115–120
52. Dhami SPS, Kappala SS, Thompson A, Szegezdi E (2016) Three-dimensional ex vivo co-culture models of the leukaemic bone marrow niche for functional drug testing. Drug Discov Today 21:1464–1471
53. Di Maggio N, Piccinini E, Jaworski M et al (2011) Toward modeling the bone marrow niche using scaffold-based 3D culture systems. Biomaterials 32:321–329
54. Choi JS, Mahadik BP, Harley BA (2015) Engineering the hematopoietic stem cell niche: frontiers in biomaterial science. Biotechnol J 10:1529–1545
55. Nelson MR, Roy K (2016) Bone-marrow mimicking biomaterial niches for studying hematopoietic stem and progenitor cells. J Mater Chem B 4:3490–3503

56. Torisawa Y, Spina CS, Mammoto T et al (2014) Bone marrow-on-a-chip replicates hematopoietic niche physiology in vitro. Nat Methods 11:663–669
57. Krupnick AS, Shaaban S, Radu A, Flake AW (2002) Bone marrow tissue engineering. Tissue Eng 8:145–155
58. Chen B, Lin H, Wang J et al (2007) Homogeneous osteogenesis and bone regeneration by demineralized bone matrix loading with collagen-targeting bone morphogenetic protein-2. Biomaterials 28:1027–1035
59. Schwartz Z, Doukarsky-Marx T, Nasatzky E et al (2008) Differential effects of bone graft substitutes on regeneration of bone marrow. Clin Oral Implants Res 19:1233–1245
60. Naveiras O, Nardi V, Wenzel PL et al (2009) Bone-marrow adipocytes as negative regulators of the haematopoietic microenvironment. Nature 460:259–263
61. Scotti C, Piccinini E, Takizawa H et al (2013) Engineering of a functional bone organ through endochondral ossification. Proc Natl Acad Sci U S A 110:3997–4002
62. Reinisch A, Etchart N, Thomas D et al (2015) Epigenetic and in vivo comparison of diverse MSC sources reveals an endochondral signature for human hematopoietic niche formation. Blood 125:249–260
63. Reinisch A, Thomas D, Corces MR et al (2016) A humanized bone marrow ossicle xenotransplantation model enables improved engraftment of healthy and leukemic human hematopoietic cells. Nat Med 22:812–821
64. Kumar S, Geiger H (2017) HSC niche biology and HSC expansion Ex Vivo. Trends Mol Med 23:799–819
65. Walasek MA, van Os R, de Haan G (2012) Hematopoietic stem cell expansion: challenges and opportunities. Ann N Y Acad Sci 1266:138–150
66. Csaszar E, Kirouac DC, Yu M et al (2012) Rapid expansion of human hematopoietic stem cells by automated control of inhibitory feedback signaling. Cell Stem Cell 10:218–229
67. Fares I, Chagraoui J, Gareau Y et al (2014) Pyrimidoindole derivatives are agonists of human hematopoietic stem cell self-renewal. Science 345:1509–1512
68. Takagi M (2005) Cell processing engineering for ex-vivo expansion of hematopoietic cells. J Biosci Bioeng 99:189–196
69. Cook MM, Futrega K, Osiecki M et al (2012) Micromarrows – three-dimensional coculture of hematopoietic stem cells and mesenchymal stromal cells. Tissue Eng Part C Methods 18:319–328
70. Leisten I, Kramann R, Ventura Ferreira MS et al (2012) 3D co-culture of hematopoietic stem and progenitor cells and mesenchymal stem cells in collagen scaffolds as a model of the hematopoietic niche. Biomaterials 33:1736–1747
71. Ferreira MS, Jahnen-Dechent W, Labude N et al (2012) Cord blood-hematopoietic stem cell expansion in 3D fibrin scaffolds with stromal support. Biomaterials 33:6987–6997
72. Nichols JE, Cortiella J, Lee J et al (2009) In vitro analog of human bone marrow from 3D scaffolds with biomimetic inverted colloidal crystal geometry. Biomaterials 30:1071–1079
73. Huang X, Zhu B, Wang X et al (2016) Three-dimensional co-culture of mesenchymal stromal cells and differentiated osteoblasts on human bio-derived bone scaffolds supports active multi-lineage hematopoiesis in vitro: functional implication of the biomimetic HSC niche. Int J Mol Med 38:1141–1151
74. Sieber S, Wirth L, Cavak N et al (2017) Bone marrow-on-a-chip: long-term culture of human haematopoietic stem cells in a three-dimensional microfluidic environment. J Tissue Eng Regen Med 12:479–489
75. Torisawa Y, Mammoto T, Jiang E et al (2016) Modeling hematopoiesis and responses to radiation countermeasures in a bone marrow-on-a-chip. Tissue Eng Part C Methods 22:509–515
76. Singh VK, Romaine PL, Newman VL (2015) Biologics as countermeasures for acute radiation syndrome: where are we now? Expert Opin Biol Ther 15:465–471
77. Singh VK, Newman VL, Berg AN et al (2015) Animal models for acute radiation syndrome drug discovery. Expert Opin Drug Discov 10:497–517

78. Cary LH, Ngudiankama BF, Salber RE et al (2012) Efficacy of radiation countermeasures depends on radiation quality. Radiat Res 177:663–675
79. Guinan EC, Barbon CM, Kalish LA et al (2011) Bactericidal/permeability-increasing protein (rBPI21) and fluoroquinolone mitigate radiation-induced bone marrow aplasia and death. Sci Transl Med 3:110ra118
80. Gagliano O, Elvassore N, Luni C (2016) Microfluidic technology enhances the potential of human pluripotent stem cells. Biochem Biophys Res Commun 473:683–687
81. Giobbe GG, Michielin F, Luni C et al (2015) Functional differentiation of human pluripotent stem cells on a chip. Nat Methods 12:637–640
82. Wang G, McCain ML, Yang L et al (2014) Modeling the mitochondrial cardiomyopathy of Barth syndrome with induced pluripotent stem cell and heart-on-chip technologies. Nat Med 20:616–623
83. Kim C (2015) iPSC technology-powerful hand for disease modeling and therapeutic screen. BMB Rep 48:256–265
84. Benam KH, Dauth S, Hassell B et al (2015) Engineered in vitro disease models. Annu Rev Pathol 10:195–262
85. Thon JN, Mazutis L, Wu S et al (2014) Platelet bioreactor-on-a-chip. Blood 124:1857–1867
86. Di Buduo CA, Wray LS, Tozzi L et al (2015) Programmable 3D silk bone marrow niche for platelet generation ex vivo and modeling of megakaryopoiesis pathologies. Blood 125:2254–2264
87. Avanzi MP, Mitchell WB (2014) Ex vivo production of platelets from stem cells. Br J Haematol 165:237–247
88. Nakamura S, Takayama N, Hirata S et al (2014) Expandable megakaryocyte cell lines enable clinically applicable generation of platelets from human induced pluripotent stem cells. Cell Stem Cell 14:535–548
89. Smith BW, Murphy GJ (2014) Stem cells, megakaryocytes, and platelets. Curr Opin Hematol 21:430–437
90. Lee HJ, Li N, Evans SM et al (2013) Biomechanical force in blood development: extrinsic physical cues drive pro-hematopoietic signaling. Differentiation 86:92–103
91. North TE, Goessling W, Peeters M et al (2009) Hematopoietic stem cell development is dependent on blood flow. Cell 137:736–748
92. Diaz MF, Li N, Lee HJ et al (2015) Biomechanical forces promote blood development through prostaglandin E2 and the cAMP-PKA signaling axis. J Exp Med 212:665–680
93. Adamo L, Naveiras O, Wenzel PL et al (2009) Biomechanical forces promote embryonic haematopoiesis. Nature 459:1131–1135
94. Wahlster L, Daley GQ (2016) Progress towards generation of human haematopoietic stem cells. Nat Cell Biol 18:1111–1117
95. Sugimura R, Jha DK, Han A et al (2017) Haematopoietic stem and progenitor cells from human pluripotent stem cells. Nature 545:432–438
96. Esch MB, King TL, Shuler ML (2011) The role of body-on-a-chip devices in drug and toxicity studies. Annu Rev Biomed Eng 13:55–72
97. Imura Y, Sato K, Yoshimura E (2010) Micro total bioassay system for ingested substances: assessment of intestinal absorption, hepatic metabolism, and bioactivity. Anal Chem 82:9983–9988
98. Kimura H, Ikeda T, Nakayama H et al (2015) An on-chip small intestine-liver model for pharmacokinetic studies. J Lab Autom 20:265–273
99. Maschmeyer I, Lorenz AK, Schimek K et al (2015) A four-organ-chip for interconnected long-term co-culture of human intestine, liver, skin and kidney equivalents. Lab Chip 15:2688–2699
100. Satoh T, Sugiura S, Shin K et al (2017) A multi-throughput multi-organ-on-a-chip system on a plate formatted pneumatic pressure-driven medium circulation platform. Lab Chip 18:115–125

Chapter 9
In Vitro Tissue Construction for Organ-on-a-Chip Applications

Yuya Morimoto, Nobuhito Mori, and Shoji Takeuchi

Abstract Functional living tissues, reconstructed in vitro, have been demanded as grafts for regenerative medicine and drug test models for pharmacokinetic study. For the reconstruction of the functional living tissues, cellular structures with the shapes of point, line and plane are attractive building blocks to form macroscopic cellular tissues. Microfluidics are suitable techniques in the preparation of the cellular structures with design flexibility and high productivity. Owing to their shape controllability, the cellular structures are able to be assembled into macroscopic tissues in various dimensions using microfluidic devices. Furthermore, the integration of in vitro constructed cellular tissues and microfluidic devices provides organ-on-a-chips to evaluate the change of cellular functions in response to applied drugs. This chapter introduces microfluidic fabrication methods for cellular structures with the shapes of point, line and plane, and methods to manipulate and assemble the cellular structures for formation of macroscopic cellular tissues. Furthermore, we show that organ-on-a-chip, combination of the cellular tissues and microfluidic devices, can find applications in biological studies and the analysis of pharmacokinetics and toxicology.

Keywords Microfluidics · Biofabrication · BioMEMS · Tissue manipulation · Functional analysis

9.1 Introduction

There is a need for in vitro reconstruction of functional living tissues as transplantable grafts for regenerative medicine and as an alternative to animal pharmacokinetic models [1, 2]. To reconstruct functional living tissues, tissue engineering methods are necessary because cells cannot form three-dimensional (3D) structures spontaneously when cultured on a culture dish. To assemble cells into 3D structures,

Y. Morimoto · N. Mori · S. Takeuchi (✉)
Institute of Industrial Science, the University of Tokyo, Tokyo, Japan
e-mail: takeuchi@iis.u-tokyo.ac.jp

M. Tokeshi (ed.), *Applications of Microfluidic Systems in Biology and Medicine*, Bioanalysis 7, https://doi.org/10.1007/978-981-13-6229-3_9

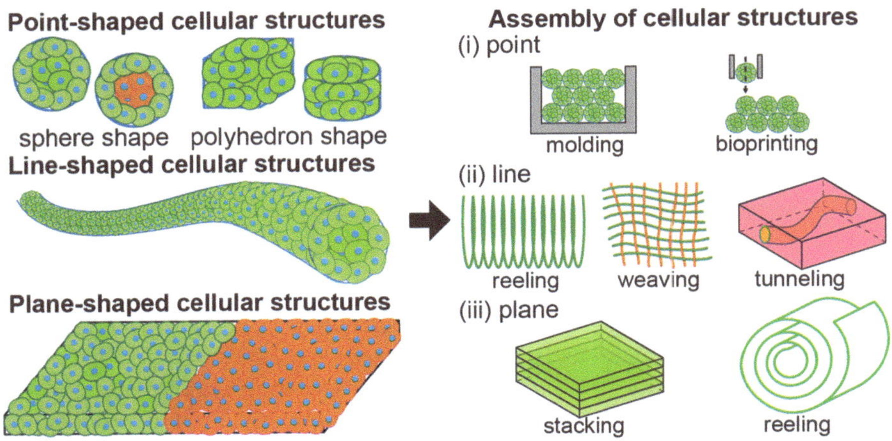

Fig. 9.1 Illustration of point-, line-, and plane-shaped cellular structures fabricated by microfluidic techniques. Microfluidic methods facilitate mass and reproducible construction of cellular structures, which are then available as building modules to form macroscopic cellular structures. (Figures are reprinted with permission from [7], ©2015 Elsevier)

cultures of cells in 3D biocompatible scaffolds are required [3, 4]. Hydrogels, especially the extracellular matrix (ECM), are commonly used as a material for 3D biocompatible scaffolds [5, 6]. The hydrogel scaffolds enable cells to adhere onto their surfaces and/or be embedded as cellular structures.

The cellular structures are categorized into three types according to their shapes: point-, line-, and plane-shaped cellular structures (Fig. 9.1) [7]. For the preparation of the various-shaped cellular structures, microfluidic devices are well suited because they can manipulate hydrogel solutions in a controlled manner by providing design flexibility and high productivity [8–10]. Moreover, microfluidic devices allow for the cell-laden hydrogel structures to be assembled into macroscopic cellular structures. The fabricated cellular structures are available as transplantable grafts for treatment of endocrine disease and repair of defective sites.

Furthermore, by integration of the cellular structures with microfluidic and microfabricated devices, we can fabricate organ-on-a-chips with in vitro constructed cellular structures [11, 12]. The organ-on-a-chips allow us to measure mechanical functions of the cellular structures such as contractile force and permeability; the chips facilitate the evaluation of cellular functions that are difficult to measure using common biological methods. Therefore, the organ-on-a-chip achieves in vitro drug testing by the evaluation of cellular functions depending on the types and concentrations of applied drugs.

In this chapter, we present an overview of (i) microfluidic fabrication methods for point-, line-, and plane-shaped cellular structures; (ii) methods for manipulating cellular structures for assembly into macroscopic cellular tissues; and (iii) organ-on-a-chip applications of the cellular structures.

9.2 Microfluidic Fabrication of Cellular Structures

For the fabrication of various-shaped cellular structures, hydrogel scaffolds with shapes corresponding to shapes of the cellular structures are necessary. Microfluidic devices allow for the flow control of hydrogel solutions, resulting in the production of hydrogel scaffolds with various configurations such as point, line, and plane shapes. In this section, we introduce microfluidic devices for the formation of hydrogel scaffolds according to the classification of each shape, and also describe the resulting cellular structures by culturing cells on and/or in the hydrogel scaffolds.

9.2.1 Point-Shaped Cellular Structures

Point-shaped cellular structures are defined as cellular beads in spherical shape and cellular blocks in a polyhedron shape. A simple fabrication method of cellular beads is the formation of cellular spheroids by aggregating cells. Because most kinds of cells tend to aggregate without their adhesion to the culture substrate [13], the size of the cellular spheroids is controllable depending on the number of aggregated cells. Therefore, the production of uniform cellular spheroids is achieved by enclosing the same number of cells in hanging-droplets [14, 15] and microwells [16, 17]. By combination with microfluidic devices, the hanging-droplets and the microwells allow the formation of cellular spheroids in a high-throughput manner, and addition of drugs to cellular spheroids and collection of secretions from cellular spheroids [18, 19]. In addition, microfluidic devices achieve aggregation of cells without enclosing cells by cell manipulations with external force [20, 21]. Usage of micro-rotational flows and magneto-Archimedes effect in a paramagnetic medium succeeded in collecting cells for the production of cellular spheroids of a desired size (Fig. 9.2). Thus, microfluidic devices are a useful tool for the mass formation of size controlled spheroids and biological evaluation by addition of reagents.

Cellular beads are also constructed by cell culture with hydrogel beads. In cell-laden hydrogel beads, their sizes are determined according to the size of the hydrogel beads. In addition, ECM can be used for the hydrogel, allowing the cells to be cultured with an appropriate culture condition. For the formation of size-controlled hydrogel beads, quasi-two-dimensional (2D) microchannels, such as T-shaped and flow-focusing channels, have been widely used [22–24]. Both microchannels allow for the formation of hydrogel droplets with high uniformity and production rates by infusing oil and a hydrogel solution. After gelation of the hydrogel droplets, hydrogel beads are formed (Fig. 9.3a, b). In this process, microfluidic devices allow for the size control of the hydrogel beads depending on the configurations of the flow channels and the flow velocities. However, the wettability of the channel surface can influence the possibility of droplet formation in the quasi-2D microchannels, and the formation of hydrogel droplets becomes difficult when the channel surface is hydrophilic. In addition, even when the channel surface is

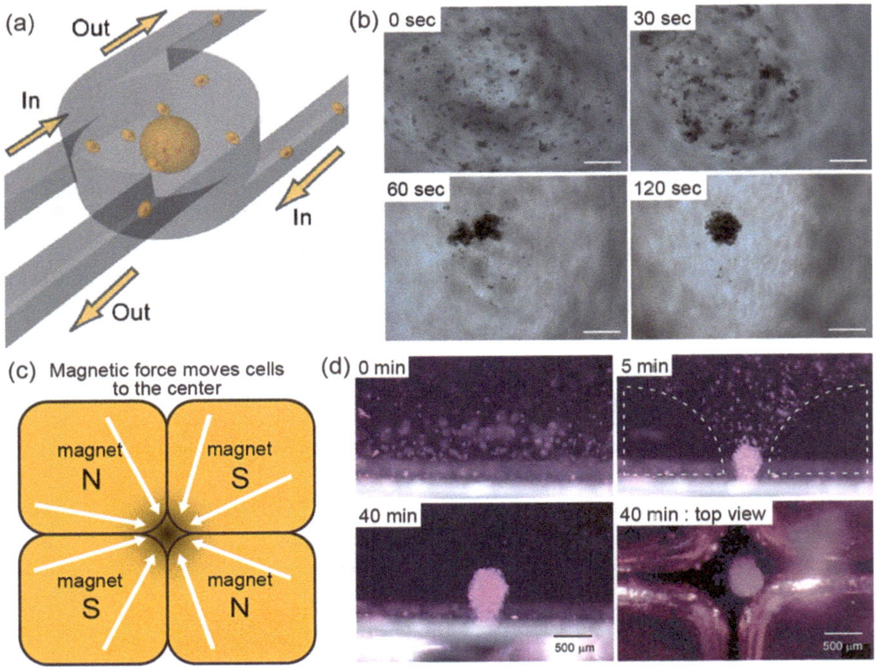

Fig. 9.2 (**a**) Conceptual illustration of cellular spheroid formation using micro-rotational flows. When cells suspended in culture medium are introduced into the device, the cells aggregate into a cellular spheroid. (**b**) Sequential images of the cellular spheroid formation using micro-rotational flows. After cell rotation occurs throughout the entire area with high-speed flow, cells start aggregating in the center of the device (~120 s) when the flow speed is reduced. Scale bars, 200 μm. (**c**) Conceptual illustration of cellular spheroid formation using the magneto-Archimedes effect. A magnetic force acts on cells within the paramagnetic medium, and then the cells aggregate to areas with a lower magnetic flux density. (**d**) Sequential images of cellular spheroid formation using the magneto-Archimedes effect. Cells gradually aggregate in the gap of the magnetic array. (Images are reprinted with (**a**, **b**) permission from [20], ©2010 Elsevier, and (**d**) permission from [21], ©2011 AIP Publishing LLC)

hydrophobic, the formation of droplets is sometimes difficult due to the adsorption of proteins and saccharides onto the channel surfaces in cases of extended usage.

Some 3D microchannels, such as the axisymmetric flow-focusing devices (AFFDs) and capillary microchannels, have been proposed as solutions for the wetting problem on the channel surface. In the AFFD, hydrogel droplets are prepared without contacts onto channel surfaces because a hydrogel solution is covered with oil (Fig. 9.3c) [25–28]. The sizes of the hydrogel beads due to the gelation of hydrogel droplets are controlled by the flow velocities and dimensions of the AFFDs, similar to the quasi-2D microchannels. Additionally, in the formation of hydrogel beads using a capillary microchannel, hydrogel droplets formed at the microchannel tip are gelated by direct injection into a gelling agent [29–31]. This method does not require pumps for liquid delivery, in contrast to AFFDs and quasi-2D microchannels. The infusion of hydrogel solution by centrifugation allows for

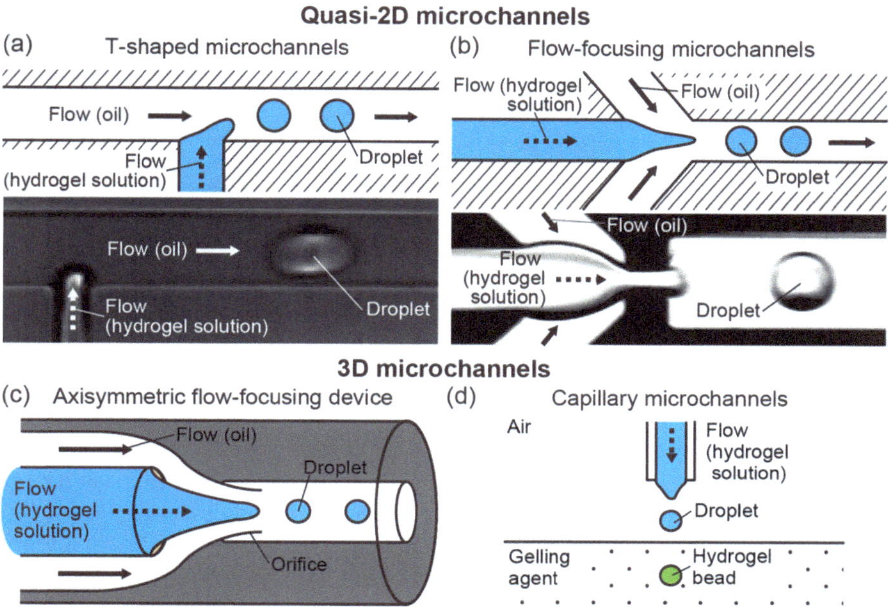

Fig. 9.3 (**a**, **b**) Schematic illustration of quasi-2D microchannels for the production of hydrogel beads. (**a**) T-shaped microchannels and (**b**) flow-focusing microchannels. (**c**, **d**) Schematic illustration of 3D microchannels for the production of hydrogel beads. (**c**) Axisymmetric flow-focusing device (AFFD) and (**d**) capillary microchannels

the formation of hydrogel beads with a uniform diameter (Fig. 9.3d). By using capillary microchannels, the compartmentalization of hydrogel beads is also achieved by the placement of multiple microchannels (i.e.: θ tube geometry) at the tip as well as through the infusion of different solutions [32]. In this method, since hydrogel beads can be formed easily without oil, it is possible to use them as culture scaffolds without any damage to the cells due to residual oil.

Depending on the microfluidic methods used for the formation of the hydrogel beads and gelation mechanisms of the hydrogel, the types of constructible hydrogel beads are determined. Favorable gelation mechanism used in microfluidic methods are ionic cross-linking used in alginate gel formation, inherent phase transition by heat used in collagen and gelatin gel formation, and reactions with photoreactive crosslinkers used in synthetic polymers such as polyethylene glycol (PEG). The AFFDs and the quasi-2D microchannels exhibit good affinity with all types of gelation mechanisms, since gelation stably occurs in oil. As a result, these microfluidic devices allow for the formation of alginate [22, 26, 27], PEG [23, 24], and collagen gel beads [33]. On the other hand, capillary microchannels are also available for the production of hydrogel beads by ionic cross-linking, and are not available for the formation of hydrogel under inherent phase transition due to the slow speed of the gelation. Therefore, the capillary microchannels have been widely used for the production of alginate gel beads.

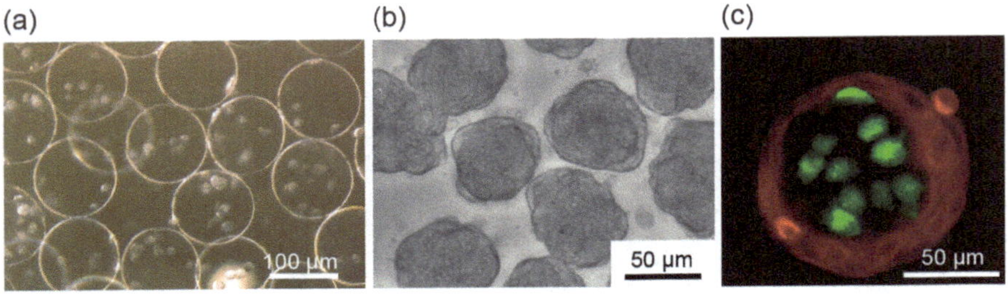

Fig. 9.4 Cellular beads fabricated from hydrogel beads. (**a**) Cell-encapsulated alginate gel beads. (**b**) Cellular beads fabricated by culturing cells on collagen beads. (**c**) The 3D hierarchic co-culture beads, where the inner cells (green) are dermal fibroblasts and outer cells (red) are epidermal keratinocytes, are fabricated by culturing the different cells with collagen beads. (Images are reprinted with (**a**) permission from [22], ©2007 John Wiley and Sons, (**b**) permission from [33], ©2011 John Wiley and Sons, and (**c**) permission from [34], ©2012 John Wiley and Sons)

Cell-laden hydrogel beads can be constructed with cells cultured on the hydrogel beads or in cell-encapsulated hydrogel beads formed by the addition of cells into the hydrogel solution before gelation. Owing to the size uniformity of the hydrogel beads, uniform cell-encapsulated alginate gel beads and cell-adhered collagen gel beads were constructed (Fig. 9.4a, b) [22, 33]. Moreover, seeding another type of cell on cell-encapsulated hydrogel beads can be performed to construct 3D hierarchical co-culture in the beads. This method has been used for the construction of fibroblast-hepatocyte co-culture beads and dermal fibroblast-epidermal keratinocyte co-culture beads (Fig. 9.4c) [34]. In both co-culture beads, cellular functions were improved due to cell-cell interactions, indicating that the 3D hierarchical co-cultured beads can be used as high-throughput experimental tools for pathological and physiological studies under 3D co-culture conditions.

Cellular blocks in polyhedron shape, another type of point-shaped cellular structure, have also been fabricated in microfluidic devices for high-throughput preparation (Fig. 9.5) [35, 36]. Light exposure to hydrogel solution containing a photoreactive crosslinking agent in the microfluidic devices, so-called flow-lithography, allows for the continuous formation of cellular blocks, the shape of which is determined according to exposure patterns controlled with photomasks or digital mirror devices. By using flow-lithography, cell-encapsulated PEG blocks can be produced in a continuous flow by controlling their shape via columns, triangle poles, and square poles.

9.2.2 Line-Shaped Cellular Structures

Cellular fibers and tubes, which are line-shaped cellular structures, are constructed by cells cultured with hydrogel fibers or tubes. Hydrogel fibers and tubes easily formed from cylindrical flows of hydrogel solution in microfluidic devices. For the

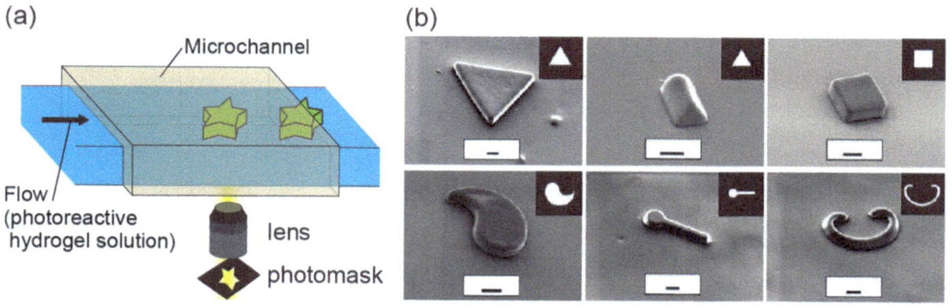

Fig. 9.5 (**a**) Illustration of the fabrication process for polyhedron-shaped cellular structures. (**b**) Images of fabricated cellular blocks with various shapes. The shapes of the cellular blocks were controlled by the shapes of the exposure patterns. Scale bars, 10 μm. (Images are reprinted with (**b**) permission from [35], ©2006 Nature Publishing Group)

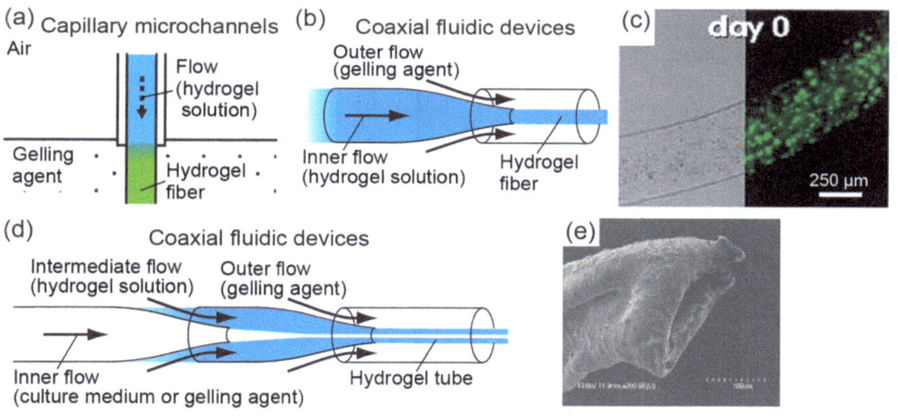

Fig. 9.6 (**a**, **b**) Fabrication process for hydrogel fibers using (**a**) capillary microchannels and (**b**) coaxial fluidic devices. (**c**) Cell-laden alginate gel fiber fabricated by the extrusion method. Green shows living cells and red shows dead cells. (**d**) Fabrication process for hydrogel tubes using coaxial fluidic devices. (**e**) Scanning electron microscopy image of an alginate gel tube. (Images are reprinted with (**c**) permission from [38], ©2011 Royal Society of Chemistry, and (**e**) permission from [43], ©2009 John Wiley and Sons)

formation of hydrogel fibers, the extrusion of hydrogel solution flows from capillary microchannels have been widely used [37–39]. By introducing the flows into a gelling agent, hydrogel fibers are constructed (Fig. 9.6a). In other fabrication methods for hydrogel fibers, microfluidic devices allow for the formation of coaxial flows with the hydrogel solution as the inner flow and gelling agent as the outer flow, whereby the hydrogel solution is gelated in the coaxial flows, and a hydrogel fiber is formed (Fig. 9.6b) [40–42]. Since the diameters of the hydrogel fibers are controllable without length limitations depending on the configuration of the microfluidic channels and the flow rate ratio, cellular fibers can be prepared with various

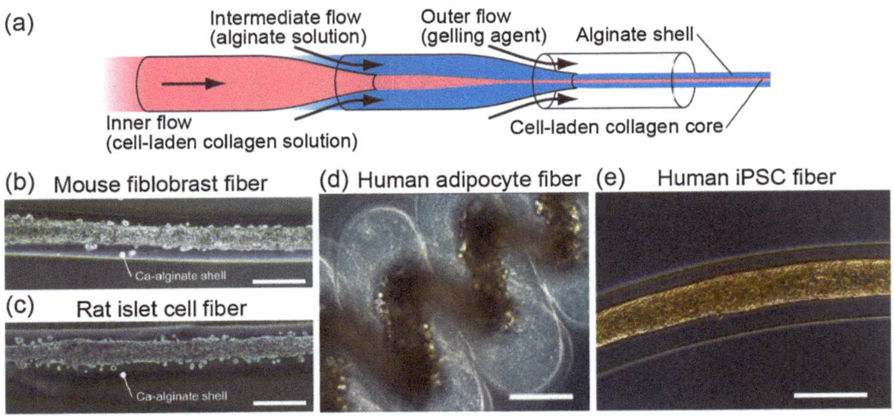

Fig. 9.7 (**a**) Fabrication process for core-shell hydrogel fibers with cell-laden collagen. (**b–e**) Images of core-shell hydrogel fibers using (**b**) mouse fibroblasts, (**c**) rat islet cells, (**d**) human adipocytes, and (**e**) human induced pluripotent stem cells (iPSCs). Scale bars, 200 μm. (Images are reprinted with (**b**, **c**) permission from [44], ©2013 Nature Publishing Group, (**d**) permission from [45], ©2015 John Wiley and Sons, and (**e**) permission from [47], 2017 licensed under Creative Commons Attribution 4.0 International License)

dimensions by culturing cells with the hydrogel fiber (Fig. 9.6c) [38]. On the other hand, hydrogel tubes are formed by using microfluidic coaxial flows with gelling agent or culture solution as the inner flow, hydrogel solution as the intermediate flow, and gelling agent as the outer flow (Fig. 9.6d). In the hydrogel tube formation, when cells are embedded in the hydrogel solution, cell-laden hydrogel tubes are constructed as a cellular tube (Fig. 9.6e) [43]. For example, the use of vascular endothelial cells enables the formation of cellular tissues. Dimensions of the cellular tubes are also variable according to the configuration of the microfluidic channels and the flow rate ratio.

The construction of fiber-shaped alginate gel structures that included cell-laden collagen solution at the core was achieved by microfluidic formation of coaxial flows with cell-laden collagen solution as the inner flow, alginate solution as the intermediate flow, and a calcium chloride solution, which is a gelling agent for alginate solution, as the outer flow (Fig. 9.7a) [44]. Here, owing to cover by the alginate gel structure, the gelation of collagen is possible regardless of the slow rate of gelation, allowing for a core-shell hydrogel fiber with cell-laden collagen as the core and alginate gel as the shell to be constructed. After the cells were cultured with the fiber, cellular adhesion to each other in the collagen to form a cellular tissue resulted in the formation of a core-shell cellular fiber. Because cells were cultured in collagen having high cell compatibility, line-shaped cellular tissues were successfully constructed using various kinds of cells such as fibroblasts [44], islet cells [44], adipocytes [45], smooth muscle cells [46] and human induced pluripotent stem cells (iPSCs) [47] (Fig. 9.7b–e).

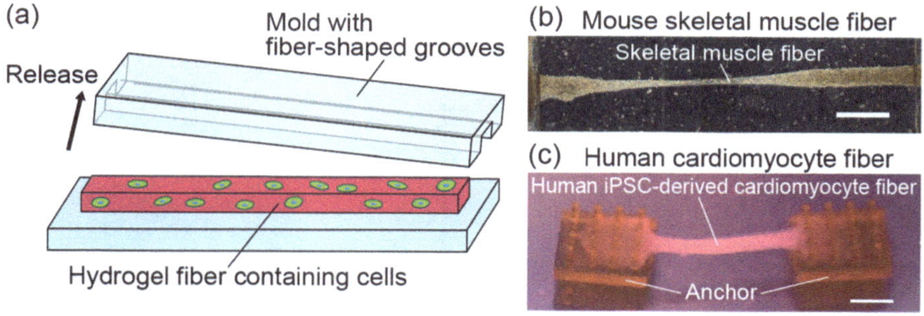

Fig. 9.8 (**a**) Fabrication process for hydrogel fibers containing cells by using a mold with fiber-shaped grooves. (**b**, **c**) Cellular fibers constructed by culturing cells in the hydrogel fibers. (**b**) Mouse skeletal muscle fiber, and (**c**) Human iPSC-derived cardiomyocyte fiber. Scale bars, 1 mm. (Images are reprinted with (**b**) permission from [48], ©2013 Elsevier, (**c**) Reprinted from [49], ©2016 Royal Society of Chemistry)

Cellular fibers can also be formed by patterning cell-laden hydrogel structures using molds with fiber-shaped grooves (Fig. 9.8a) [48, 49]. In this method, the dimensions of the cellular fibers are controlled by the geometries of the grooves in the molds. In addition, this method allows for the integration of cellular fibers with artificial components when cell-laden hydrogel structures are patterned in state of placement of the components in the hydrogel solution. In the case of embedding anchors at the ends of the hydrogel structures, the cellular fiber can be cultured without shrinkage because the anchors maintain its shape. Therefore, the cellular fiber with anchors are well suited for the formation of muscle tissues, which have strong traction force. As a result, by patterning skeletal myoblast-laden matrigel structures or human iPSC-derived cardiomyocyte-laden collagen structures with the molds and placing anchors at the ends of the structures, skeletal muscle fibers and human iPSC-derived cardiomyocyte fibers can be constructed without deformation caused by their traction force (Fig. 9.8b, c).

9.2.3 *Plane-Shaped Cellular Structures*

As a method for constructing plane-shaped cellular structures, cells cultured on temperature-responsive culture dishes have been widely used [50, 51]. In this case, cells adhered onto the culture dish are released when the dish is cooled. Due to its characteristics, by cooling the dish with the cells in a confluent state, a cellular sheet is obtained. In addition, by patterning the cells on the culture dish, a cellular sheet with a heterogeneous arrangement of cells is constructed. Although temperature-responsive culture dishes are a useful tool to form cellular sheets, their productivity and design flexibility are limited to the number and size of the dish.

To increase the productivity and design flexibility of cellular sheets, the use of microfluidic devices is an effective means. Microfluidic devices with a flat channel

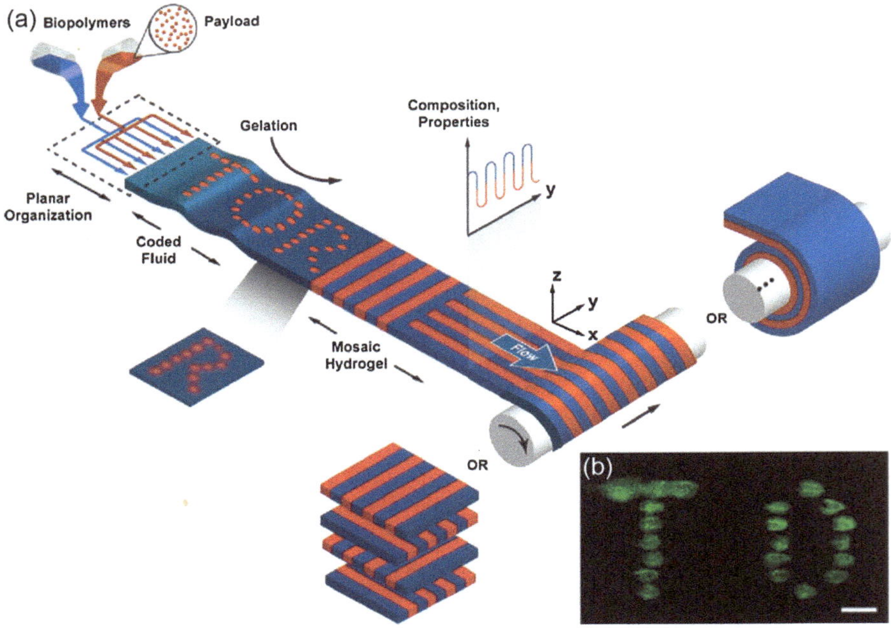

Fig. 9.9 (**a**) Conceptual illustration of the fabrication process for cell-laden hydrogel sheets using microfluidic devices. (**b**) Fluorescent image of patterned cardiomyocytes in an alginate gel-collagen sheet. Scale bar, 2 mm. (Images are reprinted with (**a**, **b**) permission from [52], ©2012 John Wiley and Sons)

allow for a planar flow of cell-laden hydrogel solution and contact to a gelling agent (Fig. 9.9). Thus, after gelation of the hydrogel solution flow, cell-laden hydrogel sheets are constructed continuously without length limitations [52]. Moreover, the microfluidic devices enable control for the placement of different cells in the planar flow, resulting in the formation of cell-laden hydrogel sheets with arrangement of different cells [53].

Furthermore, plane-shaped cellular structures are also constructed by using a special culture frame with anchors. When culturing a cell-laden hydrogel structures in the frame to fix the periphery of the structure, shrinkage of the hydrogel structures due to cell traction force could be prevented, and plane-shaped cellular structures could be constructed (Fig. 9.10) [54]. Although it is difficult to release the cellular structures from the frame, this method allows for the integration of the cellular structures with external machines such as a pump via the frame. Therefore, by making tube connectors as anchors on the frame and connecting pumps with vessel-like structures, perfusion cultures of the plane-shaped cellular structures are now available. This method is appropriate for the construction of skin tissues requiring it to be cultured at an air-liquid interface because the nutrients are supplied by the perfusion culture.

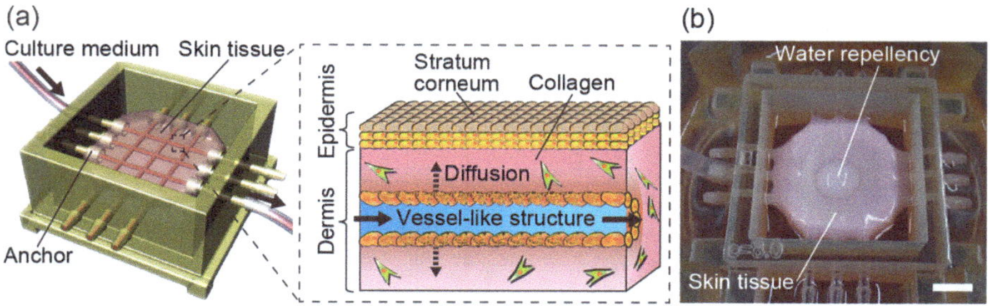

Fig. 9.10 (**a**) Conceptual illustration of the fabrication process for plane-shaped structures fixed with anchors. (**b**) Skin tissue with vessel-like structure. After perfusion culture, the skin tissue repelled water. Scale bar, 5 mm. (Images are reprinted with (**a**, **b**) permission from [54], ©2017 Elsevier)

9.3 Macroscopic Tissue Engineering by Assembling Cellular Structures

The above-mentioned various-shaped cellular structures are attractive modules for reconstructing organomimetic and homogeneous dense macroscopic cellular tissues. When assembling the cellular structures, microfluidic techniques facilitate the precise control of cell arrangements by using appropriate methods depending on the shapes of the cellular structures. In this section, we introduce the microfluidic techniques used to assemble cellular structures according to the classification of each shape.

9.3.1 Assembling Point-Shaped Cellular Structures

In the assembly of point-shaped cellular structures, especially cellular beads, the molding method has been widely used [33, 55–57]. By culturing cellular beads packed in a mold, the cellular beads are integrated by cell adhesion, resulting in the construction of macroscopic cellular tissues. In this state, since nutrients and oxygen can be supplied from the gap between the cellular beads, and central necrosis due to the depletion of nutrients and oxygen is prevented in the cellular tissues. In addition, shapes of the fabricated cellular tissues are transcribed from the mold shapes. Therefore, using various shaped molds, the molding method can be used for the construction of a millimeter to centimeter-sized cellular tissues with a human-doll shape, mouse-brain shape, or lotus root shape (Fig. 9.11a–c).

To arrange the cellular beads in the cellular tissues, bioprinting is a promising method for the construction of macroscopic cellular tissues by assembling cellular beads. In bioprinting, printers or liquid dispensers eject and stack the cellular beads [33, 58, 59]. In this process, cellular beads can be assembled by selecting a type and

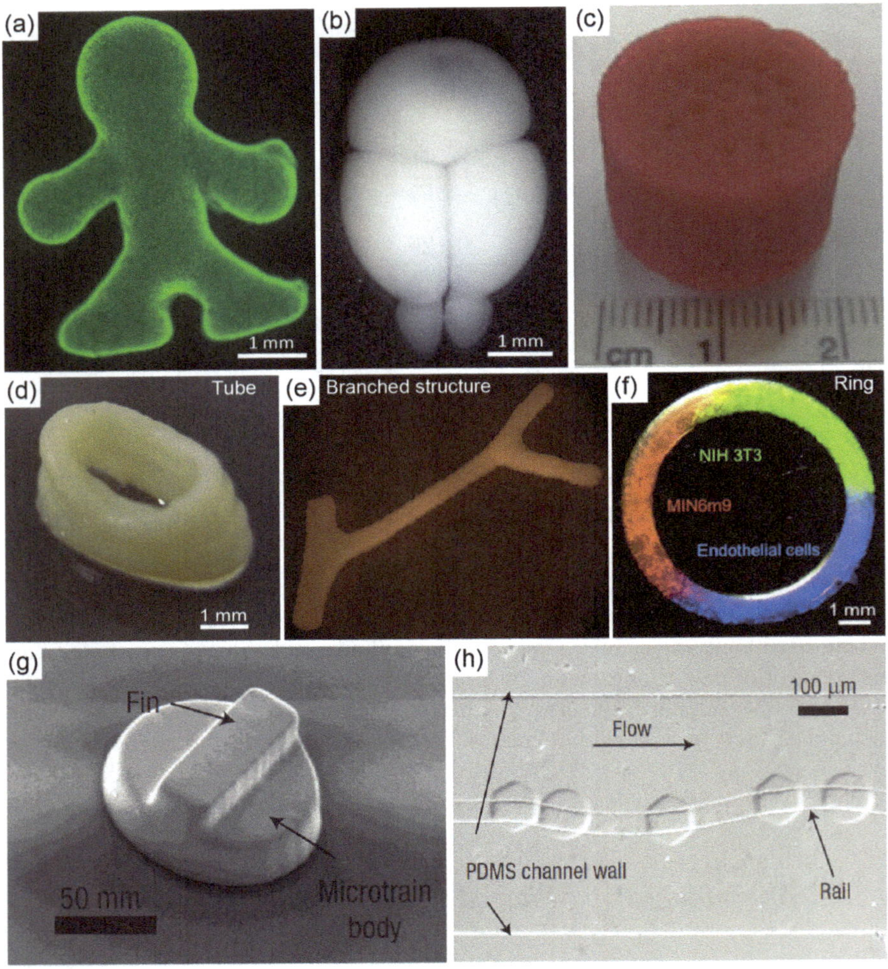

Fig. 9.11 (**a–c**) Cellular tissues fabricated by the molding method. (**a**) Fluorescent image of a human doll-shaped cellular tissue made of fibroblast-laden collagen beads. Green shows living cells and red shows dead cells. (**b**) Mouse brain-shaped cellular tissue made of neurospheres. (**c**) Lotus root-shaped cellular tissue made from fibroblast-laden cellulose beads. (**d–f**) Cellular tissue fabricated by bioprinting. (**d**) Tube-shaped tissue made of fibroblast-laden collagen beads. (**e**) Branched cellular tissue made from human skin cellular spheroids. (**f**) Ring-shaped cellular tissue made of collagen beads containing various typed cells. (**g**) Scanning electron microscopy image of the convex block. (**h**) Guided motion of the convex block along the concave rail. (Images are reprinted with (**a, d, f**) permission from [33], ©2013 John Wiley and Sons, (**c**) permission from [57], ©2012 John Wiley and Sons, (**e**) permission from [58], ©2009 Elsevier, and (**g, h**) permission from [62], ©2008 Nature Publishing Group)

placing them at arbitrary positions (Fig. 9.11d, e). Using these properties, ring-shaped tissues compartmentalized by cell types and tube-shaped tissues have been achieved with bioprinting (Fig. 9.11f).

To precisely control the arrangement of the cellular beads, microfluidic techniques capable of manipulating the cellular beads have been used. Since microfluidic devices are effective for manipulating cellular beads one by one, fluid force in microchannels can move cellular beads [60] or cell-laden PEGDA rings [61] to construct ordered cellular tissues. Alternatively, microchannels with concave rails have been proposed for assembling cell-laden convex PEGDA blocks since the concave rails can guide the motions of the convex blocks (Fig. 9.11g, h) [62]. By combining the rail networks and variable shapes of the PEGDA blocks, this approach can be used for the construction of complex-shaped and heterogeneous cellular structures.

9.3.2 Assembling Line-Shaped Cellular Structures

Reeling and weaving are suitable for the assembly of line-shaped cellular structures. A millimeter-sized cellular tissue can be easily constructed by reeling cellular fibers around a support such as a rod or plate (Fig. 9.12a–c) [39, 42–44]. In this state, reeling cellular fibers with different typed cells can provide 3D co-culture conditions in the cellular tissues (Fig. 9.12d). For assembling cellular fibers without the support, the cellular fibers can be weaved [44]. Since microfluidic operations allow for the manipulation of cellular fibers accurately, weaving the cellular fibers can be

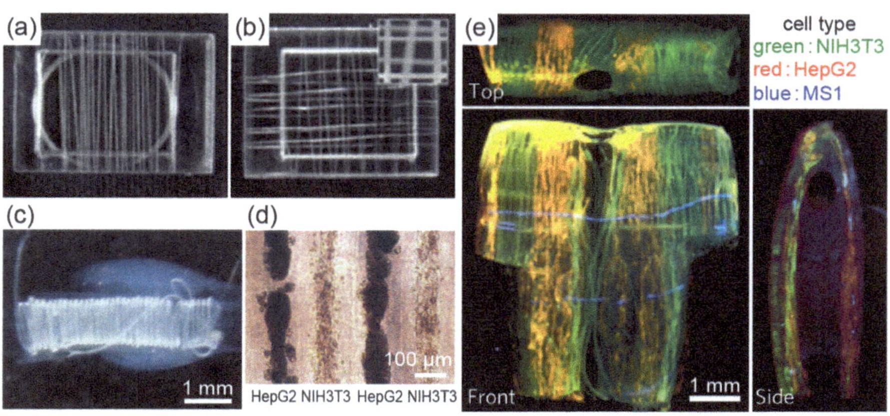

Fig. 9.12 (**a**, **b**) Reeled chitosan fibers with hepatocytes around plates. (**c**) Helical tube released from a rod after reeling it around the rod. (**d**) Co-cultured helical tube fabricated by reeling cellular fibers containing HepG2 cells and NIH3T3 cells. (**e**) T-shirt-shaped cellular tissues fabricated by weaving cellular fibers with fibroblasts, hepatocytes, and small lung carcinoma cells. (Images are reprinted with (**a**, **b**) permission from [39], ©2010 Royal Society of Chemistry, and (**c**–**e**) permission from [44], ©2013 Nature Publishing Group)

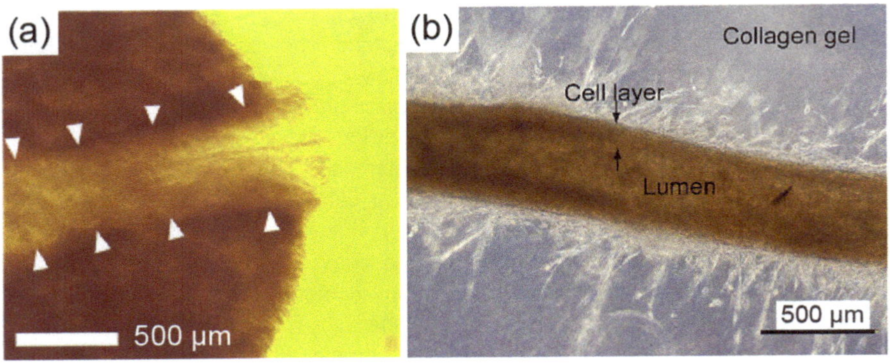

Fig. 9.13 (**a**) Formation of a channel (white arrow) fabricated by dissolving alginate gel after embedding a poly-L-lysine-coated alginate fiber in collagen gel containing human neonatal dermal fibroblasts. (**b**) Blood vessel-like structure fabricated by dissolving an alginate gel fiber with smooth muscle cells and endothelial cells in a collagen structure. (Images are reprinted with (**a**) permission from [63], ©2016 Elsevier, and (**b**) permission from [64], ©2008 American Chemical Society)

achieved by combining microfluidic channels and a loom for microsized fibers (Fig. 9.12e). Moreover, using cellular fibers with different typed cells, the method has been successfully used for the construction of centimeter-sized 3D cellular tissues with compartmentalization according to cell types.

Furthermore, the cellular fibers and cellular tubes can be used to form blood vessel-like structures for the supply of nutrients in macroscopic cellular tissues. It is well known that lumens can be formed in the cellular tissues by dissolving the hydrogel with enzymes after embedding hydrogel fibers or tubes in the tissues (Fig. 9.13a) [63]. When cellular fibers or tubes containing vascular endothelial cells are embedded and the hydrogel is dissolved, cell-covered lumens can be formed as blood vessel-like structures by cell adhesion to the luminal wall surface (Fig. 9.13b) [64]. Construction of blood vessel-like structures in macroscopic cellular tissues is important for maintaining the cell viability of the tissue for extended culture times.

9.3.3 Assembling Plane-Shaped Cellular Structures

For the assembly of plane-shaped cellular structures, stacking and reeling are suitable methods. Stacking cellular sheets have been widely used to produce macroscopic cellular tissues such as myocardial and liver [50, 51, 65]. In addition, by reeling overlapped cellular sheets with different types of cells, millimeter-sized vascular tubes hierarchized with vascular endothelial cells, smooth muscle cells, and fibroblasts have been successfully formed (Fig. 9.14) [53]. Thus, these methods are attractive for the construction of macroscopic cellular tissues with simple shapes.

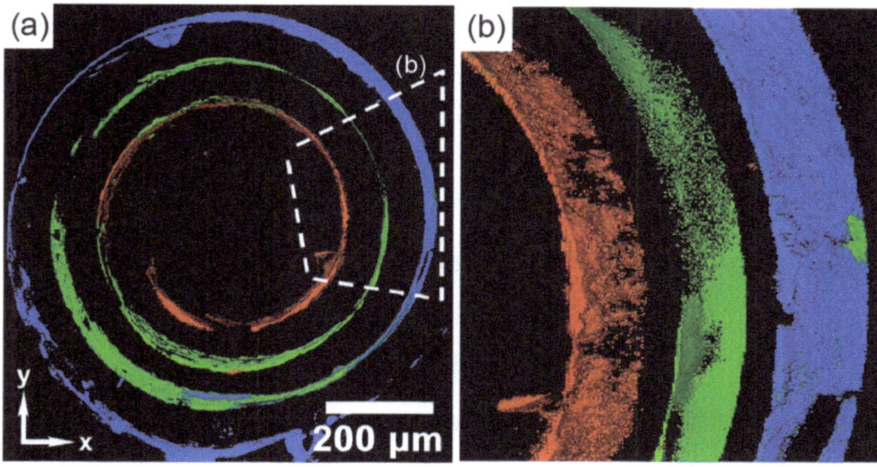

Fig. 9.14 3D macroscopic tubular tissues fabricated by reeling a cell-laden sheet with location control of various cell types. Red shows endothelial cells, green shows smooth muscle cells, and blue shows fibroblasts. (Images are reprinted with permission from [53], ©2012 John Wiley and Sons)

9.4 Organ-on-a-Chip Application

By integrating cellular tissues with microfluidic systems, devices referred to as "organ-on-a-chip" have been developed. Organ-on-a-chip systems mimic physiological conditions such as spatiotemporal chemical gradients, fluid shear stress, cyclic stretch, and interactions between various types of tissues and organs. Organ-on-a-chip systems are used not only as models of organs and pathology, but also as platforms for drug discovery since organ-on-a-chip systems can better emulate pharmacokinetics and toxicology rather than conventional 2D cultures. In this section, we introduce various organ-on-a-chip systems according to the microfluidic features.

9.4.1 Organ-on-a-Chip Based on Tissue-Trapping Structures

Many studies have integrated microfluidic devices with tissue-trapping structures such as microwells and dynamic microarrays to fabricate organ-on-a-chip systems. In these systems, cellular beads are able to be arrayed owing to the tissue-trapping structures, and are also suitable for high-throughput drug screening. As an example of a system utilizing microwells, a microfluidic device with a distributive microfluidic channel network and microwells designed to capture cancer cells and form cellular spheroids has been proposed (Fig. 9.15a) [66]. Using this device,

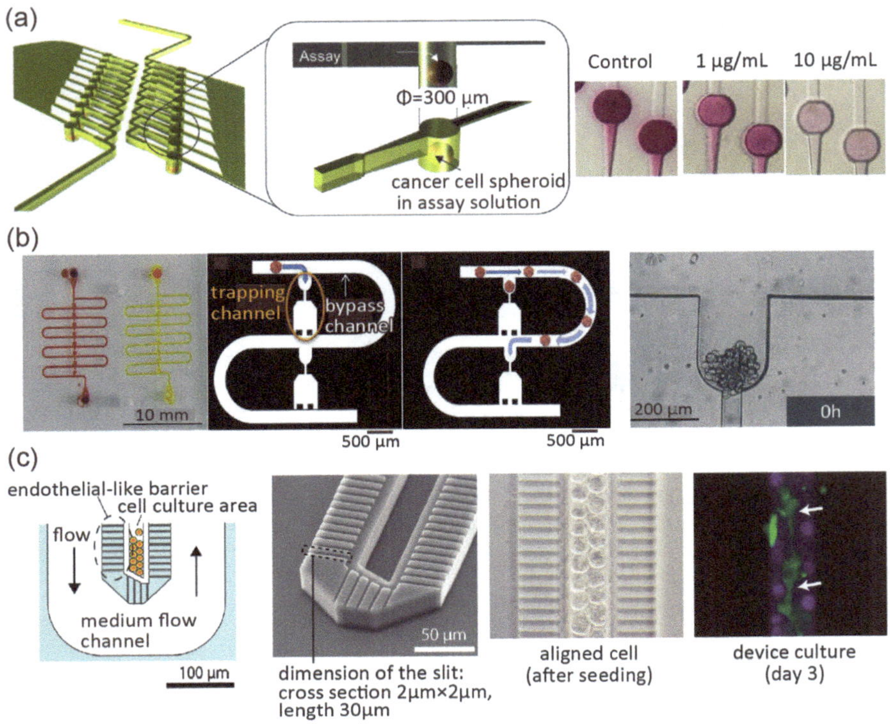

Fig. 9.15 Organ-on-a-chip systems based on a tissue-trapping structure. (**a**) System composed of a distributive microfluidic channel network and microwells designed to capture cancer spheroids. (**b**) Dynamic microarray device to trap cancer spheroids. (**c**) Liver-on-a-chip fabricated by trapping hepatocytes into two lines. (Images are reprinted with (**a**) permission from [66], ©2012 Royal Society of Chemistry, (**b**) permission from [68], 2014 licensed under Creative Commons Attribution 3.0 Unported Licence, and (**c**) permission from [69], ©2011 AIP Publishing LLC)

human breast cancer cells were distributed into 80 microwells via the microfluidic channel network and the spheroids trapped in the microwells were obtained. After formation of the spheroids, a cytotoxicity test of the anticancer drugs was performed by treating the spheroids with the drugs and performing a MTT (3-(4,5-dimethylthiazol-2-yl)-2,5-diphenyltetrazolium bromide) assay via the microfluidic channel network.

Dynamic microarray is also available to trap cellular spheroids (Fig. 9.15b) [67, 68]. The dynamic microarray is composed of a trapping channel and a bypass channel that enables trapping cellular spheroids one-by-one. Using dynamic microarray device, spheroids consisting of cancer cells were trapped and arrayed. The trapped spheroids were treated with anticancer drugs via the microchannel, and the viability of the spheroids was measured by confocal-based live/dead cell imaging [67], or by apoptosis assay based on caspase-3 activity with a normal plate reader [68]. These researches indicated that the methods to evaluate chemosensitivity can be chosen in the organ-on-a-chip systems based on dynamic microarray according to the situation; confocal imaging when spatiotemporal information is important, and apoptosis assay with plate reader when the simplicity of the procedure has high priority. The dynamic microarray was used not only for cellular spheroids but also for cell-laden hydrogel beads. For example, the dynamic microarray can trap skin cellular beads, which was composed of type-I collagen gel beads, embedded human fibroblasts, and coated by human epidermal keratinocytes [34]. Owing to the ability of continuous observation that is one of advantages of the dynamic microarray, the secretion of type-VII collagen from the skin cellular beads was analyzed with individual beads trapped. By analyzing a large amount of the beads at once using the dynamic microarray devices, this system might be a screening tool of skin drugs and cosmetics.

The trapping structure has also been used to align hepatocytes to form line-shaped liver tissues (Fig. 9.15c) [69]. The cell culture area was designed to trap and align hepatocytes into two lines, and was separated from the main perfusion channel by a wall containing a series of long narrow slits that mimics the endothelial barrier. The aligned hepatocytes formed bile canaliculi along the cell culture area, and the excretion function was evaluated by the CDFDA (5-(and-6)-carboxy-2′,7-′-dichloro-fluorescein diacetate) assay. Thus, by trapping tissues using the device with endothelial-like channels, microfluidic devices can be used for evaluating drug metabolism and excretion as an alternative to animal testing.

9.4.2 Organ-on-a-Chip Based on Membrane-Integrated Channel

In order to reconstruct epithelial structures and functions in vitro, arranging plane-shaped cellular structures in microfluidic channels provides a promising approach. As a simple method for the arrangement, porous membranes have been integrated into the microfluidic channels to separate layers of the channels. This membrane-integration approach has been widely adopted to fabricate various organ-on-a-chip systems, since most organs have epithelial tissues in vivo. The simplest way to integrate a porous membrane into the microfluidic device is to utilize commercially available cell culture inserts (Fig. 9.16a) [70, 71]. The device can be fabricated by standard soft lithography techniques and have compartments with a holder for the cell culture insert. These compartments support two culture modes: exposing tissue to medium flow, and shielding tissue from the underlying flow by the porous membrane of the cell culture insert. The latter mode is especially effective for the culture of skin tissue since the air-liquid interface is necessary for culturing and maintaining the epidermal layer. Owing to this advantage, co-cultures of skin biopsy and liver spheroid [70], and co-culture of reconstructed skin-equivalents and hair follicle [71] have been demonstrated. By co-culturing tissues, crosstalk between tissues can be established and thus improvements in their morphologies were

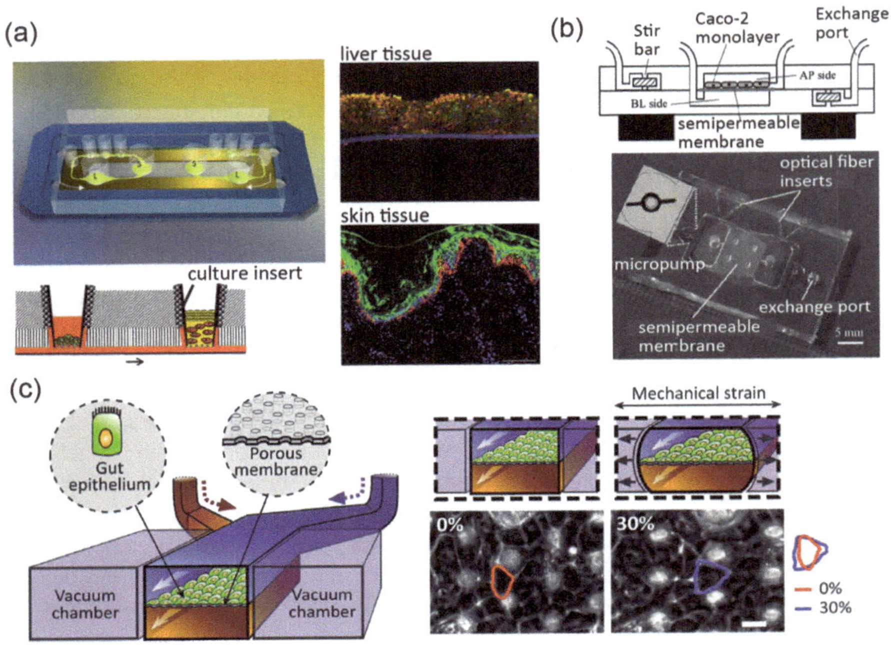

Fig. 9.16 Organ-on-a-chip system based on membrane-integrated microfluidic channels. (**a**) Skin and liver co-culture device integrated with cell culture inserts. (**b**) An intestinal model constructed in the device with microchannels separated by porous membrane. (**c**) Gut-on-a-chip system utilizing flexible porous membrane. (Images are reprinted with (**a**) permission from [70], ©2013 Royal Society of Chemistry, (**b**) permission from [72], ©2013 Royal Society of Chemistry, and (**c**) permission from [85], ©2012 Royal Society of Chemistry)

achieved. However, the extensibility of the membrane-integration using cell culture inserts is not high since the design of the microfluidic device is limited by the available size and shape of the cell culture inserts. In particular, it is difficult to set microfluidic channels on the top surface of the cell culture inserts due to the plastic walls of the membrane holder.

Bare porous membranes are used to realize more arbitrary designs than cell culture inserts. In this approach, the porous membranes are sandwiched by silicone rubber substrates with microfabricated channels. By seeding Caco-2 intestinal cells on the top surface of a semipermeable membrane separating the upper and lower microchannels, an intestinal model device was fabricated (Fig. 9.16b) [72]. The stir-bar-based micropumps in the device allowed for the perfusion of medium and the application of shear force on the cells. Furthermore, the polarized transport of rhodamine 123 from the basolateral side (lower side) to the apical side (upper side) was observed by optical fibers embedded in the device. This intestinal model device was extended to multi-organ-on-a-chip systems by adding liver and lung chambers inoculated with hepatic and lung cancer cells, respectively [73]. Orally administered or biologically active anticancer drugs were tested on the system.

Furthermore, the device named "HuMiX (human-microbial crosstalk)" used to emulate the gastrointestinal human-microbe interface was developed by extending an intestinal model [74]. The device had three layers of microfluidic channels. The first and second layers were separated by a nanoporous membrane whose top surface was inoculated with microbes, and the second and third layers were separated by a microporous membrane whose top surface was inoculated with Caco-2 cells. Owing to the layered channels, a gradient of oxygen and biomolecules was achieved. Besides the intestine, the membrane integrated in the microfluidic device can be applied for the construction of other organs such as the kidney [75, 76], blood-brain barrier [77], placenta [78], and skin [79, 80]. Owing to the microchannels separated by a porous membrane, renal inner medullary collecting duct cells on the membrane have been stimulated by fluid shear stress, and hormonal stimulation, osmotic gradient, and optimal culture conditions have been investigated [75, 76]. Similarly, the influence of the flow condition on the barrier function of the blood-brain barrier [77] and the formation of microvilli in the placenta [78] have also been investigated. Furthermore, transdermal drug permeation has been demonstrated by applying drugs on a skin model on a porous membrane and sampling the medium from the microchannel under the membrane [79].

A combination of several of the approaches stated above has also been proposed. For example, an on-chip model composed of intestinal, liver, and cancer cells has been successfully used in an assay for an anticancer drug [81]. In this device, the intestinal compartment was fabricated with a collagen-coated porous membrane, and the liver compartment was designed to trap hepatic cells cultured on carrier beads via a vertical microchannel. Similarly, a multi-organ-on-a-chip system composed of intestinal, skin, liver, and kidney equivalents has been reported [82]. In this device, the kidney compartment was constructed by seeding RPTEC/TERT-1, a proximal tubule cell line, on the membrane separating the upper and lower microchannels as stated above. The intestinal and skin compartments were established by integrating cell culture inserts into the microchannel. According to these studies, more functional organ-on-a-chip systems can be fabricated by combining several approaches such as the integration of cell culture insert, bare porous membranes, and trapping structures.

However, it is difficult to apply mechanical forces to the membrane, such as tension and compression, via the cell culture insert or bare membrane. To overcome this limitation, a microfluidic device integrated with a porous stretchable membrane has been proposed (Fig. 9.16c) [83–87]. The device consists of upper and lower microchannels separated by a polydimethylsiloxane (PDMS) porous membrane and vacuum chambers on both sides of the microchannels. By vacuuming the chambers, the porous PDMS membrane is able to be stretched. Using this device, a lung-on-a-chip was proposed [83, 84]. The alveolar epithelial cells and pulmonary endothelial cells were cultured on the upper and lower surfaces of the membrane, respectively. By applying cyclic strain, inflammatory responses of the endothelial cells to the nanoparticles were increased compared to the static condition. Moreover, as a result of the inflammatory responses, the endothelial capture of neutrophils and their transmigration from the endothelial layer to the epithelial layer were induced

[83]. Since similar results were obtained by experiments using whole mouse lung, this model could be used as alternatives to animal experiments. Furthermore, gut-on-a-chip were also reported using the similar device [85–87]. Human intestinal epithelial cells (Caco-2) were inoculated onto the PDMS porous membrane and cultured under fluid shear stress and cyclic strain. These stimulations promoted the formation of intestinal villi. Moreover, microbes were cultured on the differentiated epithelium and the influence of the cyclic stretch was investigated. As stated above, a microfluidic device integrated with a porous stretchable membrane is a promising tool for emulating in vivo conditions since real organs normally experience mechanical strain.

9.4.3 Organ-on-a-Chip Based on ECM Gel Channels

In order to investigate cell-ECM interactions in a 3D environment and/or maintaining large tissues with a nutrient supply, many studies have focused on fabricating microchannels in an ECM gel. In addition to the methods using cellular fibers and tubes (see Sect. 9.3.2), various methods have been proposed to fabricate the microchannels. The simplest method is to extract needles or wires embedded in the collagen gel (Fig. 9.17a) [88–90]. After fabricating the microchannels, endothelial cells were seeded in the microchannels to fabricate blood vessel models. The blood vessel models were used to investigate the influence of fluid shear stress on cell viability, barrier function [89], and angiogenesis [90].

However, it is difficult to fabricate complex blood vessels using this method. For the fabrication of complex blood vessels, soft lithography techniques have been employed [91]. Briefly, microstructures on collagen gels were molded from PDMS stamps. The microstructured collagen gel was sealed with flat collagen gel supported by jigs, followed by an inoculation of endothelial cells. The blood vessel model was cultured with gravity driven flow of the medium and used for permeability tests and angiogenesis studies.

To make more physiologically relevant blood vessel models, neovascularization of endothelial cells have been harnessed (Fig. 9.17b) [92, 93]. The microfluidic device for neovascularization had microchannels partitioned by microposts: the center channel was filled with fibrin gel to be seeded with endothelial cells and vascularized by the endothelial cells, medium channels flanking the center channel, and stromal cell channels next to the medium channels to induce neovascularization of the endothelial cells. Using this blood vessel model, physiological features of blood vessels were observed such as NO synthesis depending on the fluid shear stress and inflammatory response.

Recently, 3D printing technology has been utilized for the construction of microchannels with arbitrary 3D shapes in ECM gels (Fig. 9.17c) [94–97]. In this approach, ECM gel and sacrificial structures were printed by a 3D printer. The sacrificial structures were made of gelatin or Pluronic F-127 that were liquefied and removed by temperature changes. After seeding endothelial cells in the resulting

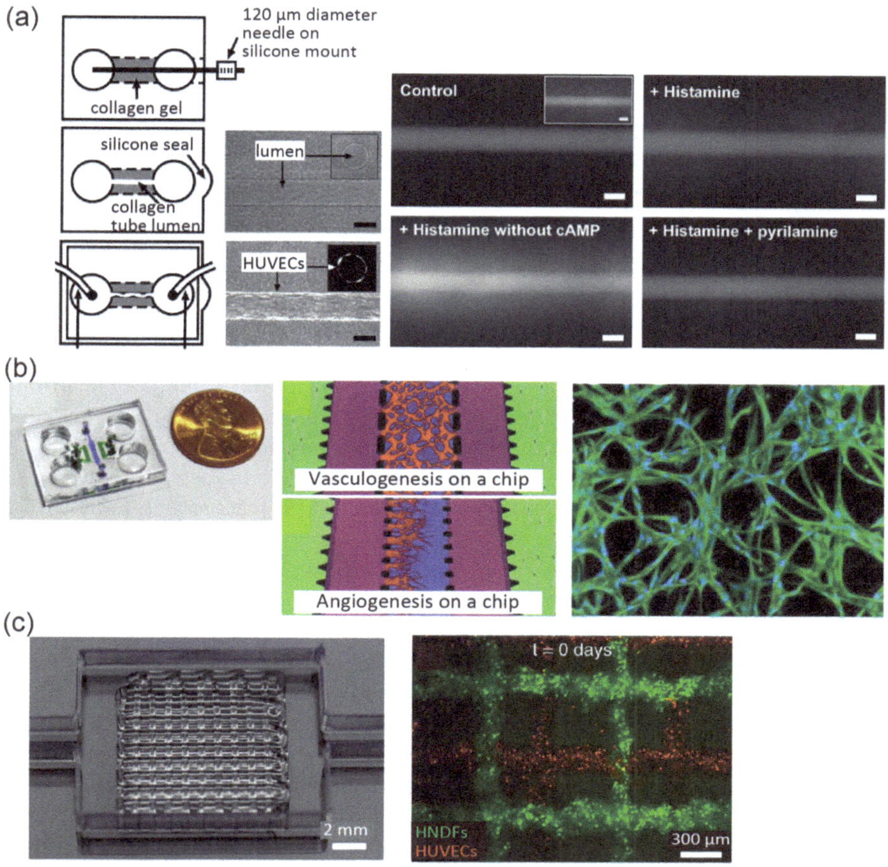

Fig. 9.17 Microchannels fabricated in ECM gel. (**a**) Blood vessel model fabricated by extracting needle and seeding endothelial cells. (**b**) Blood vessel model fabricated by neovascularization of endothelial cells. (**c**) Blood vessel model printed by 3D printer. (Images are reprinted with (**a**) permission from [88], ©2006 Elsevier, (**b**) permission from [92], ©2013 Royal Society of Chemistry, and (**c**) permission from [95], ©2014 John Wiley and Sons)

microchannels, long-term cultures (6 weeks) and permeability tests were performed [97], indicating that this system can be used as both a nutrient supply pathway for 3D tissues as well as a blood vessel model for physiological studies and drug screening.

Although many organ-on-a-chip applications using microchannels in ECM gel are blood vessel models, it is presumed that the microchannels in the ECM gel are also suitable for modeling organs with a lumen structure, such as an intestinal model on a geometrically engineered hollow channel in a porous protein scaffold [98]. The geometrically engineered hollow channel was fabricated by removing a nylon screw embedded in the silk scaffold. The channel was seeded with intestinal epithelial cells after a collagen gel containing intestinal myofibroblast was delivered into the silk scaffold. This intestinal model exhibited high accumulation of mucous secretions from the epithelium compared to a 2D culture, O_2 concentration similar to those observed in vivo, and co-culture of bacteria on the epithelium.

9.4.4 Organ-on-a-Chip Utilizing Anchored Cellular Tissue

Various cellular tissues composed of cells and ECM gel shrink and/or dynamically contract during cultivation. To maintain structure and function of these tissues against the shrinkage and contraction, anchor structures have been used. By combining the anchor structures with microfluidic devices, mechanical evaluation for tissue functions were achieved. Typical example is a heart-on-a-chip. The heart-on-a-chip was fabricated by gelling fibrin containing cardiomyocytes in a rectangular casting mold that had two flexible posts on the both edges (Fig. 9.18a) [99, 100]. The flexible post worked as anchor to keep the shape of cardiac tissue, and the force of the tissue was estimated by the deformation of the post [99]. Moreover, toxicity test of a drug (doxorubicin) was achieved by applying the drug and measuring the time course change of the force. Furthermore, microchannels were fabricated in the tissue by dissolving alginate fibers embedded in the tissue by sodium citrate or alginate lyase [100]. Since the microchannels were connected to the post with a hollow channel, medium perfusion was achieved, resulting in improved tissue morphology. However, it was difficult to precisely control the tissue shape due to shrinkage by this method. In addition, cell orientation was also difficult to be controlled since it depends on the shape of the tissue. To overcome these limitations, a heart-on-a-chip composed of fiber-shaped cardiac tissue whose ends were fixed to anchors was proposed (Fig. 9.18b) [49] (see Sect. 9.2.2). Owing to high cell density and anchors, the fiber shape was maintained, and the fiber shape aligned cardiomyocytes. Moreover, by setting anchors on a PDMS substrate with cantilevers, the force measurement was performed. Using this system, the efficacy of drugs was tested by measuring the change of contractile force and frequency.

Skin-on-a-chip systems were also fabricated by adopting anchor structure [38, 39]. Perfusable vessel-like channels were fabricated by dissolving sacrificial structures made of alginate gel and seeding endothelial cells [101]. Owing to movable posts with a hollow channel connected to the channels, perfusability was maintained even after the skin tissue had shrunk. Transdermal drug tests have been demonstrated by another skin-on-a-chip system (Fig. 9.18c) [54] (see Sect. 9.2.3). The skin tissue also had perfusable vessel-like channels fabricated by removing nylon wires embedded in the tissue and seeding endothelial cells. Since the vessel-like channels were fixed to the device channels by anchors, both the perfusability and the plane shape of the skin tissue were maintained. Owing to the perfusability and plane shape, transdermal drugs were able to be applied onto the surface of the skin tissue, and the drugs permeated into the vessel-like channels were successfully collected. From this result, this skin-on-a-chip system might be a useful tool for the development of cosmetics and drugs that are topically applied.

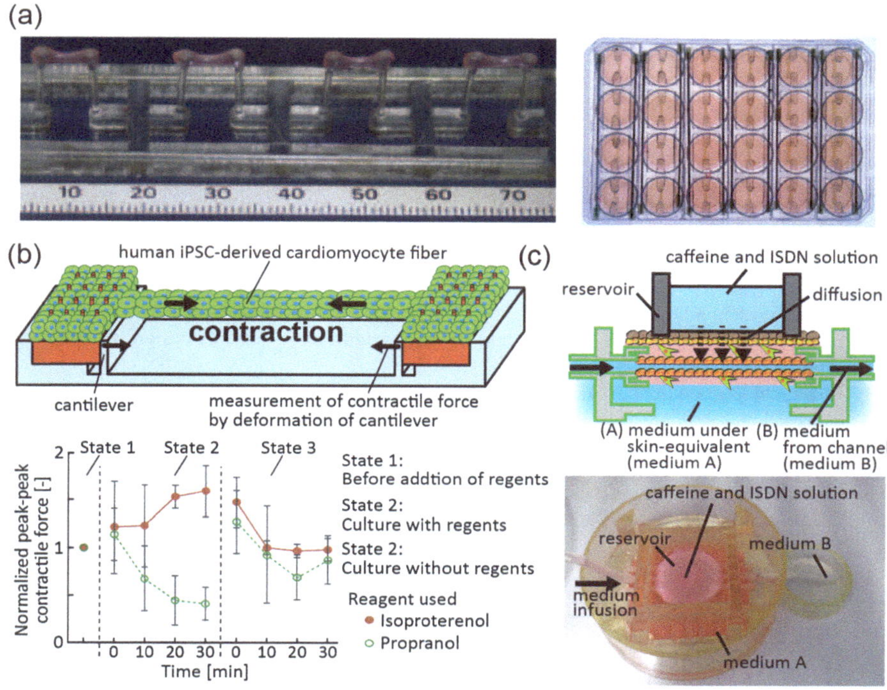

Fig. 9.18 Organ-on-a-chip systems using anchor structures. (**a**) Engineered heart tissue anchored to two posts. (**b**) Fiber-shaped cardiac tissue on a chip. (**c**) Skin-on-a-chip for percutaneous absorption test. (Images are reprinted with (**a**) permission from [99], ©2010 Wolters Kluwer Health, Inc., (**b**) permission from [49], ©2016 Royal Society of Chemistry, and (**c**) permission from [54], ©2017 Elsevier)

9.5 Summary

Microfluidic devices can construct various-shaped cellular structures with high throughput and high uniformity, while also allowing for the manipulation of the cellular structures for assembly. Therefore, using the cellular structures as building modules, macroscopic cellular tissues can be constructed. Furthermore, the integration of the cellular tissues with microfluidic devices to create organ-on-a-chip systems can be used for the analysis of pharmacokinetics and toxicology under physiological conditions. The studies introduced in the chapter will provide promising methods and tools to contribute to future improvements in this field. Furthermore, technical advances in the integration of cellular structures and microfluidic devices will help design novel devices such as biosensors and biological robots. As discussed, microfluidic technology will be widely studied with potential to explore new research fields.

References

1. Pampaloni F, Reynaud EG, Stelzer EHK (2007) The third dimension bridges the gap between cell culture and live tissue. Nat Rev Mol Cell Bio 8(10):839–845
2. Passier R, Orlova V, Mummery C (2016) Complex tissue and disease modeling using hiPSCs. Cell Stem Cell 18(3):309–321
3. Morimoto Y, Takeuchi S (2013) Three-dimensional cell culture based on microfluidic techniques to mimic living tissues. Biomater Sci 1(3):257–264
4. Groll J, Boland T, Blunk T, Burdick JA, Cho DW, Dalton PD, Derby B, Forgacs G, Li Q, Mironov VA, Moroni L, Nakamura M, Shu WM, Takeuchi S, Vozzi G, Woodfield TBF, Xu T, Yoo JJ, Malda J (2016) Biofabrication: reappraising the definition of an evolving field. Biofabrication 8(1):013001
5. Drury JL, Mooney DJ (2003) Hydrogels for tissue engineering: scaffold design variables and applications. Biomaterials 24(24):4337–4351
6. Khademhosseini A, Langer R (2007) Microengineered hydrogels for tissue engineering. Biomaterials 28(34):5087–5092
7. Morimoto Y, Hsiao AY, Takeuchi S (2015) Point-, line-, and plane-shaped cellular constructs for 3D tissue assembly. Adv Drug Deliver Rev 95:29–39
8. Chung BG, Lee KH, Khademhosseini A, Lee SH (2011) Microfluidic fabrication of microengineered hydrogels and their application in tissue engineering. Lab Chip 12(1):45–59
9. Onoe H, Takeuchi S (2015) Cell-laden microfibers for bottom-up tissue engineering. Drug Discov Today 20(2):236–246
10. Mazzitelli S, Capretto L, Quinci F, Piva R, Nastruzzi C (2013) Preparation of cell-encapsulation devices in confined microenvironment. Adv Drug Deliver Rev 65 (11–12):1533–1555
11. Bhatia SN, Ingber DE (2014) Microfluidic organs-on-chips. Nat Biotechnol 32(8):760–772
12. Perestrelo AR, Aguas ACP, Rainer A, Forte G (2015) Microfluidic organ/body-on-a-Chip devices at the convergence of biology and microengineering. Sensors 15(12):31142–31170
13. Lin RZ, Chang HY (2008) Recent advances in three-dimensional multicellular spheroid culture for biomedical research. Biotechnol J 3(9–10):1172–1184
14. Neto AI, Correia CR, Custodio CA, Mano JF (2014) Biomimetic miniaturized platform able to sustain arrays of liquid droplets for high-throughput combinatorial tests. Adv Funct Mater 24 (32):5096–5103
15. Brajsa K, Trzun M, Zlatar I, Jelic D (2016) Recent advances in three-dimensional multicellular spheroid culture for biomedical research. Period Biol 118(1):59–65
16. Kato-Negishi M, Tsuda Y, Onoe H, Takeuchi S (2010) A neurospheroid network-stamping method for neural transplantation to the brain. Biomaterials 31(34):8939–8945
17. El Assal R, Gurkan UA, Chen P, Juillard F, Tocchio A, Chinnasamy T, Beauchemin C, Unluisler S, Canikyan S, Holman A, Srivatsa S, Kaye KM, Demirci U (2016) 3-D microwell Array system for culturing virus infected tumor cells. Sci Rep 6:39144
18. Frey O, Misun PM, Fluri DA, Hengstler JG, Hierlemann A (2014) Reconfigurable microfluidic hanging drop network for multi-tissue interaction and analysis. Nat Commun 5:4250
19. Patra B, Peng CC, Liao WH, Lee CH, Tung YC (2016) Drug testing and flow cytometry analysis on a large number of uniform sized tumor spheroids using a microfluidic device. Sci Rep 6:21061
20. Ota H, Yamamoto R, Deguchi K, Tanaka Y, Kazoe Y, Sato Y, Miki N (2010) Three-dimensional spheroid-forming lab-on-a-chip using micro-rotational flow. Sens Actuat B-Chem 147(1):359–365
21. Akiyama Y, Morishima K (2011) Label-free cell aggregate formation based on the magneto-Archimedes effect. Appl Phys Lett 98(16):163702
22. Tan WH, Takeuchi S (2007) Monodisperse alginate hydrogel microbeads for cell encapsulation. Adv Mater 19:2696–2701

23. Allazetta S, Hausherr TC, Lutolf MP (2013) Microfluidic synthesis of cell-type-specific artificial extracellular matrix hydrogels. Biomacromolecules 14(4):1122–1131
24. Wieduwild R, Krishnan S, Chwalek K, Boden A, Nowak M, Drechsel D, Werner C, Zhang Y (2015) Noncovalent hydrogel beads as microcarriers for cell culture. Angew Chem Int Edit 54 (13):3962–3966
25. Takeuchi S, Garstecki P, Weibel DB, Whitesides GM (2005) An axisymmetric flow-focusing microfluidic device. Adv Mater 17(8):1067–1072
26. Morimoto Y, Tan WH, Takeuchi S (2009) Three-dimensional axisymmetric flow-focusing device using stereolithography. Biomed Microdevices 11(2):369–377
27. Morimoto Y, Tan WH, Tsuda Y, Takeuchi S (2009) Monodisperse semi-permeable microcapsules for continuous observation of cells. Lab Chip 9(15):2217–2223
28. Morimoto Y, Kuribayashi-Shigetomi K, Takeuchi S (2011) A hybrid axisymmetric flow-focusing device for monodisperse picoliter droplets. J Micromech Microeng 21(5):054031
29. Huang SB, Wu MH, Lee GB (2010) Microfluidic device utilizing pneumatic micro-vibrators to generate alginate microbeads for microencapsulation of cells. Sens Actuat B-Chem 147 (2):755–764
30. Onoe H, Inamori K, Takinoue M, Takeuchi S (2014) Centrifuge-based cell encapsulation in hydrogel microbeads using sub-microliter sample solution. RSC Adv 4(58):30480–30484
31. Morimoto Y, Onuki M, Takeuchi S (2017) Mass production of cell-laden calcium alginate particles with centrifugal force. Adv Healthc Mater 6(13):1601375
32. Maeda K, Onoe H, Takinoue M, Takeuchi S (2012) Controlled synthesis of 3D multi-compartmental particles with centrifuge-based microdroplet formation from a multi-Barrelled capillary. Adv Mater 24(10):1340–1346
33. *Matsunaga YT, *Morimoto Y, Takeuchi S (2011) Molding cell beads for rapid construction of macroscopic 3D tissue architecture. Adv Mater 23(12):H90–H94. (*equal contribution)
34. Morimoto Y, Tanaka R, Takeuchi S (2013) Construction of 3D, layered skin, microsized tissues by using cell beads for cellular function analysis. Adv Healthc Mater 2(2):261–265
35. Dendukuri D, Pregibon DC, Collins J, Hatton TA, Doyle PS (2006) Continuous-flow lithography for high-throughput microparticle synthesis. Nat Mater 5(5):365–369
36. Le Goff GC, Lee J, Gupta A, Hill WA, Doyle PS (2015) High-throughput contact flow lithography. Adv Sci 2(10):1500149
37. Sugiura S, Oda T, Aoyagi Y, Satake M, Ohkohchi N, Nakajima M (2008) Tubular gel fabrication and cell encapsulation in laminar flow stream formed by microfabricated nozzle array. Lab Chip 8(8):1255–1257
38. Mazzitelli S, Capretto L, Carugo D, Zhang X, Piva R, Nastruzzi C (2011) Optimised production of multifunctional microfibres by microfluidic chip technology for tissue engineering applications. Lab Chip 11(10):1776–1785
39. Lee KH, Shin SJ, Kim CB, Kim JK, Cho YW, Chung BG, Lee SH (2010) Microfluidic synthesis of pure chitosan microfibers for bio-artificial liver chip. Lab Chip 10(10):1328–1334
40. Jun Y, Kim MJ, Hwang YH, Jeon EA, Kang AR, Lee SH, Lee DY (2013) Microfluidics-generated pancreatic islet microfibers for enhanced immunoprotection. Biomaterials 34 (33):8122–8130
41. Yamada M, Utoh R, Ohashi K, Tatsumi K, Yamato M, Okano T, Seki M (2012) Controlled formation of heterotypic hepatic micro-organoids in anisotropic hydrogel microfibers for long-term preservation of liver-specific functions. Biomaterials 33(33):8304–8315
42. Kang E, Jeong GS, Choi YY, Lee KH, Khademhosseini A, Lee SH (2011) Digitally tunable physicochemical coding of material composition and topography in continuous microfibers. Nat Mater 10(11):877–883
43. Lee KH, Shin SJ, Park Y, Lee SH (2009) Synthesis of cell-laden alginate hollow fibers using microfluidic chips and microvascularized tissue-engineering applications. Small 5 (11):1264–1268
44. Onoe H, Okitsu T, Itou A, Kato-Negishi M, Gojo R, Kiriya D, Sato K, Mirua S, Iwanaga S, Kuribayashi-Shigetomi K, Matsunaga Y, Shimoyama Y, Takeuchi S (2013) Metre-long cell-laden Microfibres exhibit tissue morphologies and functions. Nat Mater 12:584–590

45. Hsiao AY, Okitsu T, Teramae H, Takeuchi S (2016) 3D tissue formation of Unilocular adipocytes in hydrogel microfibers. Adv Healthc Mater 5(5):548–556
46. Hsiao AY, Okitsu T, Onoe H, Kiyosawa M, Teramae H, Iwanaga S, Kazama T, Matsumoto T, Takeuchi S (2015) Smooth muscle-like tissue constructs with circumferentially oriented cells formed by the cell Fiber technology. PLoS One 10(3):e0119010
47. Ikeda K, Nagata S, Okitsu T, Takeuchi S (2017) Cell fiber-based three-dimensional culture system for highly efficient expansion of human induced pluripotent stem cells. Sci Rep 7:2850
48. Morimoto Y, Kato-Negishi M, Onoe H, Takeuchi S (2013) Three-dimensional neuron-muscle constructs with neuromuscular junctions. Biomaterials 34(37):9413–9419
49. Morimoto Y, Mori S, Sakai F, Takeuchi S (2016) Human induced pluripotent stem cell-derived fiber-shaped cardiac tissue on a chip. Lab Chip 16(12):2295–2301
50. Shimizu T, Yamato M, Isoi Y, Akutsu T, Setomaru T, Abe K, Kikuchi A, Umezu M, Okano T (2002) Fabrication of pulsatile cardiac tissue grafts using a novel 3-dimensional cell sheet manipulation technique and temperature-responsive cell culture surfaces. Circ Res 90(3):e40–e48
51. Sekine H, Shimizu T, Sakaguchi K, Dobashi I, Wada M, Yamato M, Kobayashi E, Umezu M, Okano T (2013) In vitro fabrication of functional three-dimensional tissues with perfusable blood vessels. Nat Commun 4:1399
52. Leng L, McAlister A, Zhang B, Radisic M, Cunther A (2012) Mosaic hydrogels: one-step formation of multiscale soft materials. Adv Mater 24(27):3650–3658
53. Yuan B, Jin Y, Sun Y, Wang D, Sun JS, Wang Z, Zhang W, Jiang XY (2012) A strategy for depositing different types of cells in three dimensions to mimic tubular structures in tissues. Adv Mater 24(7):890–896
54. Mori N, Morimoto Y, Takeuchi S (2017) Skin integrated with perfusable vascular channels on a chip. Biomaterials 116:48–56
55. Khan OF, Voice DN, Leung BM, Sefton MV (2015) A novel high-speed production process to create modular components for the bottom-up assembly of large-scale tissue-engineered constructs. Adv Healthc Mater 4(1):113–120
56. Kato-Negishi M, Morimoto Y, Onoe H, Takeuchi S (2013) Millimeter-sized neural building blocks for 3D heterogeneous neural network assembly. Adv Healthc Mater 2(12):1564–1570
57. Luo H, Chen M, Wang X, Mei Y, Ye Z, Zhou Y, Tan WS (2014) Fabrication of viable centimeter-sized 3D tissue constructs with microchannel conduits for improved tissue properties through assembly of cell-laden microbeads. J Tissue Eng Regen Med 8(6):493–504
58. Norotte C, Marga FS, Niklason LE, Forgacs G (2009) Scaffold-free vascular tissue engineering using bioprinting. Biomaterials 30(30):5910–5917
59. Nakamura M, Iwanaga S, Henmi C, Arai K, Nishiyama Y (2010) Biomatrices and biomaterials for future developments of bioprinting and biofabrication. Biofabrication 2(1):014110
60. Bruzewicz DA, McGuigan AP, Whitesides GM (2008) Fabrication of a modular tissue construct in a microfluidic chip. Lab Chip 8(5):663–671
61. Yue T, Nakajima M, Takeuchi M, Hu C, Huang Q, Fukuda T (2014) On-chip self-assembly of cell embedded microstructures to vascular-like microtubes. Lab Chip 14(6):1151–1161
62. Chung SE, Park W, Shin S, Lee SA, Kwon S (2008) Guided and fluidic self-assembly of microstructures using railed microfluidic channels. Nat Mater 7(7):581–587
63. Mori N, Morimoto Y, Takeuchi S (2016) Vessel-like channels supported by poly-L-lysine tubes. J Biosci Bioeng 122(6):753–757
64. Sakai S, Yamaguchi S, Takei T, Kawakami K (2008) Oxidized alginate-cross-linked alginate/gelatin hydrogel fibers for fabricating tubular constructs with layered smooth muscle cells and endothelial cells in collagen gels. Biomacromolecules 9(7):2036–2041
65. Harimoto M, Yamato M, Hirose M, Takahashi C, Isoi Y, Kikuchi A, Okano T (2002) Novel approach for achieving double-layered cell sheets co-culture: overlaying endothelial cell sheets onto monolayer hepatocytes utilizing temperature-responsive culture dishes. J Biomed Mater Res 62(3):464–470

66. Kim C, Bang JH, Kim YE, Lee SH, Kang JY (2012) On-chip anticancer drug test of regular tumor spheroids formed in microwells by a distributive microchannel network. Lab Chip 12 (20):4135–4142
67. Das T, Meunier L, Barbe L, Provencher D, Guenat O, Gervais T, Mes-Masson A-M (2013) Empirical chemosensitivity testing in a spheroid model of ovarian cancer using a microfluidics-based multiplex platform. Biomicrofluidics 7(1):011805
68. Ruppen J, Cortes-Dericks L, Marconi E, Karoubi G, Schmid RA, Peng R, Marti TM, Guenat OT (2014) A microfluidic platform for chemoresistive testing of multicellular pleural cancer spheroids. Lab Chip 14(6):1198–1205
69. Nakao Y, Kimura H, Sakai Y, Fujii T (2011) Bile canaliculi formation by aligning rat primary hepatocytes in a microfluidic device. Biomicrofluidics 5(2):022212
70. Wagner I, Materne E-M, Brincker S, Süßbier U, Frädrich C, Busek M, Sonntag F, D a S, Trushkin EV, Tonevitsky AG, Lauster R, Marx U (2013) A dynamic multi-organ-chip for long-term cultivation and substance testing proven by 3D human liver and skin tissue co-culture. Lab Chip 13(18):3538–3547
71. Ataç B, Wagner I, Horland R, Lauster R, Marx U, Tonevitsky AG, Azar RP, Lindner G (2013) Skin and hair on-a-chip: in vitro skin models versus ex vivo tissue maintenance with dynamic perfusion. Lab Chip 13(18):3555–3561
72. Kimura H, Yamamoto T, Sakai H, Sakai Y, Fujii T (2008) An integrated microfluidic system for long-term perfusion culture and on-line monitoring of intestinal tissue models. Lab Chip 8 (5):741–746
73. Kimura H, Ikeda T, Nakayama H, Sakai Y, Fujii T (2015) An on-Chip small intestine-liver model for pharmacokinetic studies. J Lab Autom 20(3):265–273
74. Shah P, Fritz JV, Glaab E, Desai MS, Greenhalgh K, Frachet A, Niegowska M, Estes M, Jäger C, Seguin-Devaux C, Zenhausern F, Wilmes P (2016) A microfluidics-based in vitro model of the gastrointestinal human-microbe interface. Nat Commun 7:11535
75. Jang K-J, Suh K-Y (2010) A multi-layer microfluidic device for efficient culture and analysis of renal tubular cells. Lab Chip 10(1):36–42
76. Jang K-J, Cho HS, Kang DH, Bae WG, Kwon T-H, Suh K-Y (2011) Fluid-shear-stress-induced translocation of aquaporin-2 and reorganization of actin cytoskeleton in renal tubular epithelial cells. Integr Biol 3(2):134–141
77. Wang YI, Abaci HE, Shuler ML (2017) Microfluidic blood-brain barrier model provides in vivo-like barrier properties for drug permeability screening. Biotechnol Bioeng 114 (1):184–194
78. Miura S, Sato K, Kato-Negishi M, Teshima T, Takeuchi S (2015) Fluid shear triggers microvilli formation via mechanosensitive activation of TRPV6. Nat Commun 6:8871
79. Abaci HE, Gledhill K, Guo Z, Christiano AM, Shuler ML (2015) Pumpless microfluidic platform for drug testing on human skin equivalents. Lab Chip 15(3):882–888
80. Ramadan Q, Ting FCW (2016) In vitro micro-physiological immune-competent model of the human skin. Lab Chip 16(10):1899–1908
81. Imura Y, Sato K, Yoshimura E (2010) Micro Total bioassay system for ingested substances: assessment of intestinal absorption, hepatic metabolism, and bioactivity. Anal Chem 82 (24):9983–9988
82. Maschmeyer I, Lorenz AK, Schimek K, Hasenberg T, Ramme AP, Hübner J, Lindner M, Drewell C, Bauer S, Thomas A, Sambo NS, Sonntag F, Lauster R, Marx U (2015) A four-organ-chip for interconnected long-term co-culture of human intestine, liver, skin and kidney equivalents. Lab Chip 15(12):2688–2699
83. Huh D, Matthews BD, Mammoto A, Montoya-Zavala M, Hsin HY, Ingber DE (2010) Reconstituting organ-level lung functions on a Chip. Science 328(5986):1662–1668
84. Huh D, Leslie DC, Matthews BD, Fraser JP, Jurek S, Hamilton GA, Thorneloe KS, McAlexander MA, Ingber DE (2012) A human disease model of drug toxicity-induced pulmonary edema in a lung-on-a-chip microdevice. Sci Transl Med 4(159):159ra147

85. Kim HJ, Huh D, Hamilton G, Ingber DE (2012) Human gut-on-a-chip inhabited by microbial flora that experiences intestinal peristalsis-like motions and flow. Lab Chip 12(12):2165–2174
86. Kim HJ, Ingber DE (2013) Gut-on-a-Chip microenvironment induces human intestinal cells to undergo villus differentiation. Integr Biol 5(9):1130–1140
87. Kim HJ, Li H, Collins JJ, Ingber DE (2016) Contributions of microbiome and mechanical deformation to intestinal bacterial overgrowth and inflammation in a human gut-on-a-chip. Proc Natl Acad Sci 113(2):E7–E15
88. Chrobak KM, Potter DR, Tien J (2006) Formation of perfused, functional microvascular tubes in vitro. Microvasc Res 71(3):185–196
89. Price GM, Wong KHK, Truslow JG, Leung AD, Acharya C, Tien J (2010) Effect of mechanical factors on the function of engineered human blood microvessels in microfluidic collagen gels. Biomaterials 31(24):6182–6189
90. Nguyen D-HT, Stapleton SC, Yang MT, Cha SS, Choi CK, P a G, Chen CS (2013) Biomimetic model to reconstitute angiogenic sprouting morphogenesis in vitro. Proc Natl Acad Sci 110(17):6712–6717
91. Zheng Y, Chen J, Craven M, Choi NW, Totorica S, Diaz-Santana A, Kermani P, Hempstead B, Fischbach-Teschl C, Lopez JA, Stroock AD (2012) In vitro microvessels for the study of angiogenesis and thrombosis. Proc Natl Acad Sci 109(24):9342–9347
92. Kim S, Lee H, Chung M, Jeon NL (2013) Engineering of functional, perfusable 3D microvascular networks on a chip. Lab Chip 13(8):1489–1500
93. Kim J, Chung M, Kim S, Jo DH, Kim JH, Jeon NL (2015) Engineering of a biomimetic Pericyte-covered 3D microvascular network. PLoS One 10(7):e0133880
94. Lee W, Lee V, Polio S, Keegan P, Lee JH, Fischer K, Park JK, Yoo SS (2010) On-demand three-dimensional freeform fabrication of multi-layered hydrogel scaffold with fluidic channels. Biotechnol Bioeng 105(6):1178–1186
95. Kolesky DB, Truby RL, Gladman AS, Busbee TA, Homan KA, Lewis JA (2014) 3D bioprinting of vascularized, heterogeneous cell-laden tissue constructs. Adv Mater 26 (19):3124–3130
96. Lee VK, Kim DY, Ngo H, Lee Y, Seo L, Yoo S-S, Vincent PA, Dai G (2014) Creating perfused functional vascular channels using 3D bio-printing technology. Biomaterials 35 (28):8092–8102
97. Kolesky DB, Homan KA, Skylar-Scott MA, Lewis JA (2016) Three-dimensional bioprinting of thick vascularized tissues. Proc Natl Acad Sci 113(12):3179–3184
98. Chen Y, Lin Y, Davis KM, Wang Q, Rnjak-Kovacina J, Li C, Isberg RR, Kumamoto CA, Mecsas J, Kaplan DL (2015) Robust bioengineered 3D functional human intestinal epithelium. Sci Rep 5(1):13708
99. Hansen A, Eder A, Bönstrup M, Flato M, Mewe M, Schaaf S, Aksehirlioglu B, Schwoerer AP, Schwörer A, Uebeler J, Eschenhagen T (2010) Development of a drug screening platform based on engineered heart tissue. Circ Res 107(1):35–44
100. Vollert I, Seiffert M, Bachmair J, Sander M, Eder A, Conradi L, Vogelsang A, Schulze T, Uebeler J, Holnthoner W, Redl H, Reichenspurner H, Hansen A, Eschenhagen T (2014) In vitro perfusion of engineered heart tissue through endothelialized channels. Tissue Eng Part A 20(3–4):854–863
101. Abaci HE, Guo Z, Coffman A, Gillette B, Lee W, Sia SK, Christiano AM (2016) Human skin constructs with spatially controlled vasculature using primary and iPSC-derived endothelial cells. Adv Healthc Mater 5(14):1800–1807

Chapter 10
Nanobiodevices for Cancer Diagnostics and Stem Cell Therapeutics

Daisuke Onoshima, Hiroshi Yukawa, and Yoshinobu Baba

Abstract This chapter describes the potential of multifunctional nanobiodevices in combination with quantum dots (QDs) to meet the requirements of diagnostic and theranostic systems. Nanobiodevices are primed to be powerful tools to provide the basis for the detection of small amounts of samples and simple operation. QDs can be utilized in these devices as bio-probes or labels for biological imaging of single molecules and cells. They have developed into new formats of biosensing to push the limits of detection. QDs has been also demonstrated to construct a multifunctional nanoplatform for stem cell transplantation and labeling with diagnostic and therapeutic modalities. The potential clinical applications of QDs have been expanded by the development of considerably low cytotoxicity QDs that do not include cadmium or selenium, as well as the development of longwave fluorescence

D. Onoshima (✉)
Institute of Innovation for Future Society, Nagoya University, Nagoya, Japan

ImPACT Research Center for Advanced Nanobiodevices, Nagoya University, Nagoya, Japan
e-mail: onoshima-d@nanobio.nagoya-u.ac.jp

H. Yukawa (✉)
Institute of Innovation for Future Society, Nagoya University, Nagoya, Japan

ImPACT Research Center for Advanced Nanobiodevices, Nagoya University, Nagoya, Japan

Department of Biomolecular Engineering, Graduate School of Engineering, Nagoya University, Nagoya, Japan
e-mail: h.yukawa@nanobio.nagoya-u.ac.jp

Y. Baba
Institute of Innovation for Future Society, Nagoya University, Nagoya, Japan

ImPACT Research Center for Advanced Nanobiodevices, Nagoya University, Nagoya, Japan

Department of Biomolecular Engineering, Graduate School of Engineering, Nagoya University, Nagoya, Japan

Health Research Institute, National Institute of Advanced Industrial Science and Technology (AIST), Takamatsu, Japan

School of Pharmacy, College of Pharmacy, Kaohsiung Medical University, Kaohsiung, Taiwan, Republic of China

M. Tokeshi (ed.), *Applications of Microfluidic Systems in Biology and Medicine*,
Bioanalysis 7, https://doi.org/10.1007/978-981-13-6229-3_10

QDs with strong permeability into the body. It provides the promising applications and further perspectives on future regenerative medicine.

Keywords Nanobiodevice · Quantum dot · Cancer diagnosis · Regenerative medicine

10.1 Introduction

Nanobiodevice is a piece of contrivance, equipment, machine, or component in the overlapping multidisciplinary activities of nanotechnology and biotechnology. The research efforts have been focused on the development of novel nanodevices for future personalized medicine and evidence based healthcare, which includes the following grand challenges: (1) multifunctional nanobiodevices for single platform bioimaging and targeting therapy; (2) ultra-high sensitive, highly specific, low invasive, and reliable multiplexed detection of disease at their inception; (3) remote disease monitoring through the ability of nanotechnology to provide on-line sensing and information relay; (4) construction of tissue-growth facilitating structured cell sheets for organ repair and replacement; (5) application of biological nanostructures in bioelectronics and environment-friendly nanofabrication processes; (6) single cell interventions and diagnostics including stem cell growth and differentiation and cellular level genomics and proteomics; and (7) stem cell differentiation and site-specific delivery. The field of multifunctional nanodevices has continued to witness strong growth during a decade. Peer reviewed publications on this field have been continuously increasing and the diversity of applications continues to expand. There is a continued healthy and necessary focus on the physical science fundamentals for both multifunctional nanodevices and an abundance of multidisciplinary biomedical applications are being built upon these fundamentals.

This chapter covers fundamentals and basic researches of nanobiodevices relating to multifunctional quantum dots (QDs)-based cancer diagnostics and stem cell therapeutics for regenerative medicine. A field of recent diagnostics and therapeutics has been advanced with QDs. QDs have developed into new formats of biomolecular sensing to push the limits of detection in biology and medicine. QDs can be also utilized as bio-probes or labels for biological imaging of living cells and tissues. More recently, QDs has been demonstrated to construct a multifunctional nanoplatform, where the QDs serve not only as an imaging agent, but also a nanoscaffold for diagnostic and therapeutic modalities. This chapter highlights the promising applications of multi-functionalized QDs as advanced nanosensors for diagnosing cancer and as innovative fluorescence probes for *in vitro* or *in vivo* stem cell imaging in regenerative medicine.

The development of semiconductor QDs has been heading toward an interdisciplinary field with bioassays and bioimaging technologies [1–3]. It offers significant advantages over those of the previous methods, which have relied on conventional organic fluorophores or fluorescent proteins. The unique and useful optical properties of QDs have been explored such as broad absorption spectra, narrow emission spectra, high quantum yields, long fluorescent lifetime, size-tunable photoluminescence, and exceptional photostability with a strong resistance to photobleaching [4–7]. The optoelectronic characteristics of QDs arise through the systematic transformation in the density distribution of the electronic energy levels as a function of the size of the QDs [8]. As a result, the nature of the QD surface plays a key role in their optical behavior. The ability to modify the surface of QDs with biomolecules or other polymers through various conjugation methods underpins their versatility.

QDs have been incorporated into various biological assays for the reversible detection and quantification of biomolecules. Such assays often focus on the use of distance-dependent fluorescence resonance energy transfer (FRET) [9, 10] andmultiplexed bioanalysis with multiple colors [11, 12]. These techniques can be introduced for applications such as immunoassays [13, 14], molecular diagnosis [15, 16], clinical assays [17–19], and cellular analysis [2, 20]. Many kinds of QD-based biological assays have been demonstrated in bulk solution or on solid substrates [21]. A more competitive platform is microfluidic device. The device is able to handle the extremely small quantities of fluid with multiple analytical samples transported through the microfabricated channels [22]. It can provide opportunities for further removal or replacement of QD-based sensing components [23].

When applying QDs to biological imaging and cellular studies, the toxic nature of cadmium-containing QDs remained a major concern. However, with the recent advances in the development of surface modifications of QDs and cadmium-free QDs, the potential toxicity of cadmiumis no longer a problem for *in vitro* and *in vivo* imaging studies. The cellular uptake of QDs can be modulated by their size [24], shape [25–27], and surface functionalization [28]. In fact, the use of multicolor QDs for stem cell imaging is probably the most important and clinically relevant application for regenerative medicine in the immediate future. Despite the enormous potential of QDs in therapeutics, the fundamental information on the interaction between QDs and therapeutic cells [29] is relatively limited. This review thus aims to outline the beneficial properties presented by QDs, along with important advances in their biological applications for cancer diagnostics and stemcell therapeutics. Particularly, the observation of the QD-mediated cellular responses, such as cellular uptake and intracellular behavior of QDs, will provide insights into the nanoparticle design and the therapeutic efficacy for regenerative medicine.

10.2 Quantum Dots-Based Sensing and Detection

10.2.1 Quantum Dots

QDs stand out as one of the most exciting research tools in chemistry, physics, and biology (Fig. 10.1). These inorganic fluorescent nanocrystals typically comprise periodic groups of CdSe and CdTe or InP and InAs semiconductor materials. For

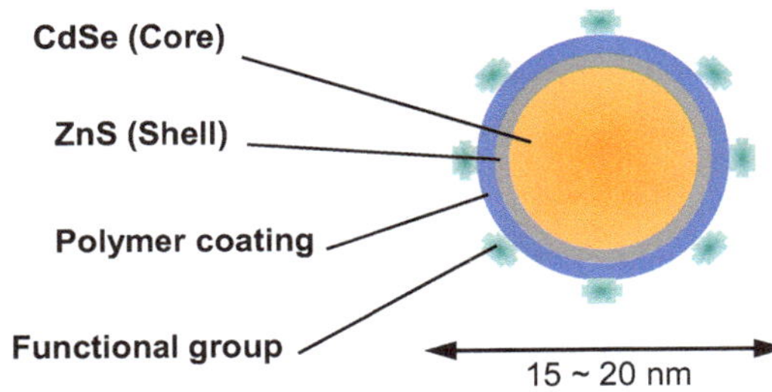

Fig. 10.1 Schematic of a major negatively charged QD. (Reprinted with permission from Ref. [137])

the semiconductor nanocrystals, the energy levels are quantized due to the quantum-confinement effect [30, 31], and their spacing can be controlled by the crystal sizes. This effect leads to the superior optical properties of QDs, such as narrow, symmetric, and size-tunable emission spectra. Also, the broad excitation spectra of QDs enable multicolor fluorescent applications. Compared with organic fluorophores or fluorescent proteins, QDs show 10–100 times brighter fluorescence and 100–1000 times higher fluorescence stability against photobleaching. These benefits of QDs facilitate long-term monitoring of intermolecular and intramolecular interactions in living cells and tissues. Consequently, much interest has been focused on the exploration of QDs for biomedical applications.

Among the array of synthetic routes devised for the preparation of biocompatible QDs, the coating of the CdSe core with the ZnS layer is indispensable [2, 3]. Passivation by the ZnS layer protects the CdSe core from oxidation, and reduces the toxicity of the CdSe core from leaching out to the surrounding solutions. It also enhances the photoluminescence yield. Although the synthesis of QDs has been performed directly in aqueous solution, QDs themselves have little specificity for the aqueous nature of the biological environment [32, 33]. For example, the ZnS-coated QDs are only soluble in nonpolar organic solvents. Altering the surface properties of QDs mainly relies on the conjugation with biological molecules such as aptamers [34], antibodies [35], oligonucleotides [36], and peptides [37] to gain the biological affinity [38].

As a result of the conjugations of QDs, they have achieved optimal stability, monodispersity, crystallinity, solubility, and biocompatibility in the fields of diagnostic and therapeutic research. For example, within a certain range of concentrations, CdSe/ZnS shows good biocompatibility with human amniotic mesenchymal stem cells [39]. The effect of QD size and poly(ethylenimine) coating on the labeling efficiency with stem cells has been also reported [40]. Furthermore, a peptide-coated QDs can mimic cellular transport mechanism in stem cells through endosomal escape [41].

10.2.2 Fluorescence Resonance Energy Transfer

In terms of the fluorescence application to monitor binding interactions or conformational changes of biomolecules, FRET phenomenon, the non-radiative energy

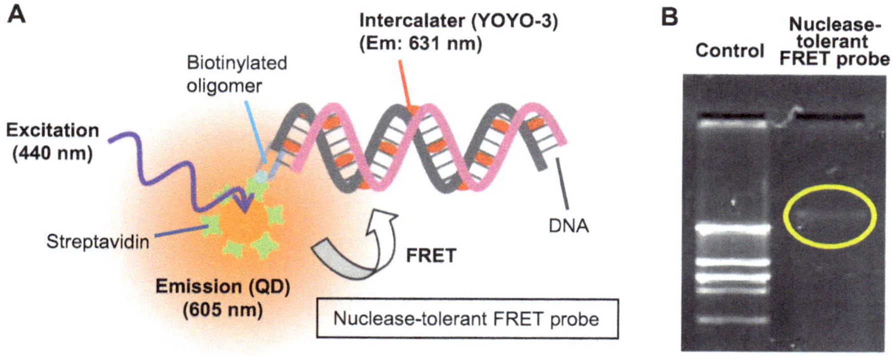

Fig. 10.2 A nuclease-tolerant FRET probe. (**a**) Schematic of the FRET between the QD donor and the DNA-intercalating dye (YOYO-3) acceptor. The QD emission overlaps the absorption spectrum of the dye, suggesting that an efficient FRET between the QD donor and the YOYO-3 dye acceptor can take place. (**b**) An electrophoresis analysis of the FRET probe for the effect on DNA digestion. A yellow circle shows the undigested FRET probe by a restriction endonuclease (EcoRI). (Reprinted with permission from Ref. [137])

transfer between an excited state fluorophore (donor) and another fluorophore (acceptor) through long-range dipole–dipole coupling, has been recognized as a quite useful detection scheme in various bioanalyses [42]. The utility of this detection scheme is explained that it generates fluorescence signals sensitive to the changes in 1–10 nm range. More specifically, the FRET-based transduction using QDs as donors has become a popular approach for assay development. Their broad absorption spectrum in the UV region can be used advantageously to avoid the direct excitation of acceptor dyes. As a result of the introduction of multiple pathways for energy transfer, the FRET efficiency can be enhanced by the QDs. The transduction strategy can be also arranged using QDs and dyes. For example, many QDs have the opportunity to interact with a single acceptor dye, or multiple dyes can interact with a single QD.

QD-based FRET probes have also been reported, including QD conjugated hybridization probes for the preliminary screening of siRNA sequences [43] and an ultrasensitive DNA nanosensor with a singleQD [36, 44],which benefit from the use of the unique photophysical properties of QDs and their conjugates. Recent work in our laboratory [45] has demonstrated the feasibility of using QDs as a nuclease tolerant FRET probe. The streptavidin-coated QD binds to the biotinylated DNA end (Fig. 10.2a), and the FRET-inducing electrostatic coupling and the structural changes to the QD-DNA conjugates can be detected (Fig. 10.2b) based on the conjugation of the QDs with YOYO-3 dye-intercalated DNA. The QD-DNA conjugate and the tolerance properties induced by dye binding are thought to be notable due to the utilization of QD conjugated functional materials in living cells or tissues. Transduction via FRET also makes it possible to employ multiple methods for data analyses. For example, ratiometric detection is often used for differences in assay preparation among multiple analyses and instrumental drift. The narrow symmetrical emission profile of QDs facilitates the deconvolution of acceptor signals for the recovery of absolute signals [46].

10.2.3 Microfluidic Devices

Microfluidic system is based on the microfabricated structures to control or manipulate fluids constrained in a small space [22, 23]. The potential to be used in biological applications is a major driving force behind the rapid development of this system [48–51]. They are primed to be powerful tools to provide the basis for the detection of biomolecules with small amounts of reagents and simple operation [47]. In addition, the large surface area-to-volume ratio and mass transport by nondiffusive means offers the transduction of analytes within seconds to minutes.

Microfluidic devices have been known to facilitate single-molecule measurements in combination with QDs [51]. Such approaches are useful to determine patterns and distributions that may otherwise be masked by ensemble averages [52]. Single-molecule studies are usually performed by monitoring a fixed volume of solution. Analytes are allowed to diffuse into this fixed volume in order to be detected. A continuous-flow system, as found in capillaries and microfluidics, ensures that multiple analytes are moved across the detection volume, increasing the probability of detection. For example, a mismatch repair protein complex that slides while maintaining continuous contact with DNA can be visualized by QDs in real time [53]. The DNA-binding proteins are engineered with epitope tags and labeled with antibody coupled QDs. The motion of the proteins is observed on the single DNA molecule that is anchored and elongated in a microfluidic device. Using similar protein labeling methods with QDs, several kinds of specific sequences related to genetic or epigenetic changes of genomic DNA molecules have been detected for optical mapping applications [54]. Moreover, the mapping method is successfully applied to the detection of DNA methylation at a single molecule level [55].

On the other hand, recent advances in a microfluidic system developed by our group (Fig. 10.3a and b) allow for the visual observation of nucleolytic cleavage for the direct screening and detection of a specific sequence in genomic DNA [56]. By using DNA molecular tagging with QD, it is possible to track the motion and position of a single DNA molecule (Fig. 10.3c and d) gaining access to a restriction enzyme on the microchannel surface. This kind of experimental approach will help enable applications for ultra-sensitive, high-speed and small-volume bioanalytical measurements.

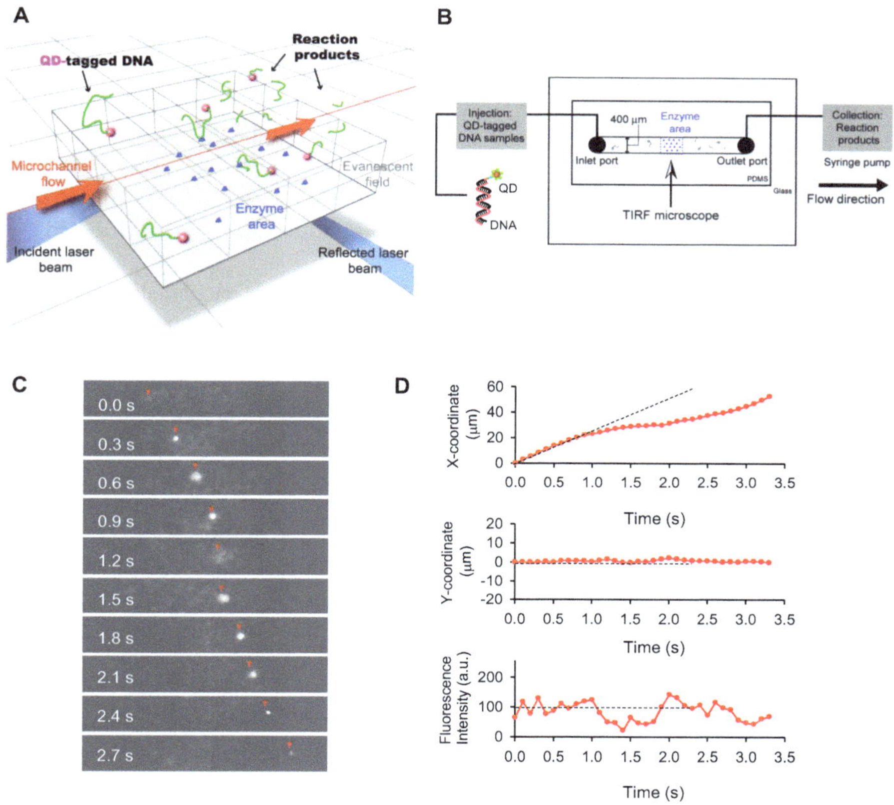

Fig. 10.3 A microfluidic molecular tracking system. (**a**) Schematic design for the time-series analysis of ongoing processes in the enzymatic reaction through the detection of single-molecular movements. (**b**) Layout of the molecular tracking chip. Enzymatic reactions are done in a conventional microfluidic device, consisting of a polydimethylsiloxane (PDMS) microstructure and a syringe pump. A single QD was connected to one DNA molecule as a tracking dye. Enzymes are immobilized on the bottom surface of a microchannel, and QD-tagged DNA molecules are flown in the channel. The fluorescent images during the event are captured by a total internal reflection fluorescence (TIRF) microscope. (**c**) Video monitoring in action at the enzyme area. The flow keeps the motion of DNA molecule constant, and a restriction enzyme (Apa I) on the surface changes the motion. (**d**) A trajectory of QD-tagged DNA represented as x coordinate, y coordinate and fluorescence intensity. Dashed line shows the control with no enzyme immobilized. (Reprinted with permission from Ref. [137])

10.3 Cancer Diagnosis by Quantum Dots

10.3.1 Protein and Nucleic Acid

Early diagnosis of cancers is at the forefront of cancer research. Cancer biomarker-based *in vitro* assays are useful for the screening and diagnosis of cancer, as well as for monitoring the response to therapy. Many proteins are considered as useful diagnostic markers of various cancers [58–60]. They are also the targets for basic biomedical research. Since the protein biomarkers are often present at very low concentrations, analytical methods with low detection limits are required for the early diagnosis of cancers [60]. QDs have proven to be applicable for the sensitive detection of cancer biomarkers such as ovarian cancer [61], breast cancer [62], and prostate cancer [63]. More recently, two tumor biomarkers, α-fetoprotein and carcinoembryonic antigen in human serum, have simultaneously been detected by a QD based nanosensor [64].

Nucleic acids are another type of tumor marker for various cancers. QDs are also very useful in the detection of nucleic acids, especially in a multiplexed format. For example, a multicolor optical coating for biological assays comprises embedding different-sized QDs into polymeric microbeads at precisely controlled ratios. These were designed as a DNA hybridization system using oligonucleotide probes and triple-color encoded beads. The coding signals can identify different DNA sequences [65]. Moreover, a solid-phase FRET assay using immobilized QDs as donors can detect two target sequences simultaneously [21]. This assay showed that over 50% of the analyte's signal readout was obtained even in bovine serum and with a large excess of non-complementary genomic DNA as background noise.

More recently, a single-quantum dot-based nanosensor for specific miRNA detection has been developed [66]. Among the numerous analytical approaches for miRNA detection, most suffer from problems such as non-specificity and low sensitivity. To overcome these problems, two-stage exponential amplification reactions and single-QD-based nanosensors were fabricated. The detection limit of 0.1 aM miRNA was achieved, and even single-nucleotide differences between miRNA family members could be distinguished.

10.3.2 Cancer Cells

Sensing the interaction between drug-carrying vehicles and cell membrane is the primary requisite for a successful diagnostic process, where the diseased cells are first located, following by subsequent cellular uptake and release of a therapeutic agent into the cytosol or nucleus of cells [68]. Especially for cancer research and therapy, QDs have been utilized as imaging probes [2, 3] for the recognition of specific cell types and tissues in the clinical settings [67]. For example, a static immunostaining of cellular targets with QDs has been demonstrated and shown to be both brighter and more photostable than comparable organic fluorophores [17]. In particular, QD-peptide conjugates can specifically target the tumor vasculature in mice [37]. A PEG coating has reduced the reticuloendothelial clearance in the *in vitro* histological results. An ABC triblock copolymer-coated QD probe has also been developed to target and image prostate cancer [35]. For *in vivo* studies, the tumor site can be actively probed by the antibody-conjugated QDs and imaged in living animals. The large size and immunogenicity of antibodies sometimes affect their pharmacological behavior. Alternatives are small ligands such as peptides [38] or aptamers [34] that are employed to conjugate QDs.

A QD nanoprobe has been developed to sense glioma cells based on the overexpression of the extracellular matrix glycoprotein, tenascin-C [69]. Tenascin-C is involved in tissue remodeling, and plays a role in the invasion of glioblastoma into the surrounding brain tissue. The QDs were targeted to tenascin-C by a single-stranded DNA aptamer. Non-cadmium-containing QDs have also been shown to be able to label cells fluorescently [70]. For example, phospholipid micelle-encapsulated silicon QDs could be conjugated to transferrin and taken up by pancreatic cancer cells. The concentration of silicon QDs applied to these cells was not toxic, because there was 95% cell viability after 24 h.

Since the cellular uptake of exogenous material generally occurs through internalization mechanisms [71], QDs are passively uptaken via non-specific endocytosis along the migratory pathway of human mammary epithelial tumor cells. Moreover, unmodified QDs have been used as an alternative marker over gold nanoparticles for phagokinetic tracking to monitor cell motility as a potential assay for cancer metastasis [72]. The use of a transfection agent such as liposomes or micelles can assist the delivery of QDs. Polymer- or ligand-modification of the QDs is usually more specific and efficient than non-specific endocytosis alone [73, 74]. A general observation is that the endocytosed QDs are often trapped in endosomes and lysosomes. They can be visualized as punctate fluorescence staining. This plays an important role in tracking the spatiotemporal behavior of the cells.

10.4 Stem Cell Labeling by Quantum Dots

10.4.1 Stem Cell Labeling

Regenerative medicine is expected to overcome the shortage of donated organs, donor site morbidity and immune reactions. Many kinds of stem cells, such as induced pluripotent stem (iPS) cells [75, 76], embryonic stem (ES) cells [77, 78], and some kinds of somatic stem cells, including bone marrow-derived stem cells [79, 80] and adipose tissue-derived stem cells [81, 82], have been discovered and may be useful in numerous applications for regenerative medicine [84–86]. To ensure the safety and maximum therapeutic effects of regenerative medicine, analyses of the behavior, accumulation and condition of transplanted stem cells *in vivo* have become increasingly important [86, 87]. However, the conventional *in vivo* imaging modalities used in clinical practice are not sufficient for the analysis of transplanted stem cells [88, 89]. Fluorescent imaging (FI) with the prominent fluorescence properties of QDs is expected to detect the transplanted stem cells *in vivo* with higher sensitivity in comparison to other imaging modalities [90, 91]. Thus, "stem cell labeling technology using QDs" and "*in vivo* fluorescent imaging technology for visualizing transplanted stem cells labeled with QDs" are essential for analyzing the transplanted stem cells.

There are generally thought to be two methods that can be used to label stem cells with QDs; the conjugation of QDs with a stem cell surface [92] and the transduction of QDs into a stem cell [90, 93]. The conjugation of QDs with the cell surface is associated with some problems, such as a reduced accumulation rate of stem cells in tissues/organs and the separation of QDs from the stem cell surface during circulation in the body. The transduction of QDs into the stem cell may overcome these problems; however, the cytotoxicity of the QDs is very high and the transduction efficiency is very low when using physical stimulus methods, such as ultrasonic transduction [91] and electroporation [74, 94]. In contrast, the labeling of stem cells with QDs using chemical modification methods, such as cationic liposomes, cell penetrating peptides, and high molecular nano-carrier (polymer micelles) was reported to be useful [96–98], because these molecules have been used in clinical applications for the transduction of DNA and proteins into cells.

10.4.2 Cationic Liposomes

QDs are unable to label stem cells with high efficiency because of their low rate of interaction with the cell membrane, so chemical modifications using cationic liposomes are expected to be useful for the stem cell labeling application of QDs [98, 99]. Some cationic liposomes such Lipofectamine® (Life technologies), COATSOME® (NOF corporation), and ScreenFect™A (Wako) are commercially available, and are mainly used for gene transfection. These cationic liposomes can interact with negatively-charged QDs (especially COOH-conjugated QDs), then the cationic liposomes rapidly enclose the negatively-charged QDs through electrostatic forces [93, 100].

In fact, when cationic liposomes and negatively-charged QDs were mixed at the optimal mixture ratio in culture medium and then were added to stem cells in the culture medium. TheQDs could be transduced into the stem cells through endocytosis within a few hours, and were maintained in the cytoplasm near the nucleus by escaping the exocytotic mechanism [101] (Fig. 10.4a). The transduction efficiency was very high, and no severe cytotoxicity was identified [102, 103] (Fig. 10.4b and c). In addition, QD labeling with cationic liposomes did not affect the stem cell characteristics, such as their potential for self-renewal and their multilineage potential [93] (Fig. 10.4d). This labeling method is simple and relatively rapid, and so appears to be useful for stem cell labeling with QDs.

However, high concentrations of cationic liposomes induce the death of stem cells. The cell membrane is generally negatively-charged, so the cell membrane structure is unstable and destroyed by the interaction with strong positively-charged cationic liposomes. The validation of the optimal concentration of cationic liposomes is important and necessary to make it possible to label various kinds of stem cells without adverse effects [99].

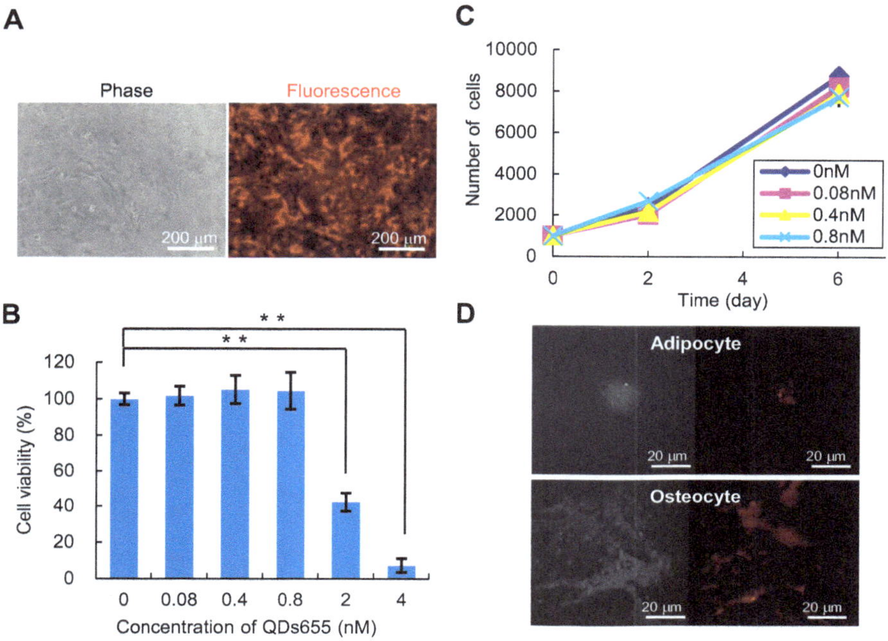

Fig. 10.4 Quantum dots labeling using poly-cationic liposome (Lipofectamine®) for imaging of adipose tissue-derived stem cells. (**a**) Morphology of ASCs and fluorescence of the QDs655 (0.8 nM) observed by phase-contrast microscopy. (**b**) Cell viability of ASCs labeled with QDs655 using Lipofectamine®. (**c**) Proliferation rate of ASCs labeled with QDs655 using Lipofectamine®. (**d**) Adipogenic and osteogenic differentiation of ASCs labeled with QDs655 (0.4 nM). (Reprinted with permission from Ref. [137])

10.4.3 Cell-Penetrating Peptides

Cell-penetrating peptides (CPPs), also known as protein transduction domains (PTDs), are expected to be useful for the stemcell labeling application of QDs, because several transduction domains can deliver a large size-independent variety of molecules into cells, including proteins, peptides, antisense oligonucleotides and large metal beads [104]. Representative CPPs include the Tat protein of human immunodeficiency virus (HIV-1) [105, 106], VP22 protein of herpes simplex virus [107] and Antennapedia (Antp) homeoprotein of Drosophila [108], and all of these CPPs possess arginine- and lysine-rich sequences. In addition, poly-arginine (PolyR), especially 8–11-arginine peptides, and poly-lysine have been shown to exhibit an even greater efficiency in the delivery of several peptides and proteins [109, 110] (Fig. 10.5a).

Most studies have reported that these CPPs were conjugated with negatively-charged QDs (CPP-QDs) through chemical or electrostatic forces, and the condition of the QD surface was dramatically changed to make it positively-charged. In a case

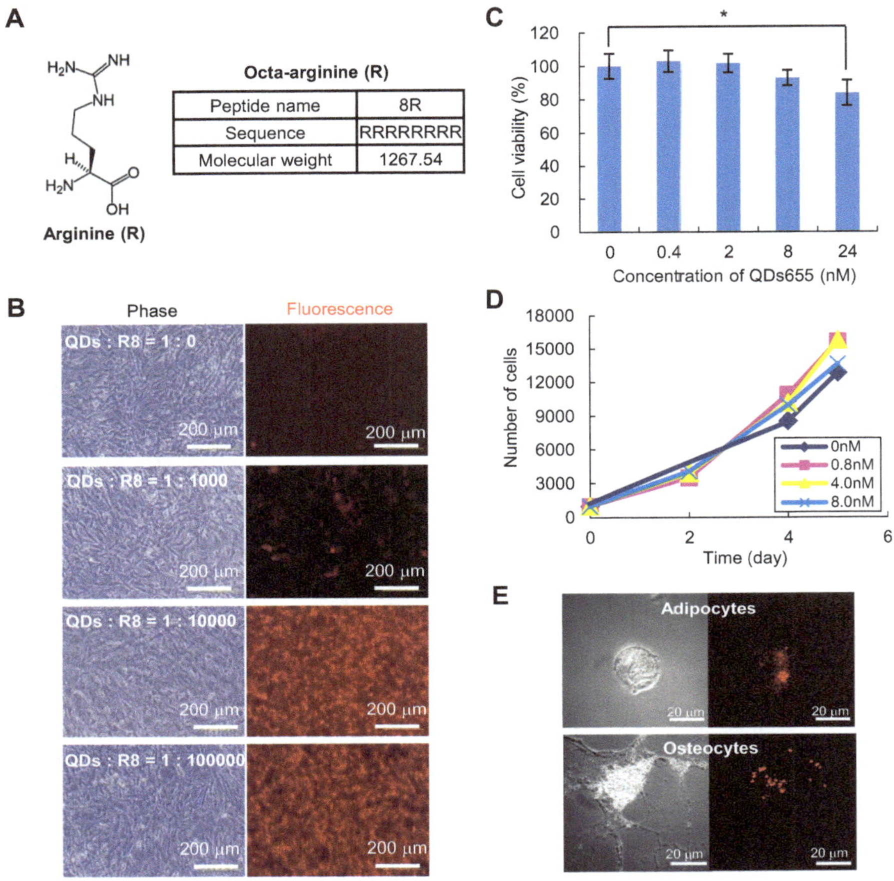

Fig. 10.5 Quantumdots labeling using cell penetrating peptides (octa-arginine: R8) for imaging of adipose tissue-derived stem cells. (**a**) Constitutional formula of arginine molecule and the information of normal chain R8. (**b**) Optimal formation of QDs and R8 for labeling ASCs. (**c**) Cell viability of ASCs labeled with QDs655 using R8. (**d**) Proliferation rate of ASCs labeled with QDs655 using R8. (**e**) Adipogenic and osteogenic differentiation of ASCs labeled with QDs655 (0.4 nM) using R8. (Reprinted with permission from Ref. [137])

of chemical binding, the CPPs and QDs were mainly chemically-bonded by the amide bonding of the amino group of CPPs with the carboxyl group of the negatively charged QDs [103, 109]. Chemically-bound CPPs–QDs are very stable; however, the synthesis and purification of CPPs–QDs involves a great deal of time and effort. On the other hand, in the case of electrostatic binding, the CPPs and QDs were conjugated by the positively-charged amino acids of CPPs binding the negatively-charged functional groups of the QDs [111, 112]. These electrostatic-bound CPPs–QDs were stable in cell culture medium, and the production of CPPs–QDs is simple and quick in comparison to chemical binding methods.

Indeed, in that case, CPPs–QDs could be transduced into stem cells through micropinocytosis within several hours, and were maintained in the cytoplasm near the nucleus by escaping exocytosis (Fig. 10.5b). Macropinocytosis occurs independent of clathrin-mediated and caveolin-mediated endocytosis, and the size of particles that can be uptake is more than 1 μm, and the process requires dynamin GTPase activity [113]. Similar to the complexation with cationic liposomes, the transduction efficiency for micropinocytosis was very high, and there was no severe cytotoxicity observed (Fig. 10.5c). The CPPs–QDs did not affect the stemcell characteristics such as the self-renewal or multilineage potential [114, 115] (Fig. 10.5d and e). In addition, the CPPs do not generally induce the death of stem cells at high concentrations, as occurs when cationic liposomes are used for transfection [90]. Thus, this

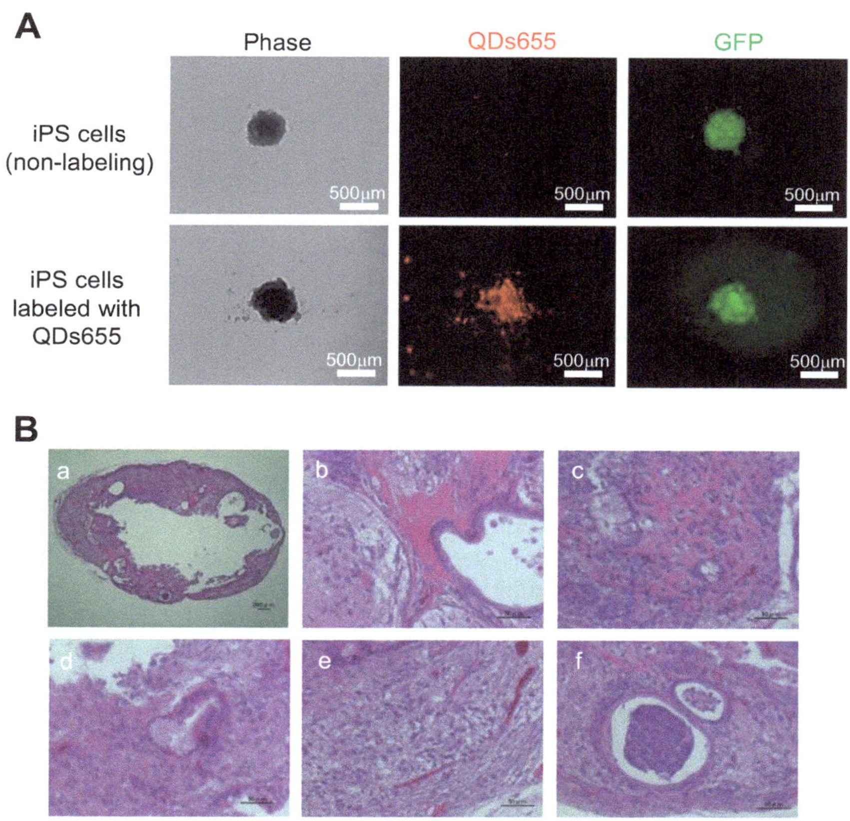

Fig. 10.6 Maintenance of undifferentiated state and multipotency of iPS cells labeled with QDs using R8. (**a**) Morphologies and fluorescence images of 1 day after EB formation of non-labeled iPS cells and iPS cells labeled with QDs using R8 (phase contrast, red fluorescence derived from QDs655, and green fluorescence derived from GFP). (**b**) Teratoma formation of iPS cells labeled with QDs using R8 after 4 weeks of the injection of labeled iPS cells to the nude mouse. *a*: Obtained histological sections were stained with hematoxylin and eosin. *b*: Nervelike structures, *c*: cartilage-like structures, *d*: gut epithelium-like structures, *e*: adipose-like structures, *f*: glomerulus of the kidney-like structures. (Reprinted with permission from Ref. [137])

labeling is also expected to be utilized for the stem cell labeling of QDs. Moreover, a previous study showed that iPS cells could be labeled with QDs using CPPs at high efficiency, and iPS cells labeled with QDs maintained their undifferentiated state and pluripotency [116] (Fig. 10.6a and b). Therefore, CPPs are expected to be useful molecules for stem cell labeling with QDs. However, validation of the optimal concentration is important and will necessary to determine the best way to label various kinds of stem cells without adverse effects [117].

10.5 *In vivo* Fluorescence Imaging of Transplanted Stem Cells Labeled with Quantum Dots

10.5.1 In Vivo Fluorescence Imaging

In vivo fluorescence imaging systems, which can detect and analyze the fluorescence or emission from the body, have been developed for small animals such as mice and rats [87]. The Maestro™ (PerkinElmer) Clair*vivo* OPT (SHIMADZU) and IVIS Imaging System (PerkinElmer) are representative instruments (Fig. 10.7a). These systems generally have an integrated ultrasensitive cooled CCD (charge-coupled device) camera, and can detect the weak fluorescence and emission *in vivo*. The number of photons from the fluorescence or emission can be counted, and the fluorescence or emission intensity can then be quantitatively determined. In addition, these fluorescence or emission data can be combined the data from other modalities, such as X-ray CT (computed tomography) [97], MRI (magnetic resonance imaging) [114, 115], SPECT (single photon emission computed tomography) and PET (positron emission computed tomography) [118, 119].

These systems can be used for *in vivo* fluorescence imaging of transplanted stem cells labeled with QDs. However, there are some major problems that prevent the successful *in vivo* fluorescence imaging of transplanted stem cells at high resolution, which include the strong scattering, absorption and autofluorescence derived from the whole body. In order to overcome these problems, fluorescence probes that can absorb the excitation light and emit strong fluorescence in the near-infrared region (about 700–900 nm) in the "Biological OpticalWindow" to decrease the scattering, absorption and autofluorescence derived from the body, are strongly desired [121–123] (Fig. 10.7b). Some QDs showing strong fluorescence in the near-infrared region have been developed and are commercially available, so QDs are expected to be useful for *in vivo* fluorescence imaging of transplanted stem cells (Fig. 10.7c).

Moreover, the autofluorescence derived from normal food must be considered [123]. In fact, autofluorescence of mice given normal feed was detected from the gastrointestinal tract by excitation in the red or near-infrared region. To diminish the impact of this effect on the results of the *in vivo* fluorescence imaging, the mice were given feed not including fluorescent components (alfalfa-free feed) for at least 1 week. The autofluorescence intensity of the alfalfa-free feed was reported to be about ten-fold lower than that of normal feed (Fig. 10.7d).

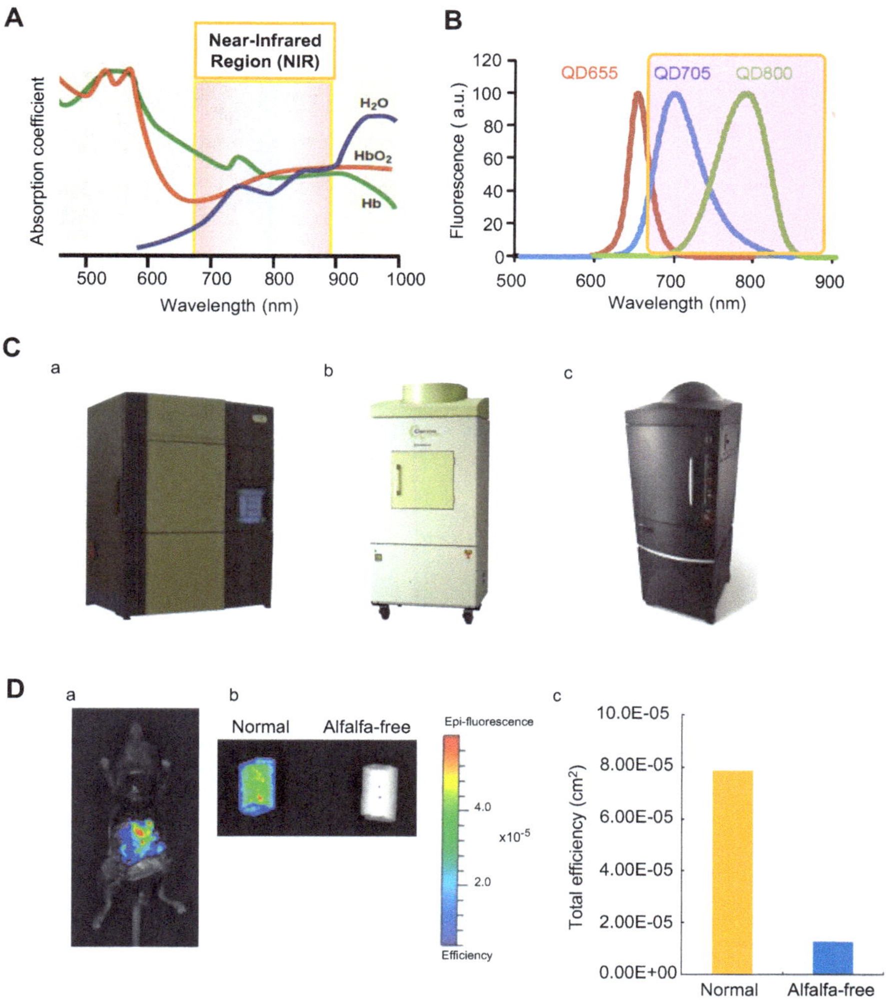

Fig. 10.7 Biological window in near-infrared region and comparison of the fluorescence intensity between normal with alfalfa-free feed. (**a**) Biological window in near-infrared region (about 700–900 nm) (**b**) Fluorescence spectra of QDs655, QDs705, and QDs800 in the culture medium. (**c**) *In vivo* imaging fluorescence system. *a*: The Maestro™ (PerkinElmer), *b*: Clair*vivo* OPT (SHIMADZU), *c*: IVIS Imaging System (PerkinElmer) (**d**) Comparison of the fluorescence intensity between normal with alfalfa-free feed. *a*: *In vivo* fluorescence image of mouse fed on normal feed. *b*: Fluorescence images of normal and alfalfa-free feed. *c*: Fluorescence intensity of normal and alfalfa-free feed. (Reprinted with permission from Ref. [137])

10.5.2 *Subcutaneous Transplantation*

In vivo fluorescence imaging of subcutaneously transplanted stem cells labeled with QDs into mice has been frequently reported [87, 124]. In fact, the fluorescence

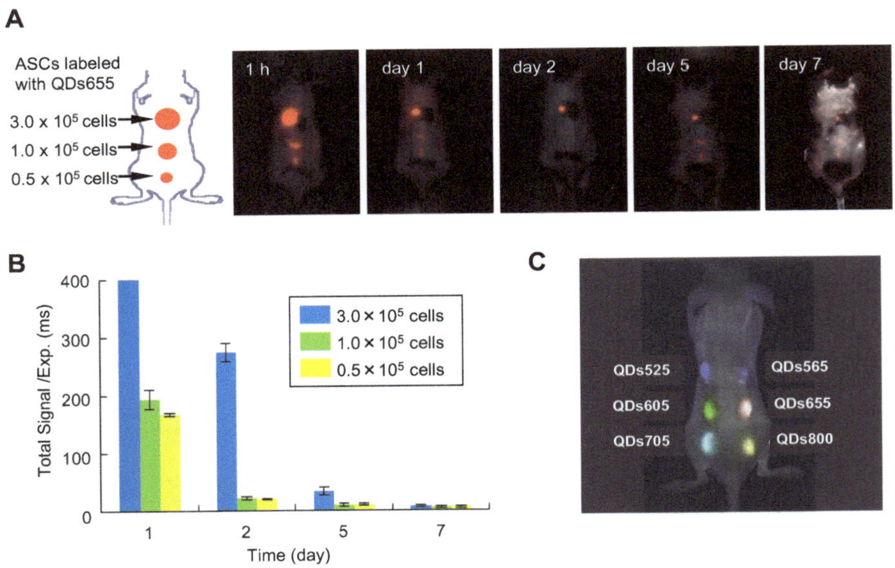

Fig. 10.8 Detection and multiplex imaging capability of QDs in subcutaneous transplantation. A: *In vivo* fluorescence images of mouse subcutaneously transplanted stem cells (0.5, 1.0 and 3.0 × 105 cells) labeled with QDs655 (0.8 nM) using R8 into the backs of the mice after 1 h, 1, 2, 5 and 7 days. B: Bar graph of the changes of fluorescence intensity at the number of stem cells labeled with QDs655 using R8 for 7 days. C: *In vivo* fluorescence Image of mouse subcutaneously transplanted stem cells labeled with QDs525, 565, 605, 655, 705 and 800 into the back of the mice after 10 min with a single excitation light source. (Reprinted with permission from Ref. [137])

intensity derived from QDs could be detected and analyzed quantitatively by using an *in vivo* fluorescence imaging system after the subcutaneous transplantation of different numbers of stem cells labeled with QDs into the back or other sites of mice. Subcutaneously transplanted stem cells labeled with QDs could be detected clearly at the level of several thousand cells over a period of about 1 week, and the fluorescence was QD dose-dependent [90] (Fig. 10.8a and b). However, the stem cells labeled with QDs could proliferate rapidly, so the sensitivity of detection was decrease in inverse proportion to the proliferation of the stem cells.

Moreover, the multiplex imaging of subcutaneously transplanted stem cells has been reported by utilizing the fluorescence characteristics of QDs [90, 125]. In fact, stem cells were labeled with different kinds of QDs emitting fluorescent light from 525 to 800 nm, then the stem cells could be detected in different colors using excitation of the same wavelength at high resolution [125] (Fig. 10.8c). The availability of QDs emitting fluorescence in the near-infrared region is relatively low for subcutaneous transplantation, because visible light can enter through the skin to some extent. This technology is expected to be useful for *in vivo* fluorescence imaging of different cell populations included in regenerative tissues and organs at the same time.

10.5.3 Intravenous and Other Transplantation

Stem cell therapy via intravenous transplantation has been expected to be useful in clinical applications for some diseases of the lungs, liver and pancreas [127–129]. Adipose tissue-derived stem cells [ASCs], bone marrow-derived stem cells, hematopoietic stem cells and progenitor cells derived from these stem cells have all been used for stem cell therapy [84–86, 129]. *In vivo* fluorescence imaging of intravenously transplanted stem cells labeled with QDs may make it possible to trace the transplanted stem cells *in vivo* and analyze their rate of accumulation into specific tissues or organs [87, 88].

In a previous study, adipose tissue-derived stem cells labeled with QDs were intravenously transplanted into mouse models of emphysema or acute liver disease, and then the location and the rate of accumulation of stem cells in major organs, such as the heart, kidneys, lungs, liver and spleen, were investigated [123, 130]. In the emphysema model, the transplanted ASCs were observed in the lungs at 1 and 4 h after transplantation, and more ASCs remained in the lungs with emphysema compared with the lungs from normal mice [130] (Fig. 10.9a). In the model of acute liver disease, the transplanted ASCs were found to accumulate more in the liver when there was simultaneous administration of heparin compared to when the ASCs were transplanted alone [123] (Fig. 10.9b, c, and d).

In these cases, QDs with near-infrared fluorescence were useful for the *in vivo* fluorescence imaging of transplanted stem cells, because near-infrared fluorescence can be seen through the skin and organs at high efficiency. In fact, intravenously transplanted ASCs labeled with QDs800 (fluorescence peak at 800 nm) could be detected in the lungs and liver even without laparotomy; whereas transplanted ASCs labeled with QDs655 (fluorescence peak at 800 nm) could not be detected [123] (Fig. 10.9b, c, and d).

On the other hand, bone marrow-derived stem cells labeled with QDs were transplanted in the ipsilateral striatum of a rat model of cerebral infarct, then the impact of the timing and stem cell dose on the therapeutic effects were determined by *in vivo* fluorescence imaging of transplanted stem cells labeled with QDs [131]. Neural stem cells labeled with QDs were transplanted into the striatum contralateral to the ischemic hemisphere, then the therapeutic effects were confirmed by *in vivo* imaging of transplanted neural stem cells labeled with QDs. The findings of that study suggest that the *in vivo* fluorescence imaging of transplanted stem cells labeled with QDs may assume increasing importance in association with the development of stem cell therapy for regenerative medicine.

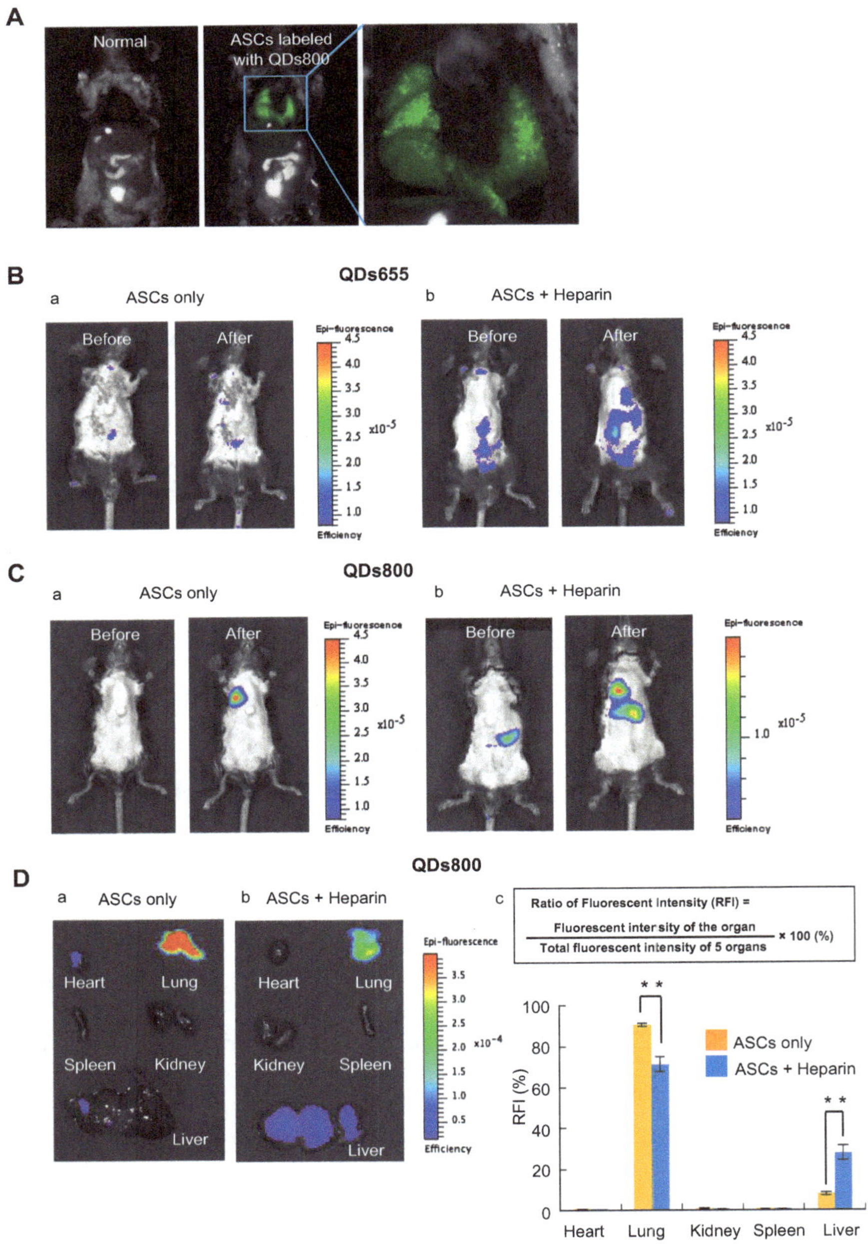

Fig. 10.9 *In vivo* and *ex vivo* imaging of stem cells labeled with QDs using R8 after intravenous injection. (**a**) *In vivo* fluorescence images of mouse transplanted stem cells labeled with QDs800 using R8 through the tail vain into mouse after 10 min. (**b**) *In vivo* fluorescence images of mice with acute liver failure after transplantation of stem cells labeled with QDs655 using R8 in combination without heparin (*a*) or with heparin (*b*) without laparotomy. (**c**) *In vivo* fluorescence images of mice with acute liver failure after transplantation of stem cells labeled with QDs800 using R8 in

10.6 Recent Progress and Views

10.6.1 Overcoming of Problems Caused by Using Quantum Dots

Three major problems caused by using QDs, such as assurance of the safety, realization of high functionality, and barrier of national regulation, appear to be resolved for the clinical application to stem cell therapy and regenerative medicine. Firstly, the assurance of safety is thought to be the most important problem. Much safer elements against several kinds of cells including iPS cells or bodies should be selected as the constituent elements of QDs to overcome this problem. Secondly, the realization of high functionality enables us to expand the extent of the application of QDs such as the detection of state changes of stem cells. Finally, the actual situation of the barrier of national regulation on regenerative medicine by using stem cells including iPS cells should be considered in each country.

CdSe and CdTe semiconductor materials have been generally used as the core of QDs, however these elements were reported to show the cytotoxicity at low concentration. Thus, the novel QDs based on silicon or $AgInS_2$ cores not including Cd, Se, and Te elements have been already developed, and these QDs were reported to have very low cytotoxicity [112, 132, 133]. On the other hand, QD-based sensing mechanisms such as FRET and multiple colors are thought to be useful for the detection of the state changes of stem cells. However, there are little information and reports on these technologies, thus the early development of these technologies is strongly expected [9–12]. Moreover, clinical research environment have been improved in recent years. New three laws associated with regenerative medicine have been established especially in Japan, so the barrier of national regulation on regenerative medicine has become lower all over the world. Collectively, QDs will actually be able to apply for stem cell therapy and regenerative medicine in the future with the further improvement and development.

10.6.2 Conclusions and Future Directions

Organic fluorophores have been widely used as fluorescent probes in conventional biometrics; however the research on QDs for biometrics has expanded greatly in recent years. The recent findings suggest that QDs can be used to detect and analyze

Fig. 10.9 (continued) combination without heparin (*a*) or with heparin (*b*) without laparotomy. (**d**) *Ex vivo* fluorescence images of five organs in mice with acute liver failure after transplantation of stem cells labeled with QDs800 using R8 without heparin (*a*) or with heparin (*b*), and ratio of the fluorescence intensity (RFI) of five organs (*c*). (Reprinted with permission from Ref. [137])

objects which conventional biometrics cannot detect. In this review, we described the fluorescence properties of QDs, and then introduced QD-based biometrics technologies, the "Bio-sensing" technology, Stem cell labeling technology and *In vivo* fluorescence imaging technology for transplanted stem cells. These technologies will likely contribute to regenerative medicine, especially stem cell therapy, which requires high-sensitivity imaging at the cellular level.

The potential clinical applications of QDs have been expanded by the development of considerably low cytotoxicity QDs that do not include cadmium (Cd) or selenium (Se) [112], as well as the development of the second near infrared region (1000–1500 nm) fluorescent QDs with strong permeability into the body [134, 135]. In fact, clinical trials based on the data obtained from basic research have been started in the U.S.A [136]., and research and development are becoming increasingly important. Moreover, the evolution of QDs including the development of hybrid materials of QDs and other functional molecules probably enables the diagnosis and therapy of stem cells *in vivo* at the same time, an approach termed "*in vivo* theranostics".

References

1. Alivisatos P (2004) The use of nanocrystals in biological detection. Nat Biotechnol 22:47–52
2. Michalet X, Pinaud FF, Bentolila LA, Tsay JM, Doose S, Li JJ, Sundaresan G, Wu AM, Gambhir SS, Weiss S (2005) Quantum dots for live cells, in vivo imaging, and diagnostics. Science 307(5709):538–544
3. Alivisatos P, Gu W, Larabell C (2005) Quantum dots as cellular probes. Annu Rev Biomed Eng 7:55–76
4. Dabbousi BO, Rodriguez-Viejo J, Mikulec FV, Heine JR, Mattoussi H, Ober R, Jensen KF, Bawendi MG (1997) (CdSe)ZnS core-shell quantum dots: synthesis and characterization of a size series of highly luminescent nanocrystallites. J Phys Chem B 101:9463–9475
5. Bruchez M, Moronne M, Gin P, Weiss S, Alivisatos AP (1998) Semiconductor nanocrystals as fluorescent biological labels. Science 281:2013–2016
6. Chan WCW, Nie S (1998) Quantum dot bioconjugates for ultrasensitive nonisotopic detection. Science 281:2016–2018
7. Sapsford KE, Pons T, Medintz IL, Mattoussi H (2006) Biosensing with luminescent semiconductor quantum dots. Sensors 6:925–953
8. Alivisatos AP (1996) Perspectives on the physical chemistry of semiconductor nanocrystals. J Phys Chem 100:13226–13239
9. Clapp AR, Medintz IL, Mattoussi H (2006) Forster resonance energy transfer investigations using quantum-dot fluorophores. ChemPhysChem 7:47–57
10. Medintz IL, Mattoussi H (2009) Quantum dot-based resonance energy transfer and its growing application in biology. Phys Chem Chem Phys 11:17–45
11. Chan WCW, Maxwell DJ, Gao X, Bailey RE, Han M, Nie S (2002) Luminescent quantum dots for multiplexed biological detection and imaging. Curr Opin Biotechnol 13:40–46
12. Algar WR, Krull UJ (2010) New opportunities in multiplexed optical bioanalyses using quantum dots and donor-acceptor interactions. Anal Bioanal Chem 398:2439–2449
13. Seydack M (2005) Nanoparticle labels in immunosensing using optical detection methods. Biosens Bioelectron 20:2454–2469

14. Deng Z, Zhang Y, Yue J, Tang F, Wei Q (2007) Green and orange CdTe quantum dots as effective pH-sensitive fluorescent probes for dual simultaneous and independent detection of viruses. J Phys Chem B 111:12024–12031
15. Algar WR, Krull UJ (2008) Quantum dots as donors in fluorescence resonance energy transfer for the bioanalysis of nucleic acids, proteins, and other biological molecules. Anal Bioanal Chem 391:1609–1618
16. Algar WR, Krull UJ (2008) Interfacial transduction of nucleic acid hybridization using immobilized quantum dots as donors in fluorescence resonance energy transfer. Langmuir 25:633–638
17. Wu X, Liu H, Liu J, Haley KN, Treadway JA, Larson JP, Ge N, Peale F, Bruchez MP (2003) Immunofluorescent labeling of cancer marker Her2 and other cellular targets with semiconductor quantum dots. Nat Biotechnol 21:41–46
18. Jain KK (2007) Applications of nanobiotechnology in clinical diagnostics. Clin Chem 53:2002–2009
19. Wagner M, Li F, Li J, Li X-F, Le X (2010) Use of quantum dots in the development of assays for cancer biomarkers. Anal Bioanal Chem 397:3213–3224
20. Liu W, Howarth M, Greytak AB, Zheng Y, Nocera DG, Ting AY, Bawendi MG (2008) Compact biocompatible quantum dots functionalized for cellular imaging. J Am Chem Soc 130:1274–1284
21. Algar WR, Krull UJ (2009) Toward a multiplexed solid-phase hybridization assay using quantum dots as donors in fluorescence resonance energy transfer. Anal Chem 81:4113–4120
22. Whitesides GM (2006) The origins and the future of microfluidics. Nature 441:368–373
23. Chen L, Algar WR, Tavares AJ, Krull UJ (2011) Toward a solid-phase nucleic acid hybridization assay within microfluidic channels using immobilized quantum dots as donors in fluorescence resonance energy transfer. Anal Bioanal Chem 399:133–141
24. Rejman J, Oberle V, Zuhorn IS, Hoekstra D (2004) Size-dependent internalization of particles via the pathways of clathrin- and caveolae-mediated endocytosis. Biochem J 377:159–169
25. Jiang W, Kim BY, Rutka JT, Chan WC (2008) Nanoparticle-mediated cellular response is size-dependent. Nat Nanotechnol 3:145–150
26. Chithrani BD, Ghazani AA, Chan WCW (2006) Determining the size and shape dependence of gold nanoparticle uptake into mammalian cells. Nano Lett 6:662–668
27. Geng Y, Dalhaimer P, Cai S, Tsai R, Tewari M, Minko T, Discher DE (2007) Shape effects of filaments versus spherical particles in flow and drug delivery. Nat Nanotechnol 2:249–255
28. Verma A, Uzun O, Hu Y, Hu Y, Han H-S, Watson N, Chen S, Irvine DJ, Stellacci F (2008) Surface-structure-regulated cell-membrane penetration by monolayerprotected nanoparticles. Nat Mater 7:588–595
29. Nel AE, Mäadler L, Velegol D, Xia T, Hoek EMV, Somasundaran P, Klaessig F, Castranova V, Thompson M (2009) Understanding biophysicochemical interactions at the nano-bio interface. Nat Mater 8:543–557
30. Nirmal M, Brus L (1999) Luminescence photophysics in semiconductor nanocrystals. Acc Chem Res 32:407–414
31. Norris DJ, Bawendi MG (1996) Measurement and assignment of the size-dependent optical spectrum in CdSe quantum dots. Phys Rev B Condens Matter 53:16338–16346
32. Correa-Duarte MA, Giersig M, Liz-Marzan LM (1998) Stabilization of CdS semiconductor nanoparticles against photodegradation by a silica coating procedure. Chem Phys Lett 286:497–501
33. Resch-Genger U, Grabolle M, Cavaliere-Jaricot S, Nitschke R, Nann T (2008) Quantum dots versus organic dyes as fluorescent labels. Nat Methods 5:763–775
34. Chu TC, Shieh F, Lavery LA, Levy M, Richards-Kortum R, Korgel BA, Ellington AD (2006) Labeling tumor cells with fluorescent nanocrystal-aptamer bioconjugates. Biosens Bioelectron 21:1859–1866
35. Gao X, Cui Y, Levenson RM, Chung LW, Nie S (2004) *In vivo* cancer targeting and imaging with semiconductor quantum dots. Nat Biotechnol 22:969–976
36. Zhang CY, Yeh HC, Kuroki MT, Wang TH (2005) Single-quantum-dot-based DNA nanosensor. Nat Mater 4:826–831

37. Akerman ME, Chan WC, Laakkonen P, Bhatia SN, Ruoslahti E (2002) Nanocrystal targeting *in vivo*. Proc Natl Acad Sci U S A 99:12617–12621
38. Lidke DS, Nagy P, Heintzmann R, Arndt-Jovin DJ, Post JN, Grecco HE, Jares- Erijman EA, Jovin TM (2004) Quantum dot ligands provide new insights into erbB/HER receptor-mediated signal transduction. Nat Biotechnol 22:198–203
40. Wang G, Zeng G, Wang C, Wang H, Yang B, Guan F, Li D, Feng X (2015) Biocompatibility of quantum dots (CdSe/ZnS) in human amnioticmembrane-derived mesenchymal stem cells *in vitro*. Biomed Pap Med Fac Univ Palacky Olomouc Czech Repub 159:227–233
41. Yang HN, Park JS, Jeon SY, Park W, Na K, Park KH (2014) The effect of quantum dot size and poly(ethylenimine) coating on the efficiency of gene delivery into human mesenchymal stem cells. Biomaterials 35:8439–8449
42. Narayanan K, Yen SK, Dou Q, Padmanabhan P, Sudhaharan T, Ahmed S, Ying JY, Selvan ST (2013) Mimicking cellular transport mechanism in stem cells through endosomal escape of new peptide-coated quantum dots. Sci Rep 3:2184
43. Jares-Erijman EA, Jovin TM (2003) FRET imaging. Nat Biotechnol 21:1387–1395
44. Bakalova R, Zhelev Z, Ohba H, Baba Y (2005) Quantum dot-conjugated hybridization probes for preliminary screening of siRNA sequences. J Am Chem Soc 127:11328–11335
45. Qian ZS, Shan XY, Chai LJ, Ma JJ, Chen JR, Feng H (2014) DNA nanosensor based on biocompatible graphene quantum dots and carbon nanotubes. Biosens Bioelectron 60:64–70
46. Onoshima D, Kaji N, Tokeshi M, Baba Y (2008) Nuclease tolerant FRET probe based on DNA-quantum dot conjugation. Anal Sci 24:181–183
47. Lee JA, Hung A, Mardyani S, Rhee A, Klostranec J, Mu Y, Li D, Chan WCW (2007) Toward the accurate read-out of quantum dot barcodes: design of deconvolution algorithms and assessment of fluorescence signals in buffer. Adv Mater 19:3113–3118
48. Liu KK, Wu RG, Chuang YJ, Khoo HS, Huang SH, Tseng FG (2010) Microfluidic systems for biosensing. Sensors 10:6623–6661
49. Yeo LY, Chang H-C, Chan PPY, Friend JR (2011) Microfluidic devices for bioapplications. Small 7:12–48
50. Choi S, Goryll M, Sin L, Wong P, Chae J (2011) Microfluidic-based biosensors toward point-of-care detection of nucleic acids and proteins. Microfluid Nanofluid 10:231–247
51. Rivet C, Lee H, Hirsch A, Hamilton S, Lu H (2011) Microfluidics for medical diagnostics and biosensors. Chem Eng Sci 66:1490–1507
52. Dittrich PS, Manz A (2005) Single-molecule fluorescence detection in microfluidic channels — the holy grail in μTAS? Anal Bioanal Chem 382:1771–1782
53. Joo C, Balci H, Ishitsuka Y, Buranachai C, Ha T (2008) Advances in single-molecule fluorescence methods for molecular biology. Annu Rev Biochem 77:51–76
54. Gorman J, Plys AJ, Visnapuu ML, Alani E, Greene EC (2010) Visualizing onedimensional diffusion of eukaryotic DNA repair factors along a chromatin lattice. Nat Struct Mol Biol 17:932–938
55. Kim S, Gottfried A, Lin RR, Dertinger T, Kim AS, Chung S, Colyer RA, Weinhold E, Weiss S, Ebenstein Y (2012) Enzymatically incorporated genomic tags for optical mapping of DNA-binding proteins. Angew Chem Int Ed 51:3578–3581
56. Michaeli Y, Shahal T, Torchinsky D, Grunwald A, Hocha R, Ebenstein Y (2013) Optical detection of epigenetic marks: sensitive quantification and direct imaging of individual hydroxymethylcytosine bases. Chem Commun 49:8599–8601
57. Onoshima D, Kaji N, Tokeshi M, Baba Y (2014) On-chip analysis of intermittent molecular encounters in nuclease digestion of specific DNA sequence. Biophys J 103:699a–700a
58. Smith AM, Dave S, Nie SM, True L, Gao XH (2006) Multicolor quantum dots for molecular diagnostics of cancer. Expert Rev Mol Diagn 6:231–244
59. Srivastava S, Srivastava R-G (2005) Proteomics in the forefront of cancer biomarker discovery. J Proteome Res 4:1098–1103
60. Cissell KA, Rahimi Y, Shrestha S, Hunt EA, Deo SK (2008) Bioluminescence-based detection of microRNA, miR21 in breast cancer cells. Anal Chem 80:2319–2325
61. Wagner M, Li F, Li JJ, Li X-F, Le XC (2010) Quantum dot based assays for cancer biomarkers. Anal Bioanal Chem 397:3213–3224

62. Wang HZ, Wang HY, Liang RQ, Ruan KC (2004) Detection of tumor marker CA125 in ovarian carcinoma using quantum dots. Acta Biochim Biophys Sin 36:681–686
63. Chen C, Peng J, Xia HS, Yang GF, Wu QS, Chen LD, Zeng LB, Zhang ZL, Pang DW, Li Y (2009) Quantumdots-based immunofluorescence technology for the quantitative determination of HER2 expression in breast cancer. Biomaterials 30:2912–2918
64. Barua S, Reqe K (2009) Cancer-cell-phenotype-dependent differential intracellular trafficking of unconjugated quantum dots. Small 5:370–376
65. Tian J, Zhou L, Zhao Y, Wang Y, Peng Y, Zhao S (2012) Multiplexed detection of tumor markers with multicolor quantum dots based on fluorescence polarization immunoassay. Talanta 92:72–77
66. Han M, Gao X, Su J, Nie S (2001) Quantum-dot-tagged microbeads for multiplexed optical coding of biomolecules. Nat Biotechnol 19:631–635
67. Zhang Y, Zhang C-Y (2012) Sensitive detection of microRNA with isothermal amplification and a single-quantum-dot-based nanosensor. Anal Chem 84:224–231
68. Medintz IL, Mattoussi H, Clapp AR (2008) Potential clinical applications of quantum dots. Int J Nanomedicine 3:151–167
69. Mulligan RC (1993) The basic science of gene therapy. Science 260:926–932
70. Chen X, Deng Y, Lin Y, Pang DW, Qing H, Qu F, Xie HY (2008) Quantum dot-labeled aptamer nanoprobes specifically targeting glioma cells. Nanotechnology 19:235105
71. Erogbogbo F, Yong K-T, Roy I, Xu G, Prasad PN, Swihart MT (2008) Biocompatible luminescent silicon quantum dots for imaging of cancer cells. ACS Nano 2:873–878
72. Kirchhausen T (2000) Three ways to make a vesicle. Nat Rev Mol Cell Biol 1:187–198
73. Parak WJ, Boudreau R, Le Gros M, Gerion D, Zanchet D, Micheel CM, Williams SC, Alivisatos AP, Larabell C (2002) Cell Motility and Metastatic Potential Studies Based on Quantum Dot Imaging of Phagokinetic Tracks. Adv Mater 14:882–885
74. Derfus AM, Chan WCW, Bhatia SN (2004) Intracellular delivery of quantum dots for live cell labeling and organelle tracking. Adv Mater 16:961–966
75. Delehanty JB, Mattoussi H, Medintz IL (2009) Delivering quantum dots into cells: strategies, progress and remaining issues. Anal Bioanal Chem 393:1091–1105
76. Dimos JT, Rodolfa KT, Niakan KK, Weisenthal LM, Mitsumoto H, Chung W, Croft GF, Saphier G, Leibel R, Goland R, Wichterle H, Henderson CE, Eggan K (2008) Induced pluripotent stem cells generated from patients with ALS can be differentiated into motor neurons. Science 321:1218–1221
77. Takahashi K, Tanabe K, Ohnuki M, Narita M, Ichisaka T, Tomoda K, Yamanaka S (2007) Induction of pluripotent stem cells from adult human fibroblasts by defined factors. Cell 131:861–872
78. Narazaki G, Uosaki H, Teranishi M, Okita K, Kim B, Matsuoka S, Yamanaka S, Yamashita JK (2008) Directed and systematic differentiation of cardiovascular cells from mouse induced pluripotent stem cells. Circulation 118:498–506
79. Okita K, Ichisaka T, Yamanaka S (2007) Generation of germline-competent induced pluripotent stem cells. Nature 448:313–317
80. Hsieh SC, Wang FF, Hung SC, Chen YJ, Wang YJ (2006) The internalized CdSe/ZnS quantum dots impair the chondrogenesis of bone marrow mesenchymal stem cells. J Biomed Mater Res B Appl Biomater 79:95–101
81. Shah BS, Clark PA, Moioli EK, Stroscio MA, Mao JJ (2007) Labeling of mesenchymal stem cells by bioconjugated quantum dots. Nano Lett 7:3071–3079
82. Zuk PA, Zhu M, Ashjian P, De Ugarte DA, Huang JI, Mizuno H, Alfonso ZC, Fraser JK, Benhaim P, Hedrick MH (2002) Human adipose tissue is a source of multipotent stem cells. Mol Biol Cell 13:4279–4295
83. Seo MJ, Suh SY, Bae YC, Jung JS (2005) Differentiation of human adipose stromal cells into hepatic lineage *in vitro* and *in vivo*. Biochem Biophys Res Commun 328:258–264
84. Amariglio N, Hirshberg A, Scheithauer BW (2009) Donor-derived brain tumor following neural stem cell transplantation in an ataxia telangiectasia patient. PLoS Med 6:e29

85. Menasche P (2009) Stem cell therapy for heart failure: are arrhythmias a real safety concern? Circulation 119:2735–2740
86. Strong M, Farrugia A, Rebulla P (2009) Stem cell and cellular therapy developments. Biologicals 37:103–107
87. Nguyen PK, Nag D, Wu JC (2010) Methods to assess stem cell lineage, fate and function. Adv Drug Deliv Rev 62:1175–1186
88. Li SC, Tachiki LML, Luo J, Dethlefs BA, Chen Z, Loudon WG (2010) A biological global positioning system: considerations for tracking stem cell behaviors in the whole body. Stem Cell Rev Rep 6:317–333
89. Wang Y, Xu C, Ow H (2013) Commercial nanoparticles for stem cell labeling and tracking. Theranostics 3:544–560
90. Rosenzweig A (2006) Cardiac cell therapy:mixed results frommixed cells. N Engl JMed 355:1274–1277
91. Yukawa H, Kagami Y, Watanabe M, Kaji N, Okamoto Y, Tokeshi M, Noguchi H, Miyamoto Y, Baba Y, Hamajima N, Hayashi S (2010) Quantum dots labeling using octaarginine peptides for imaging of adipose tissue-derived stem cells. Biomaterials 31:4094–4103
92. Slotkin JR, Chakrabarti L, Dai HN (2007) *In vivo* quantum dot labeling of mammalian stem and progenitor cells. Dev Dyn 236:3393–3401
93. Bakalova R, Ohba H, Zhelev Z, Ishikawa M, Baba Y (2004) Quantum dots as photosensitizers? Nat Biotechnol 22:1360–1361
94. Yukawa H, Mizufune S, Mamori C, Kagami Y, Oishi K, Kaji N, Okamoto Y, Tokeshi M, Noguchi H, Hamaguchi M, Hamajima N, Baba Y, Hayashi S (2009) Quantum dots for labeling adipose tissue-derived stem cells. Cell Transplant 18:591–599
95. Sun C, Cao Z, Wu M, Lu C (2014) Intracellular tracking of single native molecules with electroporation-delivered quantum dots. Anal Chem 86:11403–11409
96. Chang JC, Su HL, Hsu SH (2008) The use of peptide-delivery to protect human adiposederived adult stem cells from damage caused by the internalization of quantum dots. Biomaterials 29:925–936
97. Toita S, Hasegawa U, Koga H, Sekiya I, Muneta T, Akiyoshi K (2008) Protein-conjugated quantum dots effectively delivered into living cells by a cationic nanogel. J Nanosci Nanotechnol 8:2279–2285
98. Lagerholm BC (2007) Peptide-mediated intracellular delivery of quantum dots. Methods Mol Biol 374:105–112
99. Derfus AM, Chan WCW, Bhatia SN (2004) Intracellular delivery of quantum dots for live cell labeling and organelle tracking. Adv Mater 16:961–964
100. Zabner J, Fasbender AJ, Moninger T, Poellinger KA, Welsh MJ (1995) Cellular and molecular barriers to gene transfer by a cationic lipid. J Biol Chem 270:18997–19007
101. Biju V, Itoh T, Ishikawa M (2010) Delivering quantum dots to cells: bioconjugated quantumdots for targeted and nonspecific extracellular and intracellular imaging. Chem Soc Rev 39:3031–3056
102. Felgner PL, Gadek TR, Holm M, Roman R, Chan HW, Wenz M, Northrop JP, Ringold GM, Danielsen M (1987) Lipofection: a highly efficient, lipid-mediated DNAtransfection procedure. Proc Natl Acad Sci U S A 84:7413–7417
103. Hsieh SC, Wang FF, Lin CS, Chen YJ, Hung SC, Wang YJ (2006) The inhibition of osteogenesis with human bone marrow mesenchymal stem cells by CdSe/ZnS quantum dot labels. Biomaterials 27:1656–1664
104. Lagerholm BC, Wang MM, Ernst LA, Ly DH, Liu HJ, Bruchez MP, Waggoner AS (2004) Multicolor coding of cells with cationic peptide coated quantum dots. Nano Lett 4:2019–2022
105. Yukawa H, Noguchi H, Oishi K, Miyamoto Y, Nakase K, Futaki S, Hamaguchi M, Hamajima N, Hayashi S (2010) Transduction of cell-penetrating peptide into iPS cells. Cell Transplant 19:901–909

106. Fawell S, Seery J, Daikh Y, Moore C, Chen LL, Pepinsky B, Barsoum J (1994) Tatmediated delivery of heterologous proteins into cells. Proc Natl Acad Sci U S A 91:664–668
107. Wadia JS, Stan RV, Dowdy SF (2004) Transducible TAT–HA fusogenic peptide enhances escape of TAT-fusion proteins after lipid raft macropinocytosis. Nat Med 10:310–315
108. Elliott G, O'Hare P (1997) Intercellular trafficking and protein delivery by a herpesvirus structural protein. Cell 88:223–233
109. Schwarze SR, Hruska KA, Dowdy SF (2000) Protein transduction: unrestricted delivery into all cells? Trends Cell Biol 10:290–295
110. Silver J, Ou W (2005) Photoactivation of quantum dot fluorescence following endocytosis. Nano Lett 5:1445–1449
111. Delehanty JB, Medintz IL, Pons T, Brunel FM, Dawson PE, Mattoussi H (2006) Selfassembled quantum dot-peptide bioconjugates for selective intracellular delivery. Bioconjug Chem 17:920–927
112. Takasaki Y, Watanabe M, Yukawa H, Sabarudin A, Inagaki K, Kaji N, Okamoto Y, Tokeshi M, Miyamoto Y, Noguchi H, Umemura T, Hayashi S, Baba Y, Haraguchi H (2011) Estimation of the distribution of intravenously injected adipose tissue-derived stem cells labeled with quantum dots in mice organs through the determination of their metallic components by ICPMS. Anal Chem 83:8252–8258
113. Miyazaki Y, Yukawa H, Nishi H, Okamoto Y, Kaji N, Torimoto T, Baba Y (2013) Adipose tissue-derived stem cell imaging using cadmium-free quantum dots. Cell Med 6:91–97
114. Noguchi H, Matsumoto S (2006) Protein transduction technology: a novel therapeutic perspective. Acta Med Okayama 60:1–11
115. Lei Y, Tang H, Yao L, Yu R, Feng M, Zou B (2008) Applications of mesenchymal stem cells labeled with tat peptide conjugated quantum dots to cell tracking in mouse body. Bioconjug Chem 19:421–427
116. Chang JC, Hsu SH, Su HL (2009) The regulation of the gap junction of human mesenchymal stemcells through the internalization of quantumdots. Biomaterials 30:1937–1946
117. Yukawa H, Suzuki K, Kano Y, Yamada T, Kaji N, Ishikawa T, Baba T (2013) Induced pluripotent stemcell labeling using quantumdots. Cell Med 2013(6):83–90
118. Chen B, Liu QL, Zhang YL, Xu L, Fang XH (2008) Transmembrane delivery of the cellpenetrating peptide conjugated semiconductor quantum dots. Langmuir 24:11866–11871
119. Phelps ME (2000) Inaugural article: positron emission tomography provides molecular imaging of biological processes. Proc Natl Acad Sci U S A 97:9226–9233
120. Wu AM, Yazaki PJ, Tsai S et al (2000) High-resolution microPET imaging of carcinoembryonic antigen-positive xenografts by using a copper-64-labeled engineered antibody fragment. Proc Natl Acad Sci U S A 97:8495–8500
121. Lee H, Kim IK, Park TG (2010) Intracellular trafficking and unpacking of siRNA/quantum dot-PEI complexes modified with and without cell penetrating peptide: confocal and flow cytometric FRET analysis. Bioconjug Chem 21:289–295
122. Hsieh YH, Liu SJ, Chen HW, Lin YK, Liang KS, Lai LJ (2010) Highly sensitive rare cell detection based on quantum dot probe fluorescence analysis. Anal Bioanal Chem 396:1135–1141
123. Prasuhn DE, Blanco-Canosa JB, Vora GJ, Delehanty JB, Susumu K, Mei BC, Dawson PE, Medintz IL (2010) Combining chemoselective ligation with polyhistidinedriven self-assembly for the modular display of biomolecules on quantum dots. ACS Nano 4:267–278
124. Yukawa H, Watanabe M, Kaji N, Okamoto Y, Tokeshi M, Miyamoto Y, Noguchi H, Baba Y, Hayashi S (2012) Monitoring transplanted adipose tissue-derived stem cells combined with heparin in the liver by fluorescence imaging using quantum dots. Biomaterials 33:2177–2186
125. Biju V (2014) Chemical modifications and bioconjugate reactions of nanomaterials for sensing, imaging, drug delivery and therapy. Chem Soc Rev 43:737–962
126. Lin S, Xie X, Patel MR (2007) Quantum dot imaging for embryonic stem cells. BMC Biotechnol 7:67

127. Seleverstov O, Zabirnyk O, Zscharnack M, Bulavina L, Nowicki M, Heinrich JM (2006) Quantum dots for human mesenchymal stem cells labeling: a size-dependent autophagy activation. Nano Lett 6:2826–2832
128. Zhang T, Stilwell JL, Gerion D, Ding I, Elboudwarej O, Cooke PA (2006) Cellular effect of high doses of silica-coated quantum dot profiled with high throughput gene expression analysis and high content cellomics measurements. Nano Lett 6:800–808
129. Zhang Y, He J, Wang PN, Chen JY, Lu ZJ, Lu DR (2006) Time-dependent photoluminescence blue shift of the quantum dots in living cells: effect of oxidation by singlet oxygen. J Am Chem Soc 128:13396–13401
130. Powles R (2010) 50 years of allogeneic bone-marrow transplantation. Lancet Oncol 11:305–306
131. Kim YS, Kim JY, Shin DM, Huh JW, Lee SW, Oh YM (2014) Tracking intravenous adipose-derivedmesenchymal stemcells in a model of elastase-induced emphysema. Tuberc Respir Dis (Seoul) 77:116–123
132. Wen X, Wang Y, Zhang F, Zhang X, Lu L, Shuai X, Shen J (2014) *In vivo* monitoring of neural stem cells after transplantation in acute cerebral infarction with dual-modal MR imaging and optical imaging. Biomaterials 35:4627–4635
133. Shinchi H, Wakao M, Nagata N, Sakamoto M, Mochizuki E, Uematsu T, Kuwabata S, Suda Y (2014) Cadmium-free sugar-chain-immobilized fluorescent nanoparticles containing low-toxicity ZnS–AgInS2 cores for probing lectin and cells. Bioconjug Chem 25:286–295
134. Subramaniam P, Lee SJ, Shah S, Patel S, Starovoytov V, Lee KB (2012) Generation of a library of non-toxic quantum dots for cellular imaging and siRNA delivery. Adv Mater 24:4014–4019
135. Ozturk SS, Selcuk F, Acar HY (2010) Development of color tunable aqueous CdScysteine quantum dots with improved efficiency and investigation of cytotoxicity. J Nanosci Nanotechnol 10:2479–2488
136. Kawashima N, Nakayama K, Itoh K, Itoh T, Ishikawa M, Biju V (2010) Reversible dimerization of EGFR revealed by single-molecule fluorescence imaging using quantum dots. Chem Eur J 16:1186–1192
137. Burns AA, Vider J, Ow H, Herz E, Medina OP, Baumgart M, Larson SM, Wiesner U, Bradbury M (2008) Fluorescent silica nanoparticles with efficient urinary excretion for nanomedicine. Nano Lett 9:442–448
138. Onoshima D, Yukawa H, Baba Y (2015) Multifunctional quantum dots-based cancer diagnostics and stem cell therapeutics for regenerative medicine. Adv Drug Deliv Rev 95:2–14

Chapter 11
Nanopore Device for Single-Molecule Sensing Method and Its Application

Masateru Taniguchi and Takahito Ohshiro

Abstract Nanopore analysis is very promising for single-molecule sensing platform. The feature of nanopore platform is a simple, high-throughput single-molecule/particle detection of a wide range of analytes at low-cost at single-molecule level. The single-molecule sensing ability of nanopore device have been utilized for single-molecule DNA sequencing, protein, peptide and carbohydrates detection, and so on. Due to recent progress on the improvement of selectivity, molecular control, fabrication technique, the nanopore platform became "smarter". Therefore, the applicability of nanopore–sensing methodology are expanding not only for basic research fields but also for medical applications, such as disease diagnosis, drug screening, virus detection.

Herein this chapter, background of nanopore studies (Sect. 11.1), principle of nanopore sensing (Sect. 11.2), the nanopore-fabrication technique (Sect. 11.3), application studies of nanopore sensing (Sect. 11.4), the selective and accuracy improvement studies of nanopore-sensing (Sect. 11.5), recent novel nanopore platform studies (Sect. 11.6), and the summary and future of the nanopore-sensing method (Sect. 11.7) are introduced.

Keywords Nanopore · Single-molecule analysis · Bio-nanopore · Solid-sate nanopore

11.1 Introduction

Development of methodology of identifying molecules and/or detecting / monitoring molecular behaviors at the single-molecular level is one of the important research issues for chemistry and biology. As one of the promising methodologies, a nanopores based single molecule analysis method has been extensively studied [7, 37, 49]. A nanopore is a pore with nanometer-size diameter, which is comparable

M. Taniguchi (✉) · T. Ohshiro
Osaka University, Ibaraki, Osaka, Japan
e-mail: taniguti@sanken.osaka-u.ac.jp; toshiro@sanken.osaka-u-ac.jp

M. Tokeshi (ed.), *Applications of Microfluidic Systems in Biology and Medicine*, Bioanalysis 7, https://doi.org/10.1007/978-981-13-6229-3_11

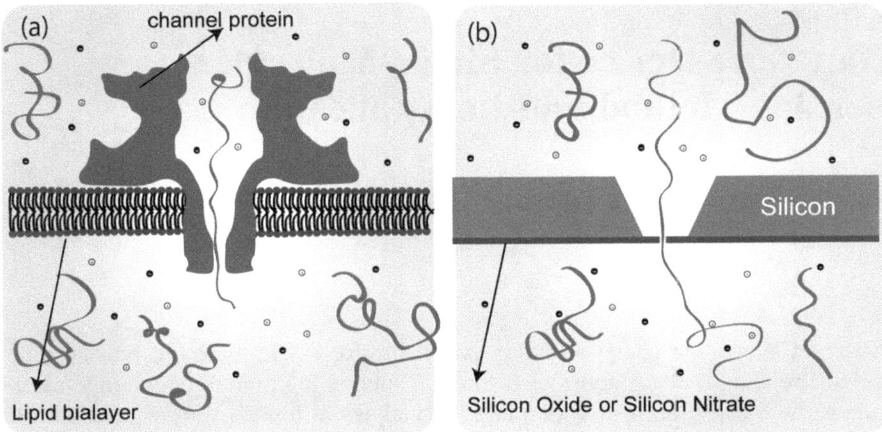

Fig. 11.1 Bio-nanopore and Solid-state nanopore: The nanopores are classified into "biological nanopores (Bio-nanopore)" (**a**) and "Solid-state nanopores" (**b**). (a) Bio-nanopore is produced by insertion of transmembrane channel-proteins in lipid membranes (See left figure). (b) Solid-state nanopore is drilled by etching, ion-beam, electron-beam lithography in an insulating thin-membrane, such as silicon, silicon oxide, silicon nitride, graphene and so on. Ideally, a single-molecule can translocate these formed nanopores across the membranes one by one. By using the nanopore-sensing, single-molecule biopolymers, such as DNA, RNA, peptide, protein, carbohydrates and sugar complexes, and so on, can be detectable

to a single-molecule size and it can be created by a pore-forming protein in lipid bilayers, "bio-nanopore", [17, 37] or as a through-hole in synthetic materials, such as silicon-nitrate or graphene, "solid-state-nanopore" [26, 74] (Fig. 11.1). These nanopores can interact with comparable-size sample molecules, and their phenomena induces some kinds of physical disturbances around the pore, resulting in detection of sample-molecular behaviors as characteristic analyte signals. The feature of nanopore based sensor is a simple, high-throughput single-molecule/particle detection of a wide range of analytes at low-cost. Therefore, the methodology become attracted not only for basic research fields but also for medical applications.

11.2 Background and Principle of Nanopore Sensing

Among nanopore based sensors, electric resistance-measurement type sensors have been well-known. As the first conventional electrical sensing method, a single-particle sensor using pores (Coulter Counter) was developed by W. H. Coulter in 1953 [22]. This is a method of measuring conductivity of pore by a change in ion current during analyte-particle translocation (Fig. 11.2). The sensing principle is as follow: This method consists of two chambers partitioned by an electrically insulated membrane having a single pore. Each of chambers is filled with an electrolyte solution and arranged with Ag/AgCl electrodes, respectively, which apply

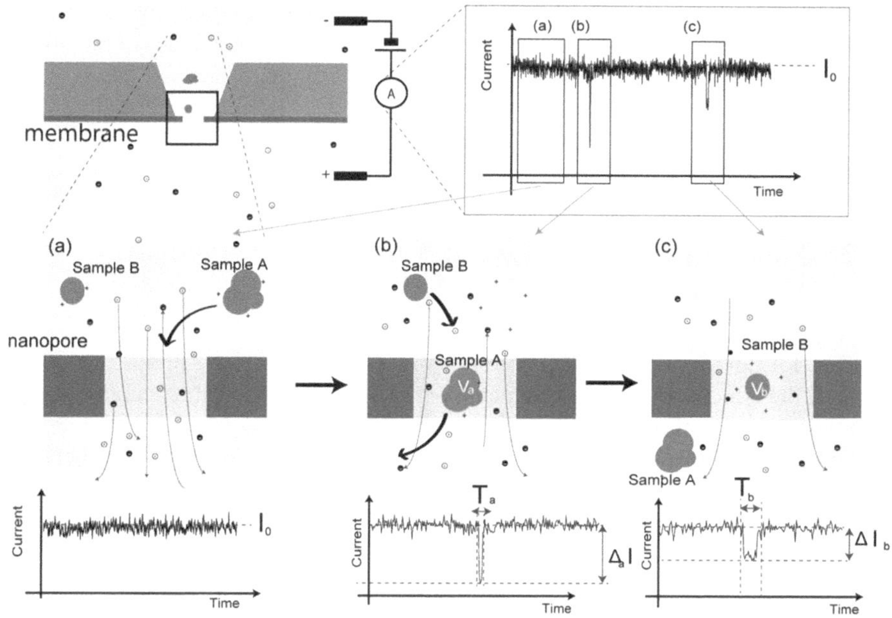

Fig. 11.2 Principles of nanopore electrical sensing. An insulated thin membrane partitions an electrolyte solution into two chambers and two Ag/AgCl electrodes are placed into each of the chamber solutions to provide dc-voltage and record electrical current through the pore. The dc-electric field causes the electrolyte ions in the solution to move only through the pore by electrophoresis. (**a**) In the "open" state, the stable background current (I_0) flows. (**b** and **c**) When a sample molecule (Sample A or B) penetrates into its pores, the ion-current flow is prevented, and current-drop signals are observed (See middle and right figures). In the "close" state, the decrement of current signal (ΔI_a for Sample A and ΔI_b for Sample B) are closely related to the excluded volume of molecules (V_a, for Sample A, and V_b for Sample B). The duration time (T_a and T_b) is time length of the sample translocation across nanopore. The difference is influenced by the sample characters such as molecular shape, orientation and the interaction between sample and side-wall of nanopore. From these information of signals can serve to identify sample molecules in a mixed sample solution

dc-electric field between chambers across the pores. The dc-electric field causes the electrolyte ions in chamber solution to move only through the pore by electrophoresis and this ion-current flow is monitored. When a single particle penetrates into its pores, the excluded volume of that particle decreases the electrical conductivity of the pore and decreases the ionic current. From the information on the magnitude, transit time, and passage frequency of the ion current-change signal of the pore generated at this time, it is possible to detect/identify the sample particle translocated through the pore. As an analysis for identification and counting of blood cells and cultured cells, this resistive measurement method is routinely used and become indispensable in the clinical uses instead of conventional time-consuming visual-type cell-examinations. Essentially, the analyzable-size of sample by using this sensing method depends on the pore-size so that the nanopore preparation/

fabrication technique is a crucial for single-molecule nanopore detection. By recent development of the nanopore preparation/fabrication technique, the use of nanopore become reproducible and reliable so that the nanopore electric resistance-measurement methods became a powerful methodology for "single-molecule" sensing method. The detail is described in Sect. 11.3.

11.3 Fabrication/Preparation of Nanopore for Nanopore Analysis

The key for development of nanopore sensors is the fabrication / production technique of the nanopores because the nanopore-shape and its robustness significantly influence the reliability and stability of signal detection. In this section, the recent nanopore fabrication studies are described. The nanopores are roughly classified into two categories. One is "biological nanopores (bio-nanopore)", which is derived from cell channel proteins, and the other is "solid-state nanopores, which are artificially created by micro−/nano-fabrication technology. Each of the techniques and their features are introduced in below.

Bio-nanopore sensors (Table 11.1) relies on the use of transmembrane channel-proteins, "porins", which are embedded in lipid membranes. These channel proteins typically have pores with nanometer-scale diameter, which can passively diffuse smaller-pore-size biomolecules across the membranes. In nanopore experiments, α-hemolysin channel protein (αHL), which is a channel protein of bacteria that causes lysis of red blood cells has, has been extensively used. Recently, the use of Mycobacterium smegmatis porin A (MspA), as bio-nanopore, is also growing because of the better ability of ion-current blockage, which is determined by the volume of the narrowest region of the pore protein. The most outstanding feature of bio-nanopore is that the uniform pores-structure can be produced by the self-organizing ability of the biomolecule. This feature serves reproducible atomic-

Table 11.1 Characteristics of bionanopores

Material	Diameter (nm)	Analytes	Spatial resolution	Refs.
α-hemolysin	1.5	ssDNA	Single base molecules	*1
		ssRNA	Single RNA molecules	*2
		microRNA	Single microRNA	*3
		Protein	Single protein	*4
MspA	1.2	ssDNA	Single base molecules	*5
		5mC, 5hmC	Single base molecules	*6
Aerolysin	1.0	Protein	Single protein	*7
		Oligosacchaaride	Single molecules	*8
N. farcinica	< 2.0	Peptide	Single peptide	*9
ClyA	3.3	ssDNA, dsDNA	Single DNA	*10

*1: [21]; *2: [23], [5], [4]; *3: [134], [112], [122]; *4: [92], [79]; *5: [71]; *6: [63]; *7: [86]; *8: [29]; *9: [101]; *10: [32]

precision geometries for bio-nanopore and reproducible current-signal by the bio-nanopore sensing. Although there is small degree of freedom in choosing the pore-size, the protein engineering technique can tune the pore-diameter, structures and charge-state by single-point mutations, resulting in improvement of detection abilities with the nanopore and/ or translocation behaviors of analytes [15, 50]. Therefore, in bio-nanopore sensing, the optimal nanopore is determined by the character of target molecules.

Solid-state nanopores (Table 11.2) are typically made of thin supporting membrane of insulated materials by top-down technique, mainly, semiconductor fabrication processing techniques. Recent development on these process techniques realize the high-reproducible mass-production of these solid-state nanopore structures. Solid-state nanopore production has been typically implemented by fabricating / sculpting a pore in an insulating thin-film materials by electron beam (EB) or an ion beam (FIB) [16, 65, 107]. They successfully demonstrated the fabrication of a pore with single-nanometer diameters. As synthetic materials of solid-state nanopore, silicon-nitrite (SiN), silicon-oxide, and aluminum oxide have been extensively studied. In addition, solid-state nanopores is recently reported using HfO_2, which is stronger than SiN and has a better chemical resistance [62]. In general, in order to achieving sufficient spatial resolution for the identification of a small molecules, thinner non-conductive film materials, i.e., 2D material membranes, are ideal because the ionic current blockade is generally determined by the volume of the molecule present in the nanopore. As the candidate of 2D materials, single and several-atomic layer material-based solid-state nanopores has also been explored. Among them, the interest for "graphene" based nanopore is increasing [33, 73, 106, 118, 129]. Graphene, a single layer of carbon atoms, is one of the ideal materials for detection of small molecules by solid-state nanopore because of atomically thinness

Table 11.2 Characteristics of solid-state nanopore

Material	Diameter (nm)	Analytes	Refs.
SiN	3.0	RNA + drug	*1
	3.7	dsDNA+ɣ-PNA	*2
$SiN/SiO_2/SiN$	20	Histone nucleosome	*3
$HfO_2/SiNx$	< 2 (HfO_2)	ssDNA, dsDNA	*4
$Graphene/SiO_2$	22 (graphene: 1 layer)	λ-DNA	*5
Graphene/SiN	5 (graphene: 1-2 layer)	10 kb DNA	*6
$Al_2O_3/graphene/Al_2O_3$	9	dsDNA, dsDNA+protein	*7
TiO_2/graphene/ SiN /SiO_2	8 (graphene: 3-15 layer)	15 kb DNA	*8
Chemical modified graphene	5-15	ssDNA	*9
$MoS_2/SiNx$	5 (MoS_2: 1-3 layer)	λ-DNA	*10
DNA origami/SiN	15	dsDNA	*11
Chemical modified SiN	25	Protein	*12
Chemical modified Au/SiN	20-25(Au)	Histagged protein IgG antibody	*13

*1: [99]; *2: [100]; *3: [104]; *4: [62]; *5: [97]; *6: [33]; *7: [118]; *8: [73]; *9: [98]; *10: [68]; *11: [11]; *12: [57]; *13: [128]

and impenetrability of ions. However, the surface of graphene is hydrophobic so that hydrophilic target molecules, such as DNA, protein, are not suitable for graphene-nanopore based sensing. To overcome this issue of the hydrophobic interactions between these biomolecules and the surface, hydrophilic thin membrane based solid-state nanopores by using material, such as chemically modified graphene [98], MoS_2 (6.5 Å-thick layers) [68], have been studied. As the other nanopore fabrication method, a bottom-up technique, such as self-folding and self-assembly methods, is also utilized. Among them, nanopore formation by using DNA origami technique, which is nanoscale folding method of DNA to create non-arbitrary two- and three-dimensional shapes at the nanoscale nanopore, have been extensively studied [11, 46, 47, 87, 127, 128]. Compared to top-down fabrication-technique, the self-assembly property and nanometer precision engineering of DNA origami nanostructures are attractive for creating hetero-structure and hybrid nanopores.

In general, compared to bio-nanopores, the structure of the solid-state pore is robust and stable in various external conditions, *i.e.*, pH, temperature, and electrolyte concentration. The feature serves to measure various analyte and the wide range of experimental condition, also serve to modify the surface of solid-sate nanopore, which can give specific interaction with analyte and/or prevent physical absorption of analyte and contaminated molecules. Therefore, solid-state nanopores are potentially suitable for a multi-target sensing platform. Until now, the multi-detection on several kinds of biopolymers target have been reported [90, 100] (Table 11.2). The detail contents on the multi-target sensing are described in Sect. 11.4. In addition, by nano-fabrication technology, solid-state nanopore can be easily integrated with various type of nanostructures, such as nano-fluid for molecular precise guidance, nano-gap electrode for sensing, and other detection sensing methods, such as optical detection, and force measurements and so on. The detail contents are described in Sect. 11.5.

11.4 Application of Nanopore Based Single-Biomolecule Sensors

11.4.1 Single-Molecule Nanopore Sequencer

Application area for nanopore-based sensors are recently growing for various bio-medical analysis because of the simple, high-throughput single-molecule detection ability and versatility at low-cost. One of the most attractive applications is a single-molecule DNA sequencing technology [13]. The first standard method for DNA sequencing is developed by Fred Sanger in the 1970s [95]. Based on the Sanger-method based sequencer, the first draft sequences of the human genome, *i.e.*, Human Genome Project, were completed in 2003 and 1000 Genomes Project has reported sequencing the entire genomes of 179 individuals [61, 111, 119]. However, those

DNA sequencing technologies were still slow and expensive to be used routinely for individual medication. Since then, new types of sequencers are being developed one after the other from companies such as Roche, Illumina, Life Technology and so on. There is a race to develop new sequencer technologies that are capable of sequencing individual genomes for a cost of $1000 or less [12]. In the race, as the new sequencer, various technologies to sequence DNA at the single-molecule level have been proposed. This category consists of various techniques, including nanopore sequencing [13], single-molecule sequencing by synthesis (SBS) [39], single-molecule real-time SBS [28], electron microscopy [10] and so on.

Among the new sequencer technologies being explored, nanopore-sensor based DNA sequencer is the most promising methodology for personal-genome uses [30, 31, 67, 123] (Fig. 11.3). Compared to other sequencers, the advantages of nanopore sequencing is potentially label-free, long reads, high throughput, and PCR-free. These features simplify the genome-reading process, resulting in

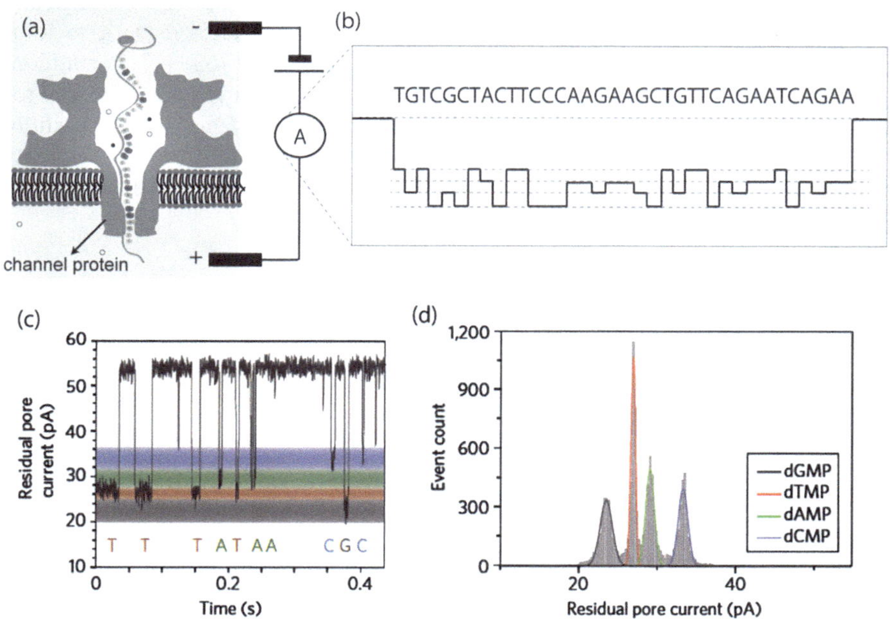

Fig. 11.3 Nanopore-based DNA sequencing. (**a**) Schematics of Bio-nanopore DNA sensing. Bio-nanopore are introduced in lipid bilayers. In the ion-current nanopore sensor, the passage of molecules induced the current blockage. (**b**) In sequencing, each of nucleotide blocks the pore differently giving a different amplitude so that this information can determine DNA sequence information. Ideally, this method can sequence long polynucleotides. (**c**) Typical ionic current-time profile through a bio-nanopore (aminocyclodextrin-modified α-haemolysin) for individual mono-nucleotides (dAMP, dCMP, dGMP, dTMP). Each mono-nucleotide causes a unique current blockage in nanopore conductance, resulting in identification of mono-nucleotide species. (**d**) Histogram of the ion-current blockage current. (Adapted with permission from [117]. Copyright 2011 Nature Publishing Group)

reduction of the cost of DNA sequencing for various research applications. First nanopore sequencing methodology was proposed around the 1990s [24]. The concept is as follow: In the ion-current nanopore sensor, when the pore is blocked due to the passage of molecules, the current flow is also blocked. In sequencing, each of nucleotide blocks the pore differently giving a different amplitude and duration of the current blockade so that this information is converted into DNA sequence information [21, 27, 71, 117] (Fig. 11.3). As the first report, the channel protein α-hemolysin was utilized for identification of homopolymers, such as poly (deoxyadenylic acid) (polydA) and poly(deoxycytidylic acid) (polydC) [2, 25]. It took 20 years to achieve the identification of single base molecules in DNA after this report of a DNA sequencer based on bionanopores. In order to obtain this ion-current signal at single-nucleotide resolution, reduction of the translocation speed of ssDNA molecules is one of the important issues. As one of the solution, this was addressed by modifying the inner nanopore surface with a cyclodextrin that had a smaller diameter than that of the nanopore [21]. In addition to this approach, DNA sequencing has been realized with α-hemolysin and a complex of Mycobacterium smegmatis porin A (MspA) with Phi29 DNA polymerase, which denatures dsDNA to form ssDNA and makes the ssDNA flow slowly into the nanopore [63, 71]. In addition, since nanopore sequencing produces fast results at low-cost, it is a powerful tool for "on-site diagnosis" of infectious agents. Indeed, nanopore sequencer successfully analyzed Ebola samples on-site in Africa [55]. The possibility of "on-site sequencing" by nanopore sequencer would potentially give a big impact on medical, clinical and health-care applications [14, 70, 121].

11.4.2 DNA/RNA Nanopore-Sensing

Bionanopore based sensing can be also applied for RNA detection and its quantification [77, 122, 125, 126] and the detection of aberrant DNA methylation, *i.e.*, N_6-methyladenosine, and N_5-methylcytosine, which is very important for cancer diagnosis and treatment. Until now, the RNA four base molecules identification is achieved by fixing the base molecules of RNA in the interior of the nanopore [4]. In order to sequence and/or identify RNA sample, fast translocation inside nanopore should be addressed for detection of single-base resolution. As one of the approaches, studies by combining an enzyme (PNPase cuts RNA one base molecule at a time) with the head of ssRNA and modifing the α-hemolysin with amino-cyclodextrin adapters is reported, and the modification enabled translocation of the single base molecules of the one molecule units [5]. Separately, the identification and/or detection of miRNA by hybridizing target miRNA with complementary oligonucleotide probes or PNA probes is being investigated [112]. In this sensing, the probe should be first translocated into a nanopore, and then, target miRNA is pulled by the probe inside the nanopore, resulting in selective detection of the target RNA molecules. The structural DNA/RNA variants, which are self-assembled nanostructures form of nucleotide such as hybridization-form and stem RNA loop-structure, can be also detected [45]. For instance, in the case of double-stranded DNA sensing, cytolysin A (ClyA), which has a minimum pore diameter of 3.3 nm and thus is larger than that of B-form dsDNA, is utilized for sensing and then found that the obtained signals have been shown to pass both ssDNA and dsDNA selectively [32]. In this way, the optimal selection of bio-nanopore especially with the optimal nanopore size is a crucial element for clear nanopore-sensing.

11.4.3 Protein/Peptide Nanopore-Sensing

The nanopore based peptide/protein sensing have been also and interested and explored. Since a protein expression and its concentration changes in the cell is closely related to biological perturbations such as disease or drug treatment, the development of single-molecule protein sensing and quantitative analysis technique can address these comprehensive proteomics analyses at single cell level. In protein/peptide analysis, most of these target molecules have various conformational variants so that the conformational control during sensing is a key issue. In general, the confinement inside nanopore can serve unfolding, denaturing, and linear translocation through a sub-nanopore structure. Using this phenomena, selective translocation and detection of the thermal unfolding of proteins with an ionic current blockade have been explored. In early studies, nanopore-sensing method exhibit the possibility of the protein identification and conformational change detections [3, 103, 133]. As one of the approach to detect specific target by nanopore-sensing, molecular recognition methods have been used. For instance, poly-γ-D-glutamic acid (γ-DPGA) detection by nanopore method is achieved by γ-DPGA monoclonal antibody modified inside the nanopore [120]. Similarly, the molecular-recognition, such as biotin/streptavidin, protein-G/immunoglobulin (IgG), and an antibody, have been also applied for nanopore sensing. However, only the use of molecule recognition agent for nanopore-sensing is not enough for the single-molecule selective detection. As another approach to selective detection of target molecules, enzyme-tethered nanopore methods are utilized, where the enzyme have the function to unfold and translocate native proteins through a nanopore sensor using the protein unfoldase. For instance, as the enzyme, ClpX was used for unfolding proteins (S1, S2 − 35, and S2 − 148) [79] and detection of a current blockade suggested that the unfolded protein was selectively translocated. In another study, a protein (V5 − C109 thioredoxin) was unfolded by placing an oligo-$(dC)_{30}$ tag on it, with the force of the tag dragging the protein into the nanopore and then the protein translocation behaviors are detected as current-blockage changes [92]. By using this method, a distinction between unphosphorylated, monophosphorylated and diphosphorylated thioredoxin variants were also identified [93]. Selective translocation of peptides has also been explored using the Nocardia farcinica channel, which has a minimum pore diameter of 2 nm or less, and it was revealed that a peptide molecule (oligoarginine) adheres to the nanopore entrance when the electrophoresis voltage is small, while it can be translocated upon the application of a large electrophoresis voltage [101].

Nanopore based sensor is also potentially applicable for protein/peptide sequencing [75, 84, 135]. In order to realize it, there are mainly two technical challenges. One is to unfold tertiary and secondary structures of target protein to allow the denatured molecules to thread through the nanopore sensor. Another is unidirectional translocation of the denatured polypeptide through the nanopore. As one of the solution for these issue, the enzyme-tethered nanopore technology can address the issues. It is well-known that some of the enzyme protein can denature protein structures. Therefore, by tethering these enzymes to bio-nanopore by protein-engineering or to solid-state nanopore by chemical modification, enzyme-assisted unfolding and translocation have been reported [79, 92]. As another solution, it is expected that the incorporation of unfolding functional unit with nanopore by nano-fabrication technology and nanopore functionalization serve to fine controls of translocation inside nanopore. The detail contents on the functional nanostructure integrated nanopore are described in Emerging nanopore of Sect. 11.6.

11.4.4 Carbohydrates and Sugar Complexes Nanopore-Sensing

Carbohydrates and sugar complexes are also a potential target of nanopore sensors. In general, carbohydrate groups are involved in various biological processes such as growth control, apoptosis, cell differentiation and proliferation, as well as physiological disorders like tumor, autoimmune diseases and inflammation. Therefore, the detection and its analysis at single-molecule level are expected to understand these important biological phenomena and its mechanisms. Until now, several studies show the nanopore-sensor can detect the carbohydrates molecules. For instance, the maltoporin (Lam B) ion channel binds sugar molecules [60] and has the ability to discriminate between different sugar species [6]. Besides, in a nanopore device based on aerolysin, which has a minimum diameter of 1 nm, the selective translocation of oligosaccharides (glycosaminoglycans) has been demonstrated for identification of saccharides with the different degrees of polymerization using an ionic current blockade and the dwell time [29].

11.5 Selectivity and Accuracy of Nanopore Electrical Sensing

Selectivity of nanopore-sensing method is a crucial parameter for the further development of nanopore-sensing. In general, nanopore based sensor can detect the difference in the volume of molecules translocating through nanopore so that it is difficult to selectively detect and/or discriminate target molecule in the analyte solution containing other molecules with the size similar to target ones. In order to overcome this issue, molecular recognition assisted nanopore can be used for the improvement of target sensing. For instance, chemical modification of solid-sate nanopore suggest can be a powerful method for this issue [57]. In this study, SiN-based solid-state nanopores covalently tethered on their surfaces to phenylalanine-glycine (FG) nucleoporins of 98 kDaA (Nup98) or FG nucleoporin of 153 kDa (Nup153) exhibited selective transportation of bovine serum albumin (BSA) and importin beta (Impβ), while both proteins passed through the unmodified nanopores at the same frequency. In a second example, selective detection is performed using solid-state nanopores modified with receptor molecules, When the IgG antibody were labeled with histidine-tag (His-tag) prior to the detection immunoglobulin G (IgG) antibody was selectively detected by using solid-state nanopore covered with nitrilotriacetic acid (NTA) receptor, which can selectively bind His-tage moiety in protein [128]. In both cases, the target analytes are identified not by the current-blockade intensity but by the current-blockade duration-time difference, which is due to the intermolecular interaction between receptor and target molecules. This approach can be applicable for any other target molecules and give the wide-range of application. However, it requires longer detection-time so that the throughput can became low. In addition, by using strong intermolecular interactions, high selectivity is achieved, but the multi-detection capacity can become lost. This trend-off relation between selectivity and throughput, and selectivity and multi-sensing should be optimized for application.

Electrical detection in nanopore devices involves the identification of picoampere-level electrical currents, and their high precision and accuracy of nanopore sensing is directly related to their ability to measure an electric current at a high signal-to-noise ratio. In order to improve the signal-to-noise ratio, it is necessary to reduce the translocation speed of the analytes [31, 58, 124]. The reason is described in following. In general, typical biopolymer analyte, such as DNA and protein, are electrically charged in electrolyte solution. Therefore, when an electrophoresis voltage is applied across a nanopore, the electrochemical bias voltage generates highly localized electrophoretic forces that are used to drive the target biopolymers in the electrolyte medium through the nanopore. Since the electrophoretic speed of negatively charged biomolecules is proportional to the strength of the electric field, single molecules pass through nanopores at very fast speeds. For instance, when an electrophoresis voltage of 0.1 V is applied, the typical thickness of a nanopore membrane is less than 50 nm, and the diameters of nanopores are less than 10 nm, an applied electric field is 0.1 V/50 nm = 20 kV/cm inside the nanopore [26, 33, 73, 94, 96, 97, 108]. For instance, in the electrical field, single DNA molecules pass through nanopores at very fast speeds of 10^{-5} s/base [117, 132]. It is expected that one method for decreasing the translocation speed is to decrease the electrophoretic voltage. However, there are no technologies currently available for the generation of 1 μV with the low voltage noises for reducing the translocation speed. Therefore, as an alternative approach, the single-base resolution at this translocation speed requires an electronic sensing system at extremely high bandwidth, and the concomitant electronic noise poses serious limitations in electrically discriminating between bases. However, the physical limit for measuring pA-level electric currents is due to the Johnson noise [54, 80], which has a maximum speed of approximately 1 MHz. At present, the maximum measuring speed of all-round measuring instruments available on the market is 250 kHz. Therefore, to be able to measure electric currents with high precision and single-molecule resolution, it is necessary to develop a method for increasing the speed of current measurements and decreasing the translocation speed of single molecules.

In bionanopores, a speed-control method has been developed mainly on the basis of the chemical modification of the nanopores by probe molecules and/or the use of enzymes [21, 71, 79, 92]. The enzyme or probe functional conditions are determined by the environment, such as the temperature, ion concentration, and so on, so that the margin for these environmental conditions can be limited. In addition, the throughput per bio-nanopore device can be low because molecular recognition based on intermolecular interactions takes time to detect analytes.

In solid-state nanopores, there are several methods for controlling the translocation speed. The passive methods use the viscosity [31], the temperature gradient [8, 42, 110], or the ion concentration gradient of the analyte solution [20, 34, 40, 43, 125, 126]. The viscosity of the solution is adjusted by adding glycerol, which has a higher viscosity than water [31]. When using this method, a five-fold reduction in the translocation speed has been obtained. Temperature and ion concentration gradients exist on both sides of the membrane; when the ion concentration gradient was adjusted, a three-fold reduction in the translocation speed was obtained [40]. Chemical modification of solid-state nanopore also serves to reduce a translocation speed inside nanopore. For instance, chemical modification of solid-state nanopores using molecules that can interact with DNA has resulted in translocations speeds 13 – 25 times slower than that observed without modification [59]. As the sophisticated molecule control method, the incorporation of functional unite for single-molecule fluid dynamics into solid-state nanopore by nano-fabrication technology have been recently emerging. The detail contents on the single-molecule speed-control nanopore technologies are described in Emerging Nanopore: Field-effect transistors (FETs) integrated Nanopore, and Speed Control Solid-State Nanopore (Sect. 11.6).

11.6 Emerging Nanopore

Besides the conventional ion-current blockage nanopore sensing shown in the previous sections, the incorporation of other functional nanostructures and/or other detection with the nanopore sensing have been recently interested. Of these studies, several promising method of nanopore-sensing, such as Nanogap nanopore, Field-effect transistors (FETs) integrated Nanopore, Speed Controllable Solid-State Nanopore, Optical-sensing Nanopore and Force-sensing Nanopore, are introduced as Emerging nanopore in this section.

11.6.1 Nanogap Nanopore

Nanogap nanopores are solid-state nanopores with the added capability of detecting specific molecular electronic structures. Nanogap nanopore-sensing have been proposed as a promising technique for single-molecule detection [102, 136, 137]. When applying bias voltage between the gap-electrodes integrated inside nanopore, the quantum-mechanical phenomenon of electron tunneling via a single-molecule is induced, which is closely related to a potential barrier via sample molecule between the nano-gape electrodes. The obtained conductance via single molecules are determined by the electronic structures of molecules, i.e., highest occupied molecular orbital (HOMO) or lowest unoccupied molecular orbital (LUMO) energy levels. Therefore, the observed electron-tunneling intensity represents the characteristic conductance of sample molecules during the translocation of gap. The ideal structure of nanogap nanopores has not yet been developed, but as a proof of the concept, which identifies single molecules via tunneling currents flowing between nanoelectrodes, the single-molecule sensing ability has been proven using scanning tunneling microscopy (STM) [18, 19, 51, 81, 109, 135] and mechanically controllable break junctions (MCBJ) [82, 84, 85, 114–116] (Table 11.3). Based on this idea, single-molecule tunnel-current based sequencing has been proposed (Fig. 11.4a, b and c). This is that, by nanogap-electrode embedded nanopore, the sequential detection of electron tunneling during translocating DNA strand can determine the DNA sequence. It is found that the electron tunneling phenomena is very sensitive to molecular orientations so that the confinement of nanopore structures can improve the stability of the molecular orientation inside nanopore. In addition, further incorporation of other functional electrodes along nanopore are proposed for precise control of molecular translocation and orientation (Fig. 11.4d) [53, 115, 116]. When applying local electric field by the incorporated electrodes to charged molecules inside nanopores, the trap and release behaviors on the pore-wall can be controlled at single-molecular level. Based on this idea, IBM group proposed that nanoelectromechanical device, "DNA transistor", can control the orientation

Table 11.3 Characteristics of Bionanopore, Solid-state Nanopore, Nanogap Nanopore

	Bio-nanopore	Solid-state nanopore	Nanogap nanopore
Detection probe	Ionic-current	Ionic-current	Tunneling-current
Analyte	dsDNA	dsDNA	ssDNA
	ssDNA	ssDNA	RNA
	RNA	RNA	Peptide
	Protein	Protein	Amino-acid
Speed control method	Chemical modification	Chemical modification	Chemical modification
	Electrophoresis	Electrophoresis	Electrophoresis
		Electroosmosis	Electroosmosis
	Temperature gradient	Temperature gradient	Temperature gradient
	Salts gradient	Salts gradient	Salts gradient

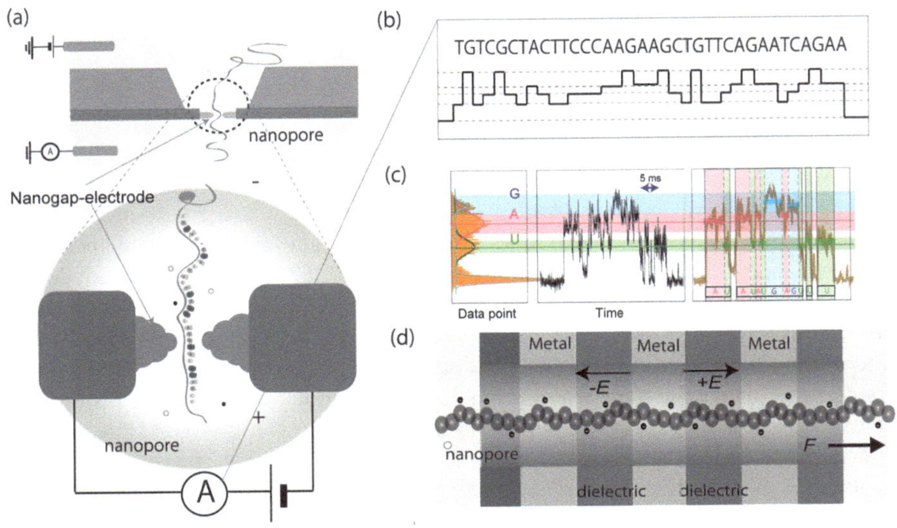

Fig. 11.4 (**a**) Schematics of nanogap nanopore sensing for single-molecule sequencing. When DNA are translocating between nanogap-electrodes inside nanopore, electron tunneling via a single-nucleotide is induced. The current intensity is closely related to a potential barrier via sample molecule between the biased nano-gape electrodes. (**b**) The sequential reading of electron tunneling during nucleotide translocation can determine the nucleotide sequence. (**c**) Typical current-time profile of UGAGGUA oligo-nucleotide. The observed electron-tunneling intensity represents the characteristic conductance of sample molecules during the translocation of gap. (Adapted with permission from [82]: Copyright 2012 Nature Publishing Group) (**d**) Schematics of DNA transistor. "DNA transistor" device have metal-dielectric structures (layers) inside the solid-state nanopore. By controlling the electric field between the metal layers, DNA can be trapped and released in the nanopore. By cyclically turning on and off these voltages, the DNA strand can move through the nanopore

and translocation speed of the single-stranded DNA inside the nanopore [38, 88]. The detail study related transistor integrated nanopore are described in Field-effect transistors (FETs) integrated Nanopore in the following section. Chemical modification on the sensing site of nanopore and/or nano-gapes can also be proposed for controlling the molecular orientation and improvement of sample identification. Several studies reported that chemical modifications of sensing tip/electrodes with a scanning tunneling microscope can improve the molecule identification, which is due to the interaction and its "recognition tunneling" between sample molecules and modified molecules [18, 19, 51, 66, 81, 135]. In early works, each of base recognition are done by each of the complementary base modified electrode [18, 81]. In the recent work, by using a 4(5)-substitutedimidazole-2-carboxamide(ICA) modified electrode, the nanogap method by using the recognition tunneling can be applicable for selective bases and amino-acid detections [19, 135].

11.6.2 Field-Effect Transistors (FETs) Integrated Nanopore

Field-effect transistors (FETs) integrated nanopore are also interested because FET-based nanopores can potentially create large electric currents by amplifying small changes in the potentials caused by sample translocation. Thus, these devices present one possible approach to overcoming the difficulty of rapidly measuring pA-level electric currents generated when single DNA molecules quickly pass through a nanopore. In early works, a design of nanopore capacitor made in a metal-oxide-semiconductor (MOS) [44] and a semiconductor-oxide-semiconductor membrane [64] were proposed. Their simulation results demonstrated that such a device can potentially sequence DNA at single base level. Recently, a progress of nano-fabrication technique and nano-materials study have accelerated FET nanopore [76, 89, 91, 113]. For instance, a fabrication of array of ionic field effect transistor (IFET) nanopores were demonstrated by combining electron beam lithography and atomic layer deposition [76]. By using the nanopore with sub-10 nm diameter on TiO_2 covered Si_3N_4, the ionic transport of KCl electrolyte can be efficiently manipulated by the embedded electrode of TiN within the nanopore. As another recent study on FET nanopore, nanopores with a diameter of 10 nm formed on Si nanowires with diameters of 30 – 50 nm can detect both changes in the ionic currents flowing through the nanopores and the channel conductance intensity of the nanowires when dsDNA passes through the nanopores [130]. Nanopores with a diameter of 10 nm fabricated on graphene nanoribbons also exhibit both ionic current blockade and changes in graphene conductance [113]. Importantly, these FET nanopore show large electric current changes on the order of 100 nA, which is greater than approximately 1000 times as large as those obtained with existing nanopore devices, and allow high-speed electric current measurements. To date, single base molecule resolution has been achieved at a measurement speed of hundreds of kHz. With these emerging nanopore technologies, however, it may be possible to realize single base molecule resolution at electric current measurement speeds of more than 1 MHz.

11.6.3 Speed Controllable Solid-State Nanopore

Single-molecule speed-control nanopore technologies have been recently explored by the incorporation of speed control unit into solid-state nanopores because solid-state nanopore can be potentially integrated with functional nanostructures by nano-fabrication technology [1, 41, 131]. In general, the fluid dynamics of single molecules within nanopore is influenced by the electrophoresis and electroosmotic flow. By voltage-application of gate electrode inside solid-state nanopore, the electroosmotic-flow can modulate the translocation speed of target molecules because, in the case of SiO_2 nanopore, cations accumulate on the surface of SiO_2 inside nanopore so that the motion of liquid are induced by an applied potential.

When a negative voltage is applied to the gate electrode, stronger upward electro-osmotic flow was induced, resulting in decrement of the translocation speed of the DNA molecules. In contrast, when a positive voltage is applied to the gate electrode, anions accumulate on the SiO_2 surface, thus resulting in a downward electroosmotic flow an increase in the translocation speed of the DNA molecules. Another speed control method of interest uses the difference in the pressures applied to both sides of the membrane. Using this method, an eight-fold reduction in the translocation speed was obtained [69]. The fluid dynamics of single molecules within nanopores has not yet been fully clarified: however, understanding the dynamics of single molecules passing through nanopores will provide important insights for improving the precision of single molecule identification using nanopores.

11.6.4 Optical-Sensing Nanopore

Optical detection method can be conjugated with solid-state nanopore platform (Fig. 11.5a) [35, 105]. In this method, optical probe tethered molecules are usually prepared and, as the optical detectors, fluorescence spectroscopy and/or total internal reflection fluorescence (TIRF) microscopy, are utilized for sample detection, and the optical detection site is set around the electrical sensing nanopore. When the labelled target molecule flow into nanopore, single-molecule optical signals are detected by these optical detectors, and simultaneously ionic-current blockage-signals are detected by current measurements. In addition, if patterning metallic membrane and/or photonic nanostructures around nanopore, the single-molecule just translocating nanopore are illuminated so the emitted optical signals are selectively enhanced inside the nanopore [9, 78]. The nanopore optical enhancement method, such as "plasmonic nanopore", can significantly suppress background signals emitted by molecules outside-nanopore so the ratio of signal to noise would be improved. These low-noise single-molecule optical nanopore method is very promising for various single-molecule biomolecule detections in real bio-samples. For instance, fluorescently labeled DNA nucleotide molecules are detected during translocating nanopore by fluorescence measurement at the nanopore opening, and identify the sample species at single-molecule level. Based on this optical method, a single molecule DNA optical sequencing technique have been proposed (Fig. 11.5b) [36, 52, 72]. In this method, the original DNA is converted to an expanded sequence by substituting every base in the sequence with a specific ordered oligonucleotides conjugated fluorescent probe. The converted DNA is hybridized with the complementary "molecular beacons" of two colors. When nanopores sequentially unzip the sample beacons during the translocation, un-quenched fluorophore at each unzipping event giving rise to a series of optical signals, resulting in the reading the original sequence. These results suggest possibility of single-molecule optical based DNA sequencing using nanopores.

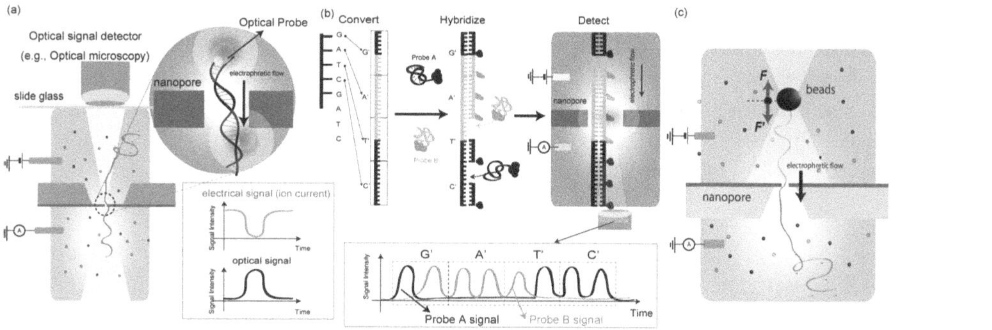

Fig. 11.5 (**a**) Schematics of optical sensing nanopore: Nanopore platform with conjugated with optical detection. First of all, optical probe tethered molecules are prepared. Next, the optical detection site is set around the electrical sensing nanopore. When the probe conjugated molecule translocate through the nanopore, the probe emits the optical signals. The emitted optical sample signals are detected by optical microscopy and simultaneously ionic-current blockage-signals are detected by current measurements. (**b**) Schematics of Single molecule DNA optical sequencing technique. First, the original DNA sequence is converted to an expanded form by substituting each and every base in the DNA sequence with a specific ordered pair of oligonucleotides. Second, the converted DNA is hybridized with complementary molecular beacons of two colors. Finally, fluorophore conjugated with the DNA sequence is sequentially detected as their optical signals, resulting in the reading the original sequence (**c**) Schematics of Nanopore force sensing: Nanopore sensing platform integrated with optical trapping systems. In this nanopore platform, the applying electrical field around the nanopore can control the exerted force on analyte (F') and the bead conjugated with sample molecule are pulled by the force of optical trapping (F). In this set-up, the observed ion-current time-profiles represent the characteristic single-molecule force signals during the translocation of molecular behaviors such as denaturing, refolding, and so on. The behaviors are strongly influenced by the inter/intra-molecular interaction between molecules and/or between molecule and wall inside a nanopore

11.6.5 Force-Sensing Nanopore

Single-molecule force-detection nanopore methods have also emerged as a powerful tool to investigate the inter/intra-molecular forces and motions of biomolecules. By conjugation of nanopore platform with force measurement techniques such as magnetic tweezers, optical trapping, and atomic force microscopy, some of nanopore force sensing platform have been reported [48, 56]. Since nanopore platform can control the exerted electrophoretic force on charged single analyte by the applying electrical field across the nanopore, the detected time-signals represent "single-molecule force signals", which are closely related to the intermolecular interaction and/or interaction between pore and analytes during the translocation (Fig. 11.5c). For instance, when the one-end of DNA tethered to polystyrene beads are trapped by optical tweezer and the other end of the DNA are pulled by the voltage-driven force of negatively charged DNA translocating through nanopore, the observed ion-current time-profiles represent the characteristic single-molecule motion behaviors, *i.e.,* self-folding of hair-pin structure, hybridization, dehybrization of DNA duplex and so on. Such "single-molecule force signals" under the nanopore-electrical field driven force can provide plenty of structural information for the interpretation of biological phenomena, *i.g.*, the enzymatic activity and biological molecules replication errors and misfolding in structure, which are closely related to disease, such as Parkinson's, Alzheimer's disease, and cancers. Therefore, nanopore based single-molecule force detection method would be widely applicable for various biological research field and medical diagnosis.

11.7 Future of Nanopore Sensor Method

Nanopore sensor have characteristic features, such as high-throughput, single-molecule measurement/capturing capability, and sample-versatility, so that the development of the nanopore platform is expanding the application interests as shown in Sect. 11.3. Of these applications, nanopore-based single-molecule sequencing would make a big impact on biomedical application such as personal medicine and drug screening. In addition, beyond conventional ion-current blockade detection, by incorporation of various emerging detection methods into nanopore platform, the nanopore-platform can detect various kinds of single-molecule information, and become "smarter", as shown in Sect. 11.4. The development of "smarter" nanopore platform are supported by the development of nano-fabrication technique, which provide the ability of mass-production, and highly-parallelization of sensors in the platform. The development of the fabrication technique would also realize more compact and miniature nanopore platforms, which can produce tremendous single-molecule related data at low cost. These features are very suitable and promising for portable sensors for the Age of Internet of Things (IoT) and Big-Data. Therefore, the development of nanopore platform fields would be still challenging and expanding on various applications, *i.g.*, future clinical, health-care application fields such as on-site diagnosis and personal medicine.

References

1. Ai Y, Liu J, Zhang BK, Qian S (2010) Field effect regulation of DNA translocation through a nanopore. Anal Chem 82:8217–8225
2. Akeson M, Branton D, Kasianowicz JJ et al (1999) Microsecond time-scale discrimination among polycytidylic acid, polyadenylic acid, and polyuridylic acid as homopolymers or as segments within single RNA molecules. Biophys J 77:3227–3233
3. Archakov AI, Ivanov YD (2007) Analytical nanobiotechnology for medicine diagnostics. Mol BioSyst 3:336–342
4. Ayub M, Bayley H (2012) Individual RNA Base Recognition in Immobilized Oligonucleotides Using a Protein Nanopore. Nano Lett 12:5637–5643
5. Ayub M, Hardwick SW, Luisi BF, Bayley H (2013) Nanopore-based identification of individual nucleotides for direct RNA sequencing. Nano Lett 13:6144–6150
6. Bacri L et al (2011) Discrimination of neutral oligosaccharides through a nanopore. Biochem Biophys Res Commun 412:561–564

7. Bayley H, Martin CR (2000) Resistive-pulse sensing from microbes to molecules. Chem Rev 100:2575
8. Belkin M, Maffeo C, Wells DB, Aksimentiev A (2013) Stretching and controlled motion of single-stranded DNA in locally heated solid-state nanopores. ACS Nano 7:6816–6824
9. Belkin M, Chao SH, Jonsson MP et al (2015) Plasmonic Nanopores for Trapping, Controlling Displacement, and Sequencing of DNA. ACS Nano 9:10598–10611
10. Bell DC, Thomas WK, Murtagh KM, Dionne CA, Graham AC, Anderson JE et al (2012a) DNA base identification by electron microscopy. Microsc Microanal 18:1049–1053
11. Bell NAW, Engst CR, Ablay M et al (2012b) DNA origami nanopores. Nano Lett 12:512–517
12. Bennett ST, Barnes C, Cox A et al (2005) Toward the $1000 human genome. Pharmacogenomics 6:373–382
13. Branton D, Deamer DW, Marziali A et al (2008) The potential and challenges of nanopore sequencing. Nat Biotechnol 26:1146–1153
14. Buermans HP, den Dunnen JT (2014) Next generation sequencing technology: advances and applications. Biochim Biophys Acta Mol basis Dis 1842:1932–1941
15. Butler TZ, Pavlenok M, Derrington IM et al (2008) Single-molecule DNA detection with an engineered MspA protein nanopore. Proc Natl Acad Sci U S A 105:20647–20652
16. Cai Q, Ledden B, Krueger E, Golovchenko JA, Li J (2006) Nanopore sculpting with noble gas ions. J Appl Phys 100:024914
17. Chan C, Yi-Tao L (2018) Biological nanopores: confined spaces for electrochemical single-molecule analysis. Acc Chem Res 51:331–341
18. Chang S, He J, Kibel A, Lee M, Sankey O, Zhang P, Lindsay S (2009) Tunneling readout of hydrogen-bonding based recognition. Nat Nanotechnol 4:297–301
19. Chang S, Huang S, He J, Liang F, Zhang P, Li S, Chen X, Sankey O, Lindsay S (2011) Electronic signatures of all four DNA nucleosides in a tunneling gap. Nano Lett 10:1070–1075
20. Chou T (2009) Enhancement of charged macromolecule capture by nanopores in a salt gradient. Chem Phys 131:034703
21. Clarke J, Wu HC, Jayasinghe L et al (2009) Continuous base identification for single-molecule nanopore DNA sequencing. Nat Nanotechnol 4:265–270
22. Coulter WH (1953) Means for conting paticles suspenedied in a fluid. US Patent No.2656508
23. Cracknell JA, Japrung D, Bayley H (2013) Translocating kilobase RNA through the staphylococcal α-hemolysin nanopore. Nano Lett 13:2500–2505
24. Deamer DW, Akeson M (2000) Nanopores and nucleic acids: prospects for ultrarapid sequencing. Trends Biotechnol 18:147–151
25. Deamer DW, Branton D (2002) Characterization of nucleic acids by nanopore analysis. Acc Chem Res 35:817–825
26. Dekker C (2007) Solid-state nanopores. Nature Nanotech 2:209–215
27. Derrington IM, Butler TZ, Collins MD et al (2010) Nanopore DNA sequencing with MspA. Proc Natl Acd Sci USA 107:16060–16065
28. Eid J, Fehr A, Gray J, Luong K, Lyle J, Otto G et al (2009) Real-time DNA sequencing from single polymerase molecules. Science 323:133–138
29. Fennouri A, Przybylski C, Pastoriza-Gallego M et al (2012) Single molecule detection of glycosaminoglycan hyaluronic acid oligosaccharides and depolymerization enzyme activity using a protein nanopore. ACS Nano 6:9672–9678
30. Fologea D, Gershow M, Ledden B et al (2005a) Detecting single stranded DNA with a solid state nanopore. Nano Lett 5:1905–1909
31. Fologea D, Uplinger J et al (2005b) Slowing DNA translocation in a solid-state nanopore. Nano Lett 5:1734–1737
32. Franceschini L, Soskine M, Biesemans A, Maglia G (2013) A nanopore machine promotes the vectorial transport of DNA across membranes. Nat Commun 4:2415
33. Garaj S, Hubbard W, Reina A et al (2010) Graphene as a subnanometre trans-electrode membrane. Nature 467:190–193

34. Ghosal S (2007) Effect of salt concentration on the electrophoretic speed of a polyelectrolyte through a nanopore. Phys Rev Lett 98:238104
35. Gilboa T, Meller A (2015) Optical sensing and analyte manipulation in solid state nanopores. Analyst 140:4733–4747
36. Gilboa T, Torfstein C, Juhasz M, Grunwald A, Ebenstein Y, Weinhold E, Meller A (2016) Single-molecule DNA methylation quantification using electro-optical sensing in solid-state nanopores. ACS Nano 10:8861–8870
37. Guihua W, Liang W, Yujing H, Shuo Z, Xiyun G (2013) Nanopore stochastic detection: diversity, sensitivity, and beyond. Acc Chem Res 46:2867–2877
38. Harrer S, Waggoner PS, Luan B, Afzali-Ardakani A, Goldfarb DL, Peng H, Martyna G, Rossnagel SM, Stolovitzky GA (2011) Electrochemical protection of thin film electrodes in solid state nanopores. Nanotechnology 22:275304
39. Harris TD, Buzby PR, Babcock H, Beer E, Bowers J, Braslavsky I et al (2008) Single-molecule DNA sequencing of a viral genome. Science 320:106–109
40. Hatlo MM, Panja D, van Roij R (2011) Translocation of DNA molecules through nanopores with salt gradients: the role of osmotic flow. Phys Rev Lett 107:068101
41. He Y, Tsutsui M, Fan C, Taniguchi M, Kawai T (2011) Controlling DNA translocation through gate modulation of nanopore wall surface charges. ACS Nano 5:5509–5518
42. He Y et al (2013a) Thermophoretic manipulation of DNA Translocation through nanopores. ACS Nano 7:538–546
43. He Y, Tsutsui M, Scheicher RH, Fan C, Taniguchi M, Kawai T (2013b) Mechanism of how salt-gradient-induced charges affect the translocation of DNA molecules through a nanopore. Biophys J 105:776–782
44. Heng JB, Aksimentiev A, Ho C, Dimitrov V, Sorsch TW, Miner JF et al (2005) Beyond the gene-chip. Bell Labs Tech J 10:5–22
45. Henley RY, Carson S, Wanunu M (2016) Studies of RNA sequence and structure using nanopores. Prog Mol Biol Transl Sci 139:73–99
46. Hernandez-Ainsa S et al (2013) DNA origami nanopores for controlling DNA translocation. ACS Nano 7:6024–6030
47. Hernandez-Ainsa S et al (2014) Voltage-dependent properties of DNA origami nanopores. Nano Lett 14:1270–1274
48. Hornblower B, Coombs A, Whitaker RD et al (2007) Single-molecule analysis of DNA-protein complexes using nanopores. Nat Methods 4:315–317
49. Howorka S, Siwy Z (2009) Nanopore analytics: sensing of single molecules. Chem Soc Rev 38:2360–2384
50. Howorka S, Cheley S, Bayley H (2001) Sequence-specific detection of individual DNA strands using engineered nanopores. Nat Biotechnol 19:636–639
51. Huang S, He J, Chang S et al (2010) Identifying single bases in a DNA oligomer with electron tunneling. Nat Nanotechnol 5:868–873
52. Huang S, Romero-Ruiz M, Castell OK, Bayley H, Wallace MI (2015) High-throughput optical sensing of nucleic acids in a nanopore array. Nat Nanotechnol 10:986–991
53. Ivanov AP, Instuli E, McGilvery CM et al (2011) DNA tunneling detector embedded in a nanopore. Nano Lett 11:279–285
54. Johnson JB (1928) Thermal agitation of electricity in conductors. Phys Rev 32:97–109
55. Joshua Q, Loman NJ, Duraffour S et al (2016) Real-time, portable genome sequencing for Ebola surveillance. Nature 530:228–232
56. Keyser UF, Koeleman BN, Van Dorp S et al (2006) Direct force measurements on DNA in a solid-state nanopore. Nature Phys 2:473–477
57. Kowalczyk SW, Kapinos L, Blosser TR, Magalhaes T, van Nies P, Lim RYH, Dekker C (2011) Single-molecule transport across an individual biomimetic nuclear pore complex. Nat Nanotechnol 6:433–438
58. Kowalczyk SW, Wells DB, Aksimentiev A, Dekker C (2012) Slowing down DNA translocation through a nanopore in lithium chloride. Nano Lett 12:1038–1044

59. Krishnakumar P, Gyarfas B, Song WS, Sen S, Zhang PM, Krstic P, Lindsay S (2013) Slowing DNA translocation through a nanopore using a functionalized electrode. ACS Nano 7:10319–10326
60. Kullman L, Winterhalter M, Bezrukov SM (2002) Transport of maltodextrins through maltoporin: a single-channel study. Biophys J 82:803–812
61. Lander ES et al (2001) Initial sequencing and analysis of the human genome. Nature 409:860–921
62. Larkin J, Henley R, Bell DC et al (2013) Slow DNA transport through nanopores in hafnium oxide membranes. ACS Nano 7:10121–10128
63. Laszlo AH, Derrington IM, Brinkerhoff H et al (2013) Detection and mapping of 5-methylcytosine and 5-hydroxymethylcytosine with nanopore MspA. Proc Natl Acad Sci U S A 110:18904–18909
64. Leroux A, Destine J, Vanderheyden B, Gracheva ME, Leburton J (2010) SPICE circuit simulation of the electrical response of a semiconductor membrane to a single-stranded DNA translocating through a nanopore. IEEE Trans Nanotechnol 9:322–329
65. Li J, Stein D, McMullan C et al (2001) Ion-beam sculpting at nanometre length scales. Nature 412:166–169
66. Lindsay S et al (2010) Recognition tunneling. Nanotechnology 21:262001
67. Liu L, Li Y, Li S et al (2012) Comparison of next-generation sequencing systems. J Biomed Biotechnol 2012:251364
68. Liu K et al (2014) Atomically thin molybdenum disulfide nanopores with high sensitivity for DNA translocation. ACS Nano 8:2504–2511
69. Lu B, Hoogerheide DP, Zhao Q, Zhang HB, Zhipeng TP, Yu DP, Goloychenko JA (2013) Pressure-controlled motion of single polymers through solid-state nanopores. Nano Lett 13:3048–3052
70. Majd S, Yusko EC, Billeh YN et al (2010) Applications of biological pores in nanomedicine, sensing, and nanoelectronics. Curr Opin Biotech 21:439–476
71. Manrao EA, Derrington IM, Laszlo AH et al (2012) Reading DNA at single-nucleotide resolution with a mutant MspA nanopore and phi29 DNA polymerase. Nat Biotechnol 30:349–353
72. McNally B, Singer A, Yu ZL, Sun YJ, Weng ZP, Meller A (2010) Optical recognition of converted DNA nucleotides for single-molecule DNA sequencing using nanopore arrays. Nano Lett 10:2237–2244
73. Merchant CA, Healy K, Wanunu M et al (2010) DNA translocation through graphene nanopores. Nano Lett 10:2915–2921
74. Miles BN, Ivanov AP, Wilson KA et al (2013) Single molecule sensing with solid-state nanopores: novel materials, methods, and applications. Chem Soc Rev 42:15–28
75. Movileanu L, Howorka S, Braha O et al (2000) Detecting protein analytes that modulate transmembrane movement of a polymer chain within a single protein pore. Nat Biotechnol 18:1091–1095
76. Nam SW, Rooks MJ, Kim KB, Rossnagel SM (2009) Ionic field effect transistors with sub-10 nm multiple nanopores. Nano Lett 9:2044–2048
77. Neely LA, Patel S, Garver J et al (2006) A single-molecule method for the quantitation of microRNA gene expression. Nat Methods 3:41–46
78. Nicoli F, Verschueren D, Klein M et al (2014) DNA translocations through solid-state plasmonic nanopores. Nano Lett 14:6917–6925
79. Nivala J, Marks DB, Akeson M (2013) Unfoldase-mediated protein translocation through an α-hemolysin nanopore. Nat Biotechnol 31:247–250
80. Nyquist H (1928) Thermal agitation of electric charge in conductors. Phys Rev 32:110–113
81. Ohshiro T, Umezawa Y (2006) Complementary base-pair-facilitated electron tunneling for electrically pinpointing complementary nucleobases. Proc Natl Acad Sci U S A 103:10–14
82. Ohshiro T, Matsubara K, Tsutsui M et al (2012) Single-molecule electrical random resequencing of DNA and RNA. Sci Rep 2:501

83. Ohshiro T, Tsutsui M, Yokota K, Furuhashi M, Taniguchi M (2014) Detection of post-translational modifications in single peptides using electron tunnelling currents. Nat Nanotechnol 9:835–840
84. Ohshiro T, Tsutsui M, Yokota K, Taniguchi M (2018) Quantitative analysis of DNA with single-molecule sequencing. Sci Rep 8:8517
85. Ohshiro T, Komoto U, Konno M et al (2019) Direct analysis of incorporation of an anticancer drug into DNA at single-molecule resolution. Sci Rep 9:3886
86. Payet L, Martinho M, Pastoriza-Gallego M, Betton JM, Auvray L, Pelta J, Mathe J (2012) Thermal unfolding of proteins probed at the single molecule level using nanopores. J Anal Chem 84:4071–4076
87. Plesa C et al (2014) Ionic permeability and mechanical properties of DNA origami nanoplates on solid-state nanopores. ACS Nano 8:35–43
88. Polonsky S, Rossnagel S, Stolovitzky G (2007) Nanopore in metal-dielectric sandwich for DNA position control. Appl Phys Lett 91:153103
89. Puster M, Rodriguez-Manzo JA, Balan A, Drndic M (2013) Toward sensitive graphene nanoribbon nanopore devices by preventing electronbeam induced damage. ACS Nano 7:11283–11289
90. Raillon C et al (2012) Nanopore detection of single molecule RNAP–DNA transcription complex. Nano Lett 12:1157–1164
91. Ren R, Zhang Y, Nadappuram BP et al (2017) Nanopore extended field-effect transistor for selective single-molecule biosensing. Nat Commun 8:586
92. Rodriguez-Larrea D, Bayley H (2013) Multistep protein unfolding during nanopore translocation. Nat Nanotechnol 8:288–295
93. Rosen CB, Rodriguez-Larrea D, Bayley H (2014) Single-molecule site-specific detection of protein phosphorylation with a nanopore. Nat Biotechnol 32:179–181
94. Rosenstein JK et al (2012) Integrated nanopore sensing platform with sub-microsecond temporal resolution. Nat Methods 9:487–492
95. Sanger F, Nicklen S, Coulson AR (1977) DNA sequencing with chain-terminating inhibitors. Proc Natl Acad Sci U S A 74:5463–5467
96. Schneider GF, Dekker C (2012) DNA sequencing with nanopores. Nat Biotechnol 30:326–328
97. Schneider GF, Kowalczyk SW, Calado VE et al (2010) DNA translocation through graphene nanopores. Nano Lett 10:3163–3167
98. Schneider GF, Xu Q, Hage S et al (2013) Tailoring the hydrophobicity of graphene for its use as nanopores for DNA translocation. Naure Commun 4:2619
99. Shasha C et al (2014) Nanopore-based conformational analysis of a viral RNA drug target. ACS Nano 8:6425–6430
100. Singer A, Rapireddy S, Ly DH, Meller A (2012) Electronic barcoding of a viral gene at the single-molecule level. Nano Lett 12:1722–1728
101. Singh PR, Barcena-Uribarri I, Modi N, Kleinekathofer U, Benz R, Winterhalter M, Mahendran KR (2012) Pulling peptides across nanochannels: resolving peptide binding and translocation through the hetero-oligomeric channel from Nocardia farcinica. ACS Nano 6:10699–10707
102. Siwy ZS, Davenport M (2010) Graphene opens up to DNA. Nat Nanotechnol 5:697–698
103. Siwy Z, Trofin L, Kohli P, Lane A, Baker LA, Trautmann C, Martin C (2005) Protein biosensors based on biofunctionalized conical gold nanotubes. J Am Chem Soc 127:5000–5001
104. Soni GV, Dekker C (2012) Detection of nucleosomal substructures using solid-state nanopores. Nano Lett 12:3180–3186
105. Soni GV et al (2010) Synchronous optical and electrical detection of biomolecules traversing through solid-state nanopores. Rev Sci Instrum 81:014301
106. Stephanie JH, Dekker C (2016) Graphene nanodevices for DNA sequencing. Nat Nanotechnol 11:127–136
107. Storm AJ, Chen JH, Ling XS, Zandbergen HW, Dekker C (2003) Fabrication of solid-state nanopores with single-nanometre precision. Nature Mater 2:537–540

108. Storm AJ et al (2005) Fast DNA translocation through a solid-state nanopore. Nano Lett 5:1193–1197
109. Tanaka H, Kawai T (2009) Partial sequencing of a single DNA molecule with a scanning tunnelling microscope. Nat Nanotechnol 4:518–522
110. Thamdrup LH, Larsen NB, Kristensen A (2010) Light-induced local heating for thermophoretic manipulation of DNA in polymer micro- and nanochannels. Nano Lett 10:826–832
111. The 1000 Genomes Project Consortium (2010) A map of human genome variation from population-scale sequencing. Nature 467:1061–1073
112. Tian K, He ZJ, Wang Y, Chen SJ, Gu LQ (2013) Designing a polycationic probe for simultaneous enrichment and detection of microRNAs in a nanopore. ACS Nano 7:3962–3969
113. Traversi F, Raillon C, Benameur SM, Liu K, Khlybov S, Tosun M, Krasnozhon D, Kis A, Radenovic A (2013) Detecting the translocation of DNA through a nanopore using graphene nanoribbons. Nat Nanotechnol 8:939–945
114. Tsutsui M, Taniguchi M, Yokota K et al (2010) Identifying single nucleotides by tunnelling current. Nature Nanotech 5:286–290
115. Tsutsui M et al (2011a) Single-molecule sensing electrode embedded in-plane nanopore. Sci Rep 1:46
116. Tsutsui M, Matsubara K, Ohshiro T, Furuhashi M, Taniguchi M, Kawai T (2011b) Electrical detection of single methylcytosines in a DNA oligomer. J Am Chem Soc 133:9124–9128
117. Venkatesan BM, Rashid B (2011) Nanopore sensors for nucleic acid analysis. Nature Nanotech 6:615–624
118. Venkatesan BM et al (2012) Stacked graphene-Al_2O_3 nanopore sensors for sensitive detection of DNA and DNA–protein complexes. ACS Nano 6:441–450
119. Venter JC et al (2001) The sequence of the human genome. Science 291:1304–1351
120. Vlassiouk I, Kozel TR, Siwy ZS (2009) Biosensing with Nanofluidic Diodes. J Am Chem Soc 131:8211–8220
121. Voelkerding KV, Dames SA, Durtschi JD (2009) Next-Generation Sequencing: From Basic Research to Diagnostics. Clin Chem 55:641–658
122. Wang Y, Zheng D, Tan Q, Wang MX, Gu LQ (2011) Nanopore-based detection of circulating microRNAs in lung cancer patients. Nat Nanotechnol 6:668–674
123. Wanunu M (2012) Nanopores: A journey towards DNA sequencing. Phys Life Rev 9:125–158
124. Wanunu M, Sutin J, McNally B, Chow A, Meller A (2008) DNA translocation governed by interactions with solid-state nanopores. Biophys J 95:4716–4725
125. Wanunu M, Dadosh T, Ray V et al (2010a) Rapid electronic detection of probe-specific microRNAs using thin nanopore sensors. Nat Nanotechnol 5:807–814
126. Wanunu M, Morrison W et al (2010b) Electrostatic focusing of unlabelled DNA into nanoscale pores using a salt gradient. Nat Nanotechnol 5:160–165
127. Wei RS, Martin TG, Rant U, Dietz H (2012a) DNA origami gatekeepers for solid-state nanopores. Angew Chem Int Ed 51:4864–4867
128. Wei RS, Gatterdam V, Wieneke R, Tampe R, Rant U (2012b) Stochastic sensing of proteins with receptor-modified solid-state nanopores. Nat Nanotechnol 7:257–263
129. Wells DB, Belkin M, Comer J et al (2012) Assessing graphene nanopores for sequencing DNA. Nano Lett 12:4117–4123
130. Xie P, Xiong QH, Fang Y, Qing Q, Lieber CM (2012) Local electrical potential detection of DNA by nanowire-nanopore sensors. Nat Nanotechnol 7:119–125
131. Yen PC, Wang CH, Hwang GJ, Chou YC (2012) Gate effects on DNA translocation through silicon dioxide nanopore. Rev Sci Instrum 83:034301
132. Yokota K, Tsutsui M, Taniguchi M (2014) Electrode-embedded nanopores for label-free single-molecule sequencing by electric currents. RSC Adv 4:15886–15899
133. Yusko E, Jay M, Johnson J, Majd S, Prangkio P, Rollings R, Li J, Yang J, Mayer M (2011) Controlling protein translocation through nanopores with bio-inspired fluid walls. Nat Nanotechnol 6:253–260

134. Zhang XY, Wang Y, Fricke BL, Gu LQ (2014) Programming nanopore ion flow for encoded multiplex microRNA detection. ACS Nano 8:3444–3450
135. Zhao Y, Ashcroft B, Zhang P et al (2014) Single-molecule spectroscopy of amino acids and peptides by recognition tunneling. Nat Nanotechnol 9:466–473
136. Zwolak M, Di Ventra M (2005) Electronic signature of DNA Nucleotides via transverse transport. Nano Lett 5:421–424
137. Zwolak M, Di Ventra M (2008) Colloquium: physical approaches to DNA sequencing and detection. Rev Mod Phys 80:141–165

Chapter 12
Paper Microfluidics for POC Testing in Low-Resource Settings

Elain Fu

Abstract Paper microfluidics is a subarea of microfluidics in which porous materials are used to create devices. Advantages of paper microfluidics include fluid transport via capillary forces, so that external pumping equipment is not necessary, and the use of less expensive materials than those commonly used in conventional microfluidic devices. Paper microfluidics enables the development of fully disposable devices that are appropriate for use in even the lowest-resource settings, and the potential for high impact improvement to human health. In this chapter, we first discuss the paper microfluidic device fabrication processes of materials selection, fluidic boundary definition, and reagent patterning. Next, we discuss tools development for manipulating fluids in paper microfluidic devices. Then, we describe specific medical applications with discussion of three promising paper microfluidic devices. Finally, we close with a general discussion of challenges in the translation of paper microfluidic devices from the lab to the field.

Keywords Paper microfluidics · Porous materials · Point-of-care testing

12.1 Introduction to Paper Microfluidics

12.1.1 Paper Microfluidics as a Subfield of Microfluidics

Paper microfluidics is a subarea of microfluidics in which porous materials are used to create devices. As in the case of conventional microfluidics-based sensor development, one can leverage key micro-scale properties. The effects of viscosity are more prominent than those of inertia, such that the dimensionless Reynolds number is smaller than one. The result is flow in the laminar regime with predictable fluid transport and mass transfer (e.g., diffusion-based mixing across adjacent streams). Surface tension and capillary forces can also be prominent. This is especially true in

E. Fu (✉)
Oregon State University, Corvallis, OR, USA
e-mail: elain.fu@oregonstate.edu

M. Tokeshi (ed.), *Applications of Microfluidic Systems in Biology and Medicine*, Bioanalysis 7, https://doi.org/10.1007/978-981-13-6229-3_12

paper microfluidics, where the dimensionless capillary number, the ratio of viscous forces to surface tension, is very small, ~1×10^{-6} for many paper systems[1].

Paper microfluidics shares several advantages with conventional microfluidics [1]. These include high surface area to volume ratios that enable fast surface reaction-based processes. Another shared advantage is the short diffusion distances (and times) that enable efficient mass transfer. For paper microfluidic systems, the diffusion distances set by the material pore diameter are typically 10 times smaller than in conventional microfluidic channels (i.e., a 10 μm characteristic pore diameter versus a 100 μm channel height). The micro-scale devices are compatible with microliter-sized sample volumes, and enable a rapid time to result, reagent savings, and reduced waste. Finally, parallelization enables the probing of multiple reaction conditions, as well as high-throughput processing in a reasonable amount of time.

Paper microfluidics has several advantages over conventional microfluidics. First, lateral flow test development over the past three decades has produced a strong knowledge base that is useful for paper microfluidic device development. Second, porous materials are generally lower cost than other materials that have traditionally been used in microfluidic devices (e.g., silicon and glass). And most importantly, fluid transport within porous materials occurs via capillary forces, so external pumping systems are not necessary. This enables the development of fully disposable devices that are appropriate for use in even the lowest-resource settings.

12.1.2 Capillary Flow in Paper Microfluidic Devices

Flow within a capillary tube is driven by capillary pressure [2],

$$P_c = \frac{2 \cdot \gamma \cdot cos\theta}{R},$$

where γ is the fluid surface tension, R is the capillary tube radius, and θ is the contact angle of the liquid-air interface with the solid. The Hagen-Poiseuille equation [3] describes fluid flow in the tube,

$$Q = \frac{\pi \cdot R^4 \cdot \Delta P}{8 \cdot \eta \cdot L},$$

where Q is the volumetric flow rate, η is the dynamic fluid viscosity, L is the fluid column length, and the pressure difference across L, ΔP, is equal to P_C. From these expressions, a relationship between the position of the fluid front L and time t, the Lucas-Washburn equation [4, 5] can be derived,

[1]The estimate is based on a surface tension of 7.2×10^{-2} N/m, a velocity of 1×10^{-4} m/s, and a viscosity of 1×10^{-3} Pa•s.

$$L = \left(\frac{\gamma \cdot R \cdot cos\theta}{2 \cdot \eta} t\right)^{\frac{1}{2}}.$$

Fluid is pulled into the tube by the capillary force, while viscous resistance counteracts the capillary force. The viscous resistance increases with the fluid column length within the capillary tube, such that the velocity of the fluid front decreases as the fluid penetrates into the capillary.

Given that a group of parallel cylindrical capillaries of varying radii can be used to describe a strip of porous material, the fluid front within the strip will follow Lucas-Washburn flow with R equal to the mean of the capillary radii, R_{mean} [4]. (See Mendez et al. [6] for a thorough discussion of this topic.) The pore structure of common porous materials used in paper microfluidics can be far from uniform, both in terms of the shape of the "pores", as well as the distribution of pore sizes. Nitrocellulose, most often used in lateral flow-based devices, has a sponge-like structure [7]. In contrast, cellulose consists of a dense network of randomly oriented fibers. Chopped glass fiber also consists of an intricate network of fibers. The wicking behavior of some common porous materials is described below.

12.1.3 *The Role of Paper Microfluidics in Point-of-Care Testing (POCT)*

A main goal of the field of POCT is to translate current high-resource laboratory-based testing into technologies that can be used in POC settings by the patient. There are many POC settings and these vary in terms of their resource levels [8]. POC settings can range from the higher-resource hospital emergency rooms, to clinics and physician's offices, to the lower-resource settings in the developed world, such as schools, workplaces, and homes, to the very lowest resource settings of the developing world. For the lowest-resource settings, there are a number of challenges that must be addressed for a technology to succeed in those settings [8–10]. These include a limited contact time with patients, a limited training of test providers, a lack of laboratory facilities (i.e., testing environments with uncontrolled temperatures and humidity levels), limited local infrastructure including a lack of cold chain for refrigeration of reagents, a lack of support for maintenance on instrumentation, and a need for a low cost per test. Finally, the performance specifications of the application must be met. These requirements are summarized by the ASSURED standards: affordable, sensitive, specific, user-friendly, rapid (and robust), equipment-free, and deliverable to the user [11].

The current standard bioassay format that is used in the lowest-resource settings is the lateral flow test (LFT) [12]. A well-known example is the home pregnancy dipstick test. It is affordable, and can be very low cost to produce [13]. The test is user-friendly and rapid. Most tests require the user to apply the sample (and a buffer) and then look for the appearance of color in the 'test' and 'control' lines within a period of 10–20 min. The test is often equipment free. An external pumping system is not required since fluid is transported via capillary forces in the nitrocellulose and cellulose substrates. There is also no need for instrumentation for detection of the signal. The test can be based on the interpretation of colored line development via human visual perception. And the test is 'deliverable to the user' with typical shelf lives of 1 or 2 years at ambient temperature [14]. Trade offs for LFT simplicity include that conventional LFTs are limited to a single delivery step of premixed sample and conjugate that can translate to limited sensitivity and specificity. Additionally, LFTs generally have a qualitative output when not used with some sort of an instrument, and they are difficult to multiplex for the simultaneous detection of multiple targets. Thus, although LFTs conform to many of the ASSURED standards, conventional LFT performance would not enable clinical relevance for a number of compelling heath conditions. Thus, there is a need to improve POCT in the lowest-resource settings, and paper microfluidics is a potential high impact solution.

One area of focus in paper microfluidics device development has been enabling testing for multiple targets from a single sample. To address this, the Whitesides group has developed paper networks for multi-analyte detection. Their 2007 article, in which they demonstrated a device consisting of a diverging network for the simultaneous detection of glucose and protein from a small volume of urine, started a resurgence of interest in paper-based devices [15]. An image series of the colorimetric results from their bi-plexed paper-based device for increasing analyte concentrations is shown in Fig. 12.1a. Following this work, there have been numerous demonstrations of multi-analyte detection in both 2- and 3-dimensional [16] formats in porous materials [17].

A second area of focus in paper microfluidics has been on providing quantitative output using optical (e.g., colorimetric, fluorescence, etc.) or electrochemical signals [18, 19]. These efforts range from developing equipment-free quantitative readout [20] to minimally-instrumented strategies based on simple readers [21]. One approach to achieving equipment-free quantitative readout is to use differences in colorimetric signal intensities or hues to indicate analyte level [20, 22]. The use of multiple indicators to produce distinct intensity- and hue-based signals is shown in Fig. 12.1b [22]. Another approach that has potential for equipment-free quantitative readout is to use the spatial distribution of signal in the detection region to indicate analyte level [20, 23]. The visible signal develops along the length of the detection region in the direction of flow and in proportion to the concentration of analyte in the sample [23]. A variation of this is the "ladder-bar" readout, in which the level of analyte is quantified by enumerating the number of discrete detection sub-regions in which signal developed [20, 24]. A demonstration of this method using an enzyme-based system is shown in Fig. 12.1c [24]. Complementary to this, much progress has been made in the area of developing analysis algorithms and simple companion reader technology to extract high quality optical [25–27], or electrochemical signals [21] that can inform on the level of target analyte within a sample.

A third area of focus in paper microfluidics device development, complementary to the previous areas, is on using paper networks to perform automated multi-step sample processing for high sensitivity testing [28]. A common theme across high-

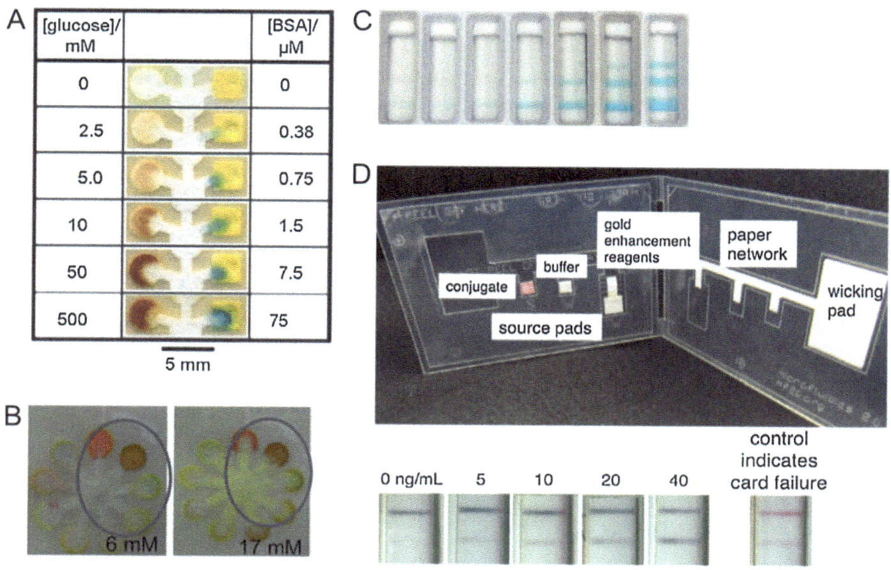

Fig. 12.1 Three areas in which paper microfluidics can improve on the performance of standard conventional lateral flow tests are in enabling multi-analyte detection, quantitative output, and higher sensitivity analysis. (**a**) 2D multi-analyte paper-based device for the simultaneous detection of glucose and protein in a urine sample. (Reproduced with permission from Ref. [15]). (**b**) Paper-based device that produces visual quantitative output based on intensity and hue through the use of multiple colorimetric indicators. (Reproduced with permission from Ref. [22]). (**c**) Paper-based device with "ladder-bar" readout where the number of bands in which signal develops increases with increasing concentration of analyte. (Reproduced with permission from Ref. [24]). (**d**) Prototype amplified malaria immunoassay in an automated paper-based device. The operator adds sample and water to pads on the card and folds the card to start the assay. The amplified assay detects clinically-relevant levels of malaria antigen in a mock sample. The limit of detection of the amplified assay was found to be four-times lower than without the rinse and amplification steps. (Reproduced with permission from Ref. [30])

performance laboratory-based tests is processing of the sample to maximize signal and minimize noise. For example, laboratory-based ELISA, or enzyme-linked immunosorbent assay, requires multiple sequential sample processing steps to achieve high sensitivity. The assay takes place in a well, coated with an antibody molecule that specifically binds to the target analyte of interest. The multiple steps of the assay include sample introduction and incubation, one or more steps to label the binding event, and a step to chemically amplify the signal due to the original label for high sensitivity detection. Also key, are the multiple wash steps used to minimize the nonspecific binding events in the system. Thus, a promising strategy for improving the performance of LFTs in low-resource settings is to implement additional sample processing to improve device performance, while still retaining the ease of use of LFTs [29]. An example of this is the amplified assay shown in Fig. 12.1d [30]. After application of the appropriate fluids on the pads of the device, the operator folds the device closed and waits until the test is complete. A concentration series indicates that the amplified assay is able to detect malaria antigen in a clinically relevant range [30]. The ability to manipulate fluids and the reagents within them in paper networks is critical to enabling automated processing for high performance. So, there is a need for robust paper microfluidic tools that serve as the analogs of pump controls and valves of conventional microfluidics [28], and compatible paper microfluidic device fabrication methods.

12.2 Paper Microfluidic Device Fabrication

Device fabrication can be divided into three processes: materials selection, fluidic boundary definition, and reagent patterning.

12.2.1 Porous Materials Selection

For materials selection, the properties of the material, such as the capacity and flow rate, are important considerations. In addition to cellulose, there are other porous substrates with properties that can be advantageous in a device, including nitrocellulose substrates, which are widely used in LFTs, fabrics which can be useful in wearable sensor applications, and glass fiber substrates, which with their high flow rates, can be useful when rapid fluid flow is desired. Flow rates can vary substantially between different porous materials. For the nitrocellulose membranes used in LFTs (characteristic pore sizes from 3 to 17 μm), capillary flow times range from 75 to 240 s per 4 cm (Millipore Hi-Flow membranes) and variability in capillary flow times can be up to 25%. For cellulose materials, the commonly used Whatman #1 filter paper has a capillary flow time of ~70 s per 2 cm, while for glass fiber, capillary flow times can be considerably shorter, e.g., 4 s and 15 s per 2 cm for Ahlstrom 8951 and 8950, respectively. The fluid capacities of different porous materials can also vary substantially (e.g., ~12 μL/cm^2 for Millipore Hi-Flow 135, ~16 μL/cm^2 for Whatman #1 filter paper, ~80 μL/cm^2 for thicker cellulose substrates [31], and ~60 μL/cm^2 for Ahlstrom 8951 glass fiber). Thus, for a given application, target processing volumes and average flow rates can be attained by judicious choice of material or a combination of materials.

12.2.2 Defining Channels in Porous Materials

Fluidic boundary definition is a critical process in the fabrication of a paper microfluidic device. Since the Whitesides group first described the use of optical lithography [15, 32] to pattern fluidic boundaries in cellulose, many methods of defining channels have been developed [33, 34], such as depositing hydrophobic PDMS barriers using a plotter [35], etching using inkjet technology [36] or a plasma [37], laser-etching hydrophobic paper [38], and contact stamping ink barriers [39]. One of the most commonly used fabrication methods has been printing wax directly onto substrates using a wax printer, followed by heating of the substrate to create functional wax barriers in the substrate. The main advantages of wax printing are its simplicity and reasonable cost with wax printers retailing for less than $600 (USD). The most commonly used porous substrate in direct wax printing is cellulose filter paper [40, 41], although nitrocellulose [42] has also been demonstrated to be compatible with direct wax printing. A severe limitation of direct wax printing is that it is only compatible with substrates that can be mechanically processed by the wax printer. Substrates that are too thick, coarse, fragile, flexible, or deviate significantly from the canonical 8.5 inch × 11 inch sheet dimensions are not recommended for use in the printer and could result in damage to the printer or substrate. Typical wax printers are limited to substrates with basis weight between 60 g/m^2 and 220 g/m^2 and dimensions between 76 mm × 127 mm and 216 mm × 610 mm. Thus, non-standard porous substrates that might have useful properties in paper microfluidics are not always compatible with the method of direct wax printing.

Several alternate wax-based patterning methods have been demonstrated including patterning using a wax pen [41], wax screen-printing [43], and wax dipping using an iron mold to selectively exclude wax in desired regions [44]. An extension of direct wax printing in porous materials is wax transfer printing [31, 45]. In this method, a transparency film is used as an intermediate template that can be mechanically processed by the wax printer and subsequently used to transfer the printed wax pattern onto a porous substrate by placing the two into contact and heating. Using the wax transfer method, fragile and flexible porous materials such as tissue and fabric can be straightforwardly patterned with wax [31]. Further, multiple wax transfer cycles can be used to pattern thicker materials (e.g., cellulose substrates with a large fluid capacity) [31]. Several wax-based methods are schematically highlighted in Figs. 12.2a–d.

Complementary to methods that define channels through patterning are methods based on physically cutting out structures with a blade [46] or laser [47]. The latter are particularly useful for creating hybrid devices composed of multiple materials. Advantages of these methods are simplicity and a relatively low cost of supporting equipment. Further, in the case of laser cutting systems, the power and speed of the laser beam can be adjusted to etch materials rather than cut through the materials. This can be useful when fabricating devices containing porous materials on a backing (e.g., nitrocellulose is often cast on a polyester backing), since the porous material can be etched away, while keeping the backing intact [47]. A potential disadvantage of cutting out structures is that they can be difficult to handle without attaching them to a support layer, which then adds a step to the device assembly process. These methods are highlighted in Figs. 12.2e–f.

One factor that may affect the choice of which channel definition method to use is the target size of the porous channel. Typical paper microfluidic devices have a channel thickness of ~100 to several hundred μm and channel widths and lengths of

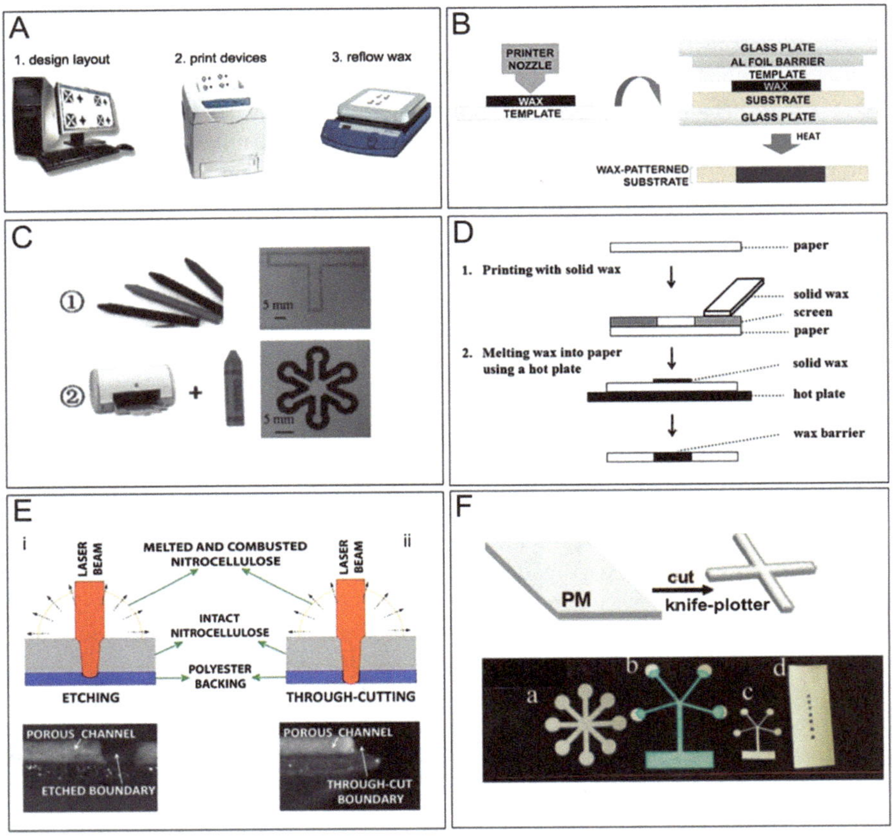

Fig. 12.2 Examples of fabrication methods for defining channels in porous materials. (**a**) The most common method is direct wax printing, in which a printer is used to deposit wax onto the porous substrate, and then the wax-coated substrate is heated to create robust hydrophobic boundaries. (Reproduced with permission from Ref. [40]). (**b**) For substrates that are not compatible with processing through the wax printer, an intermediate template, such as a polyester sheet, can be used to transfer the patterned wax to the substrate in a method called wax transfer printing. (Reproduced with permission from Ref. [31]). (**c**) Alternatives that do not require a wax printer include creating boundaries using a wax "pencil", or tracing wax boundaries onto a pre-printed ink pattern. (Reproduced with permission from Ref. [41]). (**d**) A substrate can also be patterned by applying melted wax through a screen or mask in wax screen printing. (Reproduced with permission from Ref. [43]). (**e**) Fluidic channels can be defined in porous material by cutting the porous material using a laser. Further, partial etching of materials can be useful for device assembly. (Reproduced with permission from Ref. [47]). (**f**) A knife blade can also be used to define the channels in a porous material. (Reproduced with permission from Ref. [46])

at least several mm. The mm-scale lateral dimensions are compatible with the majority of the above channel definition methods. However, in cases for which smaller dimensions are required, direct wax patterning (with demonstrated channel widths of ~600 μm [40] and ~300 μm [48]) and optical lithography [15] are the methods that would have the highest lateral patterning resolution. Additional factors that may affect the choice of channel definition method for a particular application are cost, ease of use, and the need for compatibility with other device fabrication processes such as reagent patterning and the incorporation of fluidic controls.

12.2.3 Patterning Reagents in Porous Materials

A second critical process in the fabrication of microfluidic devices is patterning reagents within the device to perform the required biochemical reactions in the field. Current patterning methods vary greatly in terms of cost and versatility.

On one end of the spectrum, manual pipetting is a low-cost method for reagent deposition [49, 50], but is constrained to mm-scale circular patterns. In addition, the spot uniformity can be highly dependent on operator skill and the method is limited to low-throughput situations. Further, in the case of lateral flow immunoassays (i.e., nonspecific adsorption of antibodies onto nitrocellulose [7]), producing different uniform reagent concentrations on the substrate is not straightforward.

At the other end of the spectrum, expensive, automated, liquid dispensing systems can produce high-resolution (~200 μm) patterns of protein in a variety of shapes and over a range of concentrations. In the context of LFTs, the most common reagent patterning system "stripes" antibodies in a narrow region across the nitrocellulose strip [12]. Other useful configurations of patterned reagents include extended regions to obtain information on the signal binding profile [51] or for continuous distance-based readout [20], and multiple discrete capture regions for distance-based readout [52] or multi-analyte patterning [53]. Additionally, patterning nanoliter drops to control rehydrated reagent concentration as a function of space and time has been demonstrated [54] and applied to an amplified assay [55]. Liquid dispensing instrumentation [56] has been used to produce spots [55, 57, 58], stripes [59, 60], and larger, mm-scale rectangular regions [51]. Disadvantages of these instruments include a high cost, between $30 k and $120 k (USD), the need for continued maintenance, and the need for operator training.

Alternative cost-effective stamping methods have been developed to complement manual pipetting of reagents. In one example, highlighted in Fig. 12.3a, cellulose stamps were used to pattern dye and protein onto cellulose substrates in high volume [61], but with reduced edge definition due to the imbibition of fluid into adjacent non-stamped substrate regions. In another example, highlighted in Fig. 12.3b, composite porous glass fiber stamps that included buffer-containing regions to restrict imbibition of the protein solution in the substrate were used to produce mm-scale rectilinear patterns of antibody on nitrocellulose [62].

Although physical adsorption works well for creating high-density regions of antibodies on nitrocellulose substrates, it may not be as effective for other molecule and substrate combinations. The latter is true for antibodies and cellulose, and a number of methods for creating high-density regions of immobilized antibody on cellulose have been demonstrated. These include crosslinking with glutaraldehyde

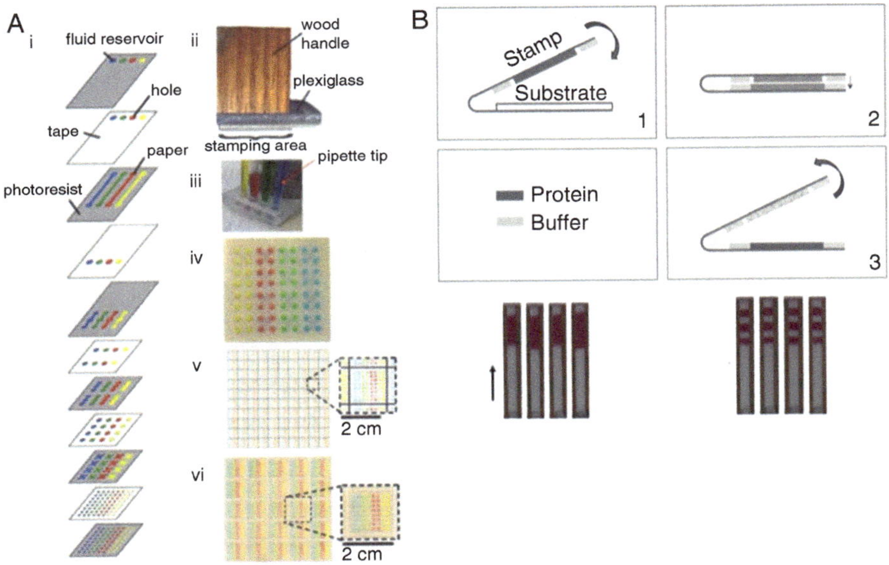

Fig. 12.3 Examples of fabrication methods for patterning reagents in porous materials. (**a**) Cellulose stamps, configured in a layered network, were used to pattern reagents onto cellulose in a high-throughput parallel process. (Reproduced with permission from Ref. [61]). (**b**) Schematic of a related method in which glass fiber stamps were used to pattern antibodies onto nitrocellulose. The stamps incorporated buffer-filled pads adjacent to the protein-filled stamps in order to improve boundary definition. The images below the schematic show antibody stamped in different patterns and subsequently labeled with gold nanoparticles. (Reproduced with permission from Ref. [62])

[63], covalent bonding using periodate oxidation [63], diazonium-based covalent functionalization [64], and photoreactive methods [65].

Adding a layer of complexity to the patterning of reagents in a paper microfluidic device is the need for the preservation of these reagents until rehydration with sample. As mentioned above, LFTs provide a benchmark for preservation timescales of 1 or 2 years at ambient temperature [14]. Methods of drying reagents, such as antibodies [66], for preservation include lyophilization and vacuum drying. One of the most commonly used additives for the preservation of proteins is the disaccharide trehalose, and it is known to stabilize protein structure [67]. As an example, in the demonstration of a paper microfluidic enzyme-amplified assay, horseradish peroxidase (HRP) was preserved by vacuum drying the enzyme with a mixture of trehalose, iron, and EDTA on glass fiber. The activity of the enzyme after rehydration (and previous dry storage at elevated temperature for 5 months) was ~80% that of fresh enzyme [68]. In a related study, SU-8 epoxy resin [69] was demonstrated to have a positive stabilizing effect on HRP storage in cellulose.

12.2.4 Manipulating Fluids in Porous Materials

The key to enabling high performance in paper microfluidic devices is the robust and accurate control of fluids within the devices. Thus, the paper microfluidics community has dedicated substantial effort to increasing the capability of paper-based devices via the development of a versatile set of methods to control fluid flow in paper-based devices or "flow tools" [70]. Flow tools have been developed to change the fluid velocity within a channel relative to a conventional porous channel, as well as function as a valve by turning on or off fluid flow within a porous channel. These flow tools use many different methods to affect fluid flow. A comprehensive review of the area of flow tools development is presented in a review by Fu and Downs [70]. The majority of the tools operate via manipulation of capillary flow in porous media, including fluid surface tension, medium pore size, contact angle, or fluid viscosity. Some of the more recent flow tools use open channels within the porous material to manipulate the overall transport of fluid within the device. Complementary to these are tools that rely on mechanical actuation, either via the operator or automated via an element responsive to a particular external stimulus. A summary of some flow tools, organized by function, is presented in Table 12.1 [70].

The ability to sequentially deliver multiple fluid volumes is required for multistep sample processing (e.g., signal amplification [30, 71]), and potentially improved analytical performance. Many of the fluid flow control tools that have been developed can be used to sequence fluids; four methods are highlighted in Fig. 12.4. A first strategy for sequencing fluids uses dissolvable barriers as shown schematically in Fig. 12.4a. Specifically, sugar barriers can be used to create delays in the transport of the fluid within a paper network [72]. Both the extent of the sugar barrier and the concentration of the sugar solution used to form the barrier within the porous material can be used to produce a delay time of seconds to an hour [72, 73]. The time delay is mainly due to an increased viscosity of the system fluid by the addition of the sugar [72]. Coefficients of variation for the hand-dipped barriers range between 11 and 24%. A second strategy for sequencing fluids is to use varying path lengths to control the arrival times of multiple fluid volumes to a common detection region [30, 73]. As shown in Fig. 12.4b, the multiple reagent volumes are delivered to the detection region in order from right to left. This device design was used to create the amplified assay shown in Fig. 12.1d. A third strategy for sequencing fluids is to use porous material shunts in which the porous material enables an alternate flow path for the system fluid. Delays between 3 and ~12 min were produced. As shown in the example in Fig. 12.4c, the shunts can be integrated into a paper network with varying path length in order to effectively sequence fluids [49]. In particular, the shunts enable the delivery of larger fluid volumes for a given network design. And a fourth strategy highlighted in Fig. 12.4d, also based on path length, uses a linear design with multiple input pads to produce the sequential delivery of multiple well-defined fluid volumes [74].

Table 12.1 Summary of some fluid flow control tools

Function	Tool
Methods of changing fluid velocity	Changes in channel width/length produce small magnitude velocity changes [73]
	Downstream region with increasing cross-sectional area produces quasi-steady flow [6]
	Porous channels enclosed by flexible films produce up to 10× increase in flow rate compared to capillary flow [75]
	Hollow channels with hydrophilic base produce up to 7× increase in flow rate [76, 77]
	Etched trenches in paper channel produce speeding up or slowing of fluid [78]
	Double layer porous channels produce an over 6× decrease in flow time (4.5 cm at 35% relative humidity) [79]
	Open channels in omniphobic paper produce fast pressure-driven flow [80]
	Hollow channels with hydrophobic base enable pressure-driven flow [81]
	Wax coatings in porous channels decrease wettability and produce time delays [82, 83]
	UV treatments of TiO_2-coated papers alter surface wettability and produce delay times [84]
	Pressed papers with reduced pore sizes produce delay times of 7.4-fold [85]
	Dissolvable species dried into channels for increased fluid viscosity upon rehydration produce delay times [72, 73]
	Porous shunts divert fluid to produce time delays [49]
	Sealing paper channels with charged plastic substrates produces time delays [86]
Methods of valving	Mechanical user actions including bridging [87], push buttons [88], folding [30], and sliding [89]
	Polarization of a hydrophobic dielectric coated on paper to be hydrophilic [90]
	Temperature affects the surfactant solubility in water-filled, hydrophobic porous channels [91]
	Corona discharge treatment switches coated paper from hydrophobic to hydrophilic [92]
	Melting of wax plug allows flow [93]
	Submerged inlet length in a well shuts off flow when fluid level drops below inlet [94]
	Dissolvable bridge shuts off flow of fluid [95, 96]
	Magnetic forces actuate to connect or disconnect a fluidic path [97]
	Expandable polymers actuate upon a fluid trigger to connect/disconnect a fluidic path [50]
	Rehydration of dried surfactant enables fluid to cross a hydrophobic region [98, 99]

(continued)

Table 12.1 (continued)

Function	Tool
Pseudo-valves	2D paper network [71, 73] and 3D paper network [89] with varying path lengths
	Linear design (1DPN) with wells or pads [74]
	Inkjet printing of barriers produces varying path lengths [100]
	SlipPAD configuration of fluid sources and channel in adjacent layers sequences the fluids in the channel upon sliding contact between the layers [101]

The table is adapted with permission from Ref. [70]

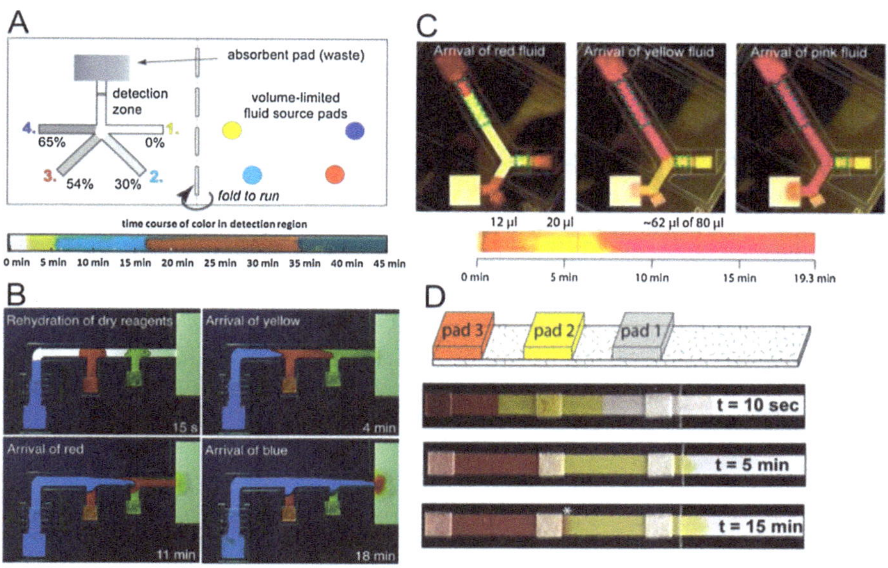

Fig. 12.4 Four methods of sequencing fluids in paper-based devices. (**a**) Dissolvable barriers consisting of sugar can be used to create time delays from seconds to 50 min. (Reproduced with permission from Ref. [72]). (**b**) Varying the path lengths of different fluid inputs into the network can be used to sequentially deliver the fluids to a common detection region. (Reproduced with permission from Ref. [30]). (**c**) Porous shunts can be integrated into a design based on varying path lengths to achieve additional control of fluid delivery. (Reproduced with permission from Ref. [49]). (**d**) Varying path lengths in a linear design with porous pads is a straightforward method of sequencing fluids. (Reproduced with permission from Ref. [74])

A main advantage of the dissolvable barriers is a large tunable range of time delays, while a potential disadvantage is the large amount of sugar released that may be incompatible with downstream assay processes. A main advantage of using the 2D network with varying path lengths for sequencing fluids is simplicity, with fluid control built into the network without additional components. However, for larger volumes, the network size (and associated dead volume) increases. The use of shunts can be used to increase the volumes delivered for a given size network, but at the expense of additional materials and fabrication steps. The 1D design is the simplest of the four and compatible with current lateral flow manufacturing processes. Additional potential benefits of integrating flow tools into a device are simplified operation and/or improved device robustness. The choice of which flow tool to use depends on the requirements of the application of interest. Compatibility with core device fabrication processes, requirements for assay time, requirements for reproducibility, or environmental constraints of the setting, can all guide the choice of flow tool.

12.3 Paper Microfluidic Devices for Medical Applications

12.3.1 Categories of Medically-Relevant Devices

There have been numerous examples of paper-based devices developed for medical applications. Paper microfluidics has the potential to decentralize sophisticated laboratory testing and reach populations in low-resource settings that would not otherwise be able to access testing. One category of medical applications for paper microfluidic devices is tumor biomarker detection. Targets include alpha-fetoprotein [102, 103], carcinoembryonic antigen [102], prostate specific antigen [102], and cancer antigen (CA) 15-3 [102, 104], CA 125 [105], and CA 19-9 [102, 106]. Another category of medical applications for paper microfluidic devices is the analysis of metabolite level to assess health status. Targets in this category include glucose [107, 108], lactate [22, 109, 110], uric acid [22, 108], cholesterol [109, 111, 112], creatinine [113], and markers of liver function [114–116]. A third category of targets for paper microfluidic devices is infectious disease detection. These include *E. coli* [117], methicillin-resistant *S. aureus [118]*, hepatitis C virus [119], *Plasmodium falciparum* [120], and influenza virus [121–123]. And a fourth category of targets is composed of analytes relevant for therapy monitoring. Targets in this category include phenylalanine for phenylketonuria patients to better manage nutritional therapy [124], theophiline for patients with asthma and other respiratory conditions to optimize their therapeutic dosages [125, 126], and lithium for patients with depression [127]. A critical review by Yamada et al. [128] includes a more comprehensive list of medically-relevant targets in paper microfluidics.

The majority of the work in the field of paper microfluidics thus far has focused on either the development of general tools/formats in model systems or proof-of-concept demonstrations of specific applications using mock samples and expert operators in well-controlled laboratory environments. This work is critical to overall progress in the field, but is also only the first step in the process of the translation of promising technology into medically-relevant, field-use devices. There are relatively few examples of paper microfluidic devices that have advanced to the stage of testing clinical specimens in the field. Three informative examples, a liver function monitor, a sickle cell disease diagnostic test, and an influenza diagnostic test, are highlighted below.

12.3.2 Paper-Based Liver Function Monitor

Liver toxicity for patients undergoing HIV and TB therapy is a significant issue. The incidence of drug-associated liver toxicity can reach ~13% for the HIV therapeutic nevirapine [129], and can be as great as 33% for the commonly used TB therapeutics isoniazid, rifampin, and pyrazinamide [130, 131]. Liver function monitoring is routine in developed countries [116]. However, access to healthcare facilities for this testing is not available to many in low-resources settings. Thus, the ability to monitor liver function on populations in need within low-resource settings is an excellent potential application of paper microfluidics.

A paper-based liver function monitor was introduced in 2011 [114]. The rapid test accepted a finger-stick sample of whole blood and produced colorimetric output for three targets related to liver function, (i) aspartate aminotransferase (AST), (ii) alkaline phosphatase (ALP), and (iii) total serum protein. The vertical flow design consisted of stacked patterned cellulose and a plasma separation membrane, encased between polyester support layers. The multiple zones for detection were patterned using wax printing of the cellulose substrate and reagents were dried into the zones. The plasma separation membrane (Pall Vivid GX) was composed of a graded pore structure that effectively separated the red blood cells from input whole blood. Device operation consisted of two user steps. First, a finger-stick sample of whole blood was obtained using a lancet and applied to the inlet of the device. Whole blood was wicked into the sample port, and the plasma that exited the opposite face of the plasma separation membrane was split between the multiple detection zones. Second, at 15 min, readout of the device was performed on the device face opposite to the sample port, either by eye or using a reader.

A subsequent stage in the development of the liver function monitor narrowed the targets to AST and alanine aminotransferase (ALT) and addressed four critical issues in device translation [116]. First, three controls were added to the device to detect improper functioning of various test components. Second, the effect of ambient temperature on assay results was assessed and indicated that higher temperatures resulted in accelerated assay signal development. The authors demonstrated that appropriate choice of read time, based on the ambient temperature, could be used to correct for temperature effects. Third, the effect of sample volume was assessed and indicated that a minimum sample volume of 25 μL was required to achieve robust function of the device, but that reasonable variation in sample volume above that minimum had a negligible effect on the assay signal. Fourth, assay interference by common substances in a blood matrix was assessed and indicated that two substances, pyruvic acid and ascorbic acid had a significant effect on the ALT signal (greater than 10% change). And finally, the ability of human readers ($N = 3$) to resolve different levels of analyte using a color intensity guide was assessed using

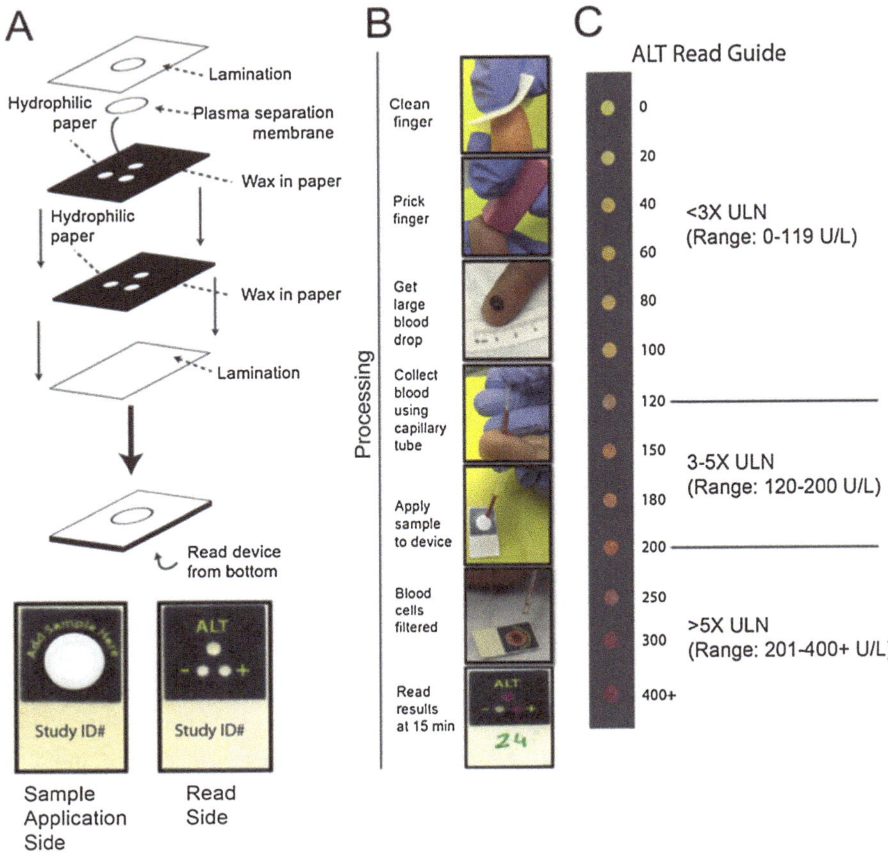

Fig. 12.5 An excellent example of the process of paper-based device technology translation is provided by the development of a liver function monitor for patients with HIV or TB. (**a**) The simple design consisted of layers of wax-patterned cellulose and a plasma separation membrane laminated together. Finger-stick blood was applied to one side of the device and readout of the device consisted of noting the intensity of color that appears in detection zones on the opposite side of the device. (**b**) The multiple operator steps are shown in the image series. Note that the device required a large blood drop (i.e., larger than for standard glucose meters). A capillary tube was used for the robust application of blood sample to the device. (**c**) A "Read Guide" was used by device operators to determine the level of ALT in the sample based on the color intensity in the device detection zone. (Reproduced with permission from Ref. [132])

clinical samples and estimated to be 90% accurate relative to gold-standard measurements on paired patient samples.

Subsequent development work on the liver function monitor further addressed issues critical to device translation, and assessed device performance in the field [132]. The device, shown in Fig. 12.5a, was modified to include only ALT detection, the assay chemistry was optimized for clinically-relevant color intensity ranges, and two controls zones were incorporated into the device to indicate inadequate sample volume, hemolysis, or nonfunctioning reagents. Operator steps, outlined in Fig. 12.5b, were further refined, and included (i) the use of a capillary tube for the transfer of the finger-stick blood sample to the device, and (ii) a more comprehensive set of assay read times between 10 and 18 min. The read guide for this version of their test is shown in Fig. 12.5c. The field evaluation was performed at a clinic in Vietnam with local nurses as the device operators on HIV patient samples ($N = 600$). ALT levels in patient samples were also assessed using a lab-based automated analyzer for comparison. The study results were promising, indicating that operator accuracy on ALT level assessment was ~84% and agreement between the two operators was ~96%. Disadvantages of the evaluated version of the device include a substantially higher limit of detection compared to a gold-standard ALT detection method, and the high-level of operator training (described as "intensive") required to perform and to interpret the test. Overall, the documented development of this device is an instructive example of multiple issues that must be addressed in creating a robust, field-use device.

12.3.3 Paper-Based Sickle Cell Diagnosis

Sickle cell disease (SCD) is a genetic disorder caused by a mutant form of hemoglobin. Under conditions of deoxygenation, the "sickle" hemoglobin forms long polymer chains that produce structural and functional changes in the red blood cells of an individual with sickle cell anemia (SCA), including cell rupture or the blockage of blood vessels [133]. SCA is most prevalent in Africa where children under 5 years old with SCA are at substantial risk of death [134, 135]. Early detection and treatment can substantially improve health outcomes and quality of life for individuals with SCA [133], but lab-based testing is not available to many populations in developing countries. Thus, there has been considerable interest in a test to diagnose SCD in low-resource settings [136].

A paper microfluidic device for the detection of SCD in children and adults was developed [137, 138]. The user steps, highlighted schematically in Fig. 12.6a, consisted of (i) mixing 20 μL of finger-stick blood and a buffer (composed of a reducing species and a hemolytic) in a small tube at a 1:10 volume ratio, (ii) waiting 10 min and pipetting a drop onto the paper test, and (iii) waiting 25 min and interpreting the result. The differences in the flow of the soluble non-sickle Hb species and the polymerized sickle Hb species resulted in distinguishable blood patterns between HbAA (wild-type alleles), HbSA (heterozygous for the sickle cell trait), and HbSS (homozygous for the sickle cell trait) individuals, as shown in Fig. 12.6b. The field evaluation was performed at a hospital in Angola on mothers of newborns ($N = 226$) and compared with gold-standard measurements (isoelectric focusing) on paired samples. Results indicated a sensitivity and specificity for HbS detection of 94% and 97%, respectively, and positive predictive and negative predictive values of 92% and 98%, respectively. Alternative readout using automated analysis of blood pattern image data indicated 100% sensitivity and specificity

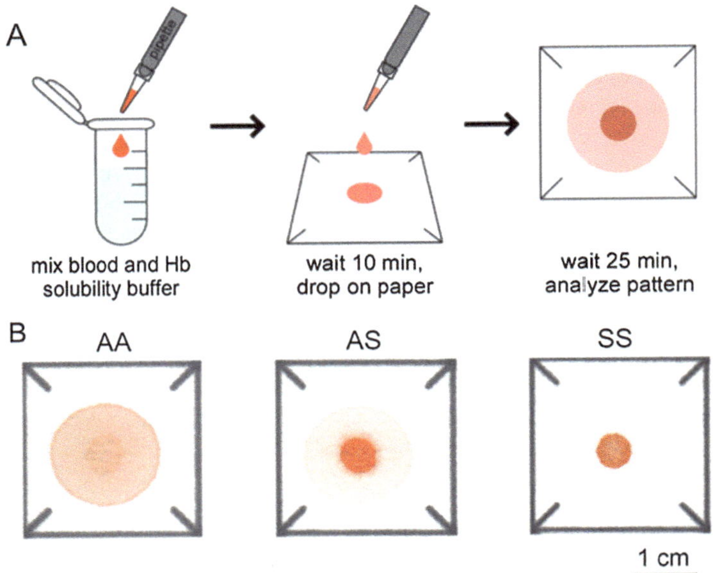

Fig. 12.6 Another informative example of paper-based device technology translation is the development of a diagnostic test for sickle cell anemia for use in low-resource settings. (**a**) User steps included pipetting a finger-stick blood sample into a tube containing a buffer and mixing, pipetting a drop of the mixture onto the device, and then interpreting the resulting blood pattern at 35 min after the start of the test. (**b**) Red blood cells that contain sickle hemoglobin have slowed transport compared to red blood cells containing normal hemoglobin, and results in distinct blood patterns for individuals with two normal alleles, one mutant and one normal allele, or two mutant alleles. (Reproduced with permission from Ref. [138])

and would be compatible with use in some higher-resource settings. Strengths of the paper-based test are it's high level of performance even in variable ambient conditions (note that quantification of the effects of temperature and humidity were not attempted in this study), and its relative ease of use that is compatible with minimal user training. Further development produced a version of the device sensitive enough for screening newborns for SCD at the expense of additional user steps to remove cellular debris from the sample [139, 140]. A limitation of the visual readout version of the device is that it did not provide information on other forms of SCD, but progress has been made on a version that makes use of automated, instrumented readout of blood pattern image data [138].

12.3.4 Paper-Based Influenza Detection

The respiratory infection influenza causes approximately 36,000 deaths and 200,000 hospitalizations each year in the U.S., and results in over S10 billion in direct

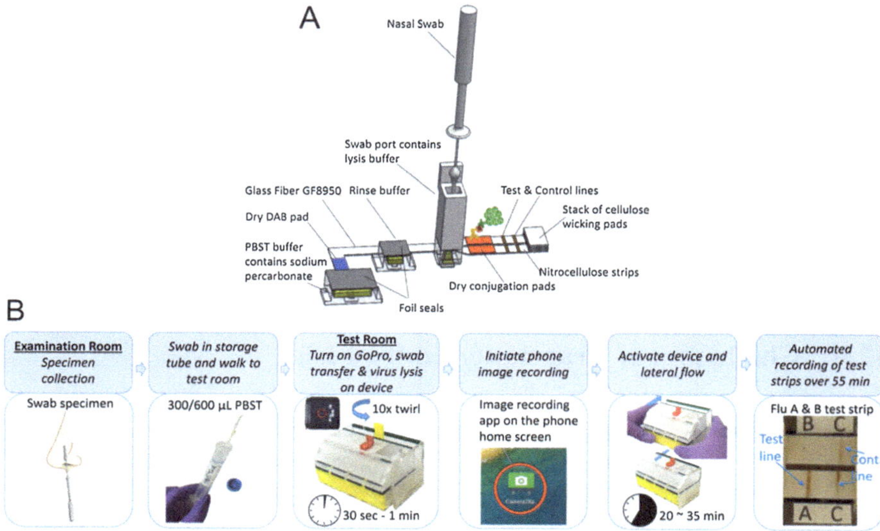

Fig. 12.7 A third informative example is the development of an influenza diagnostic. The panel summaries the device operating protocol. A strength of the device design is that operator steps have been streamlined to eliminate pipetting steps that can lead to variations in input sample volume or consistency, and downstream variability in signal. (Reproduced with permission from Ref. [123])

medical costs [141]. FDA-approved LFTs have been reported to have poor sensitivities of less than 53% [142, 143]. Thus, there is a clear need for a higher sensitivity POC test than is currently available. Potential target settings include hospital emergency rooms, clinics, military units, and the home. The potential positive impact of accurate influenza diagnosis include the use of anti-viral therapy [144], isolation of infected persons to avoid outbreaks [145], and the reduction of antibiotic use due to misdiagnosis [144]. Thus, there has been interest in developing improved POC tests for influenza detection [123, 146–149].

One of these tests advanced to the stage of preliminary clinical testing in a Seattle children's hospital [123]. The test targeted influenza virus nucleoprotein A and B via parallel sandwich immunoassays, followed by enzyme-based amplification. Automation of the multiple delivery steps was accomplished using a 2D paper network with varying path lengths (based on the work of Fig. 12.4b) for the sequential delivery of sample and label, rinse, and amplification reagents. A schematic of the 2D paper network is shown in Fig. 12.7a, and the operator steps performed by healthcare professionals, are described in Fig. 12.7b. The prototype paper-based device completed analyte detection from a nasal swab in approximately 35 min.

Device accuracy from the preliminary clinical study ($N = 25$) was ~70% for the detection of influenza A, and the device negative predictive value was 81%. Planned, targeted improvements to device sensitivity include optimizing the device flow rate and geometry, and automating a robust protocol for readout with a smartphone. The main strength of this demonstration was the high level of device integration, minimizing user steps and the potential for error. Current rapid tests that are compatible with nasal swab samples require the user to insert the swab into a tube of lysis buffer and then pipet a fraction of the lysis buffer-plus sample mixture into the device. User steps for the paper-based influenza test were significantly simplified -- the user inserted the swab into a port in the device and a well-defined volume of sample fluid was automatically input into the device. Further, optimization of the device ease of use was based on feedback generated from a user study focused on making operator steps easy to understand and perform.

12.4 Challenges of Translation to the Field

Paper microfluidics holds particular promise for enabling medical testing that is not currently available in low-resource settings. There have been a number of promising proof-of concept demonstrations of potential medical applications of paper microfluidic devices, but few field demonstrations. The above discussion of three paper-based devices that have advanced to field evaluation highlights multiple issues that must be addressed in order to create a robust field-ready device. These include (i) optimizing user steps to make the device robust to variability in operators, (ii) incorporating the appropriate positive and negative device controls and calibrators, (iii) ensuring that the device performance is robust across the range of environmental conditions that span the expected variation in field conditions, (iv) ensuring the device is robust against interference from common substances in the human matrix of interest, (v) ensuring that the device is appropriate for use by operators who have the expected level of training for the end use scenario, and (vi) ensuring that devices have robust performance across multiple manufacturing lots, and after realistic delivery and storage conditions and times.

Moving forward, the overall challenge in paper microfluidics device development for medical applications continues to be the design and implementation of devices appropriate for the intended operator and setting. The examples above highlight the tension between simplifying a device design and placing more burden on the operator, versus automating a device for ease of use, but at the expense of increasing device complexity, adding fabrication processes, and increasing cost. Following best practices, the requirements and constraints of the application should guide device design choices. The ideal device design for use by nurses at a high-resource clinic in a developed country may differ substantially from the ideal device design for use by healthcare workers at an outdoor clinic in a developing country. In particular, the device characteristics of time to result, cost, ease of use, robustness to variations in environmental conditions, and analytical performance, must match the requirements of the application and the constraints of the use scenario. By addressing these challenges of device translation to the field, the paper microfluidics community can realize the potential of paper-based devices for many precision health applications.

References

1. Beebe DJ, Mensing GA, Walker GM (2002) Physics and applications of microfluidics in biology. Annu Rev Biomed Eng 4:261–286
2. Adamson AW, Gast A (1997) Physical chemistry of surfaces, 6th edn. Wiley, New York, p 4–16
3. Bird RB, Stewart WE, Lightfoot EN (1960) Transport phenomena. Wiley, New York, p 48–52
4. Washburn EW (1921) The dynamics of capillary flow. Phys Rev 17:273–283
5. Lucas R (1918) The time law of the capillary rise of liquids. Kolloid-Zeitschrift 23:15–22
6. Mendez S, Fenton EM, Gallegos GR, Petsev DN, Sibbett SS, Stone HA, Zhang Y, Lopez GP (2010) Imbibition in porous membranes of complex shape: quasi-stationary flow in thin rectangular segments. Langmuir 26:1380–1385
7. Fridley GE, Holstein CA, Oza SB, Yager P (2013) The evolution of nitrocellulose as a material for bioassays. MRS Bull 38:326–330
8. Fu E, Yager P, Floriano P, Christodoulides N, McDevitt J (2011) Perspective on diagnostics for global health. IEEE Pulse 2:40–50
9. Urdea M, Penny LA, Olmsted SS, Giovanni MY, Kaspar P, Shepherd A, Wilson P, Dahl CA, Buchsbaum S, Moeller G, Burgess DCH (2006) Requirements for high impact diagnostics in the developing world. Nature 444:73–79
10. Yager P, Domingo GJ, Gerdes J (2008) Point-of-care diagnostics for global health. Annu Rev Biomed Eng 10:107–144
11. Peeling RW, Holmes KK, Mabey D, Ronald A (2006) Rapid tests for sexually transmitted infections (STIs): the way forward. Sex Transm Infect 82:v1–v6
12. O'Farrell B (2009) Evolution in lateral flow-based immunoassay systems. In: Wong R, Tse H (eds) Lateral flow immunoassay. Humana Press, New York
13. Posthuma-Trumpie GA, Korf J, van Amerongen A (2009) Lateral flow (immuno) assay: its strengths, weaknesses, opportunities and threats. A literature survey. Anal Bioanal Chem 393:569–582
14. O'Farrell B (2013) Lateral flow immunoassay systems: evolution from the current state of the art to the next generation of highly sensitive, quantitative rapid assays. In: Wild D (ed) The immunoassay handbook. Elsevier, Oxford
15. Martinez AW, Phillips ST, Butte MJ, Whitesides GM (2007) Patterned paper as a platform for inexpensive, low-volume, portable bioassays. Angew Chem Int Ed 46:1318–1320
16. Martinez AW, Phillips ST, Whitesides GM (2008) Three-dimensional microfluidic devices fabricated in layered paper and tape. Proc Natl Acad Sci U S A 105:19606–19611
17. Martinez AW, Phillips ST, Whitesides GM, Carrilho E (2010) Diagnostics for the developing world: microfluidic paper-based analytical devices. Anal Chem 82:3–10
18. Adkins J, Boehle K, Henry C (2015) Electrochemical paper-based microfluidic devices. Electrophoresis 36:1811–1824
19. Mettakoonpitak J, Boehle K, Nantaphol S, Teengam P, Adkins JA, Srisa-Art M, Henry CS (2016) Electrochemistry on paper-based analytical devices: a review. Electroanalysis 28:1420–1436
20. Fu E (2014) Enabling robust quantitative readout in an equipment-free model of device development. Analyst 139:4750–4757
21. Mak WC, Beni V, Turner APF (2016) Lateral-flow technology: from visual to instrumental. Trends Anal Chem 79:297–305
22. Dungchai W, Chailapakul O, Henry CS (2010) Use of multiple colorimetric indicators for paper-based microfluidic devices. Anal Chim Acta 674:227–233
23. Cate DM, Dungchai W, Cunningham JC, Volckens J, Henry CS (2013) Simple, distance-based measurement for paper analytical devices. Lab Chip 13:2397–2404

24. Fung KK, Chan CPY, Renneberg R (2009) Development of enzyme-based bar code-style lateral-flow assay for hydrogen peroxide determination. Anal Chim Acta 634:89–95
25. Coskun AF, Wong J, Khodadadi D, Nagi R, Tey A, Ozcan A (2013) A personalized food allergen testing platform on a cellphone. Lab Chip 13:636–640
26. Mudanyali O, Dimitrov S, Sikora U, Padmanabhan S, Navruz I, Ozcan A (2012) Integrated rapid-diagnostic-test reader platform on a cellphone. Lab Chip 12:2678–2686
27. Zhu HY, Yaglidere O, Su TW, Tseng D, Ozcan A (2011) Cost-effective and compact wide-field fluorescent imaging on a cell-phone. Lab Chip 11:315–322
28. Byrnes S, Thiessen G, Fu E (2013) Progress in the development of paper-based diagnostics for low-resource point-of-care settings. Bioanalysis 5:2821–2836
29. Fu E, Liang T, Houghtaling J, Ramachandran S, Ramsey SA, Lutz B, Yager P (2011) Enhanced sensitivity of lateral flow tests using a two-dimensional paper network format. Anal Chem 83:7941–7946
30. Fu E, Liang T, Spicar-Mihalic P, Houghtaling J, Ramachandran S, Yager P (2012) Two-dimensional paper network format that enables simple multistep assays for use in low-resource settings in the context of malaria antigen detection. Anal Chem 84:4574–4579
31. To A, Downs C, Fu E (2017) Wax transfer printing to enable robust barrier definition in devices based on non-standard porous materials. J Micromech Microeng 27:057001. (6pp)
32. Martinez AW, Phillips ST, Wiley BJ, Gupta M, Whitesides GM (2008) FLASH: a rapid method for prototyping paper-based microfluidic devices. Lab Chip 8:2146–2150
33. Jiang X, Fan ZH (2016) Fabrication and operation of paper-based analytical devices. In: Bohn PW, Pemberton JE (eds) Annual review of analytical chemistry, vol 9. Annual Reviews, Palo Alto, pp 203–222
34. Xia Y, Si J, Li Z (2016) Fabrication techniques for microfluidic paper-based analytical devices and their applications for biological testing: a review. Biosens Bioelectron 77:774–789
35. Bruzewicz DA, Reches M, Whitesides GM (2008) Low-cost printing of poly (dimethylsiloxane) barriers to define microchannels in paper. Anal Chem 80:3387–3392
36. Abe K, Suzuki K, Citterio D (2008) Inkjet-printed microfluidic multianalyte chemical sensing paper. Anal Chem 80:6928–6934
37. Li X, Tian JF, Nguyen T, Shen W (2008) Paper-based microfluidic devices by plasma treatment. Anal Chem 80:9131–9134
38. Chitnis G, Ding Z, Chang C-L, Savran CA, Ziaie B (2011) Laser-treated hydrophobic paper: an inexpensive microfluidic platform. Lab Chip 11:1161–1165
39. Curto VF, Lopez-Ruiz N, Capitan-Vallvey LF, Palma AJ, Benito-Lopez F, Diamond D (2013) Fast prototyping of paper-based microfluidic devices by contact stamping using indelible ink. RSC Adv 3:18811–18816
40. Carrilho E, Martinez AW, Whitesides GM (2009) Understanding wax printing: a simple micropatterning process for paper-based microfluidics. Anal Chem 81:7091–7095
41. Lu Y, Shi WW, Jiang L, Qin JH, Lin BC (2009) Rapid prototyping of paper-based microfluidics with wax for low-cost, portable bioassay. Electrophoresis 30:1497–1500
42. Lu Y, Shi WW, Qin JH, Lin BC (2010) Fabrication and characterization of paper-based microfluidics prepared in nitrocellulose membrane by wax printing. Anal Chem 82:329–335
43. Dungchai W, Chailapakul O, Henry CS (2011) A low-cost, simple, and rapid fabrication method for paper-based microfluidics using wax screen-printing. Analyst 136:77–82
44. Songjaroen T, Dungchai W, Chailapakul O, Laiwattanapaisal W (2011) Novel, simple and low-cost alternative method for fabrication of paper-based microfluidics by wax dipping. Talanta 85:2587–2593
45. Lu Y, Lin BC, Qin JH (2011) Patterned paper as a low-cost, flexible substrate for rapid prototyping of PDMS microdevices via "liquid molding". Anal Chem 83:1830–1835
46. Fenton EM, Mascarenas MR, Lopez GP, Sibbett SS (2009) Multiplex lateral-flow test strips fabricated by two-dimensional shaping. ACS Appl Mater Interfaces 1:124–129

47. Spicar-Mihalic P, Toley B, Houghtaling J, Liang T, Yager P, Fu E (2013) CO2 laser cutting and ablative etching for the fabrication of paper-based devices. J Micromech Microeng 23:067003. (6pp)
48. Tenda K, Ota R, Yamada K, Henares TG, Suzuki K, Citterio D (2016) High-resolution microfluidic paper-based analytical devices for sub-microliter sample analysis. Micromachines 7:80
49. Toley BJ, McKenzie B, Liang T, Buser JR, Yager P, Fu E (2013) Tunable-delay shunts for paper microfluidic devices. Anal Chem 85:11545–11552
50. Toley BJ, Wang JA, Gupta M, Buser JR, Lafleur LK, Lutz BR, Fu E, Yager P (2015) A versatile valving toolkit for automating fluidic operations in paper microfluidic devices. Lab Chip 15:1432–1444
51. Liang T, Robinson R, Fridley G, Yager P, Fu E (2016) Investigation of varying reagent delivery formats in a malaria system and implications on assay sensitivity. Anal Chem 88:2311–2320
52. Cho JH, Paek SH (2001) Semiquantitative, bar code version of immunochromatographic assay system for human serum albumin as model analyte. Biotechnol Bioeng 75:725–732
53. Li J, Macdonald J (2016) Multiplexed lateral flow biosensors: technological advances for radically improving point-of-care diagnoses. Biosens Bioelectron 83:177–192
54. Fridley GE, Le HQ, Fu E, Yager P (2012) Controlled release of dry reagents in porous media for tunable temporal and spatial distribution upon rehydration. Lab Chip 12:4321–4327
55. Fridley GE, Le H, Yager P (2014) Highly sensitive immunoassay based on controlled rehydration of patterned reagents in a 2-dimensional paper network. Anal Chem 86:6447–6453
56. Romanov V, Davidoff SN, Miles AR, Grainger DW, Gale BK, Brooks BD (2014) A critical comparison of protein microarray fabrication technologies. Analyst 139:1303–1326
57. Mujawar LH, Maan AA, Khan MKI, Norde W, van Amerongen A (2013) Distribution of biomolecules in porous nitrocellulose membrane pads using confocal laser scanning microscopy and high-speed cameras. Anal Chem 85:3723–3729
58. Mujawar LH, van Amerongen A, Norde W (2015) Influence of Pluronic F127 on the distribution and functionality of inkjet-printed biomolecules in porous nitrocellulose substrates. Talanta 131:541–547
59. Abe K, Kotera K, Suzuki K, Citterio D (2010) Inkjet-printed paperfluidic immuno-chemical sensing device. Anal Bioanal Chem 398:885–893
60. Teerinen T, Lappalainen T, Erho T (2014) A paper-based lateral flow assay for morphine. Anal Bioanal Chem 406:5955–5965
61. Cheng C, Mazzeo A, Gong J, Martinez A, Phillips S, Jain N, Whitesides G (2010) Millimeter-scale contact printing of aqueous solutions using a stamp made out of paper and tape. Lab Chip 10:3201–3205
62. Imdieke J, Fu E (2017) Porous stamp-based reagent patterning for lateral flow immunoassays. Anal Methods 9:2751–2756
63. Peng Y, Van Gelder V, Amaladoss A, Patel KH (2016) Covalent binding of antibodies to cellulose paper discs and their applications in naked-eye colorimetric immunoassays. J Vis Exp. https://doi.org/10.3791/54111
64. Credou J, Volland H, Dano J, Berthelot T (2013) A one-step and biocompatible cellulose functionalization for covalent antibody immobilization on immunoassay membranes. J Mater Chem B 1:3277–3286
65. Credou J, Volland H, Berthelot T (2015) Photolinker-free photoimmobilization of antibodies onto cellulose for the preparation of immunoassay membranes. J Mater Chem B 3:1079–1088
66. Stevens DY, Petri CR, Osborn JL, Spicar-Mihalic P, McKenzie KG, Yager P (2008) Enabling a microfluidic immunoassay for the developing world by integration of on-card dry reagent storage. Lab Chip 8:2038–2045

67. Jain NK, Roy I (2009) Effect of trehalose on protein structure. Protein Sci 18:24–36
68. Ramachandran S, Fu E, Lutz B, Yager P (2014) Long-term dry storage of an enzyme-based reagent system for ELISA in point-of-care devices. Analyst 139:1456–1462
69. Ganaja KA, Chaplan CA, Zhang JY, Martinez NW, Martinez AW (2017) Paper microzone plates as analytical tools for studying enzyme stability: a case study on the stabilization of horseradish peroxidase using Trehalose and SU-8 epoxy novolac resin. Anal Chem 89:5333–5341
70. Fu E, Downs C (2017) Progress in the development and integration of fluid flow control tools in paper microfluidics. Lab Chip 17:614–628
71. Fu E, Kauffman P, Lutz B, Yager P (2010) Chemical signal amplification in two-dimensional paper networks. Sens Actuators B Chem 149:325–328
72. Lutz B, Liang T, Fu E, Ramachandran S, Kauffman P, Yager P (2013) Dissolvable fluidic time delays for programming multi-step assays in instrument-free paper diagnostics. Lab Chip 13:2840–2847
73. Fu E, Lutz B, Kauffman P, Yager P (2010) Controlled reagent transport in disposable 2D paper networks. Lab Chip 10:918–920
74. Dharmaraja S, Lafleur L, Byrnes S, Kauffman P, Fu E, Buser J, Toley B, Yager P, Lutz B (2013) Programming paper networks for point of care diagnostics. Proc SPIE 8615. 10.1117/12.2006138
75. Jahanshahi-Anbuhi S, Chavan P, Sicard C, Leung V, Hossain SMZ, Pelton R, Brennan JD, Filipe CDM (2012) Creating fast flow channels in paper fluidic devices to control timing of sequential reactions. Lab Chip 12:5079–5085
76. Renault C, Koehne J, Ricco AJ, Crooks RM (2014) Three-dimensional wax patterning of paper fluidic devices. Langmuir 30:7030–7036
77. Renault C, Li X, Fosdick SE, Crooks RM (2013) Hollow-Channel paper analytical devices. Anal Chem 85:7976–7979
78. Giokas DL, Tsogas GZ, Vlessidis AG (2014) Programming fluid transport in paper-based microfluidic devices using razor-crafted open channels. Anal Chem 86:6202–6207
79. Camplisson CK, Schilling KM, Pedrotti WL, Stone HA, Martinez AW (2015) Two-ply channels for faster wicking in paper-based microfluidic devices. Lab Chip 15:4461–4466
80. Glavan AC, Martinez RV, Maxwell EJ, Subramaniam AB, Nunes RMD, Soh S, Whitesides GM (2013) Rapid fabrication of pressure-driven open-channel microfluidic devices in omniphobic R-F paper. Lab Chip 13:2922–2930
81. Shin JH, Lee GJ, Kim W, Choi S (2016) A stand-alone pressure-driven 3D microfluidic chemical sensing analytic device. Sensors Actuators B Chem 230:380–387
82. Noh N, Phillips ST (2010) Metering the capillary-driven flow of fluids in paper-based microfluidic devices. Anal Chem 82:4181–4187
83. Weng CH, Chen MY, Shen CH, Yang RJ (2014) Colored wax-printed timers for two-dimensional and three-dimensional assays on paper-based devices. Biomicrofluidics 8:066502
84. Songok J, Toivakka M (2016) Controlling capillary-driven surface flow on a paper-based microfluidic channel. Microfluid Nanofluid 20:63
85. Shin JH, Park J, Kim SH, Park JK (2014) Programmed sample delivery on a pressurized paper. Biomicrofluidics 8:054121
86. da Silva E, Santhiago M, de Souza FR, Coltro WKT, Kubota LT (2015) Triboelectric effect as a new strategy for sealing and controlling the flow in paper-based devices. Lab Chip 15:1651–1655
87. Li X, Tian JF, Shen W (2010) Progress in patterned paper sizing for fabrication of paper-based microfluidic sensors. Cellulose 17:649–659
88. Martinez AW, Phillips ST, Nie Z, Cheng C-M, Carrilho E, Wiley BJ, Whitesides GM (2010) Programmable diagnostic devices made from paper and tape. Lab Chip 10:2499–2504
89. Han KN, Choi JS, Kwon J (2016) Three-dimensional paper-based slip device for one-step point-of-care testing. Sci Rep 6:25710

90. Koo C, He F, Nugen S (2013) An inkjet-printed electrowetting valve for paper-fluidic sensors. Analyst 138:4998–5004
91. Cai LF, Zhong MH, Li HL, Xu CX, Yuan BY (2015) Defining microchannels and valves on a hydrophobic paper by low-cost inkjet printing of aqueous or weak organic solutions. Biomicrofluidics 9:046503
92. Jiang Y, Hao ZX, He QH, Chen HW (2016) A simple method for fabrication of microfluidic paper-based analytical devices and on-device fluid control with a portable corona generator. RSC Adv 6:2888–2894
93. Hong S, Kim W (2015) Dynamics of water imbibition through paper channels with wax boundaries. Microfluid Nanofluid 19:845–853
94. Lutz BR, Trinh P, Ball C, Fu E, Yager P (2011) Two-dimensional paper networks: programmable fluidic disconnects for multi-step processes in shaped paper. Lab Chip 11:4274–4278
95. Houghtaling J, Liang T, Thiessen G, Fu E (2013) Dissolvable Bridges for manipulating fluid volumes in paper networks. Anal Chem 85:11201–11204
96. Jahanshahi-Anbuhi S, Henry A, Leung V, Sicard C, Pennings K, Pelton R, Brennan JD, Filipe CDM (2013) Paper-based microfluidics with an erodible polymeric bridge giving controlled release and timed flow shutoff. Lab Chip 14:229–236
97. Li X, Zwanenburg P, Liu X (2013) Magnetic timing valves for fluid control in paper-based microfluidics. Lab Chip 13:2609–2614
98. Chen H, Cogswell J, Anagnostopoulos C, Faghri M (2012) A fluidic diode, valves, and a sequential-loading circuit fabricated on layered paper. Lab Chip 12:2909–2913
99. Gerbers R, Foellscher W, Chen H, Anagnostopoulos C, Faghri M (2014) A new paper-based platform technology for point-of-care diagnostics. Lab Chip 14:4042–4049
100. Apilux A, Ukita Y, Chikae M, Chailapakul O, Takamura Y (2013) Development of automated paper-based devices for sequential multistep sandwich enzyme-linked immunosorbent assays using inkjet printing. Lab Chip 13:126–135
101. Liu H, Li X, Crooks RM (2013) Paper-based SlipPAD for high-throughput chemical sensing. Anal Chem 85:4263–4267
102. Ge L, Wang S, Song X, Ge S, Yu J (2012) 3D origami-based multifunction-integrated immunodevice: low-cost and multiplexed sandwich chemiluminescence immunoassay on microfluidic paper-based analytical device. Lab Chip 12:3150–3158
103. Li L, Li W, Yang H, Ma C, Yu J, Yan M, Song X (2014) Sensitive origami dual-analyte electrochemical immunodevice based on polyaniline/au-paper electrode and multi-labeled 3D graphene sheets. Electrochim Acta 120:102–109
104. Wu K, Zhang Y, Wang Y, Ge S, Yan M, Yu J, Song X (2015) Paper-based analytical devices relying on visible-light-enhanced glucose/air biofuel cells. ACS Appl Mater Interfaces 7:24330–24337
105. Wang S, Ge L, Song X, Yu J, Ge S, Huang J, Zeng F (2012) Paper-based chemiluminescence ELISA: lab-on-paper based on chitosan modified paper device and wax-screen-printing. Biosens Bioelectron 31:212–218
106. Li W, Li L, Ge S, Song X, Ge L, Yan M, Yu J (2014) Multiplex electrochemical origami immunodevice based on cuboid silver-paper electrode and metal ions tagged nanoporous silver-chitosan. Biosens Bioelectron 56:167–173
107. Yang X, Forouzan O, Brown T, Shevkoplyas S (2012) Integrated separation of blood plasma from whole blood for microfluidic paper-based analytical devices. Lab Chip 12:274–280
108. Chen X, Chen J, Wang F, Xiang X, Luo M, Ji X, He Z (2012) Determination of glucose and uric acid with bienzyme colorimetry on microfluidic paper-based analysis devices. Biosens Bioelectron 35:363–368
109. Nie Z, Deiss F, Liu X, Akbulut O, Whitesides GM (2010) Integration of paper-based microfluidic devices with commercial electrochemical readers. Lab Chip 10:3163–3169
110. Dungchai W, Chailapakul O, Henry CS (2009) Electrochemical detection for paper-based microfluidics. Anal Chem 81:5821–5826

111. Deng L, Chen C, Zhu C, Dong S, Lu H (2014) Multiplexed bioactive paper based on GO@SiO2@CeO2 nanosheets for a low-cost diagnostics platform. Biosens Bioelectron 52:324–329
112. Ruecha N, Rangkupan R, Rodthongkum N, Chailapakul O (2014) Novel paper-based cholesterol biosensor using graphene/polyvinylpyrrolidone/polyaniline nanocomposite. Biosens Bioelectron 52:13–19
113. Talalak K, Noiphung J, Songjaroen T, Chailapakul O, Laiwattanapaisal W (2015) A facile low-cost enzymatic paper-based assay for the determination of urine creatinine. Talanta 144:915–921
114. Vella SJ, Beattie P, Cademartiri R, Laromaine A, Martinez AW, Phillips ST, Mirica KA, Whitesides GM (2012) Measuring markers of liver function using a micropatterned paper device designed for blood from a Fingerstick. Anal Chem 84:2883–2891
115. Pollock NR, McGray S, Colby DJ, Noubary F, Nguyen H, Nguyen TA, Khormaee S, Jain S, Hawkins K, Kumar S, Rolland JP, Beattie PD, Chau NV, Quang VM, Barfield C, Tietje K, Steele M, Weigl BH (2013) Field evaluation of a prototype paper-based point-of-care Fingerstick transaminase test. PLoS One 8:e75616
116. Pollock NR, Rolland JP, Kumar S, Beattie PD, Jain S, Noubary F, Wong VL, Pohlmann RA, Ryan US, Whitesides GM (2012) A paper-based multiplexed transaminase test for low-cost, point-of-care liver function testing. Sci Transl Med 4:152ra129
117. Connelly JT, Rolland JP, Whitesides GM (2015) "paper machine" for molecular diagnostics. Anal Chem 87:7595–7601
118. Lafleur LK, Bishop JD, Heiniger EK, Gallagher RP, Wheeler MD, Kauffman P, Zhang XH, Kline EC, Buser JR, Kumar S, Byrnes SA, Vermeulen NMJ, Scarr NK, Belousov Y, Mahoney W, Toley BJ, Ladd PD, Lutz BR, Yager P (2016) A rapid, instrument-free, sample-to-result nucleic acid amplification test. Lab Chip 16:3777–3787
119. Mu X, Zhang L, Chang S, Cui W, Zheng Z (2014) Multiplex microfluidic paper-based immunoassay for the diagnosis of hepatitis C virus infection. Anal Chem 86:5338–5344
120. Glavan AC, Christodouleas DC, Mosadegh B, Yu HD, Smith BS, Lessing J, Teresa Fernandez-Abedul M, Whitesides GM (2014) Folding analytical devices for electrochemical ELISA in hydrophobic R-H paper. Anal Chem 86:11999–12007
121. Rodriguez NM, Linnes JC, Fan A, Ellenson CK, Pollock NR, Klapperich CM (2015) Paper-based RNA extraction, in situ isothermal amplification, and lateral flow detection for low-cost, rapid diagnosis of influenza a (H1N1) from clinical specimens. Anal Chem 87:7872–7879
122. Lu J, Ge S, Ge L, Yan M, Yu J (2012) Electrochemical DNA sensor based on three-dimensional folding paper device for specific and sensitive point-of-care testing. Electrochim Acta 80:334–341
123. Huang S, Abe S, Bennett S, Liang T, Ladd PD, Yokobe L, Anderson CE, Shah K, Bishop JD, Purfield M, Kauffman P, Paul S, Welch A, Strelitz B, Follmer K, Pullar K, Sanchez-Erebia L, Gerth-Guyette E, Domingo G, Klein E, Englund J, Fu E, Yager P (2017) Disposable autonomous device for swab-to-result diagnosis of influenza. Anal Chem 89:5776–5783
124. Robinson R, Wong L, Monnat RJ Jr, Fu E (2016) Development of a whole blood paper-based device for phenylalanine detection in the context of PKU therapy monitoring. Micromachines 7:28
125. Zuk RF, Ginsberg VK, Houts T, Rabbie J, Merrick H, Ullman EF, Fischer MM, Sizto CC, Stiso SN, Litman DJ (1985) Enzyme Immunochromatography – a quantitative immunoassay requiring no instrumentation. Clin Chem 31:1144–1150
126. Vaughan LM, Milavetz G, Ellis E, Szefler SJ, Conboy K, Weinberger MM, Tillson S, Jenne J, Wiener MB, Shaughnessy T, Carrico J (1986) Multicenter evaluation of disposable visual measuring device to assay theophylline from capillary blood-sample. Lancet 1:184–186
127. Novell M, Guinovart T, Blondeau P, Xavier Rius F, Andrade FJ (2014) A paper-based potentiometric cell for decentralized monitoring of Li levels in whole blood. Lab Chip 14:1308–1314

128. Yamada K, Shibata H, Suzuki K, Citterio D (2017) Toward practical application of paper-based microfluidics for medical diagnostics: state-of-the-art and challenges. Lab Chip 17:1206–1249
129. McKoy JM, Bennett CL, Scheetz MH, Differding V, Chandler KL, Scarsi KK, Yarnold PR, Sutton S, Palella F, Johnson S, Obadina E, Raisch DW, Parada JP (2009) Hepatotoxicity associated with long-versus short-course HIV-prophylactic Nevirapine use a systematic review and meta-analysis from the research on adverse drug events and reports (RADAR) project. Drug Saf 32:147–158
130. Saukkonen JJ, Cohn DL, Jasmer RM, Schenker S, Jereb JA, Nolan CM, Peloquin CA, Gordin FM, Nunes D, Strader DB, Bernardo J, Venkataramanan R, Sterling TR, Antituberculosi ATSH (2006) An official ATS statement: hepatotoxicity of antituberculosis therapy. Am J Respir Crit Care Med 174:935–952
131. Tostmann A, Boeree MJ, Aarnoutse RE, de lange WCM, van der Ven AJAM, Dekhuijzen R (2008) Antituberculosis drug-induced hepatotoxicity: concise up-to-date review. J Gastroenterol Hepatol 23:192–202
132. Pollock NR, McGray S, Colby DJ, Noubary F, Huyen N, The Anh N, Khormaee S, Jain S, Hawkins K, Kumar S, Rolland JP, Beattie PD, Chau NV, Quang VM, Barfield C, Tietje K, Steele M, Weigl BH (2013) Field evaluation of a prototype paper-based point-of-care Fingerstick transaminase test. PLoS One 8:e75616
133. Quinn CT (2013) Sickle cell disease in childhood from newborn screening through transition to adult medical care. Pediatr Clin N Am 60:1363–1381
134. Makani J, Cox SE, Soka D, Komba AN, Oruo J, Mwamtemi H, Magesa P, Rwezaula S, Meda E, Mgaya J, Lowe B, Muturi D, Roberts DJ, Williams TN, Pallangyo K, Kitundu J, Fegan G, Kirkham FJ, Marsh K, Newton CR (2011) Mortality in sickle cell Anemia in Africa: a prospective cohort study in Tanzania. PLoS One 6:e14699
135. Modell B, Darlison M (2008) Global epidemiology of haemoglobin disorders and derived service indicators. Bull World Health Organ 86:480–487
136. Alapan Y, Fraiwan A, Kucukal E, Hasan MN, Ung R, Kim M, Odame I, Little JA, Gurkan UA (2016) Emerging point-of-care technologies for sickle cell disease screening and monitoring. Expert Rev Med Devices 13:1073–1093
137. Yang XX, Kanter J, Piety NZ, Benton MS, Vignes SM, Shevkoplyas SS (2013) A simple, rapid, low-cost diagnostic test for sickle cell disease. Lab Chip 13:1464–1467
138. Piety NZ, Yang XX, Kanter J, Vignes SM, George A, Shevkoplyas SS (2016) Validation of a low-cost paper-based screening test for sickle cell Anemia. PLoS One 11:e0144901
139. George A, Piety N, Serrano S, Lanzi M, Patel P, Nirenberg D, Airewele G, Camanda J, Shevkoplyas S (2015) Initial validation of a rapid paper-based test in screening newborns for sickle cell disease. Pediatr Blood Cancer 62:43–43
140. Piety NZ, George A, Serrano S, Lanzi MR, Patel PR, Noli MP, Kahan S, Nirenberg D, Camanda JF, Airewele G, Shevkoplyas SS (2017) A paper-based test for screening newborns for sickle cell disease. Sci Rep 7:45488
141. Molinari NAM, Ortega-Sanchez IR, Messonnier ML, Thompson WW, Wortley PM, Weintraub E, Bridges CB (2007) The annual impact of seasonal influenza in the US: measuring disease burden and costs. Vaccine 25:5086–5096
142. Vasoo S, Stevens J, Singh K (2009) Rapid antigen tests for diagnosis of pandemic (swine) influenza a/H1N1. Clin Infect Dis 49:1090–1093
143. Uyeki TM, Prasad R, Vukotich C, Stebbins S, Rinaldo CR, Ferng YH, Morse SS, Larson EL, Aiello AE, Davis B, Monto AS (2009) Low sensitivity of rapid diagnostic test for influenza. Clin Infect Dis 48:E89–E92
144. Noyola DE, Demmler GJ (2000) Effect of rapid diagnosis on management of influenza a infections. Pediatr Infect Dis J 19:303–307
145. Serwint JR, Miller RM (1993) Why diagnose influenza infections in hospitalized pediatric-patients. Pediatr Infect Dis J 12:200–204

146. Lei KF, Huang CH, Kuo RL, Chang CK, Chen KF, Tsao KC, Tsang NM (2015) Paper-based enzyme-free immunoassay for rapid detection and subtyping of influenza a H1N1 and H3N2 viruses. Anal Chim Acta 883:37–44
147. Wu D, Zhang JH, Xu FH, Wen X, Li PF, Zhang XL, Qiao S, Ge SX, Xia NS, Qian SZ, Qiu XB (2017) A paper-based microfluidic dot-ELISA system with smartphone for the detection of influenza A. Microfluid Nanofluid 21:43
148. Anderson CE, Holstein CA, Strauch E, Bennett S, Chevalier A, Nelson J, Fu E, Baker D, Yager P (2017) Rapid diagnostic assay for intact influenza virus using a high affinity Hemagglutinin binding protein. Anal Chem 89:6608–6615
149. Holstein CA, Chevalier A, Bennett S, Anderson CE, Keniston K, Olsen C, Li B, Bales B, Moore DR, Fu E, Baker D, Yager P (2016) Immobilizing affinity proteins to nitrocellulose: a toolbox for paper-based assay developers. Anal Bioanal Chem 408:1335–1346

Chapter 13
Paper-Based Microfluidics for Point-of-Care Medical Diagnostics

Kentaro Yamada and Daniel Citterio

Abstract In the last decade, the chemistry research community has witnessed an explosive growth of microfluidic devices made of paper (paper-based microfluidics). Use of paper as a substrate material brings several attractive features including extremely low cost and auxiliary pump-free liquid transportation, among others, and application of paper-based microfluidics to on-site medical diagnosis has been actively pursued. To meet the demand for medical diagnostic devices operable by end-users without expert knowledge in resource-limited settings, recent studies on paper-based microfluidics pay particular attention to simplification of user operations prior to an assay (e.g. achieving multistep enzymatic assays by single pipetting) and resulting signal readout (e.g. achieving naked eye-based analog thermometer-style result interpretation). One of the objectives of this chapter is to overview state-of-the-art research progresses in simplification of user operational procedures and development of equipment-free signal readout approaches. In addition, the basics of paper-based microfluidics including a short history of paper-based microfluidics, a comparison of paper-based and conventional plastic- or glass-based microfluidic devices and general requirements for ideal point-of-care testing devices are described. The authors believe this chapter helps researchers new to the field and researchers with different background to learn about analytical applications exclusively achieved by paper-based microfluidics and future challenges in developing "truly" practical medical diagnostic devices.

Keywords Paper-based · Microfluidic devices · Point-of-care · Medical diagnosis

---

K. Yamada · D. Citterio (✉)
Department of Applied Chemistry, Keio University, Yokohama, Japan
e-mail: citterio@applc.keio.ac.jp

M. Tokeshi (ed.), *Applications of Microfluidic Systems in Biology and Medicine*,
Bioanalysis 7, https://doi.org/10.1007/978-981-13-6229-3_13

13.1 Introduction

Being driven by advancements in micro-processing technology in the semiconductor industry, research and development of miniaturized analytical systems made from polymeric substrates has grown rapidly in the late twentieth century. Since the concept of micro-total analysis systems (μTAS, or "lab-on-a-chip" if synonymized) was introduced by Manz in 1990 [1], it has gained attention as a promising analytical platform offering a number of advantageous characteristics such as low consumption of samples and reagents, short diffusion distances in the μm to nm scale, integrated pre-conditioning steps for (bio)chemical assays (*e.g.* separation, mixing) as well as multiplexed detection with high sensitivity and spatial resolution [2–5]. Above all, the capability of sensitive and high-throughput detection from a tiny volume of sample (nL to aL [2]) gave birth to a myriad of applications of μTAS predominated by the life science field such as disease diagnostics, drug discovery, cell biology, gene engineering and proteomics [6]. The primary conformation of traditional microfluidic devices is a palm-sized polymeric substrate (most typically plastics represented by polydimethylsiloxane [2, 7, 8]) having a hollow channel inside.

On the other hand, the 2010s (and the final years of the 2000s) witnessed an explosively growing interest of researchers worldwide in microfluidic devices made of a paper material (Fig. 13.1), of which the trend was ignited by the work published in 2007 from the Whitesides group at Harvard University [9]. In this report, a branched microfluidic channel was created on a sheet of chromatography paper by photolithography. By pre-depositing the assay reagents for the detection of glucose and protein (albumin) in separate areas of the channel, multiplexed colorimetric

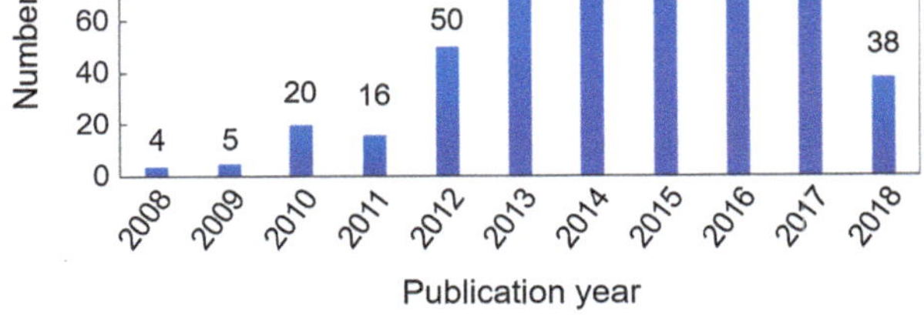

Fig. 13.1 The number of publications related to paper-based microfluidics. (Data source: Web of Science™; search criteria: "Title: paper-base*" OR "Title: patterned paper" OR "Title: paper device*" AND "microfluidic*"; search date: February 21, 2018). Note that these topical searching criteria do not necessarily cover "100%" of the publications on paper-based microfluidics, and therefore, the actual number of publications is certainly larger than shown in the graph. This data is intended to showcase the increasing number of publications related to paper-based microfluidics

assays of those analytical targets as well as a control test were demonstrated by a single application of a low-volume sample (5 μL). Microfluidic devices made of paper share some common useful capabilities with traditional microfluidics (*i.e.* low consumption of reagents and samples, multiplexed detection from a single sample application and, although not being demonstrated in the first report by Whitesides and co-workers [9], automated sample processing prior to an assay). Despite these similar benefits brought by patterning of microfluidic structures, paper-based microfluidics may not be simply described as a "paper variant" of conventional microfluidic devices. For example, the opacity of paper due to light scattering obviously makes paper-based microfluidics less compatible with microscopic observation and highly sensitive optical detection techniques relying on light transmission or luminescence emission. On the other hand, the whiteness of paper enables unaided eye-based identification of color developed on this substrate with a good contrast, making colorimetry an important detection motif for paper-based microfluidics [10, 11]. Therefore, the medical and biological application fields targeted by paper-based microfluidics are "complementary" to those of the traditional microfluidics, namely point-of-care (POC) diagnostics performable under situations lacking infrastructure and/or a skilled operator, rather than being a tool for basic life science research in central laboratories.

Besides potential application fields, the difference of device substrate materials allows discussing dissimilar characteristics of paper-based and conventional microfluidics from various standpoints. This chapter aims at describing the very basics of paper-based microfluidics (physical and chemical characteristics of the substrate material, liquid transportation principle in a microfluidic channel) as well as providing examples of state-of-the-art applications of paper-based microfluidics with particular attention on POC diagnostics by untrained users in resource-limited settings. A comparison between paper-based and traditional microfluidics has been integrated in the discussion as much as possible. Obviously, manufacturing methods of paper-based microfluidics dominated by printing techniques vary from those of the plastic-based counterparts (lithography). However, fabrication techniques of paper-based microfluidics have been comprehensively overviewed in a number of review articles [12–15], and thus, this topic is not repeatedly discussed in this chapter.

13.2 Paper as an Analytical Platform Substrate

13.2.1 Basics of Paper and Comparison with Conventional Substrates

The main component of paper is cellulose, a biopolymer consisting of a linear chain of several hundred to over ten thousand D-glucose units [16]. The fact that cellulose forms the plants' cell walls makes this material the most abundant organic polymer

Table 13.1 Property comparison of paper with conventional substrate materials for microfluidics (glass, silicon, PDMS)

	Substrate material				
Property	Glass	Silicon	PDMS	Paper	Remarks on paper
Flexibility	No	No	Yes	Yes	Less likely to be damaged during transportation and operation
Surface-to-volume ratio	Low	Low	Low	High	Provides high capacity for reagent storage
Fluid transport	Active	Active	Active	Passive	Offers capillarity-based power-free sample transport; but hard to control flow rate
Sensitivity to moisture	No	No	No	Yes	Often requires desiccant during storage
Biocompatibility	Yes	Yes	Yes	Yes	Amenable to fragile biological substances
Disposability	No	No	No	Yes	Safely disposable by incineration
High-throughput fabrication	Yes	Yes	No	Yes	Lowers manufacturing cost per device
Optical transparency	Yes	No	Yes	No	Not compatible with microscopic analyses
Material homogeneity	Yes	Yes	Yes	No	Suffers from low assay precision
Price	Moderate	High	Moderate	Low	Affordable to a wide range of users

Adapted from Ref. [21] with permission from Springer

on the earth [17–19]. The cellulose fibers composing paper are primarily originated from wood pulp in plants. Typically, paper is manufactured by passing a suspension of cellulose fibers through a mesh to remove water, followed by pressing for further water drainage and a subsequent drying process [20]. This manufacturing process results in a paper sheet having porous structures formed by randomly woven cellulosic fibers. Thanks to the abundance of natural sources and a roll-to-roll manufacturing technique, paper has become a ubiquitous substrate material of extremely low-cost. Table 13.1 compares the properties of paper to those of common polymer materials applied for conventional microfluidics [21].

13.2.2 Paper-Based Diagnostic Devices Before Emergence of Microfluidics

Among the benefits of paper shown in Table 13.1, the high surface-to-volume ratio allowing the storage of chemical assay reagents within the porous structure has been recognized since early times, and gave birth to "dip-and-read" simple paper strips represented by the litmus paper introduced in the seventeenth century [22] and urine dipsticks first emerged in 1850 [23]. Other attractive characteristics including low-cost, high manufacturing throughput, safe disposability by incineration and chemical inertness of pure cellulose (in the case of chromatography paper and filter paper) led to the success of paper as a simple diagnostic platform as seen with several commercialized relevant products. Colorimetric detection-based urine dipsticks have remained successful paper-based diagnostic devices thanks to their ability to inspect multiple urine constituents through simple "dip-and-read" user operation in a short time (~ 120 sec). Albeit information provided by the naked eye-based color inspection being semi-quantitative (*i.e.* "approximate" concentration of target analytes), simplicity and rapidity offer high-throughput screening of kidney, urogenital tract, metabolic and liver diseases, as well as hemolytic disorders. Another representative diagnostic device made of "paper" (its nitro derivative, to be exact) substrates is the lateral flow immunoassays (LFIAs). Although not being as long-established as urine dipsticks, it has been already three decades since the concept of LFIAs appeared in US Patents [24, 25]. Thanks to the use of antibodies conjugated with labeling agents (*e.g.* gold nanoparticles, dye-loaded particles), LFIAs allow visual detection of a given target antigen with high specificity. Commercially-available paper-based analytical devices relying on this technique are represented by the home pregnancy tests for detecting human chorionic gonadotropin (hCG), and the influenza testing kits for detecting the nucleoprotein of the influenza virus.

Since the emergence of LFIAs in the 1980s, there seems to be no landmark in the development of paper-based analytical devices. However, a growing number of colorimetry-based paper-made chemical assays have come into the market (*e.g.* the Merckoquant test strips [26]), reflecting a high demand for simple, portable, rapid, and disposable testing devices for various analytical targets. However, the passive liquid transport offered by the porous structure of paper substrates has been overlooked until very recent in the analytical chemistry research community, with the exception of a single report on chromatographic separation of dyes on a piece of patterned filter paper demonstrated by Müller and Clegg in 1949 [27]. In this application, the convenience of patterning of a confined separation paper channel lead to acceleration of the diffusion process and reduction of the reagent consumption. In spite of the extra-values of a patterned paper substrate, there was no response of the scientific community to this research field at that time. After more than half of a century, the American chemist George Whitesides rediscovered microfluidically patterned paper as a valuable platform to construct simple yet functional (bio) chemical sensing devices in 2007 [9]. Despite the explosively increasing number of publications, there is no unified nomenclature applied to this new class of microfluidic devices. Currently, paper-based microfluidics are also referred to as "microfluidic paper-based analytical devices (µPADs)", "lab-on-paper", "paperfluidics", and so on. In this chapter, "paper-based microfluidics" will be consistently used as a term to describe microfluidic devices made of "paper", including not only pure cellulosic filter paper and chromatography paper, but also nitrocellulose.

13.2.3 From Plastic to Paper-Based Microfluidics: Similarities and Dissimilarities

The concept of μTAS has been envisioned as a revolutionary (bio)chemical analytical platform for its efficiency, rapidity, economical use of samples and reagents and well-established manufacturing technologies. Despite a few examples of commercialized microfluidic devices [4, 28], the current state is still far from the situation where everyone monitors his/her health condition using microfluidics. The reason is primarily attributed to (1) the complexity of liquid handling systems requiring difficult start-up processes and operational expertise, and (2) the necessity of sophisticated detectors for signal acquisition. The most traditional research approaches in this field are devoted to fluidic control relying on external equipment, and hence, microfluidics have not evolved as the versatile "lab-on-a-chip" system, but partially remained a sophisticated "chip-in-a-lab" system [29].

Paper-based microfluidics are fabricated by patterning a "boundary" defining a fluidic channel region onto a paper substrate. This "boundary" is typically defined either by (1) a hydrophobic substance (*e.g.* solid wax [30, 31]) coating the surface of the cellulosic fibers throughout the paper thickness; (2) a solid substance blocking the porous structure of paper (*e.g.* photoresist [9], polystyrene [32]); or (3) the paper-air interface at the periphery of a cut paper sheet [33]. Filter paper or chromatography paper are by far the most commonly used as the substrate material because of their pure cellulose composition with a high degree of chemical inertness. More importantly, the availability of capillary action derived from μm-sized inter-fiber pores offers power-free fluid transport, addressing one of the significant bottlenecks of traditional microfluidics. Last not but least, paper is light-weight (~10 mg cm^{-2}) and affordable (~0.05 cents cm^{-2}), providing ease-of-transportation and enabling the use of paper-based microfluidics even in remote areas and resource-limited settings. Standard office paper, although being inexpensive and very accessible, is not widely utilized in the field, due to its poor wetting property originating from (1) small pore sizes (several hundred nm [34]) and (2) hydrophobic surface treatment by sizing agents to prevent ink bleeding. In addition, additives (*e.g.* fillers, optical brighteners, etc.) with unknown composition may impact the chemical reaction for detection, further discouraging developers to use this material as a substrate for chemical analyses.

As is clear from the structure of paper-based microfluidics, a bundle of numerous capillaries formed by interwoven cellulosic fibers serves as the fluidic channels on these devices, whereas the conventional microfluidics possess a hollow flow path in a substrate material such as plastic, glass and silicone. Despite their different configurations, the capability of sequential chemical reactions and multiplexed assays with small consumption of samples and reagents are common advantages. In addition, the liquid transportation in paper-based microfluidics also exhibits laminar flow [35, 36] due to the μm-sized pore radius in the paper medium and the liquid flow speeds typically at several mm per minute, making the Reynolds number (*Re*) less than 1, as expressed by the following Eq. (13.1):

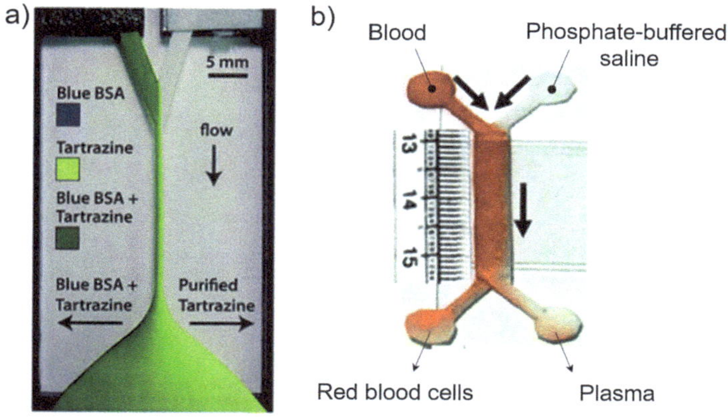

Fig. 13.2 Application examples of laminar flow on paper-based microfluidics; (**a**) separation of a dye (tartrazine) from a protein (bovine serum albumin: BSA). (Reproduced from Ref. [35] with permission from The Royal Society of Chemistry); (**b**) plasma separation from whole blood, (Adapted from Ref. [37] with permission from The Royal Society of Chemistry)

$$Re = \frac{\rho VL}{\mu} < 1, \tag{13.1}$$

where ρ is the fluid density (kg m^{-3}), V the fluid velocity (m s^{-1}), L the pore diameter (m) and μ the fluid dynamic viscosity (kg m^{-1} s^{-1} = N s m^{-2}). Various reports have demonstrated that the laminar flow phenomenon is applicable to on-paper separation purposes including dye purification from a protein solution (Fig. 13.2a) [35] and plasma separation from whole blood (Fig. 13.2b) [37].

A clear difference lies in the fact that the fluidic channels are "fully closed" in the conventional microfluidics, whereas the paper-based counterpart possesses "open-air" flow paths, unless covered by lamination films or adhesive tapes. Since detection in conventional microfluidics is carried out in a sealed channel, the generation of the detection signal is accounted for by an equilibrium between the sample phase and a second interface hosting a receptor for the given analyte (*e.g.* adjacent laminar flow, channel surface, interface of a droplet dispersed in the sample). Therefore, the signal reflects the "concentration" of the analytical target in the sample fluid introduced to the device. In the case of paper-based microfluidics, however, open-air flow paths are prone to sample evaporation. Consequently, the detection mechanism is reliant on depletion of analytes at the detection area, rather than an equilibrium state between an aqueous sample phase and a receptor immobilized onto the paper surface. In contrast to traditional microfluidics, the detection signal obtained from paper-based microfluidics reflects information on the "absolute analyte amount" transported to the detection region [38], with the presumption that sufficient ligand/receptor is available. In applications where an equilibrium state between two immiscible phases is essential (*e.g.* on-paper detection of ionic species by the ion-selective optode system [39, 40]), an additional lamination process of the paper substrate is necessary to prevent evaporative loss of the sample liquid during the assay [41–43].

13.3 Applications of Paper-Based Microfluidics to POC Diagnostics

13.3.1 Potential Target Application in the Medical and Biological Fields

Although some divided opinions may persist, the ultimate benefits of paper over the conventional substrate materials are its extremely low cost and ability of pump-free liquid transportation by capillary forces, when it comes to the use as a microfluidic device substrate. "Affordability" and the fact of "no requirement of external equipment" are of significant importance for an analytical device intended for carrying out assays in situations lacking sufficient infrastructure and economic resources, namely medical diagnostics in remote rural areas of developing countries. Even in developed countries, highly distributed medical screening could also be an important scenario where the low-cost and equipment-free characteristics of paper-based microfluidics become particularly attractive. On the other hand, more high-tech biomedical applications (*e.g.* detailed examinations requiring accurate and precise quantification of clinical targets, advanced studies in a sophisticated laboratory) less likely give paper-based microfluidics full play, because (1) the advantages of cost reduction and equipment-free operation will not pay for the reluctance to replace traditional systems, and (2) the sensitivity and spatial resolution of detection on paper-based microfluidics are generally poor compared to the advanced analytical technologies already used in fully equipped settings. Therefore, effective medical and biological application of paper-based microfluidics should be point-of-care (POC) testing for medical diagnostics in resource-limited settings and highly distributed medical screening at private homes or small clinics.

13.3.2 General Challenges in Developing Paper-Based Microfluidics

There are several landmarks in developing effective POC test devices. According to the guidelines provided by the World Health Organization (WHO), practicable POC devices should meet the following criteria summarized below under the "ASSURED" acronym [44]: **A**ffordable; **S**ensitive (no false negative results); **S**pecific (no false positive results); **U**ser-friendly (easy to operate, no use of invasive chemicals); **R**apid and **R**obust; **E**quipment-free; **D**elivered (to the end users). On the other hand, the U. S. Food & Drug Administration (FDA) formulates a guideline for *in vitro* diagnostic (IVD) tests for home or over-the-counter use known as CLIA (Clinical Laboratory Improvement Amendments) [45], where particular importance is laid on simplicity of user operation both in "sample-in" (introduction of sample to the device) and "answer-out" (interpretation of assay result) steps. Considering these criteria, practical paper-based microfluidics should balance basic analytical performance (accuracy, precision, limit of detection) and acceptance by targeted users (simplicity of sample handling, ease of result interpretation, long-term storage stability).

Despite various attractive characteristics of paper-based microfluidics and a huge number of relevant publications exceeding one thousand, they have hardly made it into the marketplace as of 2018. A large fraction of prototypes remains a proof-of-concept demonstration in an academic research paper and has yet to prove its usefulness in practical applications. The missing key to the final success of paper-based microfluidics as practical diagnostic devices has been described in recent review articles as: "*...UED (i.e. **U**ser-friendly, **E**quipment-free, **D**elivered) are important requirements that deserve more research to increase the commercialization of paper-based analytical devices.*" [46] and "*... major challenges remaining in proof-of-concept μPADs for routine health checks can be summarized as (1) necessity of complicated user operations (ass**U**red), (2) insufficient examination on long-term stability (assu**R**ed), and (3) reliance on detection equipment unfamiliar for general users (assur**E**d).*" [47] To sum up, in academic research efforts the user acceptance aspect (user-friendliness and equipment-free criteria, among others) tends to be overlooked or sacrificed to the enhancement of the analytical performance of a device. Fortunately however, the research community dealing with paper-based microfluidics is certainly becoming aware of this issue that needs to be addressed. The following sections describe the current state of the development of paper-based microfluidics with simplified user operations targeting medical and biological applications.

13.3.3 Simplification of Multi-Step Biochemical Assays

Immunoassays and nucleic acid testing are representative bioanalytical techniques with high sensitivity and specificity. They have a great potential to prevent epidemics of infectious diseases by offering rapid and accurate diagnosis. Infectious diseases (*e.g.* malaria, Ebola, Zika, tuberculosis, dengue) account for a significant mortality rate in developing countries, and thus, accurate and early diagnostic techniques are of substantial importance, because they ultimately allow adequate treatment and prevention of further spread of the diseases. With the high demand to implement immunoassays and nucleic acid testing into infectious disease marker detection in remote settings, efforts have been made to reduce the operational complexity of these analytical techniques.

13.3.3.1 Lateral Flow Immunoassays (LFIAs)

Immunoassays offer highly selective detection of an analyte of interest utilizing the very specific antigen-antibody reaction. LFIAs, prominently represented by home pregnancy tests and influenza test kits, are well-known diagnostic devices relying on this bioanalytical detection technique. They provide a "yes-no" answer of the presence of target antigens (human chorionic gonadotropin (hCG) for pregnancy testing, virus-specific protein for influenza testing) in the sample based on visualization of a line typically derived from the gold nanoparticle labels. However, the early diagnosis of infectious diseases with the conventional LFIAs is generally challenging owing to their insufficient sensitivity to detect a trace amount of the target [48, 49].

Amplification of the optical signal from the gold nanoparticle labels is achievable by adding a gold [50, 51] or silver [52–54] salt to enhance the visibility of the colloidal gold. However, the necessity of multi-step addition of sample and reagents increases the complexity of user operation. A pioneering work addressing this bottleneck utilized an automated time-lagged delivery of the assay reagents on paper-based microfluidics coined as a two-dimensional paper network (2-DPN). Figures 13.3a, b show two types of 2-DPN. Those devices allow a user to carry out a signal-amplified immunoassay of the *Pf*HRP2 malaria protein simply by (1) adding the assay components (sample, enhancement reagent for signal amplification, washing buffer) onto a specified paper area and by (2) folding the plastic card. Different path lengths (Fig. 13.3a) [55] or different flow resistance of the channels stemmed from sucrose impregnation (Fig. 13.3b) [56] result in the sequential delivery of the assay components to the detection region (indicated by a red arrow in the figures) in the desired order. In the *Pf*HRP2 malaria protein detection using each of these 2-DPN, a four-fold [55] and 2.6-fold [56] signal amplification has been achieved over the case without optical signal enhancement.

The use of the enzyme-linked immunosorbent assay (ELISA) method is another choice to enhance the sensitivity of immunoassays. In ELISAs, the secondary antibodies are labeled with enzymes (*e.g.* alkaline phosphatase, horseradish peroxidase) in place of gold nanoparticles, and subsequent addition of a chromogenic substrate generates an optical signal amplified in the enzymatic catalytic reaction. Here again, the requirement of multiple reagent addition with a strictly fixed order remains as an issue in terms of user-friendliness. In the most ideal case, the target antigen/enzyme-labeled antibody immunocomplexes and chromogenic substrates are sequentially transported to the test zone only by a single introduction of a sample liquid to the device. The use of different path lengths to the test zone in a similar manner to the 2-DPNs also works in carrying out the ELISA. The "maze-like" paper-based microfluidics schematically shown in Fig. 13.4a [57] allowed the detection of hCG only by immersing the bottom of the device into the sample with a detection limit of 8.1 mIU mL^{-1}, demonstrating a slight improvement over commercial LFIA devices for pregnancy testing (20 mIU mL^{-1}). Here, the color development of the BCIP/NBT chromogenic substrate was catalyzed by alkaline phosphatase (ALP), resulting in an improvement of the sensitivity of the immunoassay in spite of the

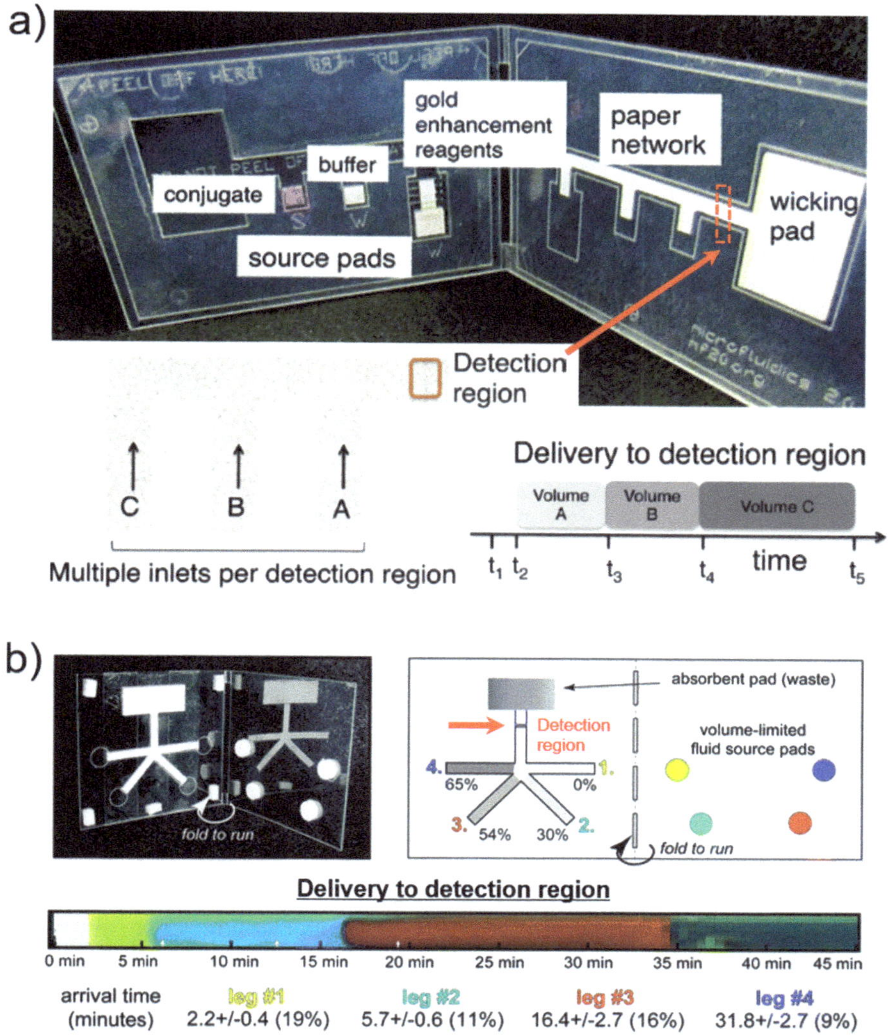

Fig. 13.3 2-DPN lateral flow immunoassay devices for automated sequential reagent delivery based on (**a**) different length of flow paths to the detection region. (Adapted with permission from Ref. [55]. Copyright 2012 American Chemical Society); (**b**) different flow resistance of sucrose-impregnated channels (concentrations of impregnated sucrose solution: 0, 30, 54, 65% from legs #1 to #4) to the detection region. (Reproduced from Ref. [56] with permission from The Royal Society of Chemistry)

much longer flow path of the sample than in the case of conventional LFIAs, potentially causing loss of analyte by adsorption onto the nitrocellulosic substrate. Very recently, an automated ELISA has been demonstrated on a paper-based microfluidics resembling the traditional LFIA devices [58]. The device shown in

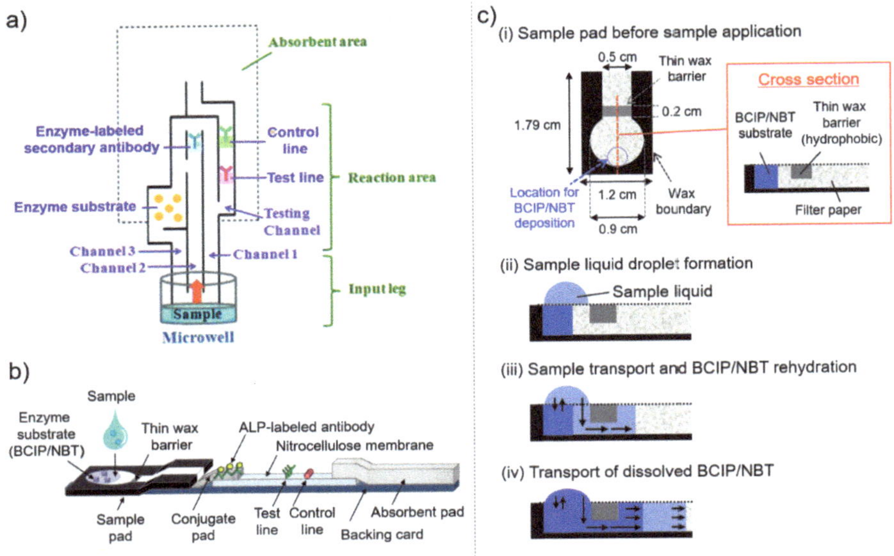

Fig. 13.4 Schematic illustrations of lateral flow devices for automated ELISA: (**a**) a "maze-like" configuration utilizing different path lengths to the reaction area. (Adapted from Ref. [57] with permission from The Royal Society of Chemistry); (**b**) a "classical format" configuration utilizing delayed transport of chromogenic enzyme substrates from a wax-modified sample pad; (**c**) detailed structure of the wax-modified sample pad and the working principle of delayed delivery of the enzyme substrates enabling the automated ELISA from single sample application. (Adapted from Ref. [58])

Fig. 13.4b shares a common architecture with the classical LFIA format, only with the exception that the sample pad is replaced by filter paper patterned by a thin wax barrier and the BCIP/NBT substrate (Fig. 13.4c, (i). The wax-modified filter paper-based sample pad allows to form a stationary sample droplet (Fig. 13.4c, (ii). While the sample liquid flows in the device by the capillary force, the pre-deposited BCIP/NBT is gradually rehydrated into the bulk sample liquid (Fig. 13.4c, (iii), and is eventually transported in a delayed manner together with the remaining sample liquid (Fig. 13.4c, (iv) toward the downstream test zone, where sandwich immunocomplexes are already present. The delayed delivery of BCIP/NBT enabled a "one-step" ELISA of mouse IgG as a model analyte. Surprisingly, the achieved limit of detection of 15.8 ng mL^{-1} was comparable to that of a commercial microtiter plate-based ELISA kit (10 ng mL^{-1}), which relies on a labor-intensive operational procedure (multistep reagent addition, incubation, washing) and absorbance measurement with a sophisticated microplate reader. Although the application to the detection of infectious disease markers is yet to be demonstrated, paper-based microfluidics for an automated ELISA are promising POC diagnostic devices providing easy-to-use yet high-performance infectious disease tests.

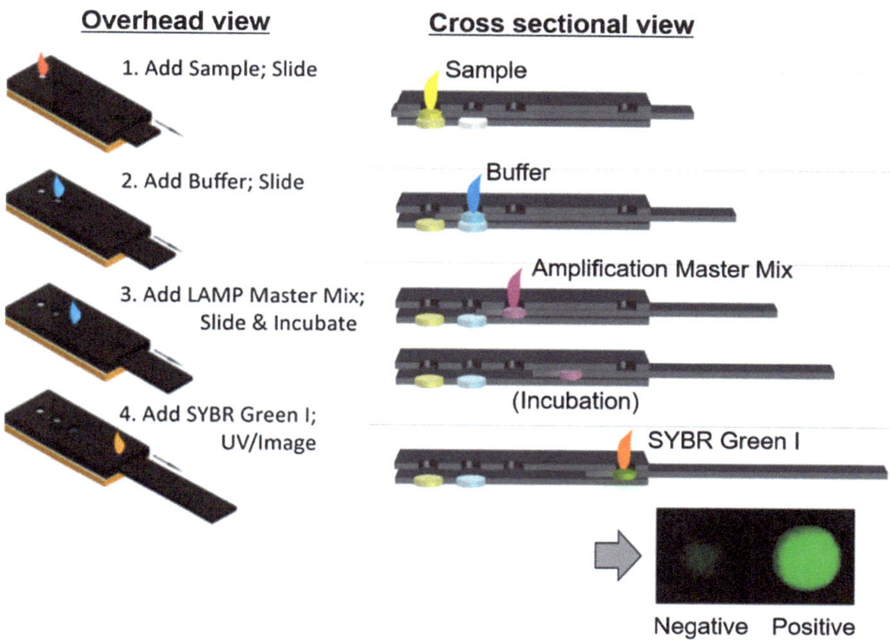

Fig. 13.5 A "paper machine" enabling nucleic acid detection by pipetting and sliding. (Adapted with permission from Ref. [62]. Copyright 2015 American Chemical Society)

13.3.3.2 Nucleic Acid Tests

The detection of nucleic acids (DNAs, RNAs) extracted from microorganisms is another powerful technique for the diagnosis of infectious diseases. To achieve high sensitivity, amplification of target nucleic acids of interest is often required prior to the detection by techniques such as the reverse transcription polymerase chain reaction (RT-PCR) [59], loop-mediated isothermal amplification (LAMP) [60], or rolling circle amplification (RCA) [61]. Although the detection of nucleic acids on paper-based microfluidics has been described in a number of reports, the current situation is that the whole detection process entails multiple steps of operation and electricity-powered heating equipment. In the most integrated and user-friendly configuration of paper-based microfluidics for nucleic acid amplification tests (NAATs), a "paper machine" developed by Diagnostics For All is able to carry out the whole blood sample processing, LAMP reaction and fluorescence-based detection in four steps of reagent deposition and handle-sliding operations (Fig. 13.5) [62]. Complexity of user operation has been minimized with the integrated "paper machine" format, however, the requirement of an incubator remained as an issue [62]. For the field use of nucleic acid detection, the development of battery-powered heating equipment requiring minimum maintenance (or elimination of its necessity) would be a key factor determining the practical feasibility in resource-limited settings. Nucleic acid detection for infectious disease diagnosis on paper-based microfluidics has been comprehensively described in a recent review article [63], and therefore, further detailed discussions will not be repeated here.

13.3.4 Simplification of Detection Signal Readout

Besides the simplicity of user operations prior to detection, interpretation of resulting detection signals is a crucial step in (bio)chemical assays for POC tests. In the case where targeted end users are non-healthcare workers (*e.g.* routine health check by monitoring clinically-relevant metabolites at private homes), intuitive signaling approaches on paper-based microfluidics are of significant importance to eliminate the risk of misinterpretation of assay results. Colorimetric urine dipsticks, a marketed paper-based medical screening tool, rely on visual comparison of color intensities or hues (*i.e.* type of color) between the sample paper and a read guide to semi-quantitatively determine the concentration of a given analyte in urine. However, readout results are inherently dependent on the color recognition capacity of the user and on the environmental light conditions. Several alternative signaling motifs have been proposed to address this issue encountered with the traditional colorimetric detection method. Table 13.2 summarizes representative works on paper-based microfluidics utilizing a simplified signal detection mode for medical and biological applications.

13.3.4.1 Analog Thermometer-Style Readout

Analog thermometers let us know the ambient temperature based on the length of an inner liquid filling a straight glass capillary. All a user has to do is to simply identify the position of the top of the inner liquid relative to adjacent scale marks. The "distance-based" quantification mode on paper-based microfluidics resembling analog thermometers is typically achieved relying on one of two distinct principles: (1) analyte depletion in a microfluidic channel (Scheme 13.1a); (2) flow resistance alteration in a microfluidic channel caused by analyte molecules (Scheme 13.1b). As shown in Table 13.2, the thermometer-style readout mode is frequently implemented on paper-based microfluidics, possibly because various chemistries are available to achieve this signaling method.

In the first approach, analyte molecules in the flowing sample liquid are continuously consumed by binding to receptors immobilized within the paper channel, eventually resulting in depletion of analyte and no further generation of analyte-receptor conjugates along the flow channel. It should be mentioned that the paper matrix itself (e.g. functional groups present on the surface of cellulose fibers) can potentially act as a "receptor" binding an analyte. At a fixed amount of receptor sites present in a microfluidic paper channel, full depletion of the analyte from the flowing sample liquid is delayed with increasing original amount of analyte, resulting in the

Table 13.2 Representative works on paper-based microfluidics utilizing simplified signal detection targeting medical applications

Signal readout mode	Analyte	Assay reagent	Detectable range	References
Analog thermometer (analyte depletion principle)	Theophylline	Antibody (receptor); glucose, GOx, HRP, 4-chloro-1-naphthol (detection)	0–40 mg L^{-1}	[64]
			0–36.6 mg L^{-1}	[65]
	Theophylline	Antibody (receptor); glucose, GOx, HRP, dicarboxidine dihydrochloride (detection)	0–40 mg L^{-1}	[66]
	Cholesterol	ChOx, HRP, MBTH, DMA (detection)	1500–4000 mg L^{-1}	[67]
	High-density lipoprotein cholesterol	Cholesterol esterase, ChOx, HRP, MBTH, DMA (detection)	250–1000 mg L^{-1}	[68]
	Glucose	GOx, HRP, DAB (detection)	11–270 mg dL^{-1}	[69]
	Lactoferrin	Terbium chloride (detection); ι-carrageenan (immobilizer)	0.05–4 mg mL^{-1}	[70]
	Cocaine	Aptamer/glucoamylase-doped hydrogel, amylose, GOx, HRP, DAB (detection)	0–400 μM	[71]
	Cocaine	Aptamer-invertase-bead conjugate, sucrose, GOx, HRP, DAB (detection)	0–500 μM	[72]
	Adenosine		0–400 μM	
	DNA	HNB, $MgCl_2$ (detection) PEI (immobilizer)	4.14×10^3–7.88×10^6 copies μL^{-1} [a]	[73]
	DNA	SYBR Green I (detection); PEG 100000 (immobilizer)	1 aM-1 fM [b]	[74]
	Carcinoembryonic antigen	HRP-labeled antibody, H_2O_2, TMB (detection)	0–40 ng mL^{-1}	[75]

(continued)

Table 13.2 (continued)

Signal readout mode	Analyte	Assay reagent	Detectable range	References
Analog thermometer (flow resistance alteration principle)	Cocaine	Aptamer-doped hydrogel (valving); food dye solution (detection)	0–100 μM	[76]
	Alkaline phosphatase	Starch (valving); food dye solution (detection)	0.075–5 U mL^{-1}	[77]
	ABO, Rh blood types	Anti-A, B, D antibodies, A-, B-cells (valving)	–	[78]
	Hematocrit	EDTA, NaCl (stabilizer)	28–57%	[79]
Time measurement	H_2O_2	Dedicated H_2O_2-degradable hydrophobic compound (valving); food dye (detection)	0–50 mM	[80]
	Alkaline phosphatase	Dedicated H_2O_2-degradable hydrophobic compound (valving); food dye (detection)	0.128–7.4 pM; 13–740 pM; 45–1480 pM; 93–1480 pM; 342–37000 pM; 56000–74000 pM [c]	[81]
	K^+	Hemin, DNA probe, H_2O_2, TMB (valving), ink solution (detection)	25–200 μM	[82]
Counting of color spots	H_2O_2	Dedicated H_2O_2-degradable hydrophobic compound (valving); food dye (detection)	0–100 mM	[80]
	Adenosine	Aptamer-GOx conjugate, glucose, $KMnO_4$ (detection)	1.5 μM–19.3 mM	[83]
Text readout	ABO, Rh blood types	Anti-A, B, D antibodies, water-insoluble red ink (detection)	—	[84]
	Albumin	Tetrabromophenol blue (detection)	0–10 mg mL^{-1}	[85]

GOx glucose oxidase, *HRP* horseradish peroxidase, *ChOx* cholesterol oxidase, *MBTH* 3-methyl-2-benzothiazolinone hydrazone, *DMA* *N,N*-dimethylaniline, *DAB* diaminobenzidine, *HNB* hydroxynaphthol blue, *PEI* polyethylenimine, *PEG* polyethylene glycol, *TMB* 3,3′,5,5-′-tetramethylbenzidine, *EDTA* ethylenediaminetetraacetic acid

[a]Initial number of genomic DNA

[b]Combined with polymerase chain reaction (PCR)

[c]Values dependent on the structure of the valving reagent

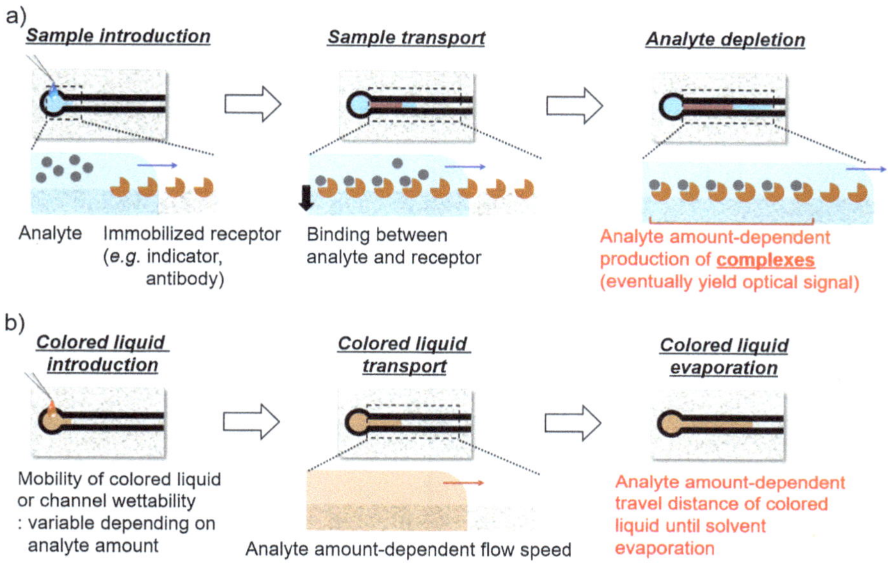

Scheme 13.1 Schematic illustration describing two major types of the working principle of an "analog thermometer-style" readout motif on paper-based microfluidics: (**a**) analyte depletion mode; (**b**) flow resistance alteration mode

observation of a distance-based signal proportional to the initial analyte content. One of the earliest examples has demonstrated the antibody-based detection of theophylline on a paper strip in the 1980s [64–66]. There, a sample containing theophylline (target analyte) is mixed with a solution containing horseradish peroxidase (HRP)-labeled theophylline (competing agent) as well as glucose oxidase, followed by wicking through a chromatography paper strip coated with theophylline-specific antibodies, consuming the free and HRP-labeled theophylline in a competitive manner. The larger the amount of initial theophylline in the sample, the longer in terms of distance of the wicking fluid the depletion of the HRP-labeled theophylline by the immobilized antibodies takes. After color development by the addition of glucose and a colorimetric substrate for HRP, the theophylline concentration in the sample can be estimated from the length of the colored part of the strip. In that work, the authors pointed out the advantage that the assay result is relatively insusceptible to the stability of the enzymatic reagents, temperature and incubation time, because the quantification signal is not reliant on the developed color intensity, but on the colored distance [64]. In the 1990s, the analog thermometer-style detection mode has been expanded to the quantification of cholesterol [67] and high-density lipoprotein cholesterol [68] in whole blood. The color band providing the distance-based signal originates from the oxidation reaction of a chromogenic peroxidase substrate (3-methyl-2-benzothiazolinone hydrazone and *N,N*-dimethylaniline) by H_2O_2 released from the reaction of the target analyte with the selectivity providing cholesterol oxidase enzyme. This reaction is catalyzed by HRP evenly immobilized along a paper strip, resulting in a gradual H_2O_2 depletion from the flowing liquid. In these cases, the analyte concentration in the sample is not detected by direct depletion of the analyte itself during the flow process, but indirectly measured through the depletion of the intermediary generated H_2O_2, which amount reflects the original analyte concentration. This mechanism is applicable to a wide range of analytes as long as a specific oxidase enzyme is available for the analyte of interest. Indeed, the "first" analog thermometer-style detection on the type of microfluidically patterned paper-based analytical devices introduced in 2007 demonstrated the quantification of glucose by using glucose oxidase (GOx), HRP and diaminobenzidine (DAB) as a colorimetric peroxidase substrate (Fig. 13.6a) [69]. The most straightforward approaches to achieve a distance-based signal make use of the direct consumption of target analyte molecules by a colorimetric or fluorometric indicator specific to the given analyte [69, 70, 86], or the migration distance modulation of a simple dye solution that indirectly represents the analytical target amount in the sample [73, 74, 76].

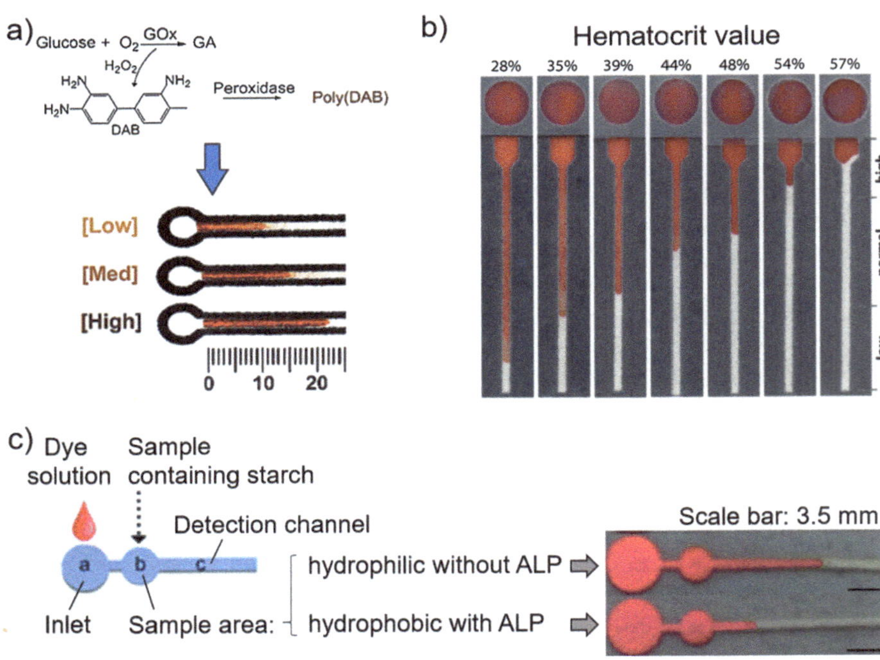

Fig. 13.6 Examples of analog thermometer-style detection on paper-based microfluidics: (a) glucose detection relying on analyte depletion within the paper channel. (Reproduced from Ref. [69] with permission from The Royal Society of Chemistry); (b) hematocrit determination relying on channel clogging by the red blood cells in the whole blood sample. (Reproduced from Ref. [79] with permission from The Royal Society of Chemistry); (c) alkaline phosphatase (ALP) detection relying on wettability change of the midstream area (zone b) of the paper channel. (Adapted from Ref. [77] with permission from The Royal Society of Chemistry)

An important condition common to all thermometer-style detection schemes relying on the analyte depletion mechanism is that the "consumption" of the analyte in the microfluidic paper channel must be continuous and rapid. Signaling compounds (e.g. a colored complex of analyte and indicator) must be produced within the paper channel in a continuous manner without being washed away by the sample liquid flow. For example, insufficient (e.g. weak binding) or slow interaction (e.g. slow reaction kinetics) of an analyte with the immobilized receptor (e.g. indicator, cellulosic substrate) does not result in a distance-based detection signal. In the case of pump-driven traditional microfluidics, reduction of the flow rate of the sample fluid allows to extend the timescale for the analyte-receptor interactions [87]. However, this strategy is not readily achievable on paper-based microfluidics, where liquid transport relies on passive wicking by capillary forces. If a water-soluble indicator specific to the given analyte is to be used for detection, it is absolutely crucial to prevent indicator molecules from being washed away by the sample liquid flow [70]. Commercialized colorimetric indicators are typically available in their sulfonated form for the sake of water-solubility, since they are designed for use in aqueous systems. A well-known strategy for immobilizing charged reagents onto a paper substrate is to deposit them together with a polymer or a particle having the opposite charge to the reagents to be immobilized [88–91]. The size or the multitopic interaction of polymer molecules or particles with the paper matrix efficiently prevents them from being transported together with the sample liquid, enabling the electrostatic immobilization of charged indicator reagents.

The second mechanism to achieve the thermometer-style detection mode is the use of microfluidic channel "clogging" to alter the flow rate of a colored solution. One early example relying on this system was demonstrated on a plastic-based microfluidic platform [92]. In this proof-of-concept work, the surface of a PDMS microfluidic channel was modified by biotin. When a sample liquid containing streptavidin is introduced into the device, it travels through the channel by capillary action, while the streptavidin-biotin interaction occurs at the channel inlet, eventually stopping the sample migration by acting as a "valve" clogging the microfluidic channel close to its inlet. Since the time for channel clogging becomes shorter with increasing concentration of streptavidin, a flow distance-based signal inversely proportional to the target analyte amount is obtained. Interestingly, this "channel clogging" mechanism successfully works on paper-based microfluidics, where the channel structure is much more complex (*i.e.* a bundle of numerous capillaries formed by randomly woven cellulosic fibers). Several literatures report blood analysis (blood typing [78], semi-quantification of the hematocrit [79]) based on a physical blocking of cellulosic paper channels. In a distance-based blood typing assay [78], blood coagulation was utilized as a "valve" to change the flow distance of red blood cells used as the color signal to identify the ABO blood type. In the semi-quantitative evaluation of the hematocrit [79], defined as the volume ratio of red blood cells to the whole blood, the analyte itself (*i.e.* red blood cells) caused clogging of the microfluidic paper channel, resulting in a blood migration length inversely proportional to the hematocrit value (Fig. 13.6b). Besides bulky blood components (e.g. red blood cells), compounds being able to undergo an analyte-dependent switching from a poorly water-soluble hydrophobic into a hydrophilic state can also serve as a valve to enable a thermometer-style detection approach. As an example, the device shown in Fig. 13.6c [77] uses starch as the valving reagent deposited together with the sample in the sample area (zone b). In the presence of alkaline phosphatase (ALP; analytical target) in the sample, the decomposition of poorly water-soluble starch into highly water-soluble glucose by an auxiliary enzymatic reaction (glucoamylase) is hindered, and thus, a dye solution deposited onto the inlet (zone a) after sample introduction migrates only a short distance in channel c. Vice versa, in the absence of ALP, starch is readily broken down into water-soluble glucose units, resulting in free flow of the dye solution into channel c. Similar kinds of "wettability changes" of a paper channel are also employed to achieve other simplified signal detection modes (timing, counting) as will be discussed in the following sections.

13.3.4.2 Time Measurement

In this detection mode, the quantification signal is the "time" between appointed events (*e.g.* sample application and color observation on a paper spot). Up to present, the timing-based quantification signal has been solely achieved by utilizing "wettability changes" of a part of the paper channel triggered by the presence of the analytical target. The first proof-of-concept work demonstrated the quantification of H_2O_2 on a vertically-assembled paper-based microfluidic device (Fig. 13.7a) [80], where a H_2O_2-degradable hydrophobic compound ("phase switching agent") is pre-deposited in the middle paper layer to regulate the sample flow-through speed depending on the H_2O_2 concentration. The penetrated sample liquid carries the green food dye to the bottom layer of the device, and green coloration of the bottom paper area is recorded as the endpoint of the time measurement. Here again, since H_2O_2 is a product of oxidase enzyme reactions, the mechanism can be straightforwardly applied to clinically-relevant metabolites converted by the respective oxidase enzyme (*e.g.* glucose, lactate, alcohol). Later on, the application of this mechanism was expanded to ALP quantification [81], where the presence of ALP generated glucose by cleaving the phosphate-substituent off a phosphorylated glucose derivative, eventually producing H_2O_2 through an auxiliary enzymatic reaction (glucose oxidase) to decompose the phase switching agent. This work has also demonstrated that the sensitivity of the time measurement-based detection can be tailored by the microfluidic channel length, the chemical structure of the phase switching agent influencing the decomposition kinetics, and the number of paper layers containing the phase switching agent. However, a bottleneck of vertically-assembled paper-based microfluidics is their relatively complex fabrication procedure (stacking of multiple layers of paper substrates and double-sided tape) requiring precise alignment and firm attachment between paper layers. The timing-based readout motif is also performable on paper-based microfluidics with a lateral flow configuration (Fig. 13.7b) [82]. There, the target potassium ions (K^+) in the sample result in the formation of a DNAzyme with a peroxidase-like catalytic

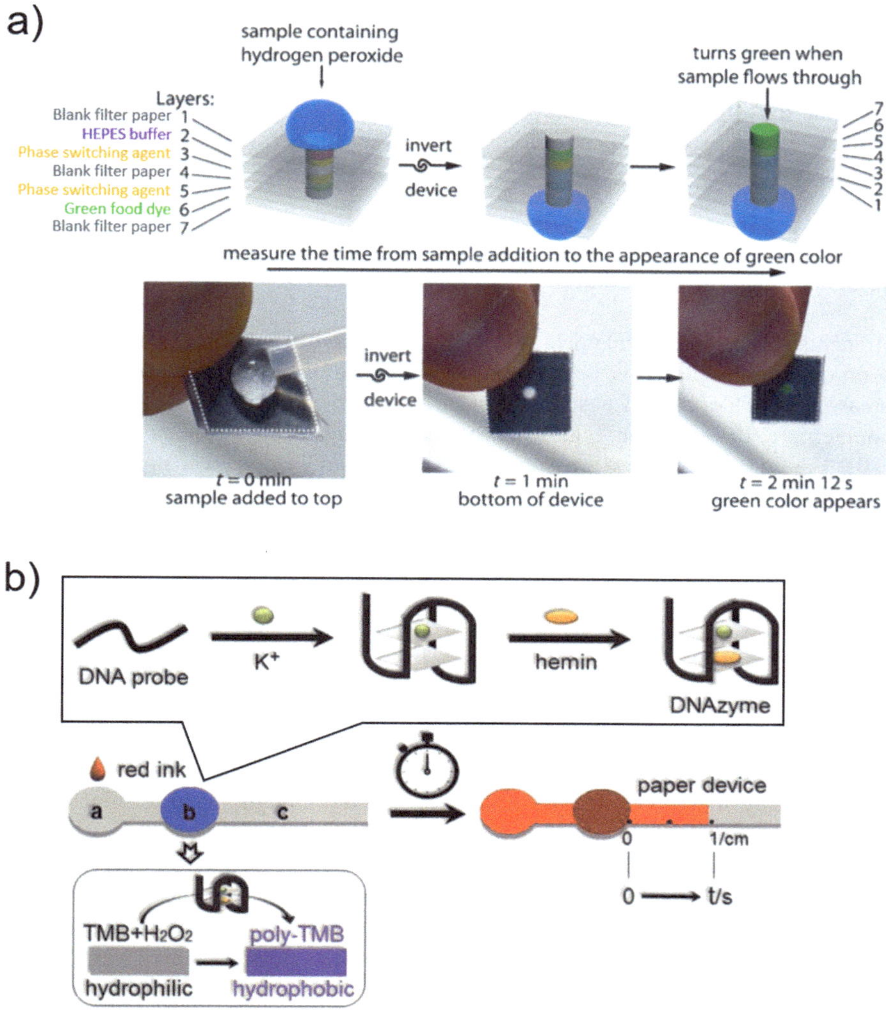

Fig. 13.7 Examples of time measurement-based detection on paper-based microfluidics: (**a**) schematic illustration of a three-dimensional device design and assay procedure for H_2O_2 quantification. (Reproduced with permission from Ref. [80]. Copyright© 2012 WILEY-VCH Verlag GmbH & Co. KGaA, Weinheim); (**b**) schematic illustration of a lateral flow-type device design and working principle of potassium ion (K^+) quantification. (Reproduced from Ref. [82], Copyright 2015, with permission from Elsevier)

activity, which is deposited into zone b in the first step. Subsequently, pre-deposited 3,3′,5,5′-tetramethylbenzidine (TMB) is oxidized into water-insoluble poly(TMB) only in the presence of K^+, resulting in increased hydrophobicity of the paper zone b. Finally, the time-based quantification of K^+ was carried out by measuring the time required for a dye solution deposited in zone a to flow a length of 1 cm in the paper channel (zone c).

Timing-based readout offers precise quantitative results simply by using an ubiquitous timer as the readout "instrument". However, one disadvantage is that this detection mode forces a user to pay attention to the device in order not to overlook the appointed endpoint phenomenon, which can be in the order of several minutes for the examples described above, making high-throughput parallel assays challenging.

13.3.4.3 Counting of Colored Spots

Branched microfluidic channels allow not only multi-analyte detection but also semi-quantitative analysis based on "counting" of the number of colored paper areas. This readout motif has been first proposed in parallel with the timing-based counterpart (illustrated in Fig. 13.7a of the previous section) in the same report [80]. Figure 13.8a shows the paper-based microfluidic device that allows semi-quantification of H_2O_2 by counting the number of green colored paper areas ("detection bars" in Fig. 13.8a) after the elapse of a certain analysis time [80]. The key idea is that each branched channel has a different amount of the H_2O_2-degradable hydrophobic compound ("phase switching agent") in the midstream. A sample containing higher concentration of H_2O_2 has greater decomposition capacity of the phase switching agent per unit of time, and therefore, is able to flow through an increasing number of the paper channels within a fixed analysis time. In contrast to the precise time measurement-based quantification mode, this "analog" counting-based approach is somewhat semi-quantitative. Nevertheless, a user is less likely to "miss" the assay endpoint in this approach as compared to the timing-based measurements, by simply setting an alarm at the required time of signal readout.

Another complementary work describes the counting-based detection of H_2O_2 by using $KMnO_4$ as an indicator [83]. The device shown in Fig. 13.8b has various amounts of $KMnO_4$ at the terminal of branched paper channels ("detection spots" in Fig. 13.8b, top), where $KMnO_4$ is decolored upon reduction by H_2O_2 in the sample. The number of decolored detection spots allows to semi-quantitatively determine the concentration of H_2O_2 in the applied sample solution. This device has been employed for the detection of adenosine in combination with an off-chip peroxidase-based enzymatic reaction (Fig. 13.8b, bottom). As is clear from the mechanism, $KMnO_4$ serves not only as an indicator, but also as a "scavenger" of H_2O_2 enabling this counting-based readout mode. Although the targeted application is environment assessment and is not directly related to the medical and biological fields, colored spot counting-based acid-base titration [93] and semi-quantification of alkali earth metals (Ca^{2+}, Mg^{2+}) [94] have also been achieved on paper-based microfluidics, where acid, base and a chelating agent serve as the scavenger of the respective analytical target (*i.e.* base, acid and alkali earth metals, respectively).

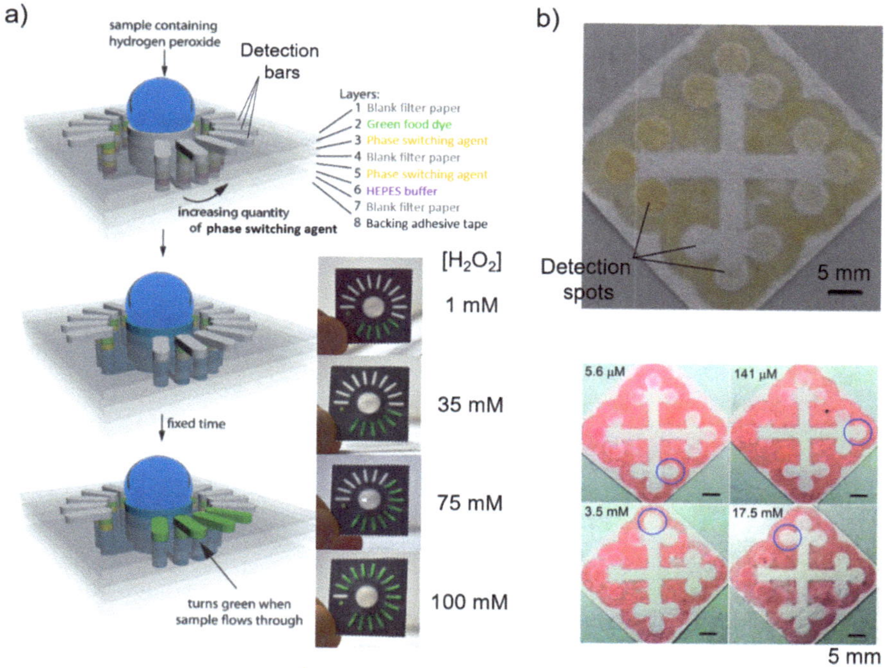

Fig. 13.8 Examples of colored spot counting-based detection on paper-based microfluidics: (**a**) schematic illustration of a three-dimensional device design, assay procedure and actual photograph of the counting-based H_2O_2 detection. (Adapted with permission from Ref. [80]. Copyright© 2012 WILEY-VCH Verlag GmbH & Co. KGaA, Weinheim); (**b**) actual photograph of a lateral flow-type device after deposition of various amounts of $KMnO_4$ (top) and processed images showing the result of application to counting-based adenosine semi-quantification. (Adapted from Ref. [83]), Copyright 2016, with permission from Elsevier)

13.3.4.4 Direct Reading of Assay Results: "Text-Displayed Signaling"

In analogy to looking at the display of a handheld blood glucose meter or a pH-meter, reading of text or a number from a piece of paper is with no doubt the most straightforward and user-friendly way to interpret the result of an analytical assay. When it comes to POC diagnostics performable under situations lacking infrastructure and/or a skilled operator, a fast and disposable "drop-and-read" or "dip-and-read" assay requiring no further user action than the application of the as-collected specimen is probably the ultimate goal that research and development should aim for. However, as compared to the other signaling approaches described above (analog thermometer-style, time measurement, counting), the "text readout" mode is scarcely implemented into paper-based tests targeting medical applications. In addition, the examples introduced in this section do not involve a sample fluid transport in a microfluidically patterned paper channel, which is deemed to be an "essence" of paper-based microfluidics. However, two examples of paper-based

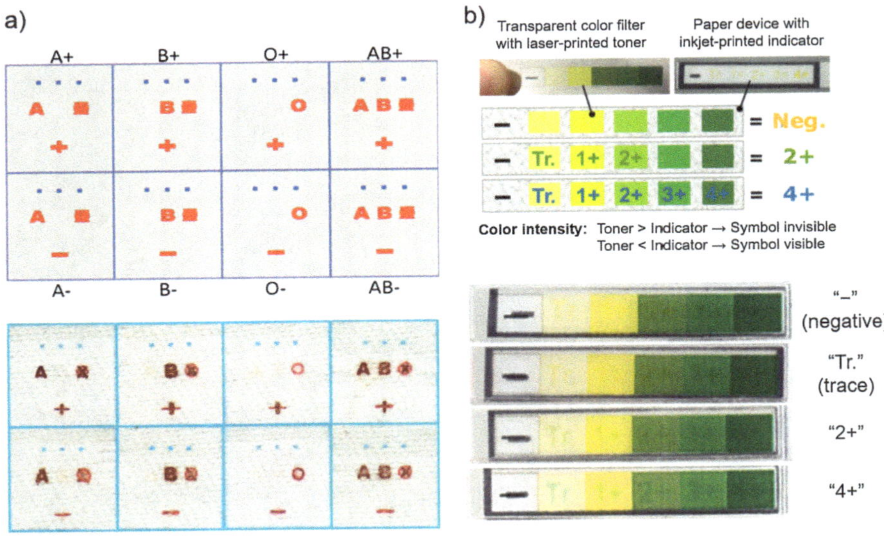

Fig. 13.9 Examples of text-displaying of assay results on paper-based devices: (**a**) ABO and Rh blood typing. (Reproduced with permission from Ref. [84]. Copyright© 2012 WILEY-VCH Verlag GmbH & Co. KGaA, Weinheim); (**b**) semi-quantitative urinary albumin detection. (Adapted with permission Ref. [85]. Copyright 2017 American Chemical Society)

devices employing the "text readout" motif are introduced here, considering the potential importance of this signaling approach allowing the most straightforward assay result interpretation. The first demonstration of a text-based signaling mode was related to ABO and Rh blood typing on a sheet of patterned paper towel [84]. The paper substrate underwent inkjet-printing of a colorless hydrophobic reagent (alkyl ketene dimer: AKD) to pattern a frame of alphabetic character- or symbol-shaped hydrophilic regions, followed by pre-deposition of anti-A, B, or D antibodies and a water-insoluble red ink. After exposure to a whole blood sample and subsequent saline washing, the sample blood type was displayed in red text consisting of coagulated blood and the red ink (Fig. 13.9a). Thanks to the highly blood type-specific agglutination reaction stemmed from the antigen-antibody reaction, this paper device achieved error-free ABO and Rh blood typing of 99 samples, including 2 weak blood samples having poor interaction capacity with the blood agglutination antibodies [84].

As is clear from the mechanism, this text-based signaling approach is limited to the blood typing application and is not straightforwardly adaptable to general (bio) chemical analysis. A recent study expanded the application of the text-based signaling on paper devices by using a classical colorimetric indicator for the detection of a given analyte. The working principle relies on "color screening" of a colorimetric indicator inkjet-deposited on paper in text form by a color-printed transparency film serving as a "mask" (Fig. 13.9b) [85]. Each text character is only visible through the mask when the analyte concentration is high enough to result in colorimetric response of the indicator whose color is more intense than the overlaid mask. This idea was applied to semi-quantitative detection of urinary albumin with combined use of tetrabromophenol blue (TBPB) as an indicator. The text-based readout result could be semi-quantitatively determined by reading the highest result symbol (−, Tr., 1+, 2+, 3+ or 4+) among the visible ones. Although the result of a comparison study with a commercial colorimetric dipstick, the gold standard for urinary albumin screening, showed similar analytical performance (accuracy rate, chance of false-positive and false-negative readouts), the text readout mode offers more intuitive result interpretation and better compatibility with users having color vision anomalies than traditional colorimetric dipsticks, where the readout relies on a sample color comparison with a read guide.

13.4 Conclusions and Outlook

The field of paper-based microfluidics showed an amazing growth since photolithographically patterned paper was rediscovered as a promising analytical platform in 2007. This new type of microfluidic devices offers several exclusive benefits including an extremely low cost, power-free sample liquid transport by capillary forces and safe disposability by incineration, among others. With these attractive characteristics and recent advancements in simplification of user operations (sample processing prior to analysis, result readout), paper-based microfluidics have a great potential to become highly functional yet affordable analytical tools enabling medical diagnostics and screening tests without the need of infrastructure and skilled operators. On the other hand, recent trends of research and development of paper-based microfluidics is also directed to more "sophisticated" applications such as cell analysis [95], tumor biomarker detection [96] and wearable sensors [97]. For these "sophisticated" applications of paper-based microfluidics, the demonstration of clear advantages over already available "high-tech" analytical systems will be a crucial factor to overcome user reluctance to replace established methods. For all targeted applications of paper-based microfluidics, be it a simple urine home test or an advanced tumor marker assay performed at POC, in-depth studies of long-term storage stability are absolutely necessary. Finally, collaboration between academia and industry is required to establish high-volume production techniques and market distribution pathways.

References

1. Manz A, Graber N, Widmer HM (1990) Miniaturized total chemical analysis systems: a novel concept for chemical sensing. Sens. Actuat. B: Chem. 1:244–248
2. Whitesides GM (2006) The origins and the future of microfluidics. Nature 442:368–373

3. Yager P, Edwards T, Fu E, Helton K, Nelson K, Tam MR, Weigl BH (2006) Microfluidic diagnostic technologies for global public health. Nature 442:412–418
4. Volpatti LR, Yetisen AK (2014) Commercialization of microfluidic devices. Trends Biotechnol 32:347–350
5. Temiz Y, Lovchik RD, Kaigala GV, Delamarche E (2015) Lab-on-a-chip devices: how to close and plug the lab? Microelectron Eng 132:156–175
6. Kurita R, Niwa O (2016) Microfluidic platforms for DNA methylation analysis. Lab Chip 16:3631–3644
7. Lim YC, Kouzani AZ, Duan W (2010) Lab-on-a-chip: a component view. Microsyst Technol 16:1995–2015
8. Tsao C-W (2016) Polymer microfluidics: simple, low-cost fabrication process bridging academic lab research to commercialized production. Micromachines 7:255
9. Martinez AW, Phillips ST, Butte MJ, Whitesides GM (2007) Patterned paper as a platform for inexpensive, low-volume, portable bioassays. Angew Chem Int Ed 46:1318–1320
10. Xu Y, Liu M, Kong N, Liu J (2016) Lab-on-paper micro- and nano-analytical devices: fabrication, modification, detection and emerging applications. Microchim Acta 183:1521–1542
11. Ruecha N, Yamada K, Suzuki K, Citterio D (2017) (Bio)chemical sensors based on paper. In: Cesar Paixão TRL, Reddy SM (eds) Materials for chemical sensing. Springer International Publishing, Cham, pp 29–74
12. Yetisen AK, Akram MS, Lowe CR (2013) Paper-based microfluidic point-of-care diagnostic devices. Lab Chip 13:2210–2251
13. Cate DM, Adkins JA, Mettakoonpitak J, Henry CS (2015) Recent developments in paper-based microfluidic devices. Anal Chem 87:19–41
14. Yamada K, Henares TG, Suzuki K, Citterio D (2015) Paper-based inkjet-printed microfluidic analytical devices. Angew Chem Int Ed 54:5294–5310
15. Xia Y, Si J, Li Z (2016) Fabrication techniques for microfluidic paper-based analytical devices and their applications for biological testing: a review. Biosens Bioelectron 77:774–789
16. Mahadeva SK, Walus K, Stoeber B (2015) Paper as a Platform for sensing applications and other devices: a review. ACS Appl Mater Interfaces 7:8345–8362
17. Klemm D, Heublein B, Fink H-P, Bohn A (2005) Cellulose: fascinating biopolymer and sustainable raw material. Angew Chem Int Ed 44:3358–3393
18. Habibi Y, Lucia LA, Rojas OJ (2010) Cellulose nanocrystals: chemistry, self-assembly, and applications. Chem Rev 110:3479–3500
19. Credou J, Berthelot T (2014) Cellulose: from biocompatible to bioactive material. J Mater Chem B 2:4767–4788
20. Bajpai P (2012) Brief description of the pulp and paper making process. In: Bajpai P (ed) Biotechnology for pulp and paper processing. Springer US, Boston, pp 7–14
21. Nery EW, Kubota LT (2013) Sensing approaches on paper-based devices: a review. Anal Bioanal Chem 405:7573–7595
22. Rodney C (2004) Scientific American inventions and discoveries. John Wiley & Sons, Inc, Hoboken
23. Maumené M (1850) On a new reagent for ascertaining the presence of sugar in certain liquids. Philos Mag Ser 3(36):482
24. Campbell RL, Wagner DB, O'Connell JP (1987) Solid phase assay with visual readout. US Patent 4703017, October 27
25. Rosenstein RW, Bloomster TG (1989) US Patent 4855240, August 8
26. Mills CK, Afable C, Candanoza C, Ferreria V, Landon S, Repaske A, Scully J, Gherna R (1989) Comparison of the merckoquant 10007 nitrite test strip with conventional reagents in the detection of nitrate reduction by bacteria. J Microbiol Methods 9:233–237
27. Müller RH, Clegg DL (1949) Automatic paper chromatography. Anal Chem 21:1123–1125
28. Chin CD, Linder V, Sia SK (2012) Commercialization of microfluidic point-of-care diagnostic devices. Lab Chip 12:2118–2134

29. Mao X, Huang TJ (2012) Microfluidic diagnostics for the developing world. Lab Chip 12:1412–1416
30. Carrilho E, Martinez AW, Whitesides GM (2009) Understanding wax printing: a simple micropatterning process for paper-based microfluidics. Anal Chem 81:7091–7095
31. Lu Y, Shi W, Jiang L, Qin J, Lin B (2009) Rapid prototyping of paper-based microfluidics with wax for low-cost, portable bioassay. Electrophoresis 30:1497–1500
32. Abe K, Suzuki K, Citterio D (2008) Inkjet-printed microfluidic multianalyte chemical sensing paper. Anal Chem 80:6928–6934
33. Fenton EM, Mascarenas MR, López GP, Sibbett SS (2009) Multiplex lateral-flow test strips fabricated by two-dimensional shaping. ACS Appl Mater Interfaces 1:124–129
34. Carson FT (1940) Some observations on determining the size of pores in paper. J Res Natl Bur Stand 24:435–442
35. Osborn JL, Lutz B, Fu E, Kauffman P, Stevens DY, Yager P (2010) Microfluidics without pumps: reinventing the T-sensor and H-filter in paper networks. Lab Chip 10:2659–2665
36. Ahmed S, Bui M-PN, Abbas A (2016) Paper-based chemical and biological sensors: Engineering aspects. Biosens Bioelectron 77:249–263
37. Kar S, Maiti TK, Chakraborty S (2015) Capillarity-driven blood plasma separation on paper-based devices. Analyst 140:6473–6476
38. Ota R, Yamada K, Suzuki K, Citterio D (2018) Quantitative evaluation of analyte transport on microfluidic paper-based analytical devices (μPADs). Analyst 143:643–653
39. Wang X, Zhang Q, Nam C, Hickner M, Mahoney M, Meyerhoff ME (2017) An ionophore-based anion-selective optode printed on cellulose paper. Angew Chem Int Ed 56:11826–11830
40. Shibata H, Henares TG, Yamada K, Suzuki K, Citterio D (2018) Implementation of a plasticized PVC-based cation-selective optode system into a paper-based analytical device for colorimetric sodium detection. Analyst 143:678–686
41. Cassano CL, Fan ZH (2013) Laminated paper-based analytical devices (LPAD): fabrication, characterization, and assays. Microfluid Nanofluid 15:173–181
42. Liu W, Cassano CL, Xu X, Fan ZH (2013) Laminated paper-based analytical devices (LPAD) with origami-enabled chemiluminescence immunoassay for cotinine detection in mouse serum. Anal Chem 85:10270–10276
43. Tenda K, Ota R, Yamada K, Henares T, Suzuki K, Citterio D (2016) High-resolution microfluidic paper-based analytical devices for sub-microliter sample analysis. Micromachines 7:80
44. Urdea M, Penny LA, Olmsted SS, Giovanni MY, Kaspar P, Shepherd A, Wilson P, Dahl CA, Buchsbaum S, Moeller G, Hay Burgess DC (2006) Requirements for high impact diagnostics in the developing world. Nature 444:73–79
45. Recommendations for Clinical Laboratory Improvement Amendments of 1988 (CLIA) (2018) Waiver applications for manufacturers of in vitro diagnostic devices. http://www.fda.gov/MedicalDevices/DeviceRegulationandGuidance/GuidanceDocuments/ucm079632.htm. Accessed 22 Feb 2018
46. Nilghaz A, Guan L, Tan W, Shen W (2016) Advances of paper-based microfluidics for diagnostics—the original motivation and current status. ACS Sens 1:1382–1393
47. Yamada K, Shibata H, Suzuki K, Citterio D (2017) Toward practical application of paper-based microfluidics for medical diagnostics: state-of-the-art and challenges. Lab Chip 17:1206–1249
48. Bai Y, Tian C, Wei X, Wang Y, Wang D, Shi X (2012) A sensitive lateral flow test strip based on silica nanoparticle/CdTe quantum dot composite reporter probes. RSC Adv 2:1778–1781
49. Xie QY, Wu YH, Xiong QR, Xu HY, Xiong YH, Liu K, Jin Y, Lai WH (2014) Advantages of fluorescent microspheres compared with colloidal gold as a label in immunochromatographic lateral flow assays. Biosens Bioelectron 54:262–265
50. Kim K, Joung H-A, Han G-R, Kim M-G (2016) An immunochromatographic biosensor combined with a water-swellable polymer for automatic signal generation or amplification. Biosens Bioelectron 85:422–428

51. Park J, Shin JH, Park JK (2016) Pressed paper-based dipstick for detection of foodborne pathogens with multistep reactions. Anal Chem 88:3781–3788
52. Shyu R-H, Shyu H-F, Liu H-W, Tang S-S (2002) Colloidal gold-based immunochromatographic assay for detection of ricin. Toxicon 40:255–258
53. Rong-Hwa S, Shiao-Shek T, Der-Jiang C, Yao-Wen H (2010) Gold nanoparticle-based lateral flow assay for detection of staphylococcal enterotoxin B. Food Chem 118:462–466
54. Chiao D-J, Shyu R-H, Hu C-S, Chiang H-Y, Tang S-S (2004) Colloidal gold-based immunochromatographic assay for detection of botulinum neurotoxin type B. J Chromatogr B 809:37–41
55. Fu E, Liang T, Spicar-Mihalic P, Houghtaling J, Ramachandran S, Yager P (2012) Two-dimensional paper network format that enables simple multistep assays for use in low-resource settings in the context of malaria antigen detection. Anal Chem 84:4574–4579
56. Lutz B, Liang T, Fu E, Ramachandran S, Kauffman P, Yager P (2013) Dissolvable fluidic time delays for programming multi-step assays in instrument-free paper diagnostics. Lab Chip 13:2840–2847
57. Apilux A, Ukita Y, Chikae M, Chailapakul O, Takamura Y (2013) Development of automated paper-based devices for sequential multistep sandwich enzyme-linked immunosorbent assays using inkjet printing. Lab Chip 13:126–135
58. Ishii M, Preechakasedkit P, Yamada K, Chailapakul O, Suzuki K, Citterio D (2018) Wax-assisted one-step enzyme-linked immunosorbent assay on lateral flow test devices. Anal Sci 34:51–56
59. Cherpillod P, Schibler M, Vieille G, Cordey S, Mamin A, Vetter P, Kaiser L (2016) Ebola virus disease diagnosis by real-time RT-PCR: a comparative study of 11 different procedures. J Clin Virol 77:9–14
60. Notomi T, Okayama H, Masubuchi H, Yonekawa T, Watanabe K, Amino N, Hase T (2000) Loop-mediated isothermal amplification of DNA. Nucleic Acids Res 28:e63–e63
61. Ali MM, Li F, Zhang Z, Zhang K, Kang D-K, Ankrum JA, Le XC, Zhao W (2014) Rolling circle amplification: a versatile tool for chemical biology, materials science and medicine. Chem Soc Rev 43:3324–3341
62. Connelly JT, Rolland JP, Whitesides GM (2015) "Paper machine" for molecular diagnostics. Anal Chem 87:7595–7601
63. Magro L, Escadafal C, Garneret P, Jacquelin B, Kwasiborski A, Manuguerra J-C, Monti F, Sakuntabhai A, Vanhomwegen J, Lafaye P, Tabeling P (2017) Paper microfluidics for nucleic acid amplification testing (NAAT) of infectious diseases. Lab Chip 17:2347–2371
64. Zuk RF, Ginsberg VK, Houts T, Rabbie J, Merrick H, Ullman EF, Fischer MM, Sizto CC, Stiso SN, Litman DJ (1985) Enzyme immunochromatography—a quantitative immunoassay requiring no instrumentation. Clin Chem 31:1144–1150
65. Vaughan L, Milavetz G, Ellis E, Szefler S, Conboy K, Weinberger M, Tillson S, Jenne J, Wiener M, Shaughnessy T, Carrico J (1986) Multicentre evaluation of disposable visual measuring device to assay theophylline from capillary blood sample. Lancet 327:184–186
66. Chen R, Li TM, Merrick H, Parrish RF, Bruno V, Kwong A, Stiso C, Litman DJ (1987) An internal clock reaction used in a one-step enzyme immunochromatographic assay of theophylline in whole blood. Clin Chem 33:1521–1525
67. Allen MP, DeLizza A, Ramel U, Jeong H, Singh P (1990) A noninstrumented quantitative test system and its application for determining cholesterol concentration in whole blood. Clin Chem 36:1591–1597
68. Liu VY, Lin TY, Schrier W, Allen M, Singh P (1993) AccuMeter noninstrumented quantitative assay of high-density lipoprotein in whole blood. Clin Chem 39:1948–1952
69. Cate DM, Dungchai W, Cunningham JC, Volckens J, Henry CS (2013) Simple, distance-based measurement for paper analytical devices. Lab Chip 13:2397–2404
70. Yamada K, Henares TG, Suzuki K, Citterio D (2015) Distance-based tear lactoferrin assay on microfluidic paper device using interfacial interactions on surface-modified cellulose. ACS Appl Mater Interfaces 7:24864–24875

71. Wei X, Tian T, Jia S, Zhu Z, Ma Y, Sun J, Lin Z, Yang CJ (2016) Microfluidic distance readout sweet hydrogel integrated paper-based analytical device (μDiSH-PAD) for visual quantitative point-of-care testing. Anal Chem 88:2345–2352
72. Tian T, An Y, Wu Y, Song Y, Zhu Z, Yang C (2017) Integrated distance-based origami paper analytical device for one-step visualized analysis. ACS Appl Mater Interfaces 9:30480–30487
73. Hongwarittorrn I, Chaichanawongsaroj N, Laiwattanapaisal W (2017) Semi-quantitative visual detection of loop mediated isothermal amplification (LAMP)-generated DNA by distance-based measurement on a paper device. Talanta 175:135–142
74. Wang AG, Dong T, Mansour H, Matamoros G, Sanchez AL, Li F (2018) Paper-based DNA reader for visualized quantification of soil-transmitted helminth infections. ACS Sens 3:205–210
75. Chen Y, Chu W, Liu W, Guo X (2018) Distance-based carcinoembryonic antigen assay on microfluidic paper immunodevice. Sens. Actuat. B: Chem. 260:452–459
76. Wei X, Tian T, Jia S, Zhu Z, Ma Y, Sun J, Lin Z, Yang CJ (2015) Target-responsive DNA hydrogel mediated "stop-flow" microfluidic paper-based analytic device for rapid, portable and visual detection of multiple targets. Anal Chem 87:4275–4282
77. Zhang L, Nie J, Wang H, Yang J, Wang B, Zhang Y, Li J (2017) Instrument-free quantitative detection of alkaline phosphatase using paper-based devices. Anal Methods 9:3375–3379
78. Noiphung J, Talalak K, Hongwarittorrn I, Pupinyo N, Thirabowonkitphithan P, Laiwattanapaisal W (2015) A novel paper-based assay for the simultaneous determination of Rh typing and forward and reverse ABO blood groups. Biosens Bioelectron 67:485–489
79. Berry SB, Fernandes SC, Rajaratnam A, DeChiara NS, Mace CR (2016) Measurement of the hematocrit using paper-based microfluidic devices. Lab Chip 16:3689–3694
80. Lewis GG, DiTucci MJ, Phillips ST (2012) Quantifying analytes in paper-based microfluidic devices without using external electronic readers. Angew Chem Int Ed 51:12707–12710
81. Lewis GG, Robbins JS, Phillips ST (2013) Point-of-care assay platform for quantifying active enzymes to femtomolar levels using measurements of time as the readout. Anal Chem 85:10432–10439
82. Zhang Y, Fan J, Nie J, Le S, Zhu W, Gao D, Yang J, Zhang S, Li J (2015) Timing readout in paper device for quantitative point-of-use hemin/G-quadruplex DNAzyme-based bioassays. Biosens Bioelectron 73:13–18
83. Zhang Y, Gao D, Fan J, Nie J, Le S, Zhu W, Yang J, Li J (2016) Naked-eye quantitative aptamer-based assay on paper device. Biosens Bioelectron 78:538–546
84. Li M, Tian J, Al-Tamimi M, Shen W (2012) Paper-based blood typing device that reports patient's blood type "in writing". Angew Chem Int Ed 51:5497–5501
85. Yamada K, Suzuki K, Citterio D (2017) Text-displaying colorimetric paper-based analytical device. ACS Sens 2:1247–1254
86. Cate DM, Noblitt SD, Volckens J, Henry CS (2015) Multiplexed paper analytical device for quantification of metals using distance-based detection. Lab Chip 15:2808–2818
87. Zhong M, Lee CY, Croushore CA, Sweedler JV (2012) Label-free quantitation of peptide release from neurons in a microfluidic device with mass spectrometry imaging. Lab Chip 12:2037–2045
88. Mentele MM, Cunningham J, Koehler K, Volckens J, Henry CS (2012) Microfluidic paper-based analytical device for particulate metals. Anal Chem 84:4474–4480
89. Rattanarat P, Dungchai W, Cate DM, Siangproh W, Volckens J, Chailapakul O, Henry CS (2013) A microfluidic paper-based analytical device for rapid quantification of particulate chromium. Anal Chim Acta 800:50–55
90. Henares TG, Yamada K, Takaki S, Suzuki K, Citterio D (2017) "Drop-slip" bulk sample flow on fully inkjet-printed microfluidic paper-based analytical device. Sens Actuat B: Chem 244:1129–1137
91. Kudo H, Yamada K, Watanabe D, Suzuki K, Citterio D (2017) Paper-based analytical device for zinc ion quantification in water samples with power-free analyte concentration. Micromachines 8:127

92. Chatterjee D, Mansfield DS, Anderson NG, Subedi S, Woolley AT (2012) "Flow valve" microfluidic devices for simple, detectorless, and label-free analyte quantitation. Anal Chem 84:7057–7063
93. Karita S, Kaneta T (2014) Acid-base titrations using microfluidic paper-based analytical devices. Anal Chem 86:12108–12114
94. Karita S, Kaneta T (2016) Chelate titrations of Ca^{2+} and Mg^{2+} using microfluidic paper-based analytical devices. Anal Chim Acta 924:60–67
95. Kenney RM, Boyce MW, Whitman NA, Kromhout BP, Lockett MR (2018) A pH-sensing optode for mapping spatiotemporal gradients in 3D paper-based cell cultures. Anal Chem 90:2376–2383
96. Wang J, Li W, Ban L, Du W, Feng X, Liu B-F (2018) A paper-based device with an adjustable time controller for the rapid determination of tumor biomarkers. Sens Actuat B: Chem 254:855–862
97. Mu X, Xin X, Fan C, Li X, Tian X, Xu KF, Zheng Z (2015) A paper-based skin patch for the diagnostic screening of cystic fibrosis. Chem Commun 51:6365–6368

FSC
www.fsc.org
MIX
Papier | Fördert gute Waldnutzung
FSC® C083411

Kolibri 360 GmbH
Ferdinand-Jühlke-Straße 7
99095 Erfurt, Deutschland
produktsicherheit@kolibri360.de